Fractional Calculus Operators and the Mittag-Leffler Function

Fractional Calculus Operators and the Mittag-Leffler Function

Editor

Maja Andrić

MDPI • Basel • Beijing • Wuhan • Barcelona • Belgrade • Manchester • Tokyo • Cluj • Tianjin

Editor
Maja Andrić
University of Split
Croatia

Editorial Office
MDPI
St. Alban-Anlage 66
4052 Basel, Switzerland

This is a reprint of articles from the Special Issue published online in the open access journal *Fractal and Fractional* (ISSN 2504-3110) (available at: https://www.mdpi.com/journal/fractalfract/special_issues/FCOMLF).

For citation purposes, cite each article independently as indicated on the article page online and as indicated below:

LastName, A.A.; LastName, B.B.; LastName, C.C. Article Title. *Journal Name* **Year**, *Volume Number*, Page Range.

ISBN 978-3-0365-5367-2 (Hbk)
ISBN 978-3-0365-5368-9 (PDF)

© 2022 by the authors. Articles in this book are Open Access and distributed under the Creative Commons Attribution (CC BY) license, which allows users to download, copy and build upon published articles, as long as the author and publisher are properly credited, which ensures maximum dissemination and a wider impact of our publications.

The book as a whole is distributed by MDPI under the terms and conditions of the Creative Commons license CC BY-NC-ND.

Contents

About the Editor vii

Maja Andrić
Editorial for Special Issue "Fractional Calculus Operators and the Mittag–Leffler Function"
Reprinted from: *Fractal Fract.* **2022**, *6*, 442, doi:10.3390/fractalfract6080442 1

Hari M. Srivastava, Abedel-Karrem N. Alomari, Khaled M. Saad and Waleed M. Hamanah
Some Dynamical Models Involving Fractional-Order Derivativeswith the Mittag-Leffler Type Kernelsand Their ApplicationsBased upon the Legendre Spectral Collocation Method
Reprinted from: *Fractal Fract.* **2021**, *5*, 131, doi:10.3390/fractalfract5030131 5

Yuri Kondratiev, José da Silva
Cesaro Limits for Fractional Dynamics
Reprinted from: *Fractal Fract.* **2021**, *5*, 133, doi:10.3390/fractalfract5040133 19

Petr A. Ryapolov and Eugene B. Postnikov
Mittag–Leffler Function as an Approximant to the Concentrated Ferrofluid's Magnetization Curve
Reprinted from: *Fractal Fract.* **2021**, *5*, 147, doi:10.3390/fractalfract5040147 31

Junesang Choi, Mohd Idris Qureshi, Aarif Hussain Bhat and Javid Majid
Reduction Formulas for Generalized Hypergeometric Series Associated with New Sequences and Applications
Reprinted from: *Fractal Fract.* **2021**, *5*, 150, doi:10.3390/fractalfract5040150 41

Bashir Ahmad and Sotiris K. Ntouyas
Hilfer–Hadamard Fractional Boundary Value Problems with Nonlocal Mixed Boundary Conditions
Reprinted from: *Fractal Fract.* **2021**, *5*, 195, doi:10.3390/fractalfract5040195 65

Dheerandra Shanker Sachan, Shailesh Jaloree and Junesang Choi
Certain Recurrence Relations of Two Parametric Mittag-Leffler Function and Their Application in Fractional Calculus
Reprinted from: *Fractal Fract.* **2021**, *5*, 215, doi:10.3390/fractalfract5040215 91

Ghulam Farid, Muhammad Yussouf and Kamsing Nonlaopon
Fejér–Hadamard Type Inequalities for $(\alpha, h\text{-}m)\text{-}p$-Convex Functions via Extended Generalized Fractional Integrals
Reprinted from: *Fractal Fract.* **2021**, *5*, 253, doi:10.3390/fractalfract5040253 109

Abdelhalim Ebaid and Hind K. Al-Jeaid
The Mittag–Leffler Functions for a Class of First-Order Fractional Initial Value Problems: Dual Solution via Riemann–Liouville Fractional Derivative
Reprinted from: *Fractal Fract.* **2022**, *6*, 85, doi:10.3390/fractalfract6020085 125

Zhiqiang Zhang, Ghulam Farid, Sajid Mehmood, Kamsing Nonlaopon and Tao Yan
Generalized k-Fractional Integral Operators Associated with Pólya-Szegö and Chebyshev Types Inequalities
Reprinted from: *Fractal Fract.* **2022**, *6*, 90, doi:10.3390/fractalfract6020090 139

Saima Naheed, Shahid Mubeen, Gauhar Rahman, Zareen A. Khan, Kottakkaran Sooppy Nisar
Certain Integral and Differential Formulas Involving the Product of Srivastava's Polynomials and Extended Wright Function
Reprinted from: *Fractal Fract.* **2022**, *6*, 93, doi:10.3390/fractalfract6020093 . . . . **151**

Abdulrahman F. Aljohani, Abdelhalim Ebaid, Ebrahem A. Algehyne, Yussri M. Mahrous, Carlo Cattani, Hind K. Al-Jeaid
The Mittag-Leffler Function for Re-Evaluating the Chlorine Transport Model: Comparative Analysis
Reprinted from: *Fractal Fract.* **2022**, *6*, 125, doi:10.3390/fractalfract6030125 . . . . **171**

Yi-Xia Li, Muhammad Samraiz, Ayesha Gul, Miguel Vivas-Cortez, Gauhar Rahman
Hermite-Hadamard Fractional Integral Inequalities via Abel-Gontscharoff Green's Function
Reprinted from: *Fractal Fract.* **2022**, *6*, 126, doi:10.3390/fractalfract6030126 . . . . **185**

Dinesh Kumar, Jeta Ram and Junesang Choi
Dirichlet Averages of Generalized Mittag-Leffler Type Function
Reprinted from: *Fractal Fract.* **2022**, *6*, 297, doi:10.3390/fractalfract6060297 . . . . **205**

Maja Andrić
Fractional Integral Inequalities of Hermite–Hadamard Type for $(h, g; m)$-Convex Functions with Extended Mittag-Leffler Function
Reprinted from: *Fractal Fract.* **2022**, *6*, 301, doi:10.3390/fractalfract6060301 . . . . **217**

Zhehao Zhang
A New Fractional Poisson Process Governed by a Recursive Fractional Differential Equation
Reprinted from: *Fractal Fract.* **2022**, *6*, 418, doi:10.3390/fractalfract6080418 . . . . **233**

About the Editor

Maja Andrić

Maja Andrić is an Associate Professor in the Faculty of Civil Engineering, Architecture and Geodesy University of Split, Croatia. In 2011 she received her Ph.D. in Mathematics from the Faculty of Natural and Mathematical Sciences, University of Zagreb. The main field of her research is Mathematical Analysis with a special focus on the Theory of Inequalities and Fractional Calculus. She published 2 scientific books (monographs) and 26 scientific papers. Since 2014, she has been a member of the Editorial Board of the international journal, *Fractional Differential Calculus*, as Associate Editor.

fractal and fractional

Editorial

Editorial for Special Issue "Fractional Calculus Operators and the Mittag–Leffler Function"

Maja Andrić

Faculty of Civil Engineering, Architecture and Geodesy, University of Split, Matice Hrvatske 15, 21000 Split, Croatia; maja.andric@gradst.hr

Citation: Andrić, M. Editorial for Special Issue "Fractional Calculus Operators and the Mittag-Leffler Function". *Fractal Fract.* **2022**, *6*, 442. https://doi.org/10.3390/fractalfract6080442

Received: 2 August 2022
Accepted: 12 August 2022
Published: 14 August 2022

Publisher's Note: MDPI stays neutral with regard to jurisdictional claims in published maps and institutional affiliations.

Copyright: © 2022 by the author. Licensee MDPI, Basel, Switzerland. This article is an open access article distributed under the terms and conditions of the Creative Commons Attribution (CC BY) license (https://creativecommons.org/licenses/by/4.0/).

Among the numerous applications of the theory of fractional calculus in almost all applied sciences, applications in numerical analysis and various fields of physics and engineering stand out. Applications of inequalities involving function integrals and their derivatives, as well as applications of fractional differentiation inequalities, have motivated many researchers to investigate extensions and generalizations using various fractional differential and integral operators. Of particular importance is the Mittag–Leffler function which, with its generalizations, appears as a solution to differential or integral equations of fractional order. This produced new results for more generalized fractional integral operators containing the Mittag–Leffler function in their kernels.

This Special Issue contains 15 published papers. In [1], the authors use derivatives of fractional order, which are based on generalized kernels of the Mittag–Leffler type, to present three models of fractional order. These models involve fractional ordinary and partial differential equations. The accuracy of the solution was checked in comparison with the exact solution and by calculating the residual error function.

The asymptotic behavior of random time changes in dynamical systems is studied in [2]. As time varies randomly, three classes are proposed that show different patterns of asymptotic decay.

In [3], the authors showed that the static magnetization curve of highly concentrated ferrofluids can be accurately approximated by the Mittag–Leffler function of the inverse external magnetic field.

By introducing two series of new numbers and their derivatives [4], and choosing to use six well-known generalized Kummer addition formulas, the authors establish six classes of generalized addition formulas and demonstrate six identities regarding finite sums.

The paper [5] deals with the existence and uniqueness of the solution for the Hilfer–Hadamard fractional differential equation, supplemented with mixed non-local boundary conditions.

In [6], the authors developed some new recurrence relations for two parametric Mittag–Leffler functions and discussed some applications of these recurrence relations. By applying Riemann–Liouville fractional integrals and differential operators, four new relations between Fox–Wright functions with certain special cases are established.

The goal of the paper [7] was to find new versions of Fejér–Hadamard-type inequalities for $(\alpha, h - m) - p$-convex functions by means of expanded fractional operators with Mittag–Leffler functions. These inequalities hold at the same time for different types of convexities and different types of fractional integrals.

In [8], a new approach is given for solving a class of first-order fractional initial value problems. The results are based on the Riemann–Liouville fractional derivative and the solutions are expressed in terms of Mittag–Leffler, exponential, and trigonometric functions.

The paper [9] aims to define an operator that contains the Mittag–Leffler function in its kernel, which leads to the deduction of many already existing operators. The results of this work reproduce Chebyshev- and Pólya-Szegö-type inequalities for various fractional integral operators.

In [10], the authors investigate certain image formulas related to the product of Srivastava polynomials and extended Wright functions using fractional integral and differential operators, including Marichev–Saigo–Maeda, Lavoie–Trottier, and Oberhettinger.

The paper [11] reviews the mathematical model of chlorine transport used as a water treatment model, when variable-order partial derivatives are included to describe the chlorine transfer system. In this paper, a new analytical solution using Mittag–Leffler functions is presented.

Hermite–Hadamard inequalities for κ-Riemann–Liouville fractional integrals are presented in [12], using a newer approach based on the Abel–Gontscharoff green function. The authors established certain integral identities and obtained new results for monotone functions.

The paper [13] investigates the Dirichlet and modified Dirichlet average of the R function, i.e., the Mittag–Leffler-type function. They are given in terms of Riemann–Liouville integrals and hypergeometric functions of several variables.

Hermite–Hadamard fractional integral inequalities for the class of $(h, g; m)$-convex functions are presented in [14]. The expanded generalized Mittag–Leffler function is contained in the kernel of applied fractional integral operators.

In [15], a new fractional Poisson process is proposed using a recursive fractional differential equation. The arrival time distribution functions are derived, while the inter-arrival times are no longer independent and identically distributed.

Funding: This research received no external funding.

Acknowledgments: As Guest Editor, I would like to express my gratitude to all the authors of this Special Issue on the contribution. I would also like to thank all reviewers for their help and efforts in improving the quality of the papers, as well as the Editorial Office for their kind cooperation and preparation toward this special collection.

Conflicts of Interest: The author declares no conflict of interest.

References

1. Srivastava, H.; Alomari, A.; Saad, K.; Hamanah, W. Some Dynamical Models Involving Fractional-Order Derivatives with the Mittag-Leffler Type Kernels and Their Applications Based upon the Legendre Spectral Collocation Method. *Fractal Fract.* **2021**, *5*, 131. [CrossRef]
2. Kondratiev, Y.; da Silva, J. Cesaro Limits for Fractional Dynamics. *Fractal Fract.* **2021**, *5*, 133. [CrossRef]
3. Ryapolov, P.; Postnikov, E. Mittag–Leffler Function as an Approximant to the Concentrated Ferrofluid's Magnetization Curve. *Fractal Fract.* **2021**, *5*, 147. [CrossRef]
4. Choi, J.; Qureshi, M.; Bhat, A.; Majid, J. Reduction Formulas for Generalized Hypergeometric Series Associated with New Sequences and Applications. *Fractal Fract.* **2021**, *5*, 150. [CrossRef]
5. Ahmad, B.; Ntouyas, S. Hilfer–Hadamard Fractional Boundary Value Problems with Nonlocal Mixed Boundary Conditions. *Fractal Fract.* **2021**, *5*, 195. [CrossRef]
6. Sachan, D.; Jaloree, S.; Choi, J. Certain Recurrence Relations of Two Parametric Mittag-Leffler Function and Their Application in Fractional Calculus. *Fractal Fract.* **2021**, *5*, 215. [CrossRef]
7. Farid, G.; Yussouf, M.; Nonlaopon, K. Fejér-Hadamard Type Inequalities for $(\alpha, h - m) - p$-Convex Functions via Extended Generalized Fractional Integrals. *Fractal Fract.* **2021**, *5*, 253. [CrossRef]
8. Ebaid, A.; Al-Jeaid, H. The Mittag-Leffler Functions for a Class of First-Order Fractional Initial Value Problems: Dual Solution via Riemann-Liouville Fractional Derivative. *Fractal Fract.* **2022**, *6*, 85. [CrossRef]
9. Zhang, Z.; Farid, G.; Mehmood, S.; Nonlaopon, K.; Yan, T. Generalized k—Fractional Integral Operators Associated with Pólya-Szegö and Chebyshev Types Inequalities. *Fractal Fract.* **2022**, *6*, 90. [CrossRef]
10. Naheed, S.; Mubeen, S.; Rahman, G.; Khan, Z.; Nisar, K. Certain Integral and Differential Formulas Involving the Product of Srivastava's Polynomials and Extended Wright Function. *Fractal Fract.* **2022**, *6*, 93. [CrossRef]
11. Aljohani, A.; Ebaid, A.; Algehyne, E.; Mahrous, Y.; Cattani, C.; Al-Jeaid, H. The Mittag-Leffler Function for Re-Evaluating the Chlorine Transport Model: Comparative Analysis. *Fractal Fract.* **2022**, *6*, 125. [CrossRef]
12. Li, Y.; Samraiz, M.; Gul, A.; Vivas-Cortez, M.; Rahman, G. Hermite-Hadamard Fractional Integral Inequalities via Abel-Gontscharoff Green's Function. *Fractal Fract.* **2022**, *6*, 126. [CrossRef]
13. Kumar, D.; Ram, J.; Choi, J. Dirichlet Averages of Generalized Mittag-Leffler Type Function. *Fractal Fract.* **2022**, *6*, 297. [CrossRef]

14. Andrić, M. Fractional Integral Inequalities of Hermite-Hadamard Type for (h,g;m)-Convex Functions with Extended Mittag-Leffler Function. *Fractal Fract.* **2022**, *6*, 301. [CrossRef]
15. Zhang, Z. A New Fractional Poisson Process Governed by a Recursive Fractional Differential Equation. *Fractal Fract.* **2022**, *6*, 418. [CrossRef]

 *fractal and fractional*

Article

Some Dynamical Models Involving Fractional-Order Derivatives with the Mittag-Leffler Type Kernels and Their Applications Based upon the Legendre Spectral Collocation Method

Hari M. Srivastava [1,2,3,4,*,†], Abedel-Karrem N. Alomari [5,†], Khaled M. Saad [6,7,†] and Waleed M. Hamanah [8,†]

1. Department of Mathematics and Statistics, University of Victoria, Victoria, BC V8W 3R4, Canada
2. Department of Medical Research, China Medical University Hospital, China Medical University, Taichung 40402, Taiwan
3. Department of Mathematics and Informatics, Azerbaijan University, 71 Jeyhun Hajibeyli Street, Baku AZ1007, Azerbaijan
4. Section of Mathematics, International Telematic University Uninettuno, I-00186 Rome, Italy
5. Department of Mathematics, Faculty of Science, Yarmouk University, Irbid 211-63, Jordan; abdomari2008@yahoo.com
6. Department of Mathematics, College of Sciences and Arts, Najran University, Najran P.O. Box 1988, Saudi Arabia; khaledma_sd@hotmail.com
7. Department of Mathematics, Faculty of Applied Science, Taiz University, Taiz P.O. Box 6803, Yemen
8. Interdisciplinary Research Center in Renewable Energy and Power Systems, King Fahd University for Petroleum and Minerals, Dhahran 31261, Saudi Arabia; g201105910@kfupm.edu.sa
* Correspondence: harimsri@math.uvic.ca
† These authors contributed equally to this work.

Citation: Srivastava, H.M.; Alomari, A.-K.N.; Saad, K.M.; Hamanah, W.M. Some Dynamical Models Involving Fractional-Order Derivatives with the Mittag-Leffler Type Kernels and Their Applications Based upon the Legendre Spectral Collocation Method. *Fractal Fract.* **2021**, *5*, 131. https://doi.org/10.3390/fractalfract5030131

Academic Editor: Maja Andrić

Received: 24 August 2021
Accepted: 17 September 2021
Published: 20 September 2021

Publisher's Note: MDPI stays neutral with regard to jurisdictional claims in published maps and institutional affiliations.

Copyright: © 2021 by the authors. Licensee MDPI, Basel, Switzerland. This article is an open access article distributed under the terms and conditions of the Creative Commons Attribution (CC BY) license (https://creativecommons.org/licenses/by/4.0/).

Abstract: Fractional derivative models involving generalized Mittag-Leffler kernels and opposing models are investigated. We first replace the classical derivative with the GMLK in order to obtain the new fractional-order models (GMLK) with the three parameters that are investigated. We utilize a spectral collocation method based on Legendre's polynomials for evaluating the numerical solutions of the pr. We then construct a scheme for the fractional-order models by using the spectral method involving the Legendre polynomials. In the first model, we directly obtain a set of nonlinear algebraic equations, which can be approximated by the Newton-Raphson method. For the second model, we also need to use the finite differences method to obtain the set of nonlinear algebraic equations, which are also approximated as in the first model. The accuracy of the results is verified in the first model by comparing it with our analytical solution. In the second and third models, the residual error functions are calculated. In all cases, the results are found to be in agreement. The method is a powerful hybrid technique of numerical and analytical approach that is applicable for partial differential equations with multi-order of fractional derivatives involving GMLK with three parameters.

Keywords: fractional derivative; generalized Mittag-Leffler kernel (GMLK); Legendre polynomials; Legendre spectral collocation method

MSC: 26A33; 41A30; 65N12; 33C45; 33E12; 65N22

1. Introduction

During the past three decades, the science of fractional calculus has attracted the interest of many scientists and researchers (see, for example, [1–6]; see also [7] for some recent developments on the subject of fractional calculus). The main motive for presenting and researching numerical and approximate methods for solving fractional-order differential equations is the scarcity and difficulty of finding analytical solutions to these equations. Therefore, many researchers resorted to deriving and presenting various numerical methods for this purpose (see [8–10]).

In order to learn more about the definitions and properties of fractional integrals and fractional derivatives, the reader can refer to (for example) [2,3]. Many researchers have drawn attention to the fractional-order modeling of a considerably wide variety of problems in mathematical, physical, chemical, biological and engineering sciences. In particular, in fluid mechanics, viscoelasticity and other applications, use has been made of fractional integrals and fractional derivatives involving non-singular kernels (see, for details, [11,12]). Such kernels as those with one parameter were considered in [13–37]). Abdeljawad (see [38,39]), Abdeljawad and Baleanu [40] considered fractional derivatives and fractional integrals with Mittag-Leffler kernels involving three parameters. For the existence of the solution to fractional-order differential equations, one of the most important advantages is in using the GMLK. This can be illustrated by noticing that the solution of the following problem:

$$^{GLMK}_{0}D^\alpha_t \Xi(\zeta) = a \quad \text{and} \quad \Xi(0) = \beta,$$

where a and β are constants, does not exist for $0 < \alpha < 1$. However, via the new definition, the solution exists (see [39]).

Our contribution in this work is to utilize the above-mentioned new definition of a fractional derivative in investigating new fractional-order models according to the associated fractional derivative operator. We also use several important and potentially useful properties of such special functions as the Legendre polynomials with a view of obtaining a special scheme through means of which we can derive the numerical solutions of the fractional-order models presented in this paper.

The format of this paper is structured as follows: In Section 2, we give some preliminaries and introduce the basic definitions and associated properties that will be used in this paper. In Section 3, the Legendre polynomials are presented together with their properties, including their fractional derivatives. In the fourth section (Section 4), the scheme and the algorithm for numerical solutions of the fractional-order models presented in this paper are constructed. In the fifth section (Section 5), the numerical results for the solutions to the presented models are discussed. Finally, in Section 6, we present our conclusions.

2. Preliminaries

In this section, we present the definition of the fractional derivative with generalized Mittag-Leffler kernels. This definition and its properties were studied by Abdeljawad and Baleanu [40] and Abdeljawad [39]. Abdeljawad [38] also undertook further study of the fundamentals and properties of these operators as well as their discrete versions.

Definition 1. *The left-sided fractional derivative with generalized Mittag-Leffler kernel $E^\gamma_{\alpha,\mu}(\lambda, t)$ is defined, for*

$$n < \alpha \leqq n+1 \quad (n \in \mathbb{N}_0 := \mathbb{N} \cup \{0\} = \{0,1,2,\cdots\}), \quad \Re(\mu) > 0,$$

$$\gamma \in \mathbb{R} \quad \text{and} \quad \lambda = -\frac{\alpha}{1-\alpha},$$

by

$$\left(^{GLMK}_{a}D^{\alpha,\mu,\gamma}f\right)(x) = \frac{M(\alpha_1)}{1-\alpha_1}\int_a^x E^\gamma_{\alpha_1,\mu}(\lambda, x-t)f^{(n+1)}(t)\,dt, \tag{1}$$

where $M(\alpha_1)$ is a normalization function such that

$$M(0) = M(1) = 1,$$

$$E^\gamma_{\alpha_1,\mu}(\lambda, t) = \sum_{k=0}^\infty \frac{(\gamma)_k}{k!\,\Gamma(\alpha_1 k + \mu)} \lambda^k\, t^{\alpha_1 k + \mu - 1}, \tag{2}$$

$$(\gamma)_0 = 1 \quad \text{and} \quad (\gamma)_k = \gamma(\gamma+1)\cdots(\gamma+k-1) \quad (k \in \mathbb{N})$$

and $\alpha_1 = \alpha - n$, that is, the fractional part of the parameter α.

Remark 1. *We note that, if $\alpha_1 = \mu = \gamma \to 1$, we obtain the ordinary derivative $f^{(n)}(x)$ of $f(x)$ of order n.*

Remark 2. *Henceforth, in this paper, we assume that the parameter μ is real and positive ($\mu > 0$).*

Theorem 1. *The fractional derivative with generalized Mittag-Leffler kernel of x^β ($\beta > n$) of order α ($n < \alpha \leq n+1$) is given, for $\mu > 0$ and $\alpha_1 = \alpha - n$ ($n \in \mathbb{N}_0$), by*

$$\,_0^{\text{GLMK}}D^{\alpha,\mu,\gamma}x^\beta = \frac{M(\alpha_1)\Gamma(\beta+1)}{1-\alpha_1} \sum_{k=0}^\infty \frac{\lambda^k(\gamma)_k x^{\alpha_1 k + \mu + \beta - n - 1}}{k!\,\Gamma(k\alpha_1 + \beta + \mu - n)} \qquad (3)$$

$$= \frac{M(\alpha_1)\Gamma(\beta+1)}{1-\alpha_1} E^\gamma_{\alpha_1, \mu+\beta-n}(\lambda, x). \qquad (4)$$

Proof. Using Definition 1 of the fractional derivative with generalized Mittag-Leffler kernel and the series of $E^\gamma_{\alpha_1, \mu}(\lambda, t)$ given by (2), we have

$$\,_0^{\text{GLMK}}D^{\alpha,\mu,\gamma}x^\beta = \frac{M(\alpha_1)}{1-\alpha_1} \int_0^x \sum_{k=0}^\infty \frac{(\gamma)_k}{k!\,\Gamma(\alpha_1 k + \mu)} \frac{\Gamma(\beta+1)}{\Gamma(\beta-n)}$$
$$\cdot \lambda^k\, t^{\beta-n-1} (x-t)^{\alpha_1 k + \mu - 1}\, dt$$

$$= \frac{M(\alpha_1)}{1-\alpha_1} \frac{\Gamma(\beta+1)}{\Gamma(\beta-n)} \sum_{k=0}^\infty \frac{(\gamma)_k}{k!\,\Gamma(\alpha_1 k + \mu)} \lambda^k$$
$$\cdot \int_0^x t^{\beta-n-1} (x-t)^{\alpha_1 k + \mu - 1}\, dt$$

$$= \frac{M(\alpha_1)\Gamma(\beta+1)}{1-\alpha_1} \sum_{k=0}^\infty \frac{(\gamma)_k}{k!\,\Gamma(k\alpha_1 + \beta + \mu - n)}$$
$$\cdot \lambda^k\, x^{\alpha_1 k + \mu + \beta - n - 1}$$

$$= \frac{M(\alpha_1)\Gamma(\beta+1)}{1-\alpha_1} E^\gamma_{\alpha_1, \mu+\beta-n}(\lambda, x),$$

which evidently proves Theorem 1. □

Definition 2 (see [40])**.** *Let f be a continuous function defined on the closed interval $[a, b]$ and assume that $0 < \alpha < 1$ and $\mu > 0$. Then the left-sided fractional integral involving the two parameters α and μ is defined by*

$$\left(\,_a^{\text{LMK}}I^{\alpha,\mu}f\right)(x) = \frac{1-\alpha}{M(\alpha)} \left(\,_aI^{1-\mu}f\right)(x) + \frac{\alpha}{M(\alpha)} \left(\,_aI^{1-\mu+\alpha}f\right)(x),$$

where

$$\left(\,_aI^\beta f\right)(x) = \frac{1}{\Gamma(\beta)} \int_a^x (x-s)^{\beta-1} f(s)\, ds$$

is the Riemann-Liouville fractional integral of the function $f(x)$ of order β.

The following remarks can be found in [39].

Remark 3. *For $\gamma \in \mathbb{N} = \{1, 2, 3, \cdots\}$, the left-sided AB fractional integral of order $\alpha > 0$ is given, for $\mu \leq 1$, by*

$$\left(\,_a^{\text{LMK}}I^{\alpha,\mu,\gamma}f\right)(x) = \sum_{i=0}^\gamma \binom{\gamma}{i} \frac{\alpha^i}{M(\alpha)(1-\alpha)^{i-1}} \left(\,_aI^{\alpha i + 1 - \mu}f\right)(x). \qquad (5)$$

Remark 4. For $0 < \alpha < 1$, $\mu > 0$ and $\gamma \in \mathbb{N}$, it is easily seen that

$$\left({}^{\text{LMK}}_{a}I^{\alpha,\mu,\gamma} \, {}^{\text{GLMK}}_{a}D^{\alpha,\mu,\gamma}f\right)(x) = f(x) - f(a).$$

3. The Shifted Legendre Polynomials and the Fractional Derivatives with Generalized Mittag-Leffler Kernel

In order to obtain the shifted Legendre polynomials on the interval $[0,1]$, we introduce the new variable $z = 2\xi - 1$. The so-shifted Legendre polynomials are defined as follows:

$$\tilde{\Theta}_s(\xi) = \Theta_s(2\xi - 1) = \Theta_{2s}(\sqrt{\xi}),$$

where the set

$$\{\Theta_s(z) \quad (s \in \mathbb{N}_0 = \{0,1,2,\cdots\})\}$$

forms a family of orthogonal Legendre polynomials on the interval $[-1,1]$.

The shifted Legendre polynomial $\tilde{\Theta}_s(\xi)$ of degree s has the expansion given by (see [41])

$$\tilde{\Theta}_s(\xi) = \sum_{k=0}^{s} \frac{(-1)^{s+k} (s+k)!}{(k!)^2 (s-k)!} \xi^k \quad (s \in \mathbb{N}_0), \tag{6}$$

so that, clearly, $\tilde{\Theta}_0(\xi) = 1$, $\tilde{\Theta}_1(\xi) = 2\xi - 1$, and so on.

For deriving approximate solutions, we can approximate the function $\Omega(\xi) \in L_2[0,1]$ as a linear combination of the following $(m+1)$ terms of $\tilde{\Theta}_s(\xi)$ given by (6):

$$\Omega(\xi) \simeq \Omega_m(\xi) = \sum_{i=0}^{m} a_i \, \tilde{\Theta}_i(\xi), \tag{7}$$

where the coefficients a_i are given by

$$a_i = (2i+1) \int_0^1 \tilde{\Theta}_i(\xi) \Omega(\xi) \, d\xi \quad (i \in \mathbb{N}_0 = \{0,1,2,\cdots\}).$$

Now, in Theorem 2 below, we construct the approximation formula for ${}^{\text{GLMK}}_{0}D^{\alpha,\mu,\gamma}(\Omega_m(\xi))$.

Theorem 2. *Given the approximation (7), it is asserted for ${}^{\text{GLMK}}_{0}D^{\alpha,\mu,\gamma}(\Omega_m(t))$ that*

$$ {}^{\text{GLMK}}_{0}D^{\alpha,\mu,\gamma}(\Omega_m(t)) = \sum_{i=\lceil\alpha\rceil}^{m} \sum_{j=\lceil\alpha\rceil}^{i} a_i \, \Pi_{i,j,\alpha_1} \, Y_{i,j}^{\alpha_1,\mu,\gamma}(t), \tag{8}$$

where

$$\Pi_{i,j,\alpha_1} = \frac{M(\alpha_1)}{1-\alpha_1} \sum_{j=\lceil\alpha\rceil}^{i} \frac{(-1)^{i+j} (i+j)! \, \Gamma(j+1)}{(j!)^2 (i-j)!} \tag{9}$$

and

$$Y_{i,j}^{\alpha_1,\mu,\gamma}(t) = \sum_{k=\lceil\alpha\rceil}^{\infty} \frac{(\gamma)_k}{k! \, \Gamma(k\alpha_1 + j + \mu - n)} \lambda^k \, t^{\alpha_1 k + \mu + j - n - 1} \tag{10}$$

with

$$\lceil\alpha\rceil := n + \lceil\alpha_1\rceil,$$

$\lceil\kappa\rceil$ being the greatest integer less than or equal to $\kappa \in \mathbb{R}$.

Proof. We apply the linearization property of the fractional derivative with generalized Mittag-Leffler kernel in (1) and Equation (7). We thus find that

$$^{\text{GLMK}}_{0}D^{\alpha,\mu,\gamma}(\Omega_m(t)) = \sum_{i=0}^{m} a_i\ {}^{\text{GLMK}}_{0}D^{\alpha,\mu,\gamma}(\tilde{\Theta}_i(t)). \qquad (11)$$

The connection between Equations (1), (6) and (11) leads us to

$$^{\text{GLMK}}_{0}D^{\alpha,\mu,\gamma}(\tilde{\Theta}_i(t)) = 0 \quad (i = 0, 1, \cdots, \lceil \alpha \rceil - 1) \qquad (12)$$

and

$$^{\text{GLMK}}_{0}D^{\alpha,\mu,\gamma}(\tilde{\Theta}_i(t)) = \sum_{j=\lceil \alpha \rceil}^{i} \frac{(-1)^{i+j}(i+j)!}{(j!)^2 (i-j)!} \frac{M(\alpha_1)\Gamma(j+1)}{1-\alpha_1}$$
$$\cdot \sum_{k=\lceil \alpha \rceil}^{\infty} \frac{(\gamma)_k}{k!\,\Gamma(k\alpha_1 + j + \mu - n)} \lambda^k\, t^{\alpha_1 k + \mu + j - n - 1} \quad (i = \lceil \alpha \rceil, \cdots, m). \qquad (13)$$

The desired result (8) follows when we combine the Equations (11)–(13). The proof of Theorem 2 is thus completed. □

4. Construction of the Schemes of the Proposed Models

In this section, we construct the schemes of the three models presented below, which are based on the spectral method and the properties of the Legendre polynomials as described in the preceding section.

Model 1

Consider the following fractional differential equation:

$$^{\text{GLMK}}_{0}D^{\alpha,\mu,\gamma}\Xi(\xi) = \xi^2 \quad \text{and} \quad \Xi(0) = 0 \quad (0 < \alpha \leq 1), \qquad (14)$$

where $^{\text{GLMK}}_{0}D^{\alpha,\mu,\gamma}\Xi(\xi)$ is the fractional derivative based on the generalized Mittag-Leffler kernel in the Liouville-Caputo sense.

The exact solution of the initial value problem (14) can be found by applying the operator $_0^{\text{LMK}}I^{\alpha,\mu,\gamma}$ on both sides of the first Equation (14) with the help of Remark 4. We thus get

$$\Xi(\xi) = \sum_{i=1}^{\gamma} \binom{\gamma}{i} \frac{2(1-\alpha)^{1-i} \alpha^i \xi^{\alpha i - \mu + 3}}{M(\alpha)\Gamma(i\alpha - \mu + 4)} + \begin{cases} \dfrac{(1-\alpha)x^2}{M(\alpha)} & (\mu \geq 1) \\[6pt] \dfrac{2(1-\alpha)x^{3-\mu}}{M(\alpha)\Gamma(4-\mu)} & (\mu < 1). \end{cases} \qquad (15)$$

For this model, we now transform Equation (14) into a system of algebraic equations. Indeed, by using a linear combination of the first $m+1$ terms of $\tilde{\Theta}_i(\xi)$, we can approximate and expand the function $\Xi(\xi)$ as follows:

$$\Xi_m(\xi) = \sum_{i=0}^{m} \Xi_i\, \tilde{\Theta}_i(\xi). \qquad (16)$$

In view of Formulas (8) and (16), and upon substituting them into Equation (14), we find that

$$\left(\sum_{i=\lceil \alpha \rceil}^{m} \sum_{j=\lceil \alpha \rceil}^{i} \Xi_i\, \Pi_{i,j,\alpha}\, Y^{\alpha}_{i,j}(\xi) \right) = \xi^2. \qquad (17)$$

Thus, in order to obtain the system of algebraic equations, we collocate Equation (17) at $m + 1 - \lceil \alpha \rceil$ points ξ_r as given below:

$$\sum_{i=\lceil\alpha\rceil}^{m}\sum_{j=\lceil\alpha\rceil}^{i}\Xi_i\,\Pi_{i,j,\alpha}\,Y_{i,j}^{\alpha}(\xi_r)=(\xi_r)^2. \tag{18}$$

If we now substitute Equation (16) into (14), we obtain the following initial condition:

$$\sum_{i=0}^{m}\Theta_i(0)\Xi_i=\Xi(0). \tag{19}$$

Since the set of Equations (18) to (19) is linear, we can solve it by using known methods and we obtain Ξ_i.

Finally, we substitute these Ξ_i into (16) and obtain the approximate solution of the initial-value problem (14).

Model 2

In this second model, we consider the following fractional-order Fisher equation (FFE) with the generalized Mittag-Leffler kernel:

$$\Xi_\eta(\xi,\eta)=^{\text{GLMK}}_{0}D_\xi^{\alpha,\mu,\gamma}\Xi(\xi,\eta)+\delta\Xi(\xi,\eta)\bigl(1-\Xi(\xi,\eta)\bigr) \tag{20}$$

$$(0<\delta<1;\ 0<\alpha\leqq 2).$$

The exact solution of the Fisher Equation (20) for $\alpha=2$ is given by (see [42])

$$\Xi(\xi,\eta)=\left[1+\exp\left(\sqrt{\frac{\delta}{6}}\xi-\frac{5}{6}\delta\eta\right)\right]^{-1}, \tag{21}$$

subject to the following initial and boundary conditions:

$$\Xi(0,\eta)=h_1(\eta)\quad\text{and}\quad\Xi(1,\eta)=h_2(\eta), \tag{22}$$

with

$$\Xi(\xi,0)=\bar{\Xi}(\xi). \tag{23}$$

Based on the following the steps, we can obtain the numerical solution of Model 2 as follows.

1. We write $\alpha=1+\alpha_1$, where $\alpha_1\in(0,1]$.
2. We can approximate and expand the function $\Xi(\xi,\eta)$ by using a linear combination of the first $m+1$ terms of $\bar{\Theta}_i(\xi)$ as given below:

$$\Xi_m(\xi,\eta)=\sum_{i=0}^{m}\Xi_i(\eta)\,\bar{\Theta}_i(\xi). \tag{24}$$

3. In view of the Formulas (8) and (24), and upon substituting them into Equation (20), we obtain

$$\sum_{i=0}^{m}\frac{d\,\Xi_i(\eta)}{d\eta}\bar{\Theta}_i(\xi)=\left(\sum_{i=\lceil\alpha\rceil}^{m}\sum_{j=\lceil\alpha\rceil}^{i}\Xi_i(\eta)\,\Pi_{i,j,\alpha_1}Y_{i,j}^{\alpha_1,\mu,\gamma}(\xi)\right)$$
$$+\left(\sum_{i=0}^{m}\Xi_i(\eta)\,\bar{\Theta}_i(\xi)\right)\left(1-\sum_{i=0}^{m}\Xi_i(\eta)\,\bar{\Theta}_i(\xi)\right). \tag{25}$$

4. We collocate Equation (25) at $m+1-\lceil\alpha\rceil$ points ξ_r in order to get the following system of first-order ODEs:

$$\sum_{i=0}^{m} \frac{d \, \Xi_i(\eta)}{d\eta} \tilde{\Theta}_i(\zeta_r) = \left(\sum_{i=\lceil \alpha \rceil}^{m} \sum_{j=\lceil \alpha \rceil}^{i} \Xi_i(\eta) \Pi_{i,j,\alpha_1} Y_{i,j}^{\alpha_1,\mu,\gamma}(\zeta_r) \right)$$
$$+ \mu \left(\sum_{i=0}^{m} \Xi_i(\eta) \tilde{\Theta}_i(\zeta_r) \right) \left(1 - \sum_{i=0}^{m} \Xi_i(\eta) \tilde{\Theta}_i(\zeta_r) \right). \quad (26)$$

5. Substituting from Equation (24) into (20), we can obtain the following boundary conditions of the system (26):

$$\sum_{i=0}^{m} \tilde{\Theta}_i(0) \Xi_i(\eta) = h_1(\eta) \quad \text{and} \quad \sum_{i=0}^{m} \tilde{\Theta}_i(1) \Xi_i(\zeta) = h_2(\zeta). \quad (27)$$

6. To obtain systems of nonlinear algebraic equations, we apply the FDM to the ODE (26) and (27), where
$$\Xi_i(\zeta) \quad (i = 0, 1, \cdots, m)$$
are unknown. We thus obtain

$$\sum_{i=0}^{m} \left(\frac{\Xi_i^s - \Xi_i^{s-1}}{\tau} \right) \tilde{\Theta}_i(\zeta_r) = \left(\sum_{i=\lceil \alpha \rceil}^{m} \sum_{j=\lceil \alpha \rceil}^{i} \Xi_i(\eta) \Pi_{i,j,\alpha_1} Y_{i,j}^{\alpha_1,\mu,\gamma}(\zeta_r) \right)$$
$$+ \mu \left(\sum_{i=0}^{m} \Xi_i(\eta) \tilde{\Theta}_i(\zeta_r) \right) \left(1 - \sum_{i=0}^{m} \Xi_i(\eta) \tilde{\Theta}_i(\zeta_r) \right) \quad (28)$$

and

$$\sum_{i=0}^{m} \left(\frac{\Xi_i^s - \Xi_i^{s-1}}{\tau} \right) \tilde{\Theta}_i(\zeta_r) = \left(\sum_{i=\lceil \alpha \rceil}^{m} \sum_{j=\lceil \alpha \rceil}^{i} \Xi_i(\eta) \Pi_{i,j,\alpha_1} Y_{i,j}^{\alpha_1,\mu,\gamma}(\zeta_r) \right)$$
$$+ \mu \left(\sum_{i=0}^{m} \Xi_i(\eta) \tilde{\Theta}_i(\zeta_r) \right) \left(1 - \sum_{i=0}^{m} \Xi_i(\eta) \tilde{\Theta}_i(\zeta_r) \right), \quad (29)$$

with

$$\sum_{i=0}^{m} \tilde{\Theta}_i(0) \Xi_i^s = h_1^s \quad \text{and} \quad \sum_{i=0}^{m} \tilde{\Theta}_i(1) \Xi_i^s = h_2^s. \quad (30)$$

7. We can explain more fully for the case when $m = 4$. By using the NIM, we obtain the following system given by (28) and (30) in the matrix form:

$$\Xi^{s+1} = \Xi^s - J^{-1}(\Xi^s) G(\Xi^s), \quad (31)$$

where
$$\Xi^s = (\Xi_0^s, \Xi_1^s, \Xi_2^s, \Xi_3^s, \Xi_4^s)^T,$$

$J^{-1}(\Xi^s)$ is the inverse of the Jacobian matrix and $G(\Xi^s)$ is the vector that represents the nonlinear equations. The initial solution Ξ^0 can be obtained by setting $s = 0$ in the initial condition (23) as detailed below.

(a) After substituting Equation (24) into the initial condition (23), we obtain

$$\Xi(\zeta, 0) = \bar{\Xi}(\zeta) \simeq \sum_{i=0}^{4} \Xi_i(0) \tilde{\Theta}_i(\zeta). \quad (32)$$

(b) Solving the following system of linear equations, we obtain the components of the initial solution Ξ^0:

$$\sum_{i=0}^{4} \Xi_i^0 \, \tilde{\Theta}_i(\zeta_r) = \bar{\Xi}(\zeta_r), \qquad r = 0,1,2,3,4, \tag{33}$$

where the points ζ_r $(r = 0,1,2,3,4)$ are the roots of $\tilde{\Theta}_5(\zeta)$.

Model 3

Now, with the third model we consider the fractional Korteweg–de Vries equation (FKdVE) with generalized Mittag-Leffler kernel

$$\Xi_\eta(\zeta,\eta) + \frac{1}{2} \, {}^{\text{GMLK}}_{\quad 0}D_\zeta^{\alpha_1,\mu,\gamma} \Xi(\zeta,\eta) + \Xi(\zeta,\eta) \, {}^{\text{GLMK}}_{\quad 0}D_\zeta^{\alpha_2,\mu,\gamma} \Xi(\zeta,\eta) = 0, \tag{34}$$

where $0 < \alpha_1 \leqq 1$ and $0 < \alpha_2 \leqq 3$.

The exact solution for the classical integer-form is given by

$$\Xi(\zeta,\eta) = 6b^2 \left[\text{sech}\left(b\zeta - 2b^3\eta\right) \right]^2, \tag{35}$$

subject to the following initial and boundary conditions:

$$\Xi(0,\eta) = f_1(\eta) \quad \text{and} \quad \Xi(1,\eta) = f_2(\eta) \tag{36}$$

and

$$\Xi(\zeta,0) = \bar{\Xi}(\zeta). \tag{37}$$

Following the same steps as we used in the cases of Model 1 and Model 2, together with

$$\alpha_2 = N + \tilde{\alpha}_2 \qquad (N = 0,1,2)$$

and $\tilde{\alpha}_2 \in (0,1]$, we can obtain the following set of algebraic equations solving Model 3:

$$\sum_{i=0}^{m} \left(\frac{\Xi_i^s - \Xi_i^{s-1}}{\tau} \right) \tilde{\Theta}_i(\zeta_r) + \frac{1}{2} \left(\sum_{i=\lceil \alpha_1 \rceil}^{m} \sum_{j=\lceil \alpha_1 \rceil}^{i} \Xi_i(\eta) \Pi_{1,i,j,\alpha_1} Y_{1,i,j}^{\alpha_1,\mu_1,\gamma_1}(\zeta_r) \right)$$

$$+ \left(\sum_{i=0}^{m} \Xi_i(\eta) \tilde{\Theta}_i(\zeta_r) \right) \left(\sum_{i=\lceil \alpha_2 \rceil}^{m} \sum_{j=\lceil \alpha_2 \rceil}^{i} \Xi_i(\eta) \Pi_{2,i,j,\tilde{\alpha}_2} Y_{2,i,j}^{\tilde{\alpha}_2,\mu_2,\gamma_2}(\zeta_r) \right) = 0. \tag{38}$$

5. Numerical Results, Graphical Illustrations and Discussions

In this section, we discuss the numerical results for the approximate solutions presented in Section 4. These results will be illustrated in a number of figures.

For Model 1, the accuracy and efficiency of the numerical approximate solution are verified by comparison with the analytical solution that we computed. For Model 2 and Model 3, the accuracy of the numerical solutions can be satisfied by using the residual error function. This will also be illustrated in various figures.

Figure 1 shows the behavior of the analytical solution (15) of Model 1 for different values of μ and γ. In this figure, we set $\alpha = 0.3$ and $m = 6$.

Figure 2 shows the absolute error between the approximate and the exact solutions of Model 1 with $\alpha = 0.3$ and $m = 6$, and for different values of μ and γ. It is clear from Figure 2 that the absolute error is very small and of the order of the error is 10^{-6}. This gives a good impression of the accuracy and efficiency of the solutions, since the comparison is made with the analytical solution. This impression is useful when numerical solutions are found for fractional-order systems that do not have an analytical solution.

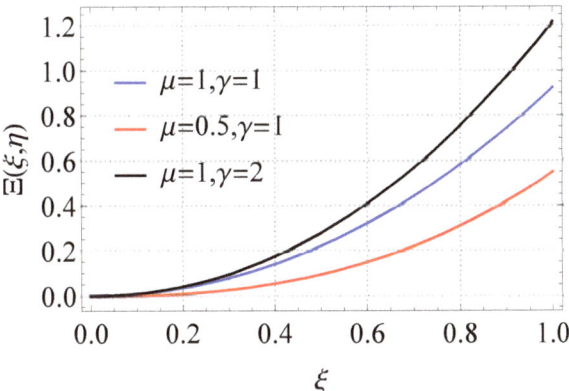

Figure 1. Graph of the exact solutions of Model 1, with $\alpha = 0.3$ and $m = 6$, and for different values of μ and γ.

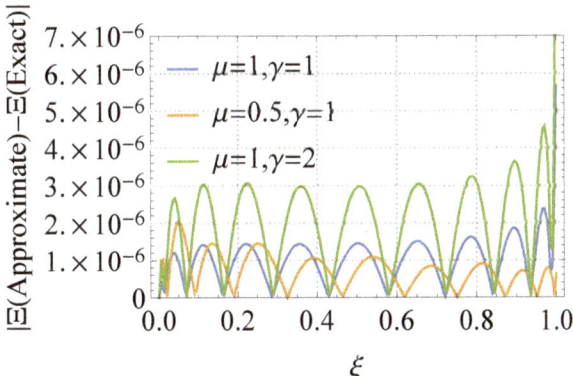

Figure 2. Graph of the absolute error between the exact and the approximate solutions of Model 1 when $\alpha = 0.3$ and $m = 6$, and for different values of μ and γ.

In Figure 3, we show the approximate solution of the FFE for different values of μ. In Model 2, we set $\alpha = 1.5$, $T = 1$, $\tau = 0.00001$, $\gamma = 1$ and $m = 10$. In Model 2, the analytical solution is unknown; thus, in order to verify the accuracy of the approximate solution, we define the residual error function (REF) as follows:

$$\text{REF}(\xi, \eta) = \left(\Xi_m(\xi, \eta)\right)_\eta - {}^{\text{GLMK}}_{0}D^{\alpha,\mu,\gamma}(\Xi_m(\xi, \eta)) \\ - \mu\, \Xi_m(\xi, \eta)(1 - \Xi_m(\xi, \eta)), \qquad (39)$$

We illustrate the REF in Figure 4 for different values of μ and γ, and for $\alpha = 1.5$, $T = 1$, $\tau = 0.00001$ and $m = 10$. We note from this illustration in Figure 4 that the order of the REF is 10^{-3}.

In the case of Model 3 for the FKdVE, we follow the same procedures and treatments that were applied for Model 2. In Model 3, we set $\alpha_2 = 2.7$, $\alpha_1 = 0.7$, $T = 1$, $\tau = 0.00001$ and $m = 5$. The numerical results of Model 3 are illustrated in Figures 5 and 6.

Remarkably, it is presumably the first time that the numerical results, which are presented in this paper, are based upon the use of the fractional derivatives and also upon the properties of the Legendre polynomials as well as GMLK.

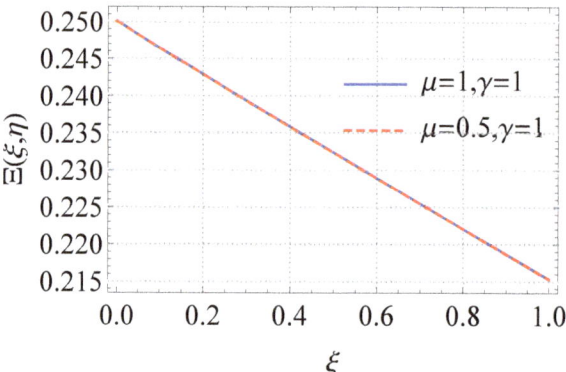

Figure 3. Graph of the solutions of Model 2, with $\alpha = 1.5$, $T = 1$, $m = 10$ and $\tau = 0.00001$, and for different values of μ and γ.

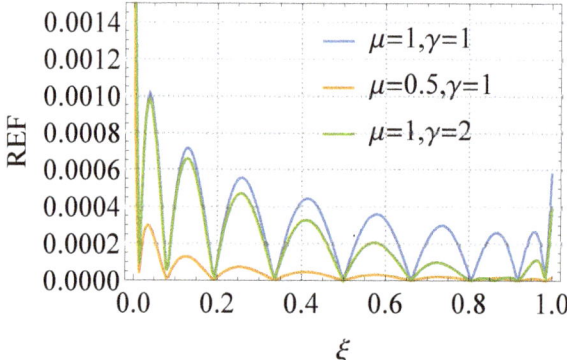

Figure 4. Graph of the residual error function (REF) of the approximate solution of Model 2, with $\alpha = 1.5$, $T = 1$, $m = 10$ and $\tau = 0.00001$, and for different values of μ and γ.

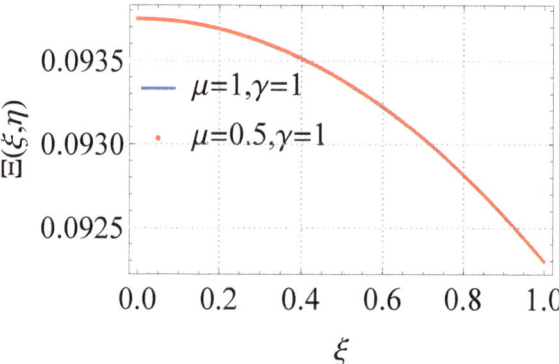

Figure 5. Graph of the solution of Model 3, with $\alpha_2 = 2.7$, $\alpha_1 = 0.7$, $T = 1$, $m = 5$ and $\tau = 0.00001$, and for different values of μ and γ.

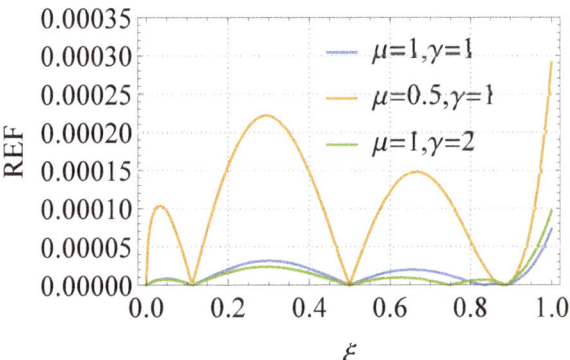

Figure 6. Graph of the of the residual error function (REF) of the approximate solution of Model 3, with $\alpha_2 = 2.7$, $\alpha_1 = 0.7$, $T = 1$, $m = 5$ and $\tau = 0.00001$, and for different values of μ and γ.

6. Conclusions

In this paper, classical (integer-order) derivatives have been replaced by some fractional-order derivatives, which are based upon generalized Mittag-Leffler type kernels. Three fractional-order models have been presented: The first model belongs to fractional ordinary differential equations, and the other two models involve fractional partial differential equations. In the first case, the treatment was achieved by directly converting the fractional-order model with the help of the Legendre polynomials to a system of algebraic equations and then finding approximate solutions by using the Newton-Raphson method, in addition to finding the solution analytically. In the second and the third fractional-order models, they were converted into differential equations, and by using the finite-difference method (FDM) to approximate the derivative with respect to time, we were led to algebraic equations. We then obtained approximate solutions as in the first model.

The accuracy of the solution was verified in the first model in comparison with the exact solution. For the second and the third models, the accuracy of the approximate solutions was verified by calculating the residual error function (REF). In all cases, the order of the error is small, and its value ranges between 10^{-3} and 10^{-6}. In all of the calculations in this paper, the *Mathematica* software package was used. Finally, we suggest that the researchers test the algorithm's efficiency using several special functions, such as Bernstein, Chebyshev, and others.

Author Contributions: H.M.S. suggested and initiated this work, performed its validation, as well as reviewed and edited the paper. A.-K.N.A. performed the formal analysis of the investigation, the methodology, the software, and wrote the first draft of the paper. K.M.S. performed the formal analysis of the investigation, the methodology, the software, and wrote the first draft of the paper. W.M.H. performed the methodology, the software, and reviewed and edited the paper. All authors have read and agreed to the published version of the manuscript.

Funding: This research received no external funding.

Institutional Review Board Statement: Not applicable.

Informed Consent Statement: Not applicable.

Data Availability Statement: Not applicable.

Conflicts of Interest: The authors declare that there are no conflict of interest regarding the publication of this paper.

References

1. Samko, S.G.; Kilbas, A.A.; Marichev, O.I. *Fractional Integrals and Derivatives: Theory and Applications*; Gordon and Breach Science Publishers: Reading, UK; Tokyo, Japan; Paris, France; Berlin, Germany; Langhorne, PA, USA, 1993.
2. Podlubny, I. Fractional Differential Equations: An Introduction to Fractional Derivatives, Fractional Differential Equations, to Methods of Their Solution and Some of Their Applications. In *Mathematics in Science and Engineering*; Academic Press: New York, NY, USA; London, UK; Sydney, NSW, Australia; Tokyo, Japan; Toronto, ON, Canada, 1999; Volume 198.
3. Kilbas, A.A.; Srivastava, H.M.; Trujillo, J.J. Theory and Applications of Fractional Differential Equations. In *North-Holland Mathematical Studies*; Elsevier (North-Holland) Science Publishers: Amsterdam, The Netherlands; London, UK; New York, NY, USA, 2006; Volume 204.
4. Tarasov, V.E. *Fractional Dynamics: Applications of Fractional Calculus to Dynamics of Particles, Fields and Media*; Springer: Berlin/Heidelberg, Germany, 2010.
5. Gorenflo, R.; Kilbas, A.A.; Mainardi, F.; Rogosin, S.V. *Mittag-Leffler Functions, Related Topics and Applications*, 2nd ed.; Springer: New York, NY, USA, 2020.
6. Srivastava, H.M. Some parametric and argument variations of the operators of fractional calculus and related special functions and integral transformations. *J. Nonlinear Convex Anal.* **2021**, *22*, 1501–1520.
7. Srivastava, H.M. Fractional-order derivatives and integrals: Introductory overview and recent developments. *Kyungpook Math. J.* **2020**, *60*, 73–116.
8. Diethelm, K. An algorithm for the numerical solution of differential equations of fractional order. *Electron. Trans. Numer. (ETNA)* **1997**, *5*, 1–6.
9. Khader, M.M.; Saad, K.M. A numerical approach for solving the problem of biological invasion (fractional Fisher equation) using Chebyshev spectral collocation method. *Chaos Solitons Fractals* **2018**, *110*, 169–177. [CrossRef]
10. Khader, M.M.; Saad, K.M. On the numerical evaluation for studying the fractional KdV, KdV-Burgers', and Burgers' equations. *Eur. Phys. J. Plus* **2018**, *133*, 1–13. [CrossRef]
11. Atangana, A.; Baleanu, D. New fractional derivatives with nonlocal and non-singular kernel: Theory and application to heat transfer model. *Therm. Sci.* **2016**, *20*, 763–769. [CrossRef]
12. Caputo, M.; Fabrizio, M. A new definition of fractional derivative without singular kernel. *Prog. Fract. Differ. Appl.* **2015**, *1*, 1–13.
13. Saad, K.M.; Khader, M.M.; Gómez-Aguilar, J.F.; Baleanu, D. Numerical solutions of the fractional Fisher's type equations with Atangana-Baleanu fractional derivative by using spectral collocation methods. *Chaos Interdiscip. J. Nonlinear Sci.* **2019**, *29*, 023116. [CrossRef]
14. Fernandez, A.; Abdeljawad, T.; Baleanu, D. Relations between fractional models with three-parameter Mittag-Leffler kernels. *Adv. Differ. Equ.* **2020**, *2020*, 1–13. [CrossRef]
15. Abdeljawad, T. A Lyapunov type inequality for fractional operators with nonsingular Mittag-Leffler kernel. *J. Inequalities Appl.* **2017**, *2017*, 1–11. [CrossRef]
16. Jarad, F.; Abdeljawad, T.; Hammouch, Z. On a class of ordinary differential equations in the frame of Atangana–Baleanu fractional derivative. *Chaos Solitons Fractals* **2018**, *117*, 16–20. [CrossRef]
17. Izadi, M.; Srivastava, H.M. Numerical approximations to the nonlinear fractional-order logistic population model with fractional-order Bessel and Legendre bases. *Chaos Solitons Fractals* **2021**, *145*, 110779. [CrossRef]
18. Morales-Delgado, V.F.; Gómez-Aguilar, J.F.; Saad, K.M.; Khan, M.A.; Agarwal, P. Analytic solution for oxygen diffusion from capillary to tissues involving external force effects: A fractional calculus approach. *Physica A Stat. Mech. Its Appl.* **2019**, *523*, 48–65. [CrossRef]
19. Srivastava, H.M.; Ahmad, H.; Ahmad, I.; Thounthong, P.; Khan, M.N. Numerical simulation of 3-D fractional-order convection-diffusion PDE by a local meshless method. *Therm. Sci.* **2021**, *25*, 347–358. [CrossRef]
20. Saad, K.M. A reliable analytical algorithm for space-time fractional cubic isothermal autocatalytic chemical system. *Pramana J. Phys.* **2018**, *91*, 1–15. [CrossRef]
21. Singh, H.; Srivastava, H.M. Numerical investigation of the fractional-order Liénard and Duffing equation arising in oscillating circuit theory. *Front. Phys.* **2020**, *8*, 120. [CrossRef]
22. Alomari, A.K. Homotopy-Sumudu transforms for solving system of fractional partial differential equations. *Adv. Differ. Equ.* **2020**, *2020*, 222. [CrossRef]
23. Kumar, S.; Pandey, R.K.; Srivastava, H.M.; Singh, G.N. A convergent collocation approach for generalized fractional integro-differential equations using Jacobi poly-fractonomials. *Mathematics* **2021**, *9*, 979. [CrossRef]
24. Alomari, A.K.; Syam, M.I.; Anakira, N.R.; Jameel, A.F. Homotopy Sumudu transform method for solving applications in physics. *Results Phys.* **2020**, *18*, 103265. [CrossRef]
25. Aljhani, S.; Noorani, M.S.; Alomari, A.K. Numerical solution of fractional-order HIV Model using homotopy method. *Discret. Dyn. Nat. Soc.* **2020**, *2020*, 2149037. [CrossRef]
26. Saad, K.M.; Al-Sharif, E.H.F. Comparative study of a cubic autocatalytic reaction via different analysis methods. *Discret. Contin. Dyn.-Syst.-S* **2019**, *12*, 665–684.
27. Saad, K.M.; Gómez-Aguilar, J.F. Coupled reaction-diffusion waves in a chemical system via fractional derivatives in Liouville-Caputo sense. *Rev. Mex. Física* **2018**, *64*, 539–547.

28. Saad, K.M.; Gómez-Aguilar J.F. Analysis of reaction-diffusion system via a new fractional derivative with non-singular kernel. *Physica A Stat. Mech. Its Appl.* **2018**, *509*, 703–716. [CrossRef]
29. Saad, K.M.; Baleanu D.; Atangana, A. New Fractional derivatives applied to the Korteweg-de Vries and Korteweg-de Vries-Burgers equations. *Comput. Appl. Math.* **2018**, *37*, 5203–5216. [CrossRef]
30. Saad, K.M.; Deniz, S.; Baleanu, D. On a new modified fractional analysis of Nagumo equation. *Int. J. Biomath.* **2019**, *12*, 1950034. [CrossRef]
31. Srivastava, H.M.; Saad, K.M. Some new and modified fractional analysis of the time-fractional Drinfeld-Sokolov-Wilson system. *Chaos Interdiscip. J. Nonlinear Sci.* **2020**, *30*, 113104. [CrossRef]
32. Srivastava, H.M.; Deniz, S.; Saad, K.M. An efficient semi-analytical method for solving the generalized regularized long wave equations with a new fractional derivative operator. *J. King Saud Univ. Sci.* **2021**, *33*, 101345. [CrossRef]
33. Srivastava, H.M.; Saad, K.M. A comparative Study of the fractional-order clock chemical model. *Mathematics* **2020**, *8*, 1436. [CrossRef]
34. Srivastava, H.M.; Saad, K.M.; Gómez-Aguilar, J.F.; Almadiy, A.A. Some new mathematical models of the fractional-order system of human immune against IAV infection. *Math. Biosci. Eng.* **2020**, *17*, 4942–4969. [CrossRef]
35. Srivastava, H.M.; Saad, K.M.; Khader, M.M. An efficient spectral collocation method for the dynamic simulation of the fractional epidemiological model of the Ebola virus. *Chaos Solitons Fractals* **2020**, *140*, 110174.
36. Saad, K.M. Comparative study on fractional isothermal chemical model. *Alex. Eng. J.* **2021**, *60*, 3265–3274. [CrossRef]
37. Saad, K.M. Comparing the Caputo, Caputo-Fabrizio and Atangana-Baleanu derivative with fractional order: Fractional cubic isothermal auto-catalytic chemical system. *Eur. Phys. J. Plus* **2018**, *133*, 94. [CrossRef]
38. Abdeljawad, T. Fractional difference operators with discrete generalized Mittag-Leffler kernels. *Chaos Solitons Fractals* **2019**, *126*, 315–324. [CrossRef]
39. Abdeljawad, T. Fractional operators with generalized Mittag-Leffler kernels and their iterated differintegrals. *Chaos Interdiscip. Nonlinear Sci.* **2019**, *29*, 023102. [CrossRef]
40. Abdeljawad, T.; Baleanu, D. On fractional derivatives with generalized Mittag-Leffler kernels. *Adv. Differ. Equ.* **2018**, *2018*, 468. [CrossRef]
41. Hesthaven, J.; Gottlieb, S.; Gottlieb, D. *Spectral Methods for Time-Dependent Problems*; Cambridge University Press: Cambridge, UK; London, UK; New York, NY, USA, 2007.
42. Feng, Z. Travelling wave solutions and proper solutions to the two-dimensional Burgers-Korteweg-de Vries. *J. Phys. A Math. Gen.* **2003**, *36*, 8817–8827. [CrossRef]

Article
Cesaro Limits for Fractional Dynamics

Yuri Kondratiev [1,2,†] and José da Silva [3,*,†]

[1] Department of Mathematics, University of Bielefeld, D-33615 Bielefeld, Germany; kondrat@math.uni-bielefeld.de
[2] Institute of Mathematics of the National Academy of Sciences of Ukraine, National Pedagogical Dragomanov University, 33615 Kiev, Ukraine
[3] CIMA, University of Madeira, Campus da Penteada, 9020-105 Funchal, Portugal
* Correspondence: joses@staff.uma.pt; Tel.: +351-291-705-185
† These authors contributed equally to this work.

Abstract: We study the asymptotic behavior of random time changes of dynamical systems. As random time changes we propose three classes which exhibits different patterns of asymptotic decays. The subordination principle may be applied to study the asymptotic behavior of the random time dynamical systems. It turns out that for the special case of stable subordinators explicit expressions for the subordination are known and its asymptotic behavior are derived. For more general classes of random time changes explicit calculations are essentially more complicated and we reduce our study to the asymptotic behavior of the corresponding Cesaro limit.

Keywords: dynamical systems; random time change; inverse subordinator; asymptotic behavior

1. Introduction

In this paper we will deal with Markov processes or dynamical systems in $\mathbb{R}^d$. These processes or dynamics starting from $x \in \mathbb{R}^d$, denote by $X^x(t)$, $t \geq 0$, have associated evolution equations on $\mathbb{R}^d$. In the Markov case we define for suitable $f : \mathbb{R}^d \longrightarrow \mathbb{R}$ the function $u(t,x) = \mathbb{E}[f(X^x(t))]$ which satisfied the Kolmogorov equation

$$\frac{\partial}{\partial t} u(t,x) = L u(t,x),$$

where L is the generator of the Markov process.

For a dynamical system we introduce $u(t,x) = f(X^x(t))$. Then this function is the solution of the Liouville equation

$$\frac{\partial}{\partial t} u(t,x) = L u(t,x),$$

where now L is the Liouville operator for the dynamical system, see e.g., Kondratiev and da Silva [1].

Let $S(t)$, $t \geq 0$ be a subordinator and $E(t)$, $t \geq 0$ denotes the inverse subordinator, that is, for each $t \geq 0$, $E(t) := \inf\{s > 0 \mid S(s) > t\}$. This random process we consider as a random time and assume to be independent of $X^x(t)$. Define a random process Y^x by

$$Y^x(t, \omega) := X^x(E(t, \omega)).$$

Then as above we may introduce

$$u^E(t,x) = \mathbb{E}[f(Y^x(t))].$$

For both Markov and dynamical system cases this function satisfies the evolution equations

$$D_t^E u^E(t,x) = L u^E(t,x)$$

where L is the Kolmogorov or Liouville operator correspondingly. Here D_t^E is a generalized fractional time derivative corresponding to the inverse subordinator $E(t)$, see Section 2 below for details, in particular the definition in (15). The main relation which is true for both cases is the following subordination formula:

$$u^E(t,x) = \int_0^\infty u(\tau,x) G_t(\tau) d\tau, \qquad (1)$$

where $G_t(\tau)$ is the density of the inverse subordinator $E(t)$, see, e.g., Toaldo [2], Kondratiev and da Silva [1] and especially the book Meerschaert and Sikorskii [3]. This formula which relates the solutions of the evolution equations with usual and fractional derivatives plays an important role in the study of dynamics with random times. Note that there exist such relations between random times, fractional equations and subordination in the framework of physical models, see, e.g., Mura et al. [4].

The goal of this paper is to study and analyze the asymptotic behavior of two elementary dynamical system after the random time change, namely $u(t,x) = e^{-at}$, $a > 0$ and $u(t,x) = t^n$, $n \geq 0$. Here the dynamical system are considered as a deterministic Markov processes. For particular classes of random times the subordination formula (1) is evaluated explicitly. This is true, for example, in the case of inverse stable subordinators. For a general inverse subordinator the properties of the density $G_t(\tau)$ are unknown and the evaluation of (1) is not possible. Actually, it is a long standing open problem in the theory of stochastic processes.

We propose an alternative approach to study the asymptotic behavior of $u^E(t,x)$. More precisely, we consider Cesaro limits (the asymptotic of the Cesaro mean of $u^E(t,x)$, see (23) below) of $u^E(t,x)$ using the subordination formula representation (1) together with the Feller–Karamata Tauberian theorem, see Theorem 1. For many classes of random times this approach leads to a precise asymptotic behavior. In this paper we investigate three classes of random time change, denote by (17)–(19), see Section 2, which exhibits different patterns of decays of the Cesaro limit of $u^E(t,x)$. We would like to emphasize that for particular classes of random times, namely inverse stable subordinators, the asymptotic of $u^E(t,x)$ which may be computed explicitly, coincides with the Cesaro limit. For other classes of random times the Cesaro limit gives one possible characteristic of the asymptotic for $u^E(t,x)$. To the best of our knowledge at the present time no other information on the asymptotic of $u^E(t,x)$ is known for a general subordinator.

The remaining of the paper is organized as follows. In Section 2 we introduce three classes (17)–(19) of subordinator processes which serves as random times. These classes are given in terms of their local behavior of the Laplace exponent at $\lambda = 0$. In addition, we state the main results of the paper. Section 3 is a preparation for the more general study of the asymptotic of the subordination in Section 4. More precisely, we investigate in detail the special case of the inverse stable subordinator where explicit expressions are known. Hence, the expression for the subordination (1) is derived (for the two dynamical systems considered above) as well as their Cesaro limit. It turns out that both asymptotic for $u^E(t,x)$ (the explicit calculations and Cesaro limit) are the same. Finally in Section 4 we study the Cesaro limit for the general classes (17)–(19) of random time changes.

2. Random Times Processes

In this section we introduce three classes of subordinators which serves as random times processes. More precisely, the random times corresponds to the inverse of subordinator processes whose Laplace exponent satisfies certain conditions, see below for details. The simplest example in class (17) below, is the well known α-stable subordinators whose inverse processes are well studied in the literature, see for example Bingham [5] or Feller [6].

The classes of processes to be introduced which serve as random times have a connection with the concept of general fractional derivatives (see Kochubei [7] for details and applications to fractional differential equations) associated to an admissible kernels $k \in L^1_{loc}(\mathbb{R}_+)$ which is characterized in terms of their Laplace transforms $\mathcal{K}(\lambda)$ as $\lambda \to 0$, see assumption (H) below.

2.1. Definitions and Main Assumptions

Let $S = \{S(t),\ t \geq 0\}$ be a subordinator without drift starting at zero, that is, an increasing Lévy process starting at zero, see Bertoin [8] for more details. The Laplace transform of $S(t), t \geq 0$ is expressed in terms of a Bernstein function $\Phi : [0, \infty) \longrightarrow [0, \infty)$ (also known as Laplace exponent) by

$$\mathbb{E}(e^{-\lambda S(t)}) = e^{-t\Phi(\lambda)}, \quad \lambda \geq 0.$$

The function Φ admits the Lévy-Khintchine representation

$$\Phi(\lambda) = \int_{(0,\infty)} (1 - e^{-\lambda \tau})\, d\sigma(\tau), \qquad (2)$$

where the measure σ (called Lévy measure) has support in $[0, \infty)$ and fulfills

$$\int_{(0,\infty)} (1 \wedge \tau)\, d\sigma(\tau) < \infty. \qquad (3)$$

In what follows we assume that the Lévy measure σ satisfy

$$\sigma\big((0, \infty)\big) = \infty. \qquad (4)$$

Using the Lévy measure σ we define the kernel k as follows

$$k : (0, \infty) \longrightarrow (0, \infty),\ t \mapsto k(t) := \sigma\big((t, \infty)\big). \qquad (5)$$

Its Laplace transform is denoted by $\mathcal{K}$, that is, for any $\lambda \geq 0$ one has

$$\mathcal{K}(\lambda) := \int_0^\infty e^{-\lambda t} k(t)\, dt. \qquad (6)$$

The relation between the function $\mathcal{K}$ and the Laplace exponent Φ is given by

$$\Phi(\lambda) = \lambda \mathcal{K}(\lambda), \quad \forall \lambda \geq 0. \qquad (7)$$

We make the following assumption on the Laplace exponent $\Phi(\lambda)$ of the subordinator S.

(H) Φ is a complete Bernstein function (more precisely, the Lévy measure σ has a completely monotone density $\rho(t)$ with respect to the Lebesgue measure, that is, $(-1)^n \rho^{(n)}(t) \geq 0$ for all $t > 0$, $n = 0, 1, 2, \ldots$) and the functions $\mathcal{K}, \Phi$ satisfy

$$\mathcal{K}(\lambda) \to \infty,\ \text{as}\ \lambda \to 0;\quad \mathcal{K}(\lambda) \to 0,\ \text{as}\ \lambda \to \infty; \qquad (8)$$

$$\Phi(\lambda) \to 0,\ \text{as}\ \lambda \to 0;\quad \Phi(\lambda) \to \infty,\ \text{as}\ \lambda \to \infty. \qquad (9)$$

Example 1. *A classical example of a subordinator S is the so-called α-stable process with index $\alpha \in (0, 1)$. Specifically, a subordinator is α-stable if its Laplace exponent is*

$$\Phi(\lambda) = \lambda^\alpha = \int_0^\infty (1 - e^{-\lambda \tau}) \frac{\alpha \tau^{-1-\alpha}}{\Gamma(1-\alpha)}\, d\tau.$$

In this case it follows that the Lévy measure is $d\sigma_\alpha(\tau) = \frac{\alpha}{\Gamma(1-\alpha)}\tau^{-(1+\alpha)}\,d\tau$. The corresponding kernel k_α has the form $k_\alpha(t) = g_{1-\alpha}(t) := \frac{t^{-\alpha}}{\Gamma(1-\alpha)}$, $t \geq 0$ and its Laplace transform is $\mathcal{K}_\alpha(\lambda) = \lambda^{\alpha-1}$, $\lambda > 0$.

Example 2. *Sum of two stable subordinators.* Let $0 < \alpha < \beta < 1$ be given and $S_{\alpha,\beta}(t)$, $t \geq 0$ the driftless subordinator with Laplace exponent given by

$$\Phi_{\alpha,\beta}(\lambda) = \lambda^\alpha + \lambda^\beta.$$

It is clear from Example 1 that the corresponding Lévy measure $\sigma_{\alpha,\beta}$ is the sum of two Lévy measures, that is,

$$d\sigma_{\alpha,\beta}(\tau) = d\sigma_\alpha(\tau) + d\sigma_\beta(\tau) = \frac{\alpha}{\Gamma(1-\alpha)}\tau^{-(1+\alpha)}\,d\tau + \frac{\beta}{\Gamma(1-\beta)}\tau^{-(1+\beta)}\,d\tau.$$

Then the associated kernel $k_{\alpha,\beta}$ is

$$k_{\alpha,\beta}(t) := g_{1-\alpha}(t) + g_{1-\beta}(t) = \frac{t^{-\alpha}}{\Gamma(1-\alpha)} + \frac{t^{-\beta}}{\Gamma(1-\beta)},\ t > 0$$

and its Laplace transform is $\mathcal{K}_{\alpha,\beta}(\lambda) = \mathcal{K}_\alpha(\lambda) + \mathcal{K}_\beta(\lambda) = \lambda^{\alpha-1} + \lambda^{\beta-1}$, $\lambda > 0$.

Let E be the inverse process of the subordinator S, that is,

$$E(t) := \inf\{s > 0 \mid S(s) > t\} = \sup\{s \geq 0 \mid S(s) \leq t\}. \tag{10}$$

For any $t \geq 0$ we denote by $G_t(\tau)$, $\tau \geq 0$ the marginal density of $E(t)$ or, equivalently

$$G_t(\tau)\,d\tau = \frac{\partial}{\partial \tau}P(E(t) \leq \tau)\,d\tau = \frac{\partial}{\partial \tau}P(S(\tau) \geq t)\,d\tau = -\frac{\partial}{\partial \tau}P(S(\tau) < t)\,d\tau.$$

The density $G_t(\tau)$ is the main object in our considerations below. Therefore, in what follows, we collect the most important properties of $G_t(\tau)$ needed in the next sections.

Remark 1. *If S is the α-stable process, $\alpha \in (0,1)$, then the inverse process $E(t)$, has Laplace transform (cf. Prop. 1(a) in Bingham [5] or Feller [6]) given by*

$$\mathbb{E}(e^{-\lambda E(t)}) = \int_0^\infty e^{-\lambda \tau} G_t(\tau)\,d\tau = \sum_{n=0}^\infty \frac{(-\lambda t^\alpha)^n}{\Gamma(n\alpha+1)} = E_\alpha(-\lambda t^\alpha), \tag{11}$$

where E_α is the Mittag-Leffler function. It follows from the asymptotic behavior of the function E_α that $\mathbb{E}(e^{-\lambda E(t)}) \sim C t^{-\alpha}$ as $t \to \infty$. It is possible to find explicitly the density $G_t(\tau)$ in this case using the completely monotonic property of the Mittag-Leffler function E_α. It is given in terms of the Wright function $W_{\mu,\nu}$, namely $G_t(\tau) = t^{-\alpha}W_{-\alpha,1-\alpha}(\tau t^{-\alpha})$, see Gorenflo et al. [9] for more details.

For a general subordinator, the following lemma determines the t-Laplace transform of $G_t(\tau)$, with k and $\mathcal{K}$ given in (5) and (6), respectively. For the proof see Kochubei [7] or Proposition 3.2 in Toaldo [2].

Lemma 1. *The t-Laplace transform of the density $G_t(\tau)$ is given by*

$$\int_0^\infty e^{-\lambda t} G_t(\tau)\,dt = \mathcal{K}(\lambda)e^{-\tau \lambda \mathcal{K}(\lambda)}. \tag{12}$$

The double (τ, t)-Laplace transform of $G_t(\tau)$ is

$$\int_0^\infty \int_0^\infty e^{-p\tau} e^{-\lambda t} G_t(\tau) \, dt \, d\tau = \frac{\mathcal{K}(\lambda)}{\lambda \mathcal{K}(\lambda) + p}. \tag{13}$$

Here we would like to make the connection of the above abstract framework with general fractional derivatives. For any $\alpha \in (0,1)$ the Caputo-Dzhrbashyan fractional derivative of order α of a function u is defined by (see e.g., Kilbas et al. [10] and references therein)

$$(\mathbb{D}_t^\alpha u)(t) = \frac{d}{dt} \int_0^t k_\alpha(t-\tau) u(\tau) \, d\tau - k_\alpha(t) u(0), \quad t > 0, \tag{14}$$

where k_α is given in Example 1, that is, $k_\alpha(t) = g_{1-\alpha}(t) = \frac{t^{-\alpha}}{\Gamma(1-\alpha)}$, $t > 0$. In general, starting with a subordinator S and the kernel $k \in L^1_{\text{loc}}(\mathbb{R}_+)$ as given in (5), we may define a differential-convolution operator by

$$(\mathbb{D}_t^{(k)} u)(t) = \frac{d}{dt} \int_0^t k(t-\tau) u(\tau) \, d\tau - k(t) u(0), \quad t > 0. \tag{15}$$

The operator $\mathbb{D}_t^{(k)}$ is also known as general fractional derivative and its applications to convolution-type differential equations was investigated in Kochubei [7].

Example 3. *Distributed order derivative. Consider the kernel k defined by*

$$k(t) := \int_0^1 g_\alpha(t) \, d\alpha = \int_0^1 \frac{t^{\alpha-1}}{\Gamma(\alpha)} \, d\alpha, \quad t > 0. \tag{16}$$

Then it is easy to see that

$$\mathcal{K}(\lambda) = \int_0^\infty e^{-\lambda t} k(t) \, dt = \frac{\lambda - 1}{\lambda \log(\lambda)}, \quad \lambda > 0.$$

The corresponding differential-convolution operator $\mathbb{D}_t^{(k)}$ is called distributed order derivative, see Atanackovic et al. [11], Daftardar-Gejji and Bhalekar [12], Hanyga [13], Kochubei [14], Gorenflo and Umarov [15], Meerschaert and Scheffler [16] for more details and applications.

We say that the functions f and g are *asymptotically equivalent at infinity*, and denote $f(x) \sim g(x)$ as $x \to \infty$, meaning that

$$\lim_{x \to \infty} \frac{f(x)}{g(x)} = 1.$$

We say that a function L is *slowly varying at infinity* (see Feller [6], Seneta [17]) if

$$\lim_{x \to \infty} \frac{L(\lambda x)}{L(x)} = 1, \quad \text{for any } \lambda > 0.$$

Below C is constant whose value is unimportant and may change from line to line.

In the following we consider three classes of admissible kernels $k \in L^1_{\text{loc}}(\mathbb{R}_+)$, characterized in terms of their Laplace transforms $\mathcal{K}(\lambda)$ as $\lambda \to 0$ (i.e., as local conditions):

$$\mathcal{K}(\lambda) \sim \lambda^{\alpha - 1}, \quad 0 < \alpha < 1. \tag{17}$$

$$\mathcal{K}(\lambda) \sim \lambda^{-1} L\left(\frac{1}{\lambda}\right), \quad L(y) := C \log(y)^{-1}, \ C > 0. \tag{18}$$

$$\mathcal{K}(\lambda) \sim \lambda^{-1} L\left(\frac{1}{\lambda}\right), \quad L(y) := C \log(y)^{-1-s}, \ s > 0, \ C > 0. \tag{19}$$

We would like to emphasize that these three classes of kernels leads to different type of differential-convolution operators. In particular, the Caputo-Djrbashian fractional derivative (17) and distributed order derivatives (18), (19). Moreover, it is simple to check that the class of subordinators from Example 2 falls into the class (17) above.

Remark 2. *The asymptotic behavior of the function $f(t)$ as $t \to \infty$ may be determined, under certain conditions, by studying the behavior of its Laplace transform $\tilde{f}(\lambda)$ as $\lambda \to 0$, and vice versa. An important situation where such a correspondence holds is described by the Feller–Karamata Tauberian (FKT) theorem.*

We state below a version of the FKT theorem which suffices for our purposes, see the monographs Bingham et al. [18] (Section 1.7) and Feller [6] (XIII, Section 1.5) for a more general version and proofs.

Theorem 1. *Feller–Karamata Tauberian. Let $U : [0, \infty) \longrightarrow \mathbb{R}$ be a monotone non-decreasing right-continuous function such that*

$$w(\lambda) := \int_0^\infty e^{-\lambda t}\, dU(t) < \infty, \quad \forall \lambda > 0.$$

If L is a slowly varying function and $C, \rho \geq 0$, then the following are equivalent

$$U(t) \sim \frac{C}{\Gamma(\rho+1)} t^\rho L(t) \quad \text{as } t \to \infty, \tag{20}$$

$$w(\lambda) \sim C\lambda^{-\rho} L\left(\frac{1}{\lambda}\right) \quad \text{as } \lambda \to 0^+. \tag{21}$$

When $C = 0$, (20) is to be interpreted as $U(t) = o(t^\rho L(t))$; similarly for (21).

2.2. Statement of the Main Results

In Section 3 and 4 we will focus our attention on deriving the asymptotic behavior of the subordination $u^E(t, x)$ given in (1) for the inverse stable subordinator as well as for the classes (17)–(19) given above. On one hand, the results concerning the inverse stable subordinator as a random time are well understood, due to the fact that the Laplace transform (in τ) of the density $G_t(\tau)$ is known (cf. Remark 1). On the other hand, for a general subordinator much less information about $G_t(\tau)$ is known and explicit results for the subordination $u^E(t, x)$ are not available. In order to get around this problem, and motivated by the results of Section 3, we study the Cesaro limit of $u^E(t, x)$ for the general classes of random times.

With the above considerations we are ready to state our main results.

Theorem 2. *Let $u^E(t, x)$ be the subordination by the density $G_t(\tau)$ associated to the inverse stable subordinator. Denote by $M_t(u^E(\cdot, x)) := \frac{1}{t} \int_0^t u^E(s, x)\, ds$ the Cesaro mean of $u^E(t, x)$.*

1. *If $u(t, x) = t^n$, $n \geq 0$, then the asymptotic behavior of $u^E(t, x)$ coincides with the Cesaro limit and is equal to*

$$Ct^{n\alpha} \quad \text{as} \quad t \to \infty.$$

2. *If $u(t, x) = e^{-at}$, $a > 0$, then the asymptotic of $u^E(t, x)$ and its Cesaro limit are equal to*

$$Ct^{-\alpha} \quad \text{as } t \to \infty.$$

The proof of Theorem 2 is essentially the contents of Section 3 while the next theorem is shown in Section 4.

Theorem 3. *Let $u^E(t, x)$ be the subordination by the density $G_t(\tau)$ associated to the classes (17)–(19) and $M_t(u^E(\cdot, x)) := \frac{1}{t} \int_0^t u^E(s, x)\, ds$ the Cesaro mean of $u^E(t, x)$.*

1. Assume that $u(t,x) = t^n$, $n \geq 0$. Then the asymptotic of the Cesaro mean for the three classes are:

   (17). $M_t(u^E(\cdot,x)) \sim C t^{\alpha n}$ as $t \to \infty$,

   (18). $M_t(u^E(\cdot,x)) \sim C \log(t)^n$ as $t \to \infty$,

   (19). $M_t(u^E(\cdot,x)) \sim C \log(t)^{(1+s)n}$ as $t \to \infty$.

2. If $u(t,x) = e^{-at}$, $a > 0$, then the asymptotic of $M_t(u^E(\cdot,x))$ for the different classes are:

   (17). $M_t(u^E(\cdot,x)) \sim C t^{-\alpha}$ as $t \to \infty$,

   (18). $M_t(u^E(\cdot,x)) \sim C \log(t)^{-1}$ as $t \to \infty$,

   (19). $M_t(u^E(\cdot,x)) \sim C \log(t)^{-1-s}$ as $t \to \infty$.

3. Inverse Stable Subordinators

In this section we consider two elementary solutions of dynamical systems, namely $u(t) = u(t,x) = t^n$, $n \geq 0$ and $u(t) = u(t,x) = e^{-at}$, $a > 0$, and investigate their subordination by the density $G_t(\tau)$ of inverse stable subordinator.

Define the function $u^E(t) = u^E(t,x)$ as the subordination of $u(t)$ (of the above type) by the kernel $G_t(\tau)$, that is,

$$u^E(t) := \int_0^\infty u(\tau) G_t(\tau) \, d\tau, \quad t \geq 0. \tag{22}$$

Our goal is to investigate the asymptotic behavior of $u^E(t)$. At first we compute explicitly the function $u^E(t)$ by solving the integral (22) and obtain the time asymptotic. Second we derive the Cesaro limit of $u^E(t)$, more precisely, the asymptotic behavior of the Cesaro mean of $u^E(t)$ defined by

$$M_t(u^E(\cdot)) := \frac{1}{t} \int_0^t u^E(s) \, ds. \tag{23}$$

It turns out that both asymptotic behaviors for the two functions $u(t)$ given above coincide. Therefore, for the random time change associated to the inverse stable subordinator $E(t)$, $t \geq 0$, the asymptotic behavior of $u^E(t)$ is the same as the Cesaro limit. On the other hand, using the Cesaro limit we may investigate a broad class of subordinators. In Section 4 we investigate the Cesaro limit for the classes (17)–(19) while in this section concentrate in the spacial case of inverse stable subordinators.

3.1. Subordination of Monomials

Let us consider at first the subordination of the function $u(t) = t^n$, $n \geq 0$. Hence, $u^E(t)$ is given by

$$u^E(t) = \int_0^\infty \tau^n G_t(\tau) \, d\tau. \tag{24}$$

It follows from (11) that $u^E(t)$ is explicitly evaluated as

$$u^E(t) = (-1)^n \frac{d^n}{d\lambda^n} E_\alpha(-\lambda t^\alpha)\big|_{\lambda=0} = \frac{n!}{\Gamma(\alpha n + 1)} t^{\alpha n}.$$

The last equality follows easily from the power series of the Mittag-Leffler function

$$E_\alpha(z) = \sum_{n=1}^\infty \frac{z^n}{\Gamma(\alpha n + 1)}.$$

In addition, the asymptotic of the Mittag-Leffler function E_α that gives

$$u^E(t) \sim C t^{n\alpha} \quad \text{as} \quad t \to \infty. \tag{25}$$

Now we turn to compute the asymptotic behavior of the Cesaro mean of $u^E(t)$ with the help of the FKT theorem. To this end we define the monotone function

$$v(t) := \int_0^t u^E(s)\,ds. \tag{26}$$

The Laplace-Stieltjes transform $w(\lambda)$ of $v(t)$ is given by

$$w(\lambda) := \int_0^\infty e^{-\lambda t}\,dv(t) = \int_0^\infty e^{-\lambda t} u^E(t)\,dt = \int_0^\infty e^{-\lambda t}\int_0^\infty \tau^n G_t(\tau)\,d\tau\,dt.$$

Using Fubini's theorem and Equation (12) we obtain

$$w(\lambda) = \int_0^\infty \tau^n \int_0^\infty e^{-\lambda t} G_t(\tau)\,dt\,d\tau = \mathcal{K}(\lambda)\int_0^\infty \tau^n e^{-\tau\lambda\mathcal{K}(\lambda)}\,d\tau.$$

The r.h.s. integral can be evaluated as

$$\int_0^\infty \tau^n e^{-\tau\lambda\mathcal{K}(\lambda)}\,d\tau = (\lambda\mathcal{K}(\lambda))^{-(1+n)} n!$$

which yields

$$w(\lambda) = n!\lambda^{-(1+n)}\mathcal{K}(\lambda)^{-n}. \tag{27}$$

On the other hand, for the stable subordinator we have $\mathcal{K}(\lambda) = \lambda^{\alpha-1}$, cf. Example 1. Thus, we obtain

$$w(\lambda) = n!\lambda^{-(1+\alpha n)} = \lambda^{-\rho} L\left(\frac{1}{\lambda}\right),$$

where $\rho = 1+\alpha n$ and $L(x) = n!$ is a trivial slowly varying function. Then Theorem 1 yields

$$v(t) \sim Ct^{1+n\alpha} \quad \text{as} \quad t \to \infty$$

and this implies the following asymptotic behavior for the Cesaro mean of $u^E(t)$

$$M_t(u^E(\cdot)) = \frac{1}{t}\int_0^t u^E(s)\,ds \sim Ct^{\alpha n} \quad \text{as} \quad t \to \infty. \tag{28}$$

Remark 3. *In conclusion, we find that the asymptotic behavior of the subordination $u^E(t)$ of any monomial by the density $G_t(\tau)$ (of the inverse stable subordinator) as well as its Cesaro limit coincides. Note also the slower decay of the subordination $u^E(t)$ compared to $u(t)$ due to $0 < \alpha < 1$.*

3.2. Subordination of Decaying Exponentials

Now we consider the solution $u(t) = e^{-at}$, $a > 0$ and proceed to study the asymptotic behavior of its subordination $u^E(t)$ by the kernel $G_t(\tau)$. Again a direct computation is possible in that case as well as the Cesaro mean.

Hence, the subordination $u^E(t)$ is given by

$$u^E(t) = \int_0^\infty u(\tau) G_t(\tau)\,d\tau = \int_0^\infty e^{-a\tau} G_t(\tau)\,d\tau. \tag{29}$$

It follows from Equation (11) that

$$u^E(t) = E_\alpha(-at^\alpha) \sim Ct^{-\alpha} \quad \text{as} \quad t \to \infty. \tag{30}$$

On the other hand, to derive the asymptotic behavior for the Cesaro mean of $u^E(t)$ (with the help of Theorem 1) we define the monotone function

$$v(t) := \int_0^t u^E(s)\,ds. \tag{31}$$

The Laplace-Stieltjes transform $w(\lambda)$ of $v(t)$ is equal to

$$w(\lambda) := \int_0^\infty e^{-\lambda t} dv(t) = \int_0^\infty e^{-\lambda t} u^E(t) \, dt = \int_0^\infty e^{-\lambda t} \int_0^\infty e^{-a\tau} G_t(\tau) \, d\tau \, dt$$

and using Fubini's theorem and Equation (11) we obtain

$$w(\lambda) = \mathcal{K}(\lambda) \int_0^\infty e^{-\tau(a + \lambda \mathcal{K}(\lambda))} \, d\tau = \frac{\mathcal{K}(\lambda)}{a + \lambda \mathcal{K}(\lambda)}. \tag{32}$$

As $\mathcal{K}(\lambda) = \lambda^{\alpha-1}$ for the class (17) we may write $\tilde{v}(\lambda)$ as

$$w(\lambda) = \lambda^{-(1-\alpha)} \frac{1}{a + \lambda^\alpha} = \lambda^{-\rho} L\left(\frac{1}{\lambda}\right), \quad \rho = 1 - \alpha, \quad L(t) := \frac{1}{a + t^{-\alpha}}.$$

It is simple to verify that L is a slowly varying function so that we may use the FKT theorem to obtain

$$v(t) \sim C t^{1-\alpha} \frac{1}{a + t^{-\alpha}} \quad \text{as} \quad t \to \infty.$$

Dividing both sides by t leads to the asymptotic behavior of the Cesaro mean of $u^E(t)$, that is,

$$M_t(u^E(\cdot)) = \frac{1}{t} \int_0^t u^E(s, x) \, ds \sim C \frac{t^{-\alpha}}{a + t^{-\alpha}} \sim C t^{-\alpha} \quad \text{as } t \to \infty. \tag{33}$$

Remark 4. *We conclude that the asymptotic behavior $u^E(t)$ given in (30) coincides with the Cesaro limit of $u^E(t, x)$. In addition, we notice that the starting function $u(t) = e^{-at}$ has an exponential decay and its subordination has a slower decay, namely polynomial decay.*

4. Cesaro Limit for General Classes of Subordinators

In this section we study the asymptotic behavior of the subordination by the density $G_t(\tau)$ associated to the classes (17)–(19). Note that Examples 1 and 2 belong to the class (17). As pointed out in Section 3 here we only study the Cesaro limit of the subordination function $u^E(t)$.

As in Section 3, $u^E(t)$ is defined by

$$u^E(t) := \int_0^\infty \tau^n G_t(\tau) \, d\tau \tag{34}$$

or

$$u^E(t) := \int_0^\infty e^{-a\tau} G_t(\tau) \, d\tau \tag{35}$$

while $v(t)$ is defined by

$$v(t) := \int_0^t u^E(s) \, ds.$$

The density $G_t(\tau)$ in (34) and (35) is associated to each class (17)–(19) described above. We study the Cesaro limit of $u^E(t)$ for each class separately.

4.1. Subordination by the Class (17)

At first we study the asymptotic behavior of $u^E(t)$ given by (34). To this end we use equality (27) to obtain the Laplace-Stieltjes transform $w(\lambda)$ of the function $v(t)$ as

$$w(\lambda) := \int_0^\infty e^{-\lambda t} dv(t) = \lambda^{-(1+n)} (\mathcal{K}(\lambda))^{-n} n!.$$

It follows from the behavior of $\mathcal{K}(\lambda)$ at $\lambda = 0$ of the class (17) that

$$w(\lambda) \sim \lambda^{-(1+\alpha n)} n! = \lambda^{-\rho} L\left(\frac{1}{\lambda}\right),$$

where $\rho = 1 + \alpha n$ and $L(x) = n!$ is a slowly varying function. It follows from the FKT theorem that
$$\tilde{v}(t) \sim Ct^\rho L(t) = Ct^{1+\alpha n} \quad \text{as} \quad t \to \infty.$$
This implies the Cesaro limit of $u^E(t)$ as
$$M_t(u^E(\cdot)) \sim Ct^{\alpha n} \quad \text{as} \quad t \to \infty.$$
Note that this asymptotic is similar to the analogous for the inverse stable subordinator, cf. (28).

Let us now study the Cesaro limit of the function $u^E(t)$ given in (35). Using the equality (32) the Laplace-Stieltjes transform $v(t)$ has the form
$$\tilde{v}(\lambda) = \frac{\mathcal{K}(\lambda)}{a + \lambda \mathcal{K}(\lambda)}.$$
Replacing the local behavior of $\mathcal{K}(\lambda)$ at $\lambda = 0$ for the class (17) gives
$$\tilde{v}(\lambda) \sim \frac{\lambda^{\alpha-1}}{a + \lambda^\alpha} = \lambda^{-\rho} L\left(\frac{1}{\lambda}\right),$$
where $\rho = 1 - \alpha$ and $L(x) = \frac{1}{1+ax^{-\alpha}}$. An applications of the FKT theorem yields the asymptotic for $v(t)$, namely $v(t) \sim Ct^\rho L(t)$ as $t \to \infty$. Finally dividing both sides by t gives the Cesaro limit of $u^E(t)$, that is,
$$M_t(u^E(\cdot)) \sim C \frac{t^{-\alpha}}{1 + at^{-\alpha}} \sim Ct^{-\alpha} \quad \text{as} \quad t \to \infty.$$
Again, we obtain the same asymptotic as for the inverse stable subordinator, see (33). In any case, since $0 < \alpha < 1$, the time decaying is slower than the initial function $u(t)$.

4.2. Subordination by the Class (18)

Assume that $u^E(t)$ is the subordination given in (34). The Laplace-Stieltjes transform $w(\lambda)$ of $v(t)$ (cf. equality (27)) has the form
$$w(\lambda) := \int_0^\infty e^{-\lambda t} dv(t) = \lambda^{-(1+n)} (\mathcal{K}(\lambda))^{-n} n!.$$
Using the behavior of $\mathcal{K}(\lambda)$ near $\lambda = 0$ for the class (18) we obtain
$$w(\lambda) \sim \lambda^{-1} L\left(\frac{1}{\lambda}\right),$$
where $L(x) = C \log(x)^n$, $C > 0$, is a slowly varying function. Then it follows from the FKT theorem that
$$v(t) \sim Ct \log(t)^n$$
and as a result the asymptotic behavior for the Cesaro mean of $u^E(t)$ follows
$$M_t(u^E(\cdot)) \sim C \log(t)^n \quad \text{as} \quad t \to \infty.$$
A similar analysis may be applied to study the asymptotic behavior for the subordination $u^E(t)$ given in (35). The Laplace-Stieltjes transform $w(\lambda)$ of the monotone function $v(t)$ may be evaluated using equality (32) to find the following expression
$$w(\lambda) = \frac{\mathcal{K}(\lambda)}{a + \lambda \mathcal{K}(\lambda)}.$$

Using the local behavior of $\mathcal{K}(\lambda)$ near $\lambda = 0$ from class (18) yields

$$w(\lambda) \sim \lambda^{-1} L\left(\frac{1}{\lambda}\right),$$

where $L(x) = C \frac{\log(x)^{-1}}{a + C \log(x)^{-1}}$ which is a slowly varying function. Using the FKT theorem we obtain the longtime behavior for the Cesaro mean of $u^E(t)$ as

$$M_t(u^E(\cdot)) \sim C \frac{\log(t)^{-1}}{a + C \log(t)^{-1}} \sim C \log(t)^{-1} \quad \text{as} \quad t \to \infty.$$

4.3. Subordination by the Class (19)

At first we study the subordination $u^E(t)$ given in (34) for the class (19). The Laplace-Stieltjes transform $w(\lambda)$ of the corresponding $v(t)$ is computed using equality (27) and we obtain

$$w(\lambda) := \int_0^\infty e^{-\lambda t} dv(t) = \lambda^{-(1+n)} (\mathcal{K}(\lambda))^{-n} n!.$$

Using the behavior of $\mathcal{K}(\lambda)$ near $\lambda = 0$ for the class (19) yields

$$w(\lambda) \sim \lambda^{-1} L\left(\frac{1}{\lambda}\right),$$

where $L(x) = C \log(x)^{(1+s)n}$, $C > 0$, is a slowly varying function. Then it follows from Theorem 1 that

$$v(t) \sim Ct \log(t)^{(1+s)n}$$

and dividing both sides by t gives the asymptotic behavior for the Cesaro mean of $u^E(t)$, namely

$$M_t(u^E(\cdot)) \sim C \log(t)^{(1+s)n} \quad \text{as} \quad t \to \infty.$$

Let $u^E(t)$ be the subordination by $u(t) = e^{-at}$, $a > 0$, that is, equality (35) with $G_t(\tau)$ from the class (19). It follows from equality (32) that the Laplace-Stieltjes transform $w(\lambda)$ of $v(t)$ has the form

$$w(\lambda) = \frac{\mathcal{K}(\lambda)}{a + \lambda \mathcal{K}(\lambda)}.$$

Using the local behavior of $\mathcal{K}(\lambda)$ near $\lambda = 0$ from class (19) yields

$$w(\lambda) \sim \lambda^{-1} L\left(\frac{1}{\lambda}\right), \qquad L(x) = C \frac{\log(x)^{-1-s}}{a + C \log(x)^{-1-s}},$$

where $C, s > 0$. As the function L is slowly varying at infinity, then by the FKT theorem we obtain the asymptotic behavior for the Cesaro mean of $u^E(t)$ as

$$M_t(u^E(\cdot)) \sim C \frac{\log(t)^{-1-s}}{a + C \log(t)^{-1-s}} \sim C \log(t)^{-1-s} \quad \text{as} \quad t \to \infty.$$

Author Contributions: Writing—original draft preparation, Y.K. and J.d.S. All authors have read and agreed to the published version of the manuscript.

Funding: This research was funded by Fundação para a Ciência e a Tecnologia, Portugal grant number UIDB/MAT/04674/2020 and Ministry of Education and Science of Ukraine grant number 0119U002583.

Institutional Review Board Statement: Not applicable.

Informed Consent Statement: Not applicable.

Conflicts of Interest: The authors declare no conflict of interest.

References

1. Kondratiev, Y.; da Silva, J. Random Time Dynamical Systems I: General Structures. *arXiv* **2020**, arXiv:2012.15201v1.
2. Toaldo, B. Convolution-type derivatives, hitting-times of subordinators and time-changed C_0-semigroups. *Potential Anal.* **2015**, *42*, 115–140. [CrossRef]
3. Meerschaert, M.M.; Sikorskii, A. *Stochastic Models for Fractional Calculus*; Walter de Gruyter & Co.: Berlin, Germany, 2012; Volume 43.
4. Mura, A.; Taqqu, M.S.; Mainardi, F. Non-Markovian diffusion equations and processes: analysis and simulations. *Phys. A* **2008**, *387*, 5033–5064. [CrossRef]
5. Bingham, N.H. Limit theorems for occupation times of Markov processes. *Z. Wahrscheinlichkeitstheorie Verwandte Geb.* **1971**, *17*, 1–22. [CrossRef]
6. Feller, W. *An Introduction to Probability Theory and Its Applications*, 2nd ed.; John Wiley & Sons Inc.: New York, NY, USA, 1971; Volume II.
7. Kochubei, A.N. General Fractional Calculus, Evolution Equations, and Renewal Processes. *Integral Equ. Oper. Theory* **2011**, *71*, 583–600. [CrossRef]
8. Bertoin, J. *Lévy Processes; Cambridge Tracts in Mathematics*; Cambridge University Press: Cambridge, UK, 1996; Volume 121, p. x+265.
9. Gorenflo, R.; Luchko, Y.; Mainardi, F. Analytical properties and applications of the Wright function. *Fract. Calc. Appl. Anal.* **1999**, *2*, 383–414.
10. Kilbas, A.A.; Srivastava, H.M.; Trujillo, J.J. *Theory and Applications of Fractional Differential Equations; North-Holland Mathematics Studies*; Elsevier Science B.V.: Amsterdam, The Netherlands, 2006; Volume 204.
11. Atanackovic, T.M.; Pilipovic, S.; Zorica, D. Time distributed-order diffusion-wave equation. I., II. Proceedings of the Royal Society of London A: Mathematical, Physical and Engineering Sciences. *R. Soc.* **2009**, *465*, 869–1891.
12. Daftardar-Gejji, V.; Bhalekar, S. Boundary value problems for multi-term fractional differential equations. *J. Math. Anal. Appl.* **2008**, *345*, 754–765. [CrossRef]
13. Hanyga, A. Anomalous diffusion without scale invariance. *J. Phys. A Mat. Theor.* **2007**, *40*, 5551. [CrossRef]
14. Kochubei, A.N. Distributed order calculus and equations of ultraslow diffusion. *J. Math. Anal. Appl.* **2008**, *340*, 252–281. [CrossRef]
15. Gorenflo, R.; Umarov, S. Cauchy and nonlocal multi-point problems for distributed order pseudo-differential equations, Part one. *Z. Anal. Anwend.* **2005**, *24*, 449–466. [CrossRef]
16. Meerschaert, M.M.; Scheffler, H.P. Stochastic model for ultraslow diffusion. *Stoch. Process. Appl.* **2006**, *116*, 1215–1235. [CrossRef]
17. Seneta, E. *Regularly Varying Functions; Lect. Notes Math.*; Springer: Berlin/Heidelberg, Germany, 1976; Volume 508.
18. Bingham, N.H.; Goldie, C.M.; Teugels, J.L. *Regular Variation; Encyclopedia of Mathematics and its Applications*; Cambridge University Press: Cambridge, UK, 1987; Volume 27.

 *fractal and fractional*

Article

Mittag–Leffler Function as an Approximant to the Concentrated Ferrofluid's Magnetization Curve

Petr A. Ryapolov [1] and Eugene B. Postnikov [2,*]

[1] Faculty of Natural Sciences, Southwest State University, 50 Let Oktyabrya St., 94, 305048 Kursk, Russia; r-piter@yandex.ru
[2] Department of Theoretical Physics, Kursk State University, Radishcheva St., 33, 305000 Kursk, Russia
* Correspondence: postnikov@kursksu.ru

Abstract: In this work, we show that the static magnetization curve of high-concentrated ferrofluids can be accurately approximated by the Mittag–Leffler function of the inverse external magnetic field. The dependence of the Mittag–Leffler function's fractional index on physical characteristics of samples is analysed and its growth with the growing degree of system's dilution is revealed. These results provide a certain background for revealing mechanisms of hindered fluctuations in concentrated solutions of strongly interacting of the magnetic nanoparticles as well as a simple tool for an explicit specification of macroscopic force fields in ferrofluid-based technical systems.

Keywords: Mittag–Leffler function; data fitting; magnetization; magnetic fluids

1. Introduction

The magnetization of ferrofluids under realistic conditions of a highly concentrated suspension of magnetic particles covered by envelopes preventing the aggregation, and taking into account the possible polydispersity of these particles, is a complicated problem of condensed matter physics still far from its final resolution [1–6].

While the qualitative picture of superparamagnetic phenomena in ideal diluted media of magnetic dipoles is well-established [7], the effects of multiparticle interactions, aggregation of nanoparticles and their hindered rotation, a wide spectrum of possible relaxation times under such conditions do not allow practically applicable straightforward calculations.

Among the most accepted approaches, one can note the second-order modified mean-field (MMF2) theory proposed in the work [8], which treats the macroscopic magnetization M of ferrofluid as a function of the applied magnetic field intensity H in the form

$$M(H) = \rho \left\langle \mu(x) L\left(\frac{\mu_0 \mu(x) H_{\text{eff}}}{k_B T} \right) \right\rangle, \quad (1)$$

where angle brackets denote averaging over the ensemble of microscopic magnetic moments $\mu(x)$ and the particle size distribution, μ_0 and k_B are is the vacuum permeability and Boltzmann's constant, respectively, ρ is the density and

$$H_{\text{eff}} = H + \frac{1}{3} M_L(H) + \frac{1}{144} M_L(H) \frac{dM_L(H)}{dH} \quad (2)$$

is the effective magnetic field in a medium with

$$M_L(H) = \rho \left\langle \mu(x) L\left(\frac{\mu_0 \mu(x) H}{k_B T} \right) \right\rangle, \quad (3)$$

where $L(z) = \coth(z) - z^{-1}$ is the Langevin function.

Although this method and its further improvements, which take into account higher-order in the dipolar coupling constant for polydisperse concentrated ferrofluids [9], reasonably reproduce the magnetization curve, and especially the initial susceptibility

$$\chi = \chi_L \left(1 + \frac{1}{3}\chi_L + \frac{1}{144}\chi_L^2\right) \qquad (4)$$

with

$$\chi_L = \left(\frac{\partial M_L}{\partial H}\right)_{H=0} = \frac{\mu_0 \rho \langle \mu^2(x) \rangle}{3 k_B T} \qquad (5)$$

that makes it useful for magnetic granulometry of ferrofluids, it requires assumptions on the statistic properties of magnetic nanoparticles and rather complicated numerical computation of integrals when calculating integrals with the respective probability density functions.

On the other hand, a variety of practical problems require the knowledge of an accurate shape of the magnetization curve not around the zero field (4) but in the range of strong external magnetic field, where this line is highly curved, and further toward the saturation but not reaching the latter.

Most traditional technical applications of the magnetic fluid-based systems such as sensors, sealers, acoustic systems [10,11] as well as modern applications in microfluidics [12], controlled magnetophoresis [13], self-assembly [14], and separation [15] use the range of magnetic fields in which the working fluid is placed is in the middle region of the magnetization curve far from both the interval of initial magnetization and saturation. Therefore, predicting the shape of the magnetization curve in the region of its significant curvature is also an important technical task.

This situation induces the emergence of several alternative approaches aimed at an efficient fitting of ferrofluid's magnetisation stated as an explicit function of the external magnetic field by an appropriate set of approximating functions [16–19] adjusted to experimental conditions.

In this work, we explore the Mittag–Leffler function as a promising universal approximant chosen for two reasons: (i) this function has high flexibility for fitting data with quite various behaviour [20] and (ii) the Mittag–Leffler function is involved in the description of fluctuational and relaxational processes in ferrofluids [21,22] that may affect the stationary magnetization in an external field.

2. Experimental Data and Their Processing

2.1. Measurements of Ferrofluid's Magnetization

As an example of the practical analysis of ferrofluid's magnetisation, we consider the set of samples specified previously in the work [23], which are obtained by the sequential dilutions of the initial magnetic liquid denoted as MF-1, see the parameters in Table 1. This procedure assures the same structure of magnetic nanoparticles in all samples, which differ by their concentration only.

To determine the magnetisation of samples, the ballistic method was used. Its essence is that the measuring cell containing magnetic fluid is placed between the poles of the electromagnet connected to a micromagnetometer (the relative uncertainty of measurements is estimated as 2.5%). The change of the magnetic flux after the cell's rotation was registered allows calculating the magnetization, see [24] for technical details.

The maximal intensity of the magnetic field allowed by the setup is equal to 800 kA/m in the region of measurements. During the experiment, the tending of ferrofluid's magnetization to the saturated state in the strong magnetic field was controlled by plotting the magnetization as a function of the inverse magnetic field as follows from the definition $M_s = M(H)$ for $H \to \infty$ that implies $H^{-1} \to 0$ as well as by monitoring the change of magnetization at subsequent steps of the magnetic field elevation. It is noted that the relative change of the registered $M(H)$ for H close to 800 kA/m tends to limits of experimental uncertainty that means that the obtained experiments data satisfy the conditions of "ferrofluid's magnetization in a strong external magnetic field".

Table 1. Physical properties of the studied ferrofluids: the density ρ, the relative volume concentration of nanoparticles ϕ and the relative volume concentration of its magnetic fraction ϕ_m as well as the parameter of their magnetization curve determined by the Mittag–Leffler function-based approximation: the saturation magnetization M_s, the inverse fractional magnetic demagnetizing susceptibility a, and the fractional index α. The last row lists relative average absolute deviations between experimental and fitted data.

	MF-1	MF-2	MF-3	MF-4
ρ, kg·m^{-3}	1245	1058	952	870
ϕ, %	11.02	6.62	4.11	2.18
ϕ_m, %	9.08	4.34	2.70	1.93
M_s, kA·m^{-1}	47.6	21.7	13.5	8.69
a, (kA·m^{-1})$^\alpha$	4.56	4.92	4.77	5.33
α	0.53	0.65	0.69	0.71
AAD, %	2.1	1.2	2.0	2.5

2.2. The Mittag–Leffler Function as an Approximant for the Static Magnetization Curve

Being based on the asymptotic expansion of the Langevin function, the conventional way to consider the magnetization curve, when it tends to the state of saturation in a strong external magnetic field, is considering the function $M(H^{-1})$, which has the asymptotic form

$$M \cong M_s\left(1 - \frac{k_B T}{\mu_0 \tilde{\mu} H}\right), \tag{6}$$

where $\tilde{\mu}$ is some effective magnetic moment obtained via the procedure of an appropriate statistical averaging. In particular, for diluted systems it is argued [17] that it is related to the harmonic mean of elementary magnetic moments; MMF2 theory gives more complicated expression, which depends on the chosen statistical distribution, but does not change the principal functional form. As a consequence, the saturated magnetization is operationally defined as the limit $M_s = M(H^{-1})$ when $H^{-1} = 0$ with the usage of the least mean square fitting experimental data.

Note however, that certain precautions related to the direct usage of Equation (6) were noted even in early works on the superparamagnetism in ferrofluids [25,26] argued to the free energy difference between initial and final states and different magnetization routes for very small and relatively larger magnetic particles. The latter was also noted recently in ref. [17].

In fact, experimental data showed in Figure 1A follow a curved path resembling some stretched exponential rather than a straight line as a function of the inverse magnetic field. This argues in favour of searching a more relevant approximation than Equation (6).

The promising candidate is the form

$$M(H^{-1}) = M_s E_\alpha(aH^{-\alpha}) \tag{7}$$

expressed via the Mittag–Leffler function defined as [27,28]

$$E_\alpha(-z^\alpha) := \sum_{n=0}^{\infty}(-1)^n \frac{z^{\alpha n}}{\Gamma(\alpha n+1)}, \tag{8}$$

where $\Gamma(\cdot)$ is the Gamma function, $\alpha > 0$ and can be fractional, and a is a parameter whose physical meaning will be discussed below.

The function (8) tends asymptotically [28] at $z \to 0$ to

$$E_\alpha(-z^\alpha) \sim \exp\left(-\frac{z^\alpha}{\Gamma(1+\alpha)}\right) \sim 1 - \frac{z^\alpha}{\Gamma(1+\alpha)}, \qquad (9)$$

i.e., to the shape visible in Figure 1A.

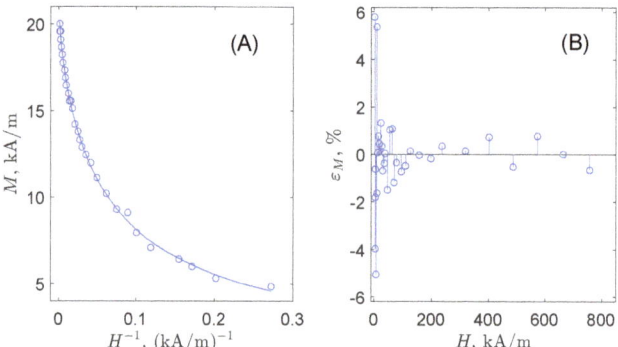

Figure 1. (**A**) The magnetization of the ferrofluid MF-2 as a function of the inverse external magnetic field, experimental values $M(H^{-1})$ (circles) and their approximation by the Mittag–Leffler function M_{fit} (curve); (**B**) the relative deviation between the data and their approximation $\varepsilon_M = 100\%\left(M - M_{fit}\right)M^{-1}$ plotted as a function of the external magnetic field applied to the sample.

At the same time, $E_1(z) = \exp(z)$, and the limiting case for $z \ll 1$

$$E_1(-z) = 1 - z$$

reduces this representation to Equation (6) in the classic superparamagnetic case. In this case $a = k_\text{B}T/\mu_0\tilde{\mu}$; in the general case it is an indefinite parameter to be determined by the fitting procedure.

In application to magnetization curve data for all dilutions, the fitting procedure and the subsequent computation of the Mittag–Leffler function with the parameters determined by this fitting were carried out using the packages [29,30] for MATLAB. Figure 2 clearly shows that the solid curves representing the functional form (7) accurately reproduce the experimental data not only in the asymptotic region of strong magnetic fields but practically over the whole range. The double logarithmic scale is used to better distinguish between curves, which otherwise go too close to each other in the region of small H^{-1}.

The values provided in Table 1 indicate that the relative average absolute deviations defined as

$$\text{AAD} = \frac{100\%}{N}\sum_{i=1}^{N}\left|\frac{M(H_i) - M_{fit}(H_i)}{M(H_i)}\right|$$

do not exceed a few percents. Figure 1B illustrates the distribution of these deviations in more details for MF-2 (for the rest of ferrofluids the picture is principally the same). One can see that they are distributed symmetrically over zero, i.e., this is connected with the experimental uncertainty. They are sufficiently small over the great majority of the external margetic filed range; the larger deviations are revealed in the close vicinity of the demagnetized state ($H = 0$) only.

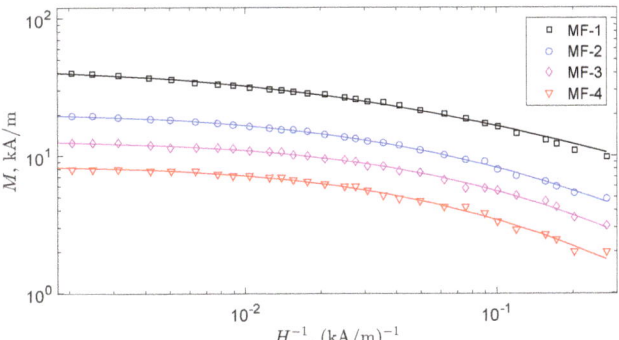

Figure 2. The magnetization curves in double logarithmic scale for all ferrofluids listed in Table 1: experimental data (markers) and their the Mittag–Leffler function-based approximations (curves).

3. Discussion

Thus, the Mittag–Leffler function-based expression (7) reproduces the magnetization curve with practically acceptable accuracy. Moreover, plotting the values of parameters listed in Table 1, one can see a certain regularity in their dependence on the magnetic phase concentration as shown in Figure 3 when a fluid's state is far from a very diluted system, which should exhibit the classic superparamegnetic behaviour, and the Langevin function does not reduces to the Mittag–Leffler representation.

On the contrary, the dependences for concentrated systems can be connected to physical mechanisms. Looking at Table 1 and Figure 3B it is seen that the α-index of the diminishes with the growing concentration of magnetic nanoparticles in a regular way that means that anomalous long-range effects accompany the more concentrated systems.

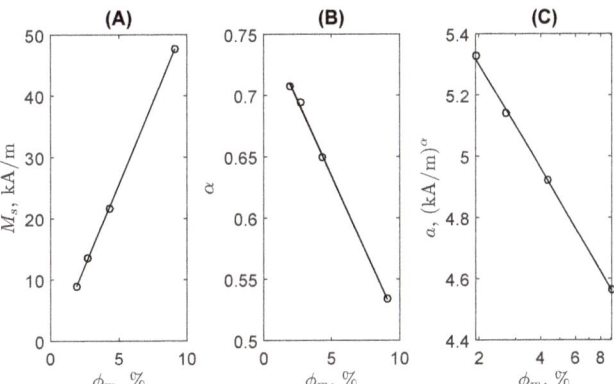

Figure 3. Parameters (circles) of Equation (7) as functions of the magnetic phase concentration. The fitting equations shown as solid straight lines are $M_s = 5.3977\phi_m - 1.4426$ (**A**), $\alpha = -0.0246\phi_m + 0.7575$ (**B**), and $a = -0.4875\ln(\phi_m) + 5.6377$ (**C**).

Note that the function (7) is a solution to the fractional differential equation

$$^C D_{H^{-1}}^\alpha M(H^{-1}) = -aM(H^{-1}) \tag{10}$$

with the initial condition $M_s = M(0^+)$, where $^CD^{\alpha}_{H^{-1}}$ the fractional derivative of order α in the Caputo sense,

$$^CD^{\alpha}_{H^{-1}}M(H^{-1}) = -\frac{1}{\Gamma(1-\alpha)}\int_0^{H^{-1}}\frac{(H^{-1})^2\frac{dM}{dH}}{(H^{-1}-h)^{\alpha}}dh. \qquad (11)$$

The integer-order derivative within the integrand in Equation (11) is especially written explicitly as $dM/dH \equiv \chi$ to highlight its meaning as the usual differential magnetic susceptibility. Therefore, it follows from Equations (10) and (11) that the parameter a can be considered to be a kind of the inverse fractional magnetic susceptibility, which is taken into account when the system is demagnetized to some state from the completely ordered one ($H \to \infty$, $M = M_s$). The higher concentrations lead to smaller values of α that implies wider integration kernels in Equation (11), i.e., more expressed effects of retarded magnetic restructuring due to many-particle interactions.

In the concentrated systems, the growth of the applied magnetic field can induce chain formation resulting in the magnetisation response of the mixture of elementary magnetic dipoles of single (even interacting particles) and multicore aggregates [31].

Another feature related to the experimental procedure of measurements of concentrated ferrofluids magnetization in a strong magnetic field was revealed in the works [32,33], where it has been noted that a relatively long measurement procedure may lead to the system's slow restructuring, which changes its macroscopic magnetic properties. Such a process is characterized by the relaxation times significantly exceeding those for Neel and Brown mechanisms. Moreover, the stretched exponential time dependence argues in favour of anomalous kinetics that unavoidably requires the usage of the Mittag–Leffler function.

Thus, we can hypothesize that the revealed fitting dependence on the external magnetic field may emerge as a fractional-order response to the switching-on external magnetic field accompanied by a hindered alignment of elementary magnetic dipoles in concentrated ferrofluids, their anomalous kinetic-based fluctuations and, additionally, a specific field shielding. As a kind of analogy, one can mention that Equation (1) contains H_{eff} as its argument, which, in turn, also contains the Langevin function. Thus, combining both (1) and (2) under the strong-field approximation (6), we obtain the nested structure similar to the first terms of the continued fraction representation of the electric ladder circuit that is a known example [34] of the system leading to the fractional-order dynamics with solutions expressed via the Mittag–Leffler function.

As a kind of argument supporting this hypothesis, we can consider dependencies shown in Figure 3. One can note that the dilution of the system, i.e., diminished concentration of the magnetic phase ϕ_m, leads to the growing α-index. This realizes a transition from Equations (8) and (9) to the classic hyperbolic law (6) in the asymptotic limit. However, there is also a caveat: this Mittag–Leffler function-based consideration seems not be applicable to the completely demagnetized state when it results in diverging $\chi_0 = (dM/dH)H = 0$. This is also supported by Figure 1B where the deviations from the approximant grow approaching this state and Figure 3A where the linear approximation of the saturation magnetization does not go through the point $M(0) = 0$. However, this conclusion does not affect the accuracy of approximation and regularities at moderate and high values of the applied external magnetic fields. On the contrary, large concentrations ϕ_m result in effects of the hindered rotation of magnetic nanoparticles preventing their alignment in the magnetic field, viscoelastic retardation of the alignment, agglomeration, etc, see the discussion above. All these factors lead to the emergence of memory effects that is reflected by the diminishing of the fractional index α in the respective Mittag–Leffler function-based representation. At the same time, the linear dependence of the magnitude of the saturated magnetization shown in Figure 3A is the completely classic effect reflected even in the expression (1): larger concentrations of the magnetic phase directly proportionally result in the larger saturation magnetizations. The last of three dependencies, see Figure 3C is a phenomenological one but its logarithmic character is qualitatively expectable because the amplitude parameter a is a multiplier but not an additive term.

Finally, we should stress that the reasons described above have physically qualitative character, while the dependences shown in Figure 3 can find a direct practical application: when several experimental data points follow linear regularity for different concentrations of the magnetic phase, one can use the obtained regression lines to predict the magnetic response (the magnetization curve) of ferrofluids at intermediate concentrations without carrying out additional time-consuming measurements.

4. Conclusions

In this work, we demonstrated that the Mittag–Leffler function can be used as an efficient approximant for the representation of the ferrofluid magnetization curve at moderate and strong external magnetic field as an explicit function of the latter. This approach has an advantage as using only a small number of parameters to be fitted. Among them, the saturated magnetization is determined as better corresponding to the way of change of experimental data within the experimentally accessible range of applied magnetic fields. In addition, it is shown, see Figure 3 that there exits a regularity in the change of these parameters in response to the change of the concentration of the magnetic nanoparticles while the latter is not very small. As a result, one can predict the change magnetic properties of a ferrofluid due to its dilution, i.e., a small number of reference dilutions used to determine the coefficients of linear fits of these parameters. In turn, they provide an opportunity to plan dilutions leading to desired magnetic properties. Thus, it can be easily used in applied problems, which demand the phenomenological high-accurate analytic representation of ferrofluid's magnetization when the controlling configuration of the external magnetic field is stated by the system's construction. Among such applications, one can mention different microfluidic devices operating with microparticles and biological cells, e.g., [12,35], devices based on the magneto-Archimedes effect, e.g., [15,16,23], ferrofluid-based measuring devices [19,36], etc.

In addition, the static field-dependent magnetic susceptibility $\xi(H) = dM(H)/dH$ (except the close vicinity of the state $H = 0$) also has a simple analytical representation in this case since the derivative of the Mittag–Leffler function (8) is known [28]

$$\frac{d}{dz}E_\alpha(-z^\alpha) = -z^{-(1-\alpha)}E_{\alpha,\alpha}(-z^\alpha)$$

and can be accurately computed numerically with the existing software [30].

Finally, the revealed mathematical dependence poses some outlooks for future more detailed investigations of possible physical mechanisms, which may lead to such a representation of static magnetization in the form close to typical for anomalous kinetics processes. In particular, does it mean the existence of fractional-order fluctuations in the case of high concentrations of magnetic nanoparticles that possibly results in trapping their rotation, etc.?

Author Contributions: Conceptualization, E.B.P.; methodology, E.B.P. and P.A.R.; software, E.B.P.; formal analysis, E.B.P.; investigation, P.A.R. and E.B.P.; resources, P.A.R.; data curation, P.A.R.; writing—original draft preparation, E.B.P. and P.A.R.; writing—review and editing, P.A.R. and E.B.P.; visualization, E.B.P.; supervision, E.B.P. All authors have read and agreed to the published version of the manuscript.

Funding: P.A.R. is supported by the Ministry of Science and Higher Education of the Russian Federation, Russia within the research project of the state task No. 0851-2020-0035.

Data Availability Statement: The raw magnetization data used for illustrating the model reported in this study by figures are available on the reasonable request from the first author (P.A.R.).

Conflicts of Interest: The authors declare no conflict of interest.

References

1. Bedanta, S.; Kleemann, W. Supermagnetism. *J. Phys. D Appl. Phys.* **2008**, *42*, 013001. [CrossRef]
2. Wu, K.; Tu, L.; Su, D.; Wang, J.P. Magnetic dynamics of ferrofluids: Mathematical models and experimental investigations. *J. Phys. D Appl. Phys.* **2017**, *50*, 085005. [CrossRef]
3. Lebedev, A.V.; Stepanov, V.I.; Kuznetsov, A.A.; Ivanov, A.O.; Pshenichnikov, A.F. Dynamic susceptibility of a concentrated ferrofluid: The role of interparticle interactions. *Phys. Rev. E* **2019**, *100*, 032605. [CrossRef]
4. Usov, N.A.; Serebryakova, O.N. Equilibrium properties of assembly of interacting superparamagnetic nanoparticles. *Sci. Rep.* **2020**, *10*, 13677. [CrossRef]
5. Devi, E.C.; Singh, S.D. Tracing the Magnetization Curves: A Review on Their Importance, Strategy, and Outcomes. *J. Supercond. Nov. Magn.* **2020**, *34*, 15–25. [CrossRef]
6. Dikansky, Y.I.; Ispiryan, A.G.; Arefyev, I.M.; Kunikin, S.A. Effective fields in magnetic colloids and features of their magnetization kinetics. *Eur. Phys. J. E* **2021**, *44*, 1–13. [CrossRef] [PubMed]
7. Rosensweig, R.E. *Ferrohydrodynamics*; Cambridge University Press: Cambridge, UK, 1985.
8. Ivanov, A.O.; Kuznetsova, O.B. Magnetic properties of dense ferrofluids: An influence of interparticle correlations. *Phys. Rev. E* **2001**, *64*, 041405. [CrossRef] [PubMed]
9. Solovyova, A.Y.; Elfimova, E.A.; Ivanov, A.O.; Camp, P.J. Modified mean-field theory of the magnetic properties of concentrated, high-susceptibility, polydisperse ferrofluids. *Phys. Rev. E* **2017**, *96*, 052609. [CrossRef]
10. Rosensweig, R.E. Magnetic fluids. *Annu. Rev. Fluid Mech.* **1987**, *19*, 437–461. [CrossRef]
11. Vékás, L. Ferrofluids and Magnetorheological Fluids. *Adv. Sci. Technol.* **2008**, *54*, 127–136. [CrossRef]
12. Zhao, W.; Cheng, R.; Miller, J.R.; Mao, L. Label-free microfluidic manipulation of particles and cells in magnetic liquids. *Adv. Funct. Mater.* **2016**, *26*, 3916–3932. [CrossRef]
13. Ivanov, A.S.; Pshenichnikov, A.F. Magnetophoresis and diffusion of colloidal particles in a thin layer of magnetic fluids. *J. Magn. Magn. Mater.* **2010**, *322*, 2575–2580. [CrossRef]
14. Wang, L.; Wang, J. Self-assembly of colloids based on microfluidics. *Nanoscale* **2019**, *11*, 16708–16722. [CrossRef]
15. Gao, Q.H.; Zhang, W.M.; Zou, H.X.; Li, W.B.; Yan, H.; Peng, Z.K.; Meng, G. Label-free manipulation via the magneto-Archimedes effect: fundamentals, methodology and applications. *Mater. Horizons* **2019**, *6*, 1359–1379. [CrossRef]
16. Lee, J.H.; Nam, Y.J.; Yamane, R.; Park, M.K. Position feedback control of a nonmagnetic body levitated in magnetic fluid. *J. Phys. Conf. Ser.* **2009**, *149*, 012107. [CrossRef]
17. Rehberg, I.; Richter, R.; Hartung, S.; Lucht, N.; Hankiewicz, B.; Friedrich, T. Measuring magnetic moments of polydisperse ferrofluids utilizing the inverse Langevin function. *Phys. Rev. B* **2019**, *100*, 134425. [CrossRef]
18. Rehberg, I.; Richter, R.; Hartung, S. Graphical Magnetogranulometry of EMG909. *J. Magn. Magn. Mater.* **2020**, *508*, 166868. [CrossRef]
19. Ryapolov, P.A.; Sokolov, E.A.; Bashtovoi, V.G.; Reks, A.G.; Postnikov, E.B. Equilibrium configurations in a magnetic fluid-based field mapping and gas pressure measuring system: Experiment and simulations. *AIP Adv.* **2021**, *11*, 015206. [CrossRef]
20. Podlubny, I.; Petráš, I.; Škovránek, T. Fitting of experimental data using Mittag–Leffler function. In Proceedings of the IEEE 13th International Carpathian Control Conference (ICCC), High Tatras, Slovakia, 28–31 May 2012; pp. 578–581. [CrossRef]
21. Fannin, P.C. On the use of dielectric formalism in the representation of ferrofluid data. *J. Mol. Liq.* **1996**, *69*, 39–51. [CrossRef]
22. Kalmykov, Y.P. Fractional rotational Brownian motion in a uniform dc external field. *Phys. Rev. E* **2004**, *70*, 051106. [CrossRef]
23. Ryapolov, P.A.; Polunin, V.M.; Postnikov, E.B.; Bashtovoi, V.G.; Reks, A.G.; Sokolov, E.A. The behaviour of gas inclusions in a magnetic fluid in a non-uniform magnetic field. *J. Magn. Magn. Mater.* **2020**, *497*, 165925. [CrossRef]
24. Polunin, V.M. *Acoustics of Nanodispersed Magnetic Fluids*; CRC Press: Boca Raton, FL, USA, 2015.
25. Mayer, A.; Vogt, E. Magnetische Messungen an Eisenamalgam zur Frage: Ferromagnetismus und Korngröße. *Z. Naturforschung A* **1952**, *7*, 334–340. [CrossRef]
26. Bean, C.P.; Jacobs, I.S. Magnetic granulometry and super-paramagnetism. *J. Appl. Phys.* **1956**, *27*, 1448–1452. [CrossRef]
27. Podlubny, I. *Fractional Differential Equations: An Introduction to Fractional Derivatives, Dractional Differential Equations, to Methods of Their Solution and Some of Their Applications*; Academic Press: San Diego, CA, USA, 1998.
28. Mainardi, F. Why the Mittag–Leffler function can be considered the Queen function of the Fractional Calculus? *Entropy* **2020**, *22*, 1359. [CrossRef]
29. Podlubny, I. Fitting Data Using the Mittag–Leffler Function. Available online: https://www.mathworks.com/matlabcentral/fileexchange/32170-fitting-data-using-the-mittag-leffler-function (accessed on 30 September 2021).
30. Podlubny, I. Mittag-Leffler Function. Available online: https://www.mathworks.com/matlabcentral/fileexchange/8738-mittag-lefler-function (accessed on 30 September 2021).
31. Li, J.; Lin, Y.; Liu, X.; Lin, L.; Zhang, Q.; Fu, J.; Chen, L.; Li, D. The quasi-magnetic-hysteresis behavior of polydisperse ferrofluids with small coupling constant. *Phys. B Condens. Matter* **2012**, *407*, 4638–4642. [CrossRef]
32. Lukashevich, M.V.; Naletova, V.A.; Turkov, V.A.; Nazarenko, A.V. A new method of measuring magnetization of a magnetic fluid not affecting its dispersive composition and calculation of volume distribution function. *J. Magn. Magn. Mater.* **1993**, *122*, 139–141. [CrossRef]
33. Jin, D.; Kim, H. Magnetization of magnetite ferrofluid studied by using a magnetic balance. *Bull. Korean Chem. Soc.* **2013**, *34*, 1715–1721. [CrossRef]

34. Sierociuk, D.; Podlubny, I.; Petras, I. Experimental evidence of variable-order behavior of ladders and nested ladders. *IEEE Trans. Control Syst. Technol.* **2012**, *21*, 459–466. [CrossRef]
35. Myklatun, A.; Cappetta, M.; Winklhofer, M.; Ntziachristos, V.; Westmeyer, G.G. Microfluidic sorting of intrinsically magnetic cells under visual control. *Sci. Rep.* **2017**, *7*, 6942. [CrossRef] [PubMed]
36. Smolyaninov, I.I.; Smolyaninova, V.N. Fine tuning and MOND in a metamaterial "multiverse". *Sci. Rep.* **2017**, *7*, 8023. [CrossRef]

Article

Reduction Formulas for Generalized Hypergeometric Series Associated with New Sequences and Applications

Junesang Choi [1,*], **Mohd Idris Qureshi** [2], **Aarif Hussain Bhat** [2] **and Javid Majid** [2]

1. Department of Mathematics, Dongguk University, Gyeongju 38066, Korea
2. Department of Applied Sciences and Humanities, Faculty of Engineering and Technology, Jamia Millia Islamia (A Central University), New Delhi 110025, India; miqureshi_delhi@yahoo.co.in (M.I.Q.); aarifsaleem19@gmail.com (A.H.B.); javidmajid375@gmail.com (J.M.)
* Correspondence: junesangchoi@gmail.com; Tel.: +82-010-6525-2262

Abstract: In this paper, by introducing two sequences of *new* numbers and their derivatives, which are closely related to the Stirling numbers of the first kind, and choosing to employ six known generalized Kummer's summation formulas for $_2F_1(-1)$ and $_2F_1(1/2)$, we establish six classes of generalized summation formulas for $_{p+2}F_{p+1}$ with arguments -1 and $1/2$ for any positive integer p. Next, by differentiating both sides of six chosen formulas presented here with respect to a specific parameter, among numerous ones, we demonstrate six identities in connection with finite sums of $_4F_3(-1)$ and $_4F_3(1/2)$. Further, we choose to give simple particular identities of some formulas presented here. We conclude this paper by highlighting a potential use of the newly presented numbers and posing some problems.

Keywords: Gamma function; Psi function; Pochhammer symbol; hypergeometric function $_2F_1$; generalized hypergeometric functions $_tF_u$; Gauss's summation theorem for $_2F_1(1)$; Kummer's summation theorem for $_2F_1(-1)$; generalized Kummer's summation theorem for $_2F_1(-1)$; Stirling numbers of the first kind

MSC: 11B73; 11B83; 33C05; 33C20; 11B68; 33B15

1. Introduction and Preliminaries

The Pochhammer symbol $(\xi)_\eta$ $(\xi, \eta \in \mathbb{C})$ is defined, in terms of Gamma function Γ (see, e.g., [1], p. 2 and p. 5), by

$$(\xi)_\eta = \frac{\Gamma(\xi+\eta)}{\Gamma(\xi)} \quad (\xi+\eta \in \mathbb{C} \setminus \mathbb{Z}_0^-, \eta \in \mathbb{C} \setminus \{0\}; \xi \in \mathbb{C} \setminus \mathbb{Z}_0^-, \eta = 0)$$

$$= \begin{cases} 1 & (\eta = 0), \\ \xi(\xi+1)\cdots(\xi+n-1) & (\eta = n \in \mathbb{N}), \end{cases} \quad (1)$$

it accepted that $(0)_0 = 1$. Here and throughout, let $\mathbb{C}$, $\mathbb{R}^+$, $\mathbb{Z}$, and $\mathbb{N}$ represent, respectively, the sets of complex numbers, positive real numbers, integers, and positive integers. Furthermore, let $\mathbb{N}_0 := \mathbb{N} \cup \{0\}$, $\mathbb{Z}^- := \mathbb{Z} \setminus \mathbb{N}_0$ and $\mathbb{Z}_0^- := \mathbb{Z} \setminus \mathbb{N}$. Further, throughout this article, it is assumed that an empty sum and an empty product are read as 0 and 1, respectively. The generalized hypergeometric series (or function) $_pF_q$ $(p, q \in \mathbb{N}_0)$, which is a parametric and logical extension of the Gaussian hypergeometric series $_2F_1$, is defined by (see, e.g., [1–9])

$$_pF_q\left[\begin{matrix}\mu_1, \ldots, \mu_p;\\ \nu_1, \ldots, \nu_q;\end{matrix} w\right] = \sum_{\ell=0}^{\infty} \frac{\prod_{j=1}^{p}(\mu_j)_\ell}{\prod_{j=1}^{q}(\nu_j)_\ell} \frac{w^\ell}{\ell!} \quad (2)$$

$$= {}_pF_q(\mu_1, \ldots, \mu_p; \nu_1, \ldots, \nu_q; w).$$

Here it is supposed that the variable w, the numerator parameters $\mu_1, \ldots, \mu_p$, and the denominator parameters $\nu_1, \ldots, \nu_q$ take on complex values, provided that

$$(\nu_j \in \mathbb{C} \setminus \mathbb{Z}_0^-; \ j = 1, \ldots, q). \tag{3}$$

Then, if a numerator parameter is in $\mathbb{Z}_0^-$, the series ${}_pF_q$ is found to terminate and becomes a polynomial in w.

With none of the numerator and denominator parameters being zero or a negative integer, the series ${}_pF_q$ in (2)

(i) diverges for all $w \in \mathbb{C} \setminus \{0\}$, if $p > q+1$;
(ii) converges for all $w \in \mathbb{C}$, if $p \leq q$;
(iii) converges for $|w| < 1$ and diverges for $|w| > 1$ if $p = q+1$;
(iv) converges absolutely for $|w| = 1$, if $p = q+1$ and $\Re(\omega) > 0$;
(v) converges conditionally for $|w| = 1$ $(w \neq 1)$, if $p = q+1$ and $-1 < \Re(\omega) \leq 0$;
(vi) diverges for $|w| = 1$, if $p = q+1$ and $\Re(\omega) \leq -1$.

where

$$\omega := \sum_{j=1}^{q} \nu_j - \sum_{j=1}^{p} \mu_j \tag{4}$$

which is called the parametric excess of the series.

Gauss's famous summation formula [10]:

$$_2F_1(\kappa, \lambda; \mu; 1) = \frac{\Gamma(\mu)\Gamma(\mu - \kappa - \lambda)}{\Gamma(\mu - \kappa)\Gamma(\mu - \lambda)} \tag{5}$$

$$\left(\Re(\mu - \kappa - \lambda) > 0, \ \mu \in \mathbb{C} \setminus \mathbb{Z}_0^- \right)$$

has been a significant, pioneering, and essential identity, especially in the theories of hypergeometric and generalized hypergeometric functions, as well as related special functions. Formula (5) can be proved by using Euler's integral representation for $_2F_1(z)$ (see, e.g., [6], pp. 44–49) or telescoping (see, e.g., [11], pp. 181–182). Since (5) appeared, a number of researchers have devoted their arduous, intrigued and penetrated endeavors to getting summation formulas for the generalized hypergeometric series in (2). As a result, the generalized hypergeometric series in (2) of the case $p = q+1$ have been found to be classified as follows: $_{q+1}F_q$ in (2) is said to be ω-balanced if the parametric excess equals ω and balanced if $\omega = 1$. Further, if $\omega = 1$ and one of the numerator parameters is a negative integer, it is called Saalschützian. It is well-poised if the parameters μ_j, ν_j can be separately permuted so that

$$1 + \mu_1 = \mu_2 + \nu_1 = \cdots = \mu_{q+1} + \nu_q$$

and very well-poised if the condition $\mu_2 = 1 + \frac{\mu_1}{2}$ holds true, along with the above condition for the well-poised nature. Consequently, a large number of summation and transformation formulas for ${}_pF_q$ have been established by means of diverse techniques. In fact, usually, certain mixed techniques are used in getting a summation formula or a transformation formula for ${}_pF_q$. Here we recall only several representative techniques which are employed in deriving some summation and transformation formulas for ${}_pF_q$:

(i) Contiguous function relations (and computer programs) [12–25].
(ii) The idea of partition of the set of nonnegative integers into its terms modulo N applied to a series involving functions Ψ_n ($n \in \mathbb{N}_0$) displayed by

$$\sum_{n=0}^{\infty} \Psi_n = \sum_{r=0}^{N-1} \sum_{n=0}^{\infty} \Psi_{nN+r} \tag{6}$$

is ubiquitously employed (see, e.g., [26,27]). In particular, partition of the series into even and odd terms gives

$$\sum_{n=0}^{\infty} \Psi_n = \sum_{n=0}^{\infty} \Psi_{2n} + \sum_{n=0}^{\infty} \Psi_{2n+1}. \tag{7}$$

The (6) and (7) have been used to get certain identities involving generalized hypergeometric series and their extensions (see, e.g., [28–37]). Exton [30] considered the following two combinations

$$_{q+1}F_q\left[\begin{array}{c} a_1, a_2, \ldots, a_{q+1}; \\ b_1, b_2, \ldots, b_q; \end{array} 1\right] + {}_{q+1}F_q\left[\begin{array}{c} a_1, a_2, \ldots, a_{q+1}; \\ b_1, b_2, \ldots, b_q; \end{array} -1\right]$$

$$= 2\,{}_{2q+2}F_{2q+1}\left[\begin{array}{c} \frac{a_1}{2}, \frac{a_1+1}{2}, \ldots, \frac{a_{q+1}}{2}, \frac{a_{q+1}+1}{2}; \\ \frac{1}{2}, \frac{b_1}{2}, \frac{b_1+1}{2}, \ldots, \frac{b_q}{2}, \frac{b_q+1}{2}; \end{array} 1\right] \tag{8}$$

and

$$_{q+1}F_q\left[\begin{array}{c} a_1, a_2, \ldots, a_{q+1}; \\ b_1, b_2, \ldots, b_q; \end{array} 1\right] - {}_{q+1}F_q\left[\begin{array}{c} a_1, a_2, \ldots, a_{q+1}; \\ b_1, b_2, \ldots, b_q; \end{array} -1\right]$$

$$= 2\,\frac{\prod_{j=1}^{q+1} a_j}{\prod_{j=1}^{q} b_j}\,{}_{2q+2}F_{2q+1}\left[\begin{array}{c} \frac{a_1+1}{2}, \frac{a_1}{2}+1, \ldots, \frac{a_{q+1}+1}{2}, \frac{a_{q+1}}{2}+1; \\ \frac{3}{2}, \frac{b_1+1}{2}, \frac{b_1}{2}+1, \ldots, \frac{b_q+1}{2}, \frac{b_q}{2}+1; \end{array} 1\right]. \tag{9}$$

If the summation formulas for $_{q+1}F_q(1)$ and $_{q+1}F_q(-1)$ are known, then summation formulas for $_{2q+2}F_{2q+1}(1)$ in (8) and (9) can be derived. Obviously, the reverse process can work.

(iii) The method in (ii) is to obtain summation formulas for certain generalized hypergeometric functions of higher order from those of lower order. Conversely, reduction formulas of generalized hypergeometric and their extended special functions are to reduce those of higher order to some other ones of lower order (see, e.g., [14,38–45]).

In connection with the method (iii), for a generalized hypergeometric function $_pF_q(z)$ with positive integral differences between certain numerator and denominator parameters, Karlsson [39] provided a formula expressing the $_pF_q(z)$ as a finite sum of lower-order functions as follows (see also [42,46,47]):

$$_pF_q\left[\begin{array}{c} b_1+\ell_1, \ldots, b_n+\ell_n, a_{n+1}, \ldots, a_p; \\ b_1, \ldots, b_n, b_{n+1}, \ldots, b_q; \end{array} z\right]$$

$$= \sum_{j_1=0}^{\ell_1} \cdots \sum_{j_n=0}^{\ell_n} A(j_1, \ldots, j_n)\, z^{J_n} \tag{10}$$

$$\times\, {}_{p-n}F_{q-n}\left[\begin{array}{c} a_{n+1}+J_n, \ldots, a_p+J_n; \\ b_{n+1}+J_n, \ldots, b_q+J_n; \end{array} z\right],$$

where $J_n = j_1 + \cdots + j_n$ and

$$A(j_1, \ldots, j_n) = \prod_{r=1}^{n} \binom{\ell_r}{j_r} \cdot \frac{\prod_{r=2}^{n} (b_r+\ell_r)_{J_{r-1}} \cdot \prod_{r=n+1}^{p} (a_r)_{J_n}}{\prod_{r=1}^{n} (b_r)_{J_r} \cdot \prod_{r=n+1}^{q} (b_r)_{J_n}}. \tag{11}$$

Here the following constraints are assumed that, with suitable permutation of parameters, $a_r = b_r + \ell_r$, $\ell_r \in \mathbb{N}$ ($r = 1, \ldots, n$), $n \leq \min\{p, q\}$, $p \leq q + 1$, $b_r \in \mathbb{C} \setminus \mathbb{Z}_0^-$ ($r = 1, \ldots, q$); if $a_r \in \mathbb{Z}_0^-$ for some $r \in \{1, \ldots, p\}$, the condition $p \leq q + 1$ is cancelled.

Using (10), Minton's two summation theorems in [43] for $p = q + 1$, $z = 1$ are derived. Srivastava [45] gave a simpler proof of (10). Gottschalk and Maslen [38] provided a good account of reduction formulas for the generalized hypergeometric functions of one variable with some useful comments on (10) and listed certain transformation formulas for generalized hypergeometric functions in [38].

The content of this paper would be derived from the reduction formula (10). Yet, in this paper, by introducing two sequences of *new* numbers and their derivatives as in Section 2 and using the six generalized summation formulas (15)–(20), we aim to establish families of generalized summation formulas for $_{t+2}F_{t+1}$ ($t \in \mathbb{N}$) with their arguments -1 and $1/2$ as in Sections 4 and 5. Furthermore, we select to give simple particular identities of some formulas presented here. By differentiating both sides of two chosen formulas presented here with respect to a specific parameter, among numerous ones, further, we demonstrate two identities associated with finite sums of $_4F_3(-1)$. We close this article by emphasizing some of the possible applications for the newly introduced numbers and presenting certain problems.

For our purpose, we also recall three basic and useful summation formulas for $_2F_1$ due to Kummer [48], p. 134, Entries 1, 2 and 3 (see also [49], Equations (1.3), (1.4) and (1.5); see further [50]) (the interested reader may refer to [49], p. 853 for clarifications on the first and true contributors to the following three summation formulae):

Summation Formula 1 due to Kummer:

$$_2F_1\left[\begin{matrix} \kappa, \lambda; \\ 1 + \kappa - \lambda; \end{matrix} -1\right] = \frac{\Gamma(1 + \kappa - \lambda)\Gamma\left(1 + \frac{\kappa}{2}\right)}{\Gamma\left(1 + \frac{\kappa}{2} - \lambda\right)\Gamma(1 + \kappa)} \tag{12}$$

$$(\kappa - \lambda \in \mathbb{C} \setminus \mathbb{Z}^-,\ \Re(\lambda) < 1).$$

Summation Formula 2 due to Kummer:

$$_2F_1\left[\begin{matrix} \kappa, \lambda; \\ \frac{1}{2}(\kappa + \lambda + 1); \end{matrix} \frac{1}{2}\right] = \frac{\Gamma\left(\frac{1}{2}\right)\Gamma\left(\frac{1}{2}\kappa + \frac{1}{2}\lambda + \frac{1}{2}\right)}{\Gamma\left(\frac{1}{2}\kappa + \frac{1}{2}\right)\Gamma\left(\frac{1}{2}\lambda + \frac{1}{2}\right)} \tag{13}$$

$$\left(\frac{\kappa + \lambda + 1}{2} \in \mathbb{C} \setminus \mathbb{Z}_0^-\right).$$

Summation Formula 3 due to Kummer:

$$_2F_1\left[\begin{matrix} \kappa, 1 - \kappa; \\ \lambda; \end{matrix} \frac{1}{2}\right] = \frac{2^{1-\lambda}\Gamma\left(\frac{1}{2}\right)\Gamma(\lambda)}{\Gamma\left(\frac{1}{2}\lambda + \frac{1}{2}\kappa\right)\Gamma\left(\frac{1}{2}\lambda - \frac{1}{2}\kappa + \frac{1}{2}\right)} \quad (\lambda \in \mathbb{C} \setminus \mathbb{Z}_0^-). \tag{14}$$

Further a number of generalizations and contiguous extensions of the above-mentioned Kummer's summation theorems have been given (see, e.g., [16,49,51–54] and the references therein). Amid this trend, Choi et al. [51], Equations (2.2) and (2.3) presented the following extensions of (12) (see also [53], Theorems 3 and 4):

$$_2F_1\left[\begin{matrix} \kappa, \lambda; \\ 1 + \kappa - \lambda + p; \end{matrix} -1\right]$$

$$= \frac{\Gamma(1 + \kappa - \lambda + p)}{2\,\Gamma(\kappa)\,(1 - \lambda)_p} \sum_{r=0}^{p} \binom{p}{r} \frac{(-1)^r \Gamma\left(\frac{r+\kappa}{2}\right)}{\Gamma\left(\frac{r+\kappa}{2} + 1 - \lambda\right)} \tag{15}$$

$$\left(p \in \mathbb{N}_0,\ \kappa - \lambda + p \in \mathbb{C} \setminus \mathbb{Z}^-,\ \Re(\lambda) < 1 + \frac{p}{2}\right).$$

and
$$_2F_1\left[\begin{matrix}\kappa,\lambda;\\1+\kappa-\lambda-p;\end{matrix}-1\right]$$
$$=\frac{\Gamma(1+\kappa-\lambda-p)}{2\Gamma(\kappa)}\sum_{r=0}^{p}\binom{p}{r}\frac{\Gamma(\frac{r+\kappa}{2})}{\Gamma(\frac{r+\kappa}{2}+1-\lambda-p)} \quad (16)$$
$$\left(p\in\mathbb{N}_0,\ \kappa-\lambda-p\in\mathbb{C}\setminus\mathbb{Z}^-,\ \Re(\lambda)<1-\frac{p}{2}\right).$$

Rakha and Rathie [53], Theorem 1 gave the following generalization of (13):
$$_2F_1\left[\begin{matrix}\kappa,\lambda;\\ \frac{1+\kappa+\lambda+p}{2};\end{matrix}\frac{1}{2}\right]=\frac{2^{\kappa-1}\Gamma(\frac{1+\kappa+\lambda+p}{2})\Gamma(\frac{1-\kappa+\lambda-p}{2})}{\Gamma(\kappa)\Gamma(\frac{1-\kappa+\lambda+p}{2})}$$
$$\times\sum_{r=0}^{p}\binom{p}{r}\frac{(-1)^r\Gamma(\frac{\kappa+r}{2})}{\Gamma(\frac{1+\lambda+r-p}{2})} \quad (17)$$
$$\left(p\in\mathbb{N}_0,\ \frac{1+\kappa+\lambda+p}{2}\in\mathbb{C}\setminus\mathbb{Z}_0^-\right).$$

The following extension of (13) is recorded in [9], p. 491, Entry 7.3.7-2 (see also [53], Theorem 2):
$$_2F_1\left[\begin{matrix}\kappa,\lambda;\\ \frac{1+\kappa+\lambda-p}{2};\end{matrix}\frac{1}{2}\right]=\frac{2^{\lambda-1}\Gamma(\frac{1+\kappa+\lambda-p}{2})}{\Gamma(\lambda)}\sum_{r=0}^{p}\binom{p}{r}\frac{\Gamma\left(\frac{\lambda+r}{2}\right)}{\Gamma(\frac{1+\kappa+r-p}{2})} \quad (18)$$
$$\left(p\in\mathbb{N}_0,\ \frac{1+\kappa+\lambda-p}{2}\in\mathbb{C}\setminus\mathbb{Z}_0^-\right).$$

Rakha and Rathie ([53], Theorems 5 and 6) provided two generalizations of (14) which, with the aid of Legendre's duplication formula for the Gamma function (e.g., [1], p. 6, Equation (29)), are slightly modified as follows:
$$_2F_1\left[\begin{matrix}\kappa,1-\kappa+p;\\ \lambda;\end{matrix}\frac{1}{2}\right]=\frac{2^{p-\kappa}\,\Gamma(\kappa-p)\,\Gamma(\lambda)}{\Gamma(\kappa)\,\Gamma(\lambda-\kappa)}$$
$$\times\sum_{r=0}^{p}(-1)^r\binom{p}{r}\frac{\Gamma(\frac{\lambda-\kappa+r}{2})}{\Gamma\left(\frac{\lambda+\kappa+r}{2}-p\right)} \quad (19)$$
$$\left(\lambda\in\mathbb{C}\setminus\mathbb{Z}_0^-,\ p\in\mathbb{N}_0\right)$$

and
$$_2F_1\left[\begin{matrix}\kappa,1-\kappa-p;\\ \lambda;\end{matrix}\frac{1}{2}\right]=\frac{2^{-p-\kappa}\,\Gamma(\lambda)}{\Gamma(\lambda-\kappa)}\sum_{r=0}^{p}\binom{p}{r}\frac{\Gamma\left(\frac{\lambda-\kappa+r}{2}\right)}{\Gamma\left(\frac{\lambda+\kappa+r}{2}\right)} \quad (20)$$
$$\left(\lambda\in\mathbb{C}\setminus\mathbb{Z}_0^-,\ p\in\mathbb{N}_0\right),$$
which is a corrected version of [53], Theorem 6.

In addition, we recall the Psi (or digamma) function $\psi(z)$ (see, e.g., [1], pp. 24–33) defined by
$$\psi(z):=\frac{d}{dz}\{\log\Gamma(z)\}=\frac{\Gamma'(z)}{\Gamma(z)}\quad(z\in\mathbb{C}\setminus\mathbb{Z}_0^-). \quad (21)$$

We recall one of the many identities involving the Psi function
$$\psi(z+n)-\psi(z)=\sum_{j=1}^{n}\frac{1}{z+j-1}\quad(n\in\mathbb{N}). \quad (22)$$

Remark 1. *Magnus Gösta Mittag–Leffler (1846–1927), a Swedish mathematician (see [55]; see also [56,57]), invented the function $E_\alpha(z)$ (23) in conjunction with the summation technique for divergent series, which is eponymously referred to as the Mittag–Leffler function and represented by the following convergent power series across the whole complex plane:*

$$E_\alpha(z) = \sum_{k=0}^{\infty} \frac{z^k}{\Gamma(\alpha k + 1)} \quad (\Re(\alpha) > 0, z \in \mathbb{C}). \tag{23}$$

The two parameterized Mittag–Leffler function $E_{\alpha,\beta}(z)$ is defined by (see, e.g., [58,59])

$$E_{\alpha,\beta}(z) = \sum_{k=0}^{\infty} \frac{z^k}{\Gamma(\alpha k + \beta)} \quad (\Re(\alpha) > 0, \beta \in \mathbb{C}). \tag{24}$$

There have been a variety of extensions of the Mittag–Leffler functions (23) and (24), most of which belong to certain special cases of the following Fox-Wright function (see [60–63], [64], p. 21):

$$_p\Psi_q\left[\begin{matrix}(\alpha_1, A_1), \ldots, (\alpha_p, A_p); \\ (\beta_1, B_1), \ldots, (\beta_q, B_q);\end{matrix} z\right] = \sum_{k=0}^{\infty} \frac{\prod_{\ell=1}^{p} \Gamma(\alpha_\ell + A_\ell k)}{\prod_{j=1}^{q} \Gamma(\beta_j + B_j k)} \frac{z^k}{k!}, \tag{25}$$

where $z \in \mathbb{C}$, $\alpha_\ell, \beta_j \in \mathbb{C}$ ($\ell = 1, \ldots, p$, $j = 1, \ldots, q$), the coefficients $A_1, \ldots, A_p \in \mathbb{R}^+$ and $B_1, \ldots, B_q \in \mathbb{R}^+$ such that $\alpha_\ell + A_\ell k \in \mathbb{C} \setminus \mathbb{Z}_0^-$ ($k \in \mathbb{N}_0$) and

$$1 + \sum_{j=1}^{q} B_j - \sum_{j=1}^{p} A_j \geqq 0. \tag{26}$$

A particular case of (25) is

$$_p\Psi_q\left[\begin{matrix}(\alpha_1, 1), \ldots, (\alpha_p, 1); \\ (\beta_1, 1), \ldots, (\beta_q, 1);\end{matrix} z\right] = \frac{\prod_{\ell=1}^{p} \Gamma(\alpha_\ell)}{\prod_{j=1}^{q} \Gamma(\beta_j)} {}_pF_q\left[\begin{matrix}\alpha_1, \ldots, \alpha_p; \\ \beta_1, \ldots, \beta_q;\end{matrix} z\right]. \tag{27}$$

In light of (27), the topic of this article may be regarded to be Mittag–Leffler type functions.

Indeed, owing to the range of its applications in fractional calculus, some scholars have nicknamed the Mittag–Leffler function the "Queen Function of the Fractional Calculus" in the past (see, e.g., [65]).

2. Sequences of New Numbers

Numerous polynomials, numbers, their extensions, degenerations, and new polynomials and new numbers have been developed and studied, owing primarily to their potential applications and use in a diverse variety of research fields (see, e.g., [66–71] and the references therein). For example, Bernoulli polynomials and numbers are among most important and useful ones (see, e.g., [5], pp. 35–40, [1], Sections 1.7 and 1.8). As with Definitions 1 and 2, this section introduces two sequences of new numbers and their derivatives that are and will be useful (at the very least) for our current and related study topics.

Definition 1. A sequence of new numbers $\{\mathcal{A}_j(\alpha,\ell)\}_{j=0}^{\ell}$ is defined by

$$(\alpha+k)_\ell = (\alpha+k)(\alpha+k+1)\cdots(\alpha+k+\ell-1)$$
$$:= \sum_{j=0}^{\ell} \mathcal{A}_j(\alpha,\ell)\, k(k-1)\cdots(k-j+1) \tag{28}$$
$$(k\in\mathbb{N}_0,\, \ell\in\mathbb{N},\, \alpha\in\mathbb{C})$$

and

$$\mathcal{A}_0(\alpha,0) := 1 \quad (\alpha\in\mathbb{C}). \tag{29}$$

Definition 2. A sequence of new numbers $\{\mathcal{B}_j(\alpha,\ell)\}_{j=0}^{\ell}$ is defined by

$$\mathcal{B}_j(\alpha,\ell) := \frac{d}{d\alpha}\mathcal{A}_j(\alpha,\ell) \quad (\ell\in\mathbb{N},\, \alpha\in\mathbb{C}) \tag{30}$$

and

$$\mathcal{B}_0(\alpha,0) := 0 \quad (\alpha\in\mathbb{C}). \tag{31}$$

Both of the following lemmas may be used to represent the numbers in Definitions 1 and 2 explicitly.

Lemma 1. Let $\alpha\in\mathbb{C}$ and $\ell\in\mathbb{N}_0$. Then

$$\sum_{j=\nu}^{\ell} \mathcal{A}_j(\alpha,\ell)\, s(j,\nu) = \sum_{j=\nu}^{\ell}(-1)^{\ell+j} s(\ell,j)\binom{j}{\nu}\alpha^{j-\nu} \quad (\nu=0,1,\ldots,\ell). \tag{32}$$

Also

$$\mathcal{A}_j(\alpha,\ell) = \binom{\ell}{j}\frac{(\alpha)_\ell}{(\alpha)_j} = \binom{\ell}{j}(\alpha+j)_{\ell-j} \quad (j=0,1,\ldots,\ell). \tag{33}$$

Lemma 2. Let $\alpha\in\mathbb{C}$ and $\ell\in\mathbb{N}_0$. Then

$$\sum_{j=\nu}^{\ell} \mathcal{B}_j(\alpha,\ell)\, s(j,\nu) = \sum_{j=\nu}^{\ell}(-1)^{\ell+j} j\, s(\ell,j)\binom{j-1}{\nu}\alpha^{j-1-\nu} \tag{34}$$
$$(\nu=0,1,\ldots,\ell).$$

Also

$$\mathcal{B}_j(\alpha,\ell) = \binom{\ell}{j}(\alpha+j)_{\ell-j}\sum_{k=j}^{\ell-1}\frac{1}{\alpha+k} \quad (j=0,1,\ldots,\ell). \tag{35}$$

Proof of Lemma 1. The Stirling numbers $s(m,r)$ of the first kind are recalled and defined by the generating function (see, e.g., [1], Section 1.6)

$$\omega(\omega-1)\cdots(\omega-m+1) = \sum_{r=0}^{m} s(m,r)\,\omega^r. \tag{36}$$

We use (36) to expand the Pochhammer symbol (1) as follows:

$$(\omega)_m = \omega(\omega+1)\cdots(\omega+m-1) = \sum_{r=0}^{m}(-1)^{m+r} s(m,r)\,\omega^r, \tag{37}$$

where $(-1)^{m+r} s(m,r)$ indicates the number of permutations of m symbols, which possesses exactly r cycles.

Applying (36) and (37) to (28), we obtain

$$\sum_{j=0}^{\ell}(-1)^{\ell+j}s(\ell,j)\sum_{\nu=0}^{j}\binom{j}{\nu}\alpha^{j-\nu}k^{\nu} = \sum_{j=0}^{\ell}\mathcal{A}_j(\alpha,\ell)\sum_{\nu=0}^{j}s(j,\nu)k^{\nu}. \tag{38}$$

Using a series rearrangement technique (see, e.g., [72], Equation (1.24))

$$\sum_{j=0}^{\ell}\sum_{\nu=0}^{j}f(j,\nu) = \sum_{\nu=0}^{\ell}\sum_{j=\nu}^{\ell}f(j,\nu) \tag{39}$$

in (38), we get

$$\sum_{\nu=0}^{\ell}\sum_{j=\nu}^{\ell}(-1)^{\ell+j}s(\ell,j)\binom{j}{\nu}\alpha^{j-\nu}k^{\nu} = \sum_{\nu=0}^{\ell}\sum_{j=\nu}^{\ell}\mathcal{A}_j(\alpha,\ell)s(j,\nu)k^{\nu}. \tag{40}$$

Now the desired identity (32) follows from (40).

The identity (33) can be obtained by matching the right-handed members of (10) and (50) when $n=1$. □

Proof of Lemma 2. Differentiating both sides of (32) and (33) yields (44) and (35), respectively. □

We recall the following identities (see, e.g., [1], Section 1.6):

$$s(m,0) = \begin{cases}1 & (m=0) \\ 0 & (m\in\mathbb{N}),\end{cases} \quad s(m,m)=1,$$

$$s(m,1) = (-1)^{m+1}(m-1)!, \quad s(m,m-1) = -\binom{m}{2} \tag{41}$$

and

$$\sum_{r=1}^{m}s(m,r) = 0 \quad (m\in\mathbb{N}\setminus\{1\}); \quad \sum_{r=0}^{m}(-1)^{m+r}s(m,r) = m!;$$

$$\sum_{j=r}^{m}s(m+1,j+1)m^{j-r} = s(m,r). \tag{42}$$

The identity (32), with the aid of (41) and (42), or the identity (33) can give explicit expressions for any $\ell\in\mathbb{N}$ with $0\leq j\leq\ell$ and $\alpha\in\mathbb{C}$. For example,

$$\mathcal{A}_\ell(\alpha,\ell) = 1 \quad (\ell\in\mathbb{N}). \tag{43}$$

$$\mathcal{A}_0(\alpha,1) = \alpha, \quad \mathcal{A}_0(\alpha,2) = \alpha+\alpha^2, \quad \mathcal{A}_1(\alpha,2) = 2+2\alpha,$$
$$\mathcal{A}_0(\alpha,3) = 2\alpha+3\alpha^2+\alpha^3, \quad \mathcal{A}_1(\alpha,3) = 6+9\alpha+3\alpha^2, \quad \mathcal{A}_2(\alpha,3) = 6+3\alpha,$$
$$\mathcal{A}_0(\alpha,4) = 6\alpha+11\alpha^2+6\alpha^3+\alpha^4, \quad \mathcal{A}_1(\alpha,4) = 24+44\alpha+24\alpha^2+4\alpha^3,$$
$$\mathcal{A}_2(\alpha,4) = 36+30\alpha+6\alpha^2, \quad \mathcal{A}_3(\alpha,4) = 12+4\alpha.$$

Differentiating both sides of (32) with respect to α, we get

$$\sum_{j=\nu}^{\ell}\mathcal{B}_j(\alpha,\ell)s(j,\nu) = \sum_{j=\nu}^{\ell}(-1)^{\ell+j}js(\ell,j)\binom{j-1}{\nu}\alpha^{j-1-\nu} \tag{44}$$

$$(\nu=0,1,\ldots,\ell).$$

Likewise, the relation (44), with the aid of (41) and (42), or the identity (35) can give explicit expressions for any $\ell \in \mathbb{N}$ with $0 \leq j \leq \ell$ and $\alpha \in \mathbb{C}$. For example,

$$\mathcal{B}_\ell(\alpha, \ell) = 0 \quad (\ell \in \mathbb{N}). \tag{45}$$

$$\mathcal{B}_0(\alpha, 1) = 1, \quad \mathcal{B}_0(\alpha, 2) = 1 + 2\alpha, \quad \mathcal{B}_1(\alpha, 2) = 2,$$
$$\mathcal{B}_0(\alpha, 3) = 2 + 6\alpha + 3\alpha^2, \quad \mathcal{B}_1(\alpha, 3) = 9 + 6\alpha, \quad \mathcal{B}_2(\alpha, 3) = 3,$$
$$\mathcal{B}_0(\alpha, 4) = 6 + 22\alpha + 18\alpha^2 + 4\alpha^3, \quad \mathcal{B}_1(\alpha, 4) = 44 + 48\alpha + 12\alpha^2,$$
$$\mathcal{B}_2(\alpha, 4) = 30 + 12\alpha, \quad \mathcal{B}_3(\alpha, 4) = 4.$$

Remark 2.
(i) $(\alpha + k)_\ell$ is a polynomial in both α and k of the same degree ℓ.
(ii) $\mathcal{A}_j(\alpha, \ell)$ is a polynomial in α of degree $\ell - j$.
(iii) $\mathcal{B}_j(\alpha, \ell)$ is a polynomial in α of degree $\ell - j - 1$.
(iv) The generalized harmonic numbers $H_n^{(s)}(\alpha)$ are defined by (see, e.g., [73], Equation (1.3))

$$H_n^{(s)}(\alpha) := \sum_{k=1}^n \frac{1}{(k+\alpha)^s} \quad (n \in \mathbb{N}, s \in \mathbb{C}, \alpha \in \mathbb{C} \setminus \mathbb{Z}^-), \tag{46}$$

where $H_n^{(1)}(\alpha) := H_n(\alpha)$ and $H_n^{(s)}(0) := H_n^{(s)}$ are the harmonic numbers of order s (see, e.g., [73], Equation (1.2))

$$H_n^{(s)} := \sum_{k=1}^n \frac{1}{k^s} \quad (n \in \mathbb{N}, s \in \mathbb{C}) \tag{47}$$

and $H_n^{(1)} := H_n$ are the harmonic numbers (see, e.g., [73], Equation (1.1))

$$H_n := \sum_{k=1}^n \frac{1}{k} \quad (n \in \mathbb{N}). \tag{48}$$

It follows from (35) and (46) that

$$\mathcal{B}_j(\alpha, \ell) = \binom{\ell}{j} (\alpha + j)_{\ell - j} \left(H_\ell(\alpha - 1) - H_j(\alpha - 1) \right). \tag{49}$$

3. Reduction Theorems in Terms of the Sequence in Definition 1

In this section, using the sequence in Definition 1, we present certain reduction formulas for $_pF_q$.

Theorem 1. *Let $\ell \in \mathbb{N}$, $1 \leq \min\{p, q\}$, $p \leq q + 1$, $b, b_r \in \mathbb{C} \setminus \mathbb{Z}_0^-$ ($r = 2, \ldots, q$); if $b + \ell, a_r \in \mathbb{Z}_0^-$ for some $r \in \{2, \ldots, p\}$, the condition $p \leq q + 1$ is cancelled. Then*

$$_pF_q \begin{bmatrix} b + \ell, a_2, \ldots, a_p; \\ b, b_2, b_3, \ldots, b_q; \end{bmatrix} z = \frac{1}{(b)_\ell} \sum_{j=0}^\ell \mathcal{A}_j(b, \ell) z^j$$
$$\times \frac{\prod_{r=2}^p (a_r)_j}{\prod_{r=2}^q (b_r)_j} {}_{p-1}F_{q-1} \begin{bmatrix} a_2 + j, \ldots, a_p + j; \\ b_2 + j, b_3 + j, \ldots, b_q + j; \end{bmatrix} z. \tag{50}$$

Proof. Let $\mathcal{L}_1$ be the left member of (50). Then using the identity

$$\frac{(b+\ell)_k}{(b)_k} = \frac{(b)_{k+\ell}}{(b)_\ell (b)_k} = \frac{(b+k)_\ell}{(b)_\ell} \tag{51}$$

to expand $\mathcal{L}_1$ gives

$$\mathcal{L}_1 = \frac{1}{(b)_\ell} \sum_{k=0}^{\infty} \frac{(b+k)_\ell (a_2)_k \cdots (a_p)_k}{k! (b_2)_k \cdots (b_q)_k} z^k. \tag{52}$$

Employing (28) in (52), we obtain

$$\mathcal{L}_1 = \frac{1}{(b)_\ell} \sum_{j=0}^{\ell} \mathcal{A}_j(b,\ell) \sum_{k=j}^{\infty} \frac{(a_2)_k \cdots (a_p)_k}{(k-j)! (b_2)_k \cdots (b_q)_k} z^k.$$

Setting $k - j = k'$ and dropping the prime on k yields

$$\mathcal{L}_1 = \frac{1}{(b)_\ell} \sum_{j=0}^{\ell} \mathcal{A}_j(b,\ell) \sum_{k=0}^{\infty} \frac{(a_2)_{k+j} \cdots (a_p)_{k+j}}{k! (b_2)_{k+j} \cdots (b_q)_{k+j}} z^{k+j}$$

$$= \frac{1}{(b)_\ell} \sum_{j=0}^{\ell} \mathcal{A}_j(b,\ell) z^j \frac{\prod_{r=2}^{p}(a_r)_j}{\prod_{r=2}^{q}(b_r)_j} \sum_{k=0}^{\infty} \frac{(a_2+j)_k \cdots (a_p+j)_k}{k! (b_2+j)_k \cdots (b_q+j)_k} z^k,$$

which is instantly apparent to be equivalent to the right-handed component of (50). □

Theorem 2. *Let* $a_r = b_r + \ell_r$, $\ell_r \in \mathbb{N}$ $(r = 1, \ldots, n)$, $n \leq \min\{p, q\}$, $p \leq q + 1$, $b_r \in \mathbb{C} \setminus \mathbb{Z}_0^-$ $(r = 1, \ldots, q)$; *if* $a_r \in \mathbb{Z}_0^-$ *for some* $r \in \{1, \ldots, p\}$, *the condition* $p \leq q + 1$ *is cancelled. Then*

$$
{}_pF_q\!\left[\begin{array}{c} b_1 + \ell_1, \ldots, b_n + \ell_n, a_{n+1}, \ldots, a_p; \\ b_1, \ldots, b_n, b_{n+1}, \ldots, b_q; \end{array} z\right]
$$
$$
= \frac{1}{\prod_{r=1}^{n}(b_r)_{\ell_r}} \sum_{j_1=0}^{\ell_1} \cdots \sum_{j_n=0}^{\ell_n} \mathcal{A}(j_1,\ldots,j_n) z^{J_n} \tag{53}
$$
$$
\times {}_{p-n}F_{q-n}\!\left[\begin{array}{c} a_{n+1} + J_n, \ldots, \ldots, a_p + J_n; \\ b_{n+1} + J_n, \ldots, b_q + J_n; \end{array} z\right],
$$

where $J_n = j_1 + \cdots + j_n$ *and* $J_0 = 0$ *and*

$$\mathcal{A}(j_1,\ldots,j_n) = \prod_{r=1}^{n} \mathcal{A}_{j_r}(b_r + J_{r-1}, \ell_r) \frac{\prod_{r=n+1}^{p}(a_r)_{J_n}}{\prod_{r=n+1}^{q}(b_r)_{J_n}}. \tag{54}$$

Proof. We may proceed with induction on n in order to demonstrate (53). This may be accomplished by applying the proof of Theorem 1 repeatedly. We omit specifics. □

Remark 3. *we have*

$$\prod_{r=1}^{n} \mathcal{A}_{j_r}(b_r + J_{r-1}, \ell_r) = \prod_{r=1}^{n} \binom{\ell_r}{j_r} \frac{(b_r)_{\ell_r + J_{r-1}}}{(b_r)_{J_r}}. \tag{55}$$

The case $n = 1$ *of* (55) *is easily found to yield the equivalent relation* (33).

4. Generalized Summation Theorems 5 mm Based on (16), (18) and (20)

The following theorems provide generalized summation formulae for the $_{t+2}F_{t+1}$ ($t \in \mathbb{N}$) with its arguments -1 and $\frac{1}{2}$.

4.1. Generalized Summation Formulas Based on (16)

Theorem 3. Let $\ell, m \in \mathbb{N}_0$ with $\ell \leq m$, and $\alpha, \beta \in \mathbb{C}$. Furthermore, let $\alpha - \beta - m \in \mathbb{C} \setminus \mathbb{Z}^-$ and $c \in \mathbb{C} \setminus \mathbb{Z}_0^-$. Further let $\Re(\beta) < \frac{2-m-\ell}{2}$. Then

$$
{}_3F_2\left[\begin{array}{c} \alpha, \beta, c+\ell; \\ 1+\alpha-\beta-m, c; \end{array} -1\right] = \frac{\Gamma(1+\alpha-\beta-m)}{2\,(c)_\ell\,\Gamma(\alpha)}
$$

$$
\times \sum_{j=0}^{\ell} (-1)^j (\beta)_j \, \mathcal{A}_j(c,\ell) \sum_{r=0}^{m-j} \binom{m-j}{r} \frac{\Gamma\left(\frac{r+j+\alpha}{2}\right)}{\Gamma\left(\frac{r+j+\alpha}{2}+1-\beta-m\right)}. \quad (56)
$$

Proof. In view of (29), the case $\ell = 0$ of (56) is found to become the identity (16). Without loss of generality, assume that ℓ is a positive integer. Let $\mathcal{L}_1$ be the left-handed member of (56). Then using the identity

$$
\frac{(c+\ell)_k}{(c)_k} = \frac{(c)_{k+\ell}}{(c)_\ell\,(c)_k} = \frac{(c+k)_\ell}{(c)_\ell} \quad (57)
$$

to expand $\mathcal{L}_1$ gives

$$
\mathcal{L}_1 = \frac{1}{(c)_\ell} \sum_{k=0}^{\infty} \frac{(\alpha)_k\,(\beta)_k\,(c+k)_\ell}{k!\,(1+\alpha-\beta-m)_k} (-1)^k. \quad (58)
$$

Employing (28) in (58), we obtain

$$
\mathcal{L}_1 = \frac{1}{(c)_\ell} \sum_{j=0}^{\ell} \mathcal{A}_j(c,\ell) \sum_{k=j}^{\infty} \frac{(\alpha)_k\,(\beta)_k\,(-1)^k}{(k-j)!\,(1+\alpha-\beta-m)_k}.
$$

Setting $k - j = k'$ and dropping the prime on k yields

$$
\mathcal{L}_1 = \frac{1}{(c)_\ell} \sum_{j=0}^{\ell} \mathcal{A}_j(c,\ell) \sum_{k=0}^{\infty} \frac{(\alpha)_{j+k}\,(\beta)_{j+k}\,(-1)^{j+k}}{k!\,(1+\alpha-\beta-m)_{j+k}}
$$

$$
= \frac{1}{(c)_\ell} \sum_{j=0}^{\ell} \mathcal{A}_j(c,\ell) \frac{(-1)^j(\alpha)_j\,(\beta)_j}{(1+\alpha-\beta-m)_j} \sum_{k=0}^{\infty} \frac{(\alpha+j)_k\,(\beta+j)_k\,(-1)^k}{k!\,(j+1+\alpha-\beta-m)_k}
$$

$$
= \frac{1}{(c)_\ell} \sum_{j=0}^{\ell} \mathcal{A}_j(c,\ell) \frac{(-1)^j(\alpha)_j\,(\beta)_j}{(1+\alpha-\beta-m)_j} {}_2F_1\left[\begin{array}{c} \alpha+j, \beta+j; \\ j+1+\alpha-\beta-m; \end{array} -1\right].
$$

For the last $_2F_1(-1)$, replacing α, λ, and p by $\alpha + j$, $\beta + j$, and $m - j$, respectively, in (16), we obtain the desired summation formula (56). □

Theorem 4. Let $\ell, \rho, m \in \mathbb{N}_0$ with $\ell + \rho \leq m$, and $\alpha, \beta \in \mathbb{C}$. Furthermore, let $\alpha - \beta - m \in \mathbb{C} \setminus \mathbb{Z}^-$ and $c, d \in \mathbb{C} \setminus \mathbb{Z}_0^-$. Further let $\Re(\beta) < \frac{2-m-\ell-\rho}{2}$. Then

$$_4F_3\left[\begin{array}{c} \alpha, \beta, c+\ell, d+\rho; \\ 1+\alpha-\beta-m, c, d; \end{array} -1\right] = \frac{\Gamma(1+\alpha-\beta-m)}{2(c)_\ell (d)_\rho \Gamma(\alpha)}$$

$$\times \sum_{\nu=0}^{\rho} \sum_{j=0}^{\ell} (-1)^{\nu+j} (\beta)_{\nu+j}\, \mathcal{A}_j(c+\nu,\ell) \mathcal{A}_\nu(d,\rho) \qquad (59)$$

$$\times \sum_{r=0}^{m-\nu-j} \binom{m-\nu-j}{r} \frac{\Gamma\left(\frac{r+\nu+j+\alpha}{2}\right)}{\Gamma\left(\frac{r+\nu+j+\alpha}{2}+1-\beta-m\right)}.$$

Proof. As in the beginning of the proof of Theorem 3, here also let assume that ℓ and ρ are positive integers. Let $\mathcal{L}_2$ be the left member of (59). Then, using (51), we have

$$\mathcal{L}_2 = \sum_{k=0}^{\infty} \frac{(\alpha)_k (\beta)_k (c+\ell)_k (d+\rho)_k}{k!\,(1+\alpha-\beta-m)_k (c)_k (d)_k} (-1)^k$$

$$= \frac{1}{(d)_\rho} \sum_{k=0}^{\infty} \frac{(\alpha)_k (\beta)_k (c+\ell)_k (d+k)_\rho}{k!\,(1+\alpha-\beta-m)_k (c)_k} (-1)^k.$$

Employing (28) in the last sum, here, with the aid of (56), as in the similar process of proof of Theorem 3, we can prove the identity (59). We omit the details. □

Theorem 5. Let $t, \ell_1, \ldots, \ell_t, m \in \mathbb{N}_0$ with $j_t \leq m$. Furthermore, let $\alpha, \beta \in \mathbb{C}$, and $\alpha - \beta - m \in \mathbb{C} \setminus \mathbb{Z}^-$, and $c_1, \ldots, c_t \in \mathbb{C} \setminus \mathbb{Z}_0^-$. Further let $\Re(\beta) < \frac{2-m-l_t}{2}$. Then

$$_{t+2}F_{t+1}\left[\begin{array}{c} \alpha, \beta, c_1+\ell_1, \ldots, c_t+\ell_t; \\ 1+\alpha-\beta-m, c_1, \ldots, c_t; \end{array} -1\right] = \frac{\Gamma(1+\alpha-\beta-m)}{2\,\Gamma(\alpha) \prod_{j=1}^{t}(c_j)_{\ell_j}}$$

$$\times \sum_{j_t=0}^{\ell_t} \cdots \sum_{j_1=0}^{\ell_1} (-1)^{j_t} (\beta)_{j_t} \prod_{k=1}^{t} \mathcal{A}_{j_k}(c_k+j_t-j_k, \ell_k) \qquad (60)$$

$$\times \sum_{r=0}^{m-j_t} \binom{m-j_t}{r} \frac{\Gamma\left(\frac{r+j_t+\alpha}{2}\right)}{\Gamma\left(\frac{r+j_t+\alpha}{2}+1-\beta-m\right)},$$

where

$$l_t := \sum_{\eta=1}^{t} \ell_\eta \quad (t \in \mathbb{N}) \quad \text{and} \quad j_k := \sum_{\eta=1}^{k} j_\eta \quad (k \in \mathbb{N}). \qquad (61)$$

Proof. By using mathematical induction on $t \in \mathbb{N}$, we may replicate the procedure used to establish Theorem 4 and therefore show the conclusion here. The specifics are avoided. □

4.2. Generalized Summation Formulas Based on (18)

Theorem 6. Let $\ell, m \in \mathbb{N}_0$, and $\alpha, \beta \in \mathbb{C}$. Furthermore, let $c, \frac{1+\alpha+\beta-m}{2} \in \mathbb{C} \setminus \mathbb{Z}_0^-$. Then

$$_3F_2\left[\begin{array}{c} \alpha, \beta, c+\ell; \\ \frac{1+\alpha+\beta-m}{2}, c; \end{array} \frac{1}{2}\right] = \frac{2^{\beta-1}\, \Gamma\left(\frac{1+\alpha+\beta-m}{2}\right)}{\Gamma(\beta)\,(c)_\ell}$$

$$\times \sum_{j=0}^{\ell} (\alpha)_j\, \mathcal{A}_j(c,\ell) \sum_{r=0}^{m} \binom{m}{r} \frac{\Gamma\left(\frac{\beta+j+r}{2}\right)}{\Gamma\left(\frac{1+\alpha-m+j+r}{2}\right)}. \qquad (62)$$

Proof. The proof would run in parallel with that of Theorem 3 with the aid of (18). The details are omitted. □

Theorem 7. *Let $\rho, \ell, m \in \mathbb{N}_0$, and $\alpha, \beta \in \mathbb{C}$. Furthermore, let $c, d, \frac{1+\alpha+\beta-m}{2} \in \mathbb{C} \setminus \mathbb{Z}_0^-$. Then*

$$
{}_4F_3\left[\begin{array}{c}\alpha, \beta, c+\ell, d+\rho; \\ \frac{1+\alpha+\beta-m}{2}, c, d;\end{array} \frac{1}{2}\right] = \frac{2^{\beta-1}\Gamma\left(\frac{1+\alpha+\beta-m}{2}\right)}{(c)_\ell (d)_\rho \Gamma(\beta)}
$$

$$
\times \sum_{\nu=0}^{\rho}\sum_{j=0}^{\ell}(\alpha)_{\nu+j}\, \mathcal{A}_\nu(d,\rho)\, \mathcal{A}_j(c+\nu,\ell) \sum_{r=0}^{m}\binom{m}{r}\frac{\Gamma\left(\frac{\beta+\nu+j+r}{2}\right)}{\Gamma\left(\frac{1+\alpha-m+\nu+j+r}{2}\right)}.
$$
(63)

Proof. The proof would continue in the same manner as that of Theorem 4, aided by (62). We omit specifics. □

Theorem 8. *Let $t, \ell_1, \ldots, \ell_t, m \in \mathbb{N}_0$, and $\alpha, \beta \in \mathbb{C}$. Furthermore, let $c, d, \frac{1+\alpha+\beta-m}{2} \in \mathbb{C} \setminus \mathbb{Z}_0^-$. Then*

$$
{}_{t+2}F_{t+1}\left[\begin{array}{c}\alpha, \beta, c_1+\ell_1, \ldots, c_t+\ell_t; \\ \frac{1+\alpha+\beta-m}{2}, c_1, \ldots, c_t;\end{array} \frac{1}{2}\right] = \frac{2^{\beta-1}\Gamma\left(\frac{1+\alpha+\beta-m}{2}\right)}{\Gamma(\beta)\prod_{k=1}^{t}(c_k)_{\ell_k}}
$$

$$
\times \sum_{j_t=0}^{\ell_t}\cdots\sum_{j_1=0}^{\ell_1}(\alpha)_{j_t}\, \mathcal{A}_{j_k}(c_k+j_t-j_k,\ell_k) \sum_{r=0}^{m}\binom{m}{r}\frac{\Gamma\left(\frac{\beta+r+j_t}{2}\right)}{\Gamma\left(\frac{1+\alpha-m+r+j_t}{2}\right)},
$$
(64)

where j_k is the same as in (61).

Proof. The proof would be accomplished by following the lines of that of Theorem 5. The involved details are omitted. □

4.3. Generalized Summation Formulas Based on (20)

Theorem 9. *Let $\ell, m \in \mathbb{N}_0$ with $2\ell \leq m$. Furthermore, let $\alpha \in \mathbb{C}$, and $\beta, c \in \mathbb{C} \setminus \mathbb{Z}_0^-$. Then*

$$
{}_3F_2\left[\begin{array}{c}\alpha, 1-\alpha-m, c+\ell; \\ \beta, c;\end{array} \frac{1}{2}\right] = \frac{2^{-\alpha-m}\Gamma(\beta)}{(c)_\ell\, \Gamma(\beta-\alpha)}
$$

$$
\times \sum_{j=0}^{\ell}(\alpha)_j\,(1-\alpha-m)_j\, \mathcal{A}_j(c,\ell) \sum_{r=0}^{m-2j}\binom{m-2j}{r}\frac{\Gamma\left(\frac{\beta-\alpha+r}{2}\right)}{\Gamma\left(\frac{\beta+\alpha+r}{2}+j\right)}.
$$
(65)

Proof. The proof would run in parallel with that of Theorem 3 with the aid of (20). The details are omitted. □

Theorem 10. *Let $\rho, \ell, m \in \mathbb{N}_0$ with $2(\ell+\rho) \leq m$. Furthermore, let $\alpha \in \mathbb{C}$, and $\beta, c, d \in \mathbb{C} \setminus \mathbb{Z}_0^-$. Then*

$$
{}_4F_3\left[\begin{array}{c}\alpha, 1-\alpha-m, c+\ell, d+\rho; \\ \beta, c, d;\end{array} \frac{1}{2}\right] = \frac{2^{-\alpha-m}\Gamma(\beta)}{(c)_\ell (d)_\rho \Gamma(\beta-\alpha)}
$$

$$
\times \sum_{\nu=0}^{\rho}\sum_{j=0}^{\ell}(\alpha)_{\nu+j}\,(1-\alpha-m)_{\nu+j}\, \mathcal{A}_\nu(d,\rho)\, \mathcal{A}_j(c+\nu,\ell)
$$
(66)

$$
\times \sum_{r=0}^{m-2\nu-2j}\binom{m-2\nu-2j}{r}\frac{\Gamma\left(\frac{\beta-\alpha+r}{2}\right)}{\Gamma\left(\frac{\beta+\alpha+r}{2}+\nu+j\right)}.
$$

Proof. The proof would run in line with that of Theorem 4 with the help of (65). We omit the details. □

Theorem 11. *Let $t, \ell_1, \ldots, \ell_t, m \in \mathbb{N}_0$ with $2j_t \leq m$. Furthermore, let $\alpha \in \mathbb{C}$ and $\beta, c_1, \ldots, c_t \in \mathbb{C} \setminus \mathbb{Z}_0^-$. Then*

$$_{t+2}F_{t+1}\left[\begin{matrix}\alpha, 1-\alpha-m, c_1+\ell_1, \ldots, c_t+\ell_t; \\ \beta, c_1, \ldots, c_t;\end{matrix} \frac{1}{2}\right] = \frac{2^{-\alpha-m}\Gamma(\beta)}{\Gamma(\beta-\alpha)\prod_{k=1}^{t}(c_k)_{\ell_k}}$$

$$\times \sum_{j_t=0}^{\ell_t} \cdots \sum_{j_1=0}^{\ell_1} (\alpha)_{j_t}(1-\alpha-m)_{j_t} \prod_{k=1}^{t} \mathcal{A}_{j_k}(c_k+j_t-j_k, \ell_k) \quad (67)$$

$$\times \sum_{r=0}^{m-2j_t} \binom{m-2j_t}{r} \frac{\Gamma\left(\frac{\beta-\alpha+r}{2}\right)}{\Gamma\left(\frac{\beta+\alpha+r}{2}+j_t\right)},$$

where j_k is the same as in (61).

Proof. The proof would flow along the lines of that of Theorem 5. The involved details are omitted. □

5. Generalized Summation Theorems 5 mm Based on (15), (17) and (19)

The following theorems offer generalized summation formulae for the $_{t+2}F_{t+1}$ ($t \in \mathbb{N}$) and its arguments -1 and $\frac{1}{2}$. The proofs of each theorem are skipped here, principally because they can be checked in the same manner as the preceding section's counterpart.

5.1. Generalized Summation Formulas Based on (15)

Theorem 12. *Let $\ell, m \in \mathbb{N}_0$, and $\alpha, \beta \in \mathbb{C}$. Furthermore, let $\alpha - \beta + m \in \mathbb{C} \setminus \mathbb{Z}^-$ and $c \in \mathbb{C} \setminus \mathbb{Z}_0^-$. Further let $\Re(\beta) < \frac{2+m-\ell}{2}$. Then*

$$_3F_2\left[\begin{matrix}\alpha, \beta, c+\ell; \\ 1+\alpha-\beta+m, c;\end{matrix} -1\right] = \frac{\Gamma(1+\alpha-\beta+m)}{2(c)_\ell \Gamma(\alpha)(1-\beta)_m}$$

$$\times \sum_{j=0}^{\ell} \mathcal{A}_j(c, \ell) \sum_{r=0}^{m+j} \binom{m+j}{r} \frac{(-1)^r \Gamma\left(\frac{\alpha+r+j}{2}\right)}{\Gamma\left(1-\beta+\frac{\alpha+r-j}{2}\right)}. \quad (68)$$

Theorem 13. *Let $\ell, \rho, m \in \mathbb{N}_0$, and $\alpha, \beta \in \mathbb{C}$. Furthermore, let $\alpha - \beta + m \in \mathbb{C} \setminus \mathbb{Z}^-$ and $c, d \in \mathbb{C} \setminus \mathbb{Z}_0^-$. Further let $\Re(\beta) < \frac{2+m-\ell-\rho}{2}$. Then*

$$_4F_3\left[\begin{matrix}\alpha, \beta, c+\ell, d+\rho; \\ 1+\alpha-\beta+m, c, d;\end{matrix} -1\right]$$

$$= \frac{\Gamma(1+\alpha-\beta+m)}{2\Gamma(\alpha)(d)_\rho(c)_\ell(1-\beta)_m} \sum_{v=0}^{\rho} \sum_{j=0}^{\ell} \mathcal{A}_v(d,\rho)\mathcal{A}_j(c+v,\ell) \quad (69)$$

$$\times \sum_{r=0}^{m+v+j} \binom{m+v+j}{r} \frac{(-1)^r \Gamma\left(\frac{\alpha+r+v+j}{2}\right)}{\Gamma\left(1-\beta+\frac{\alpha+r-v-j}{2}\right)}.$$

Theorem 14. *Let $t, \ell_1, \ldots, \ell_t, m \in \mathbb{N}_0$. Furthermore, let $\alpha, \beta \in \mathbb{C}$, and $\alpha - \beta + m \in \mathbb{C} \setminus \mathbb{Z}^-$, and $c_1, \ldots, c_t \in \mathbb{C} \setminus \mathbb{Z}_0^-$. Further let $\Re(\beta) < \frac{2+m-l_t}{2}$. Then*

$$_{t+2}F_{t+1}\left[\begin{array}{c}\alpha, \beta, c_1+\ell_1, \ldots, c_t+\ell_t; \\ 1+\alpha-\beta+m, c_1, \ldots, c_t;\end{array} -1\right] = \frac{\Gamma(1+\alpha-\beta+m)}{2\Gamma(\alpha)(1-\beta)_m \prod_{j=1}^{t}(c_j)_{\ell_j}}$$

$$\times \sum_{j_t=0}^{\ell_t}\cdots\sum_{j_1=0}^{\ell_1}\prod_{k=1}^{t}\mathcal{A}_{j_k}(c_k+j_t-j_k,\ell_k) \tag{70}$$

$$\times \sum_{r=0}^{m+j_t}\binom{m+j_t}{r}\frac{\Gamma\left(\frac{\alpha+r+j_t}{2}\right)}{\Gamma\left(1-\beta+\frac{\alpha+r-j_t}{2}\right)},$$

where l_t and j_k $(t, k \in \mathbb{N})$ are the same as in (61).

5.2. Generalized Summation Formulas Based on (17)

Theorem 15. *Let $\ell, m \in \mathbb{N}_0$, and $\alpha, \beta \in \mathbb{C}$. Furthermore, let $c, \frac{1+\alpha+\beta+m}{2} \in \mathbb{C}\setminus\mathbb{Z}_0^-$. Then*

$$_3F_2\left[\begin{array}{c}\alpha, \beta, c+\ell; \\ \frac{1+\alpha+\beta+m}{2}, c;\end{array} \frac{1}{2}\right] = \frac{2^{\alpha-1}}{\Gamma(\alpha)(c)_\ell}\frac{\Gamma(\frac{1+\alpha+\beta+m}{2})\Gamma(\frac{1-\alpha+\beta-m}{2})}{\Gamma(\frac{1-\alpha+\beta+m}{2})}$$

$$\times \sum_{j=0}^{\ell}(\beta)_j \mathcal{A}_j(c,\ell) \sum_{r=0}^{m}\binom{m}{r}\frac{(-1)^r\Gamma(\frac{\alpha+j+r}{2})}{\Gamma(\frac{1+\beta+j+r-m}{2})}. \tag{71}$$

Theorem 16. *Let $\rho, \ell, m \in \mathbb{N}_0$, and $\alpha, \beta \in \mathbb{C}$. Furthermore, let $c, d, \frac{1+\alpha+\beta+m}{2} \in \mathbb{C}\setminus\mathbb{Z}_0^-$. Then*

$$_4F_3\left[\begin{array}{c}\alpha, \beta, c+\ell, d+\rho; \\ \frac{1+\alpha+\beta+m}{2}, c, d;\end{array} \frac{1}{2}\right] = \frac{2^{\alpha-1}\Gamma(\frac{1+\alpha+\beta+m}{2})\Gamma(\frac{1-\alpha+\beta-m}{2})}{\Gamma(\alpha)\Gamma(\frac{1-\alpha+\beta+m}{2})(d)_\rho(c)_\ell}$$

$$\times \sum_{v=0}^{\rho}\sum_{j=0}^{\ell}\mathcal{A}_j(c+v,\ell)\mathcal{A}_v(d,\rho)(\beta)_{v+j}\sum_{r=0}^{m}\binom{m}{r}\frac{(-1)^r\Gamma(\frac{\alpha+v+j+r}{2})}{\Gamma(\frac{1+\beta+v+j+r-m}{2})}. \tag{72}$$

Theorem 17. *Let $t, \ell_1, \ldots, \ell_t, m \in \mathbb{N}_0$, and $\alpha, \beta \in \mathbb{C}$. Furthermore, let $c, d, \frac{1+\alpha+\beta+m}{2} \in \mathbb{C}\setminus\mathbb{Z}_0^-$. Then*

$$_{t+2}F_{t+1}\left[\begin{array}{c}\alpha, \beta, c_1+\ell_1, \ldots, c_t+\ell_t; \\ \frac{1+\alpha+\beta+m}{2}, c_1, \ldots, c_t;\end{array} \frac{1}{2}\right] = \frac{2^{\alpha-1}\Gamma(\frac{1+\alpha+\beta+m}{2})\Gamma(\frac{1-\alpha+\beta-m}{2})}{\Gamma(\alpha)\Gamma(\frac{1-\alpha+\beta+m}{2})\prod_{k=1}^{t}(c_k)_{\ell_k}}$$

$$\times \sum_{j_t=0}^{\ell_t}\cdots\sum_{j_1=0}^{\ell_1}(\beta)_{j_t}\mathcal{A}_{j_k}(c_k+j_t-j_k,\ell_k)\sum_{r=0}^{m}\binom{m}{r}\frac{(-1)^r\Gamma(\frac{\alpha+r+j_t}{2})}{\Gamma(\frac{1+\beta+r-m+j_t}{2})}, \tag{73}$$

where j_k is the same as in (61).

5.3. Generalized Summation Formulas Based on (19)

Theorem 18. *Let $\ell, m \in \mathbb{N}_0$. Furthermore, let $\alpha \in \mathbb{C}$, and $\beta, c \in \mathbb{C}\setminus\mathbb{Z}_0^-$. Then*

$$_3F_2\left[\begin{array}{c}\alpha, 1-\alpha+m, c+\ell; \\ \beta, c;\end{array} \frac{1}{2}\right] = \frac{2^{m-\alpha}\,\Gamma(\alpha-m)\,\Gamma(\beta)}{\Gamma(\alpha)\,\Gamma(\beta-\alpha)\,(c)_\ell}$$

$$\times \sum_{j=0}^{\ell}(-1)^j \mathcal{A}_j(c,\ell)\sum_{r=0}^{m+2j}(-1)^r\binom{m+2j}{r}\frac{\Gamma(\frac{\beta-\alpha+r}{2})}{\Gamma(\frac{\beta+\alpha+r}{2}-m-j)}. \tag{74}$$

Theorem 19. *Let $\rho, \ell, m \in \mathbb{N}_0$. Furthermore, let $\alpha \in \mathbb{C}$, and $\beta, c, d \in \mathbb{C}\setminus\mathbb{Z}_0^-$. Then*

$$
{}_4F_3\left[\begin{array}{c}\alpha,\ 1-\alpha+m,\ c+\ell,\ d+\rho;\ 1\\ \beta,\ c,\ d;\ 2\end{array}\right] = \frac{2^{m-\alpha}\,\Gamma(\alpha-m)\,\Gamma(\beta)}{\Gamma(\alpha)\,\Gamma(\beta-\alpha)\,(c)_\ell (d)_\rho}
$$
$$
\times \sum_{\nu=0}^{\rho}\sum_{j=0}^{\ell}(-1)^{\nu+j}\mathcal{A}_j(c+\nu,\ell)\mathcal{A}_\nu(d,\rho) \tag{75}
$$
$$
\times \sum_{r=0}^{m+2(\nu+j)}(-1)^r\binom{m+2(\nu+j)}{r}\frac{\Gamma\!\left(\frac{\beta-\alpha+r}{2}\right)}{\Gamma\!\left(\frac{\beta+\alpha+r}{2}-m-\nu-j\right)}.
$$

Theorem 20. *Let $t, \ell_1, \ldots, \ell_t, m \in \mathbb{N}_0$ Furthermore, let $\alpha \in \mathbb{C}$ and $\beta, c_1, \ldots, c_t \in \mathbb{C}\setminus \mathbb{Z}_0^-$. Then*

$$
{}_{t+2}F_{t+1}\left[\begin{array}{c}\alpha,\ 1-\alpha+m,\ c_1+\ell_1,\ \ldots,\ c_t+\ell_t;\ 1\\ \beta,\ c_1,\ \ldots,\ c_t;\ 2\end{array}\right]
$$
$$
= \frac{2^{m-\alpha}\,\Gamma(\alpha-m)\,\Gamma(\beta)}{\Gamma(\alpha)\,\Gamma(\beta-\alpha)\,\prod_{k=1}^{t}(c_k)_{\ell_k}} \tag{76}
$$
$$
\times \sum_{j_t=0}^{\ell_t}\cdots\sum_{j_1=0}^{\ell_1}(-1)^{j_t}\prod_{k=1}^{t}\mathcal{A}_{j_k}(c_k+j_t-j_k,\ell_k)
$$
$$
\times \sum_{r=0}^{m+2j_t}(-1)^r\binom{m+2j_t}{r}\frac{\Gamma\!\left(\frac{\beta-\alpha+r}{2}\right)}{\Gamma\!\left(\frac{\beta+\alpha+r}{2}-m-j_t\right)},
$$

where j_k is the same as in (61).

6. Formulas Involving Finite Sums of ${}_{t+1}F_t$

We provide formulae for finite sums of ${}_{t+1}F_t$ by using two identities in Theorems 3, 6, 9, 12, 15 and 18, which are stated in the following six theorems. This section contains just the proof of Theorem 21. The proofs of the other theorems are omitted since they would run concurrently with the proof of Theorem 21.

Theorem 21. *Let $\ell \in \mathbb{N}$, $m \in \mathbb{N}_0$ with $\ell \leq m$, and $\alpha, \beta \in \mathbb{C}$. Furthermore, let $\alpha - \beta - m \in \mathbb{C}\setminus \mathbb{Z}^-$ and $c \in \mathbb{C}\setminus \mathbb{Z}_0^-$. Further let $\Re(\beta) < \frac{2-m-\ell}{2}$. Then*

$$
\sum_{j=1}^{\ell}\frac{1}{c+j-1}\,{}_4F_3\left[\begin{array}{c}\alpha,\ \beta,\ c+\ell,\ c+j-1;\\ 1+\alpha-\beta-m,\ c,\ c+j;\end{array}-1\right] = \frac{\Gamma(1+\alpha-\beta-m)}{2\,(c)_\ell\,\Gamma(\alpha)}
$$
$$
\times \sum_{j=0}^{\ell-1}(-1)^j(\beta)_j\,\mathcal{B}_j(c,\ell)\sum_{r=0}^{m-j}\binom{m-j}{r}\frac{\Gamma\!\left(\frac{r+j+\alpha}{2}\right)}{\Gamma\!\left(\frac{r+j+\alpha}{2}+1-\beta-m\right)}. \tag{77}
$$

Proof. Multiplying both sides of (56) by $(c)_\ell$, we get

$$
\sum_{k=0}^{\infty}\frac{(\alpha)_k(\beta)_k(c)_{\ell+k}}{k!\,(1+\alpha-\beta-m)_k\,(c)_k}(-1)^k = \frac{\Gamma(1+\alpha-\beta-m)}{2\,\Gamma(\alpha)}
$$
$$
\times \sum_{j=0}^{\ell}(-1)^j(\beta)_j\,\mathcal{A}_j(c,\ell)\sum_{r=0}^{m-j}\binom{m-j}{r}\frac{\Gamma\!\left(\frac{r+j+\alpha}{2}\right)}{\Gamma\!\left(\frac{r+j+\alpha}{2}+1-\beta-m\right)}. \tag{78}
$$

Differentiating $(c)_{\ell+k}/(c)_k = \Gamma(c+k+\ell)/\Gamma(c+k)$ with respect to c, we obtain

$$
\frac{d}{dc}\frac{(c)_{\ell+k}}{(c)_k} = \frac{(c)_{\ell+k}}{(c)_k}\{\psi(c+k+\ell)-\psi(c+k)\}.
$$

Using (22), we find

$$\frac{d}{dc}\frac{(c)_{\ell+k}}{(c)_k} = \frac{(c)_{\ell+k}}{(c)_k}\sum_{j=1}^{\ell}\frac{1}{c+j-1+k}.$$

Employing the fundamental identity $\Gamma(z+1) = z\Gamma(z)$, in view of (1), we have

$$\sum_{j=1}^{\ell}\frac{1}{c+j-1+k} = \sum_{j=1}^{\ell}\frac{\Gamma(c+j-1+k)}{\Gamma(c+j+k)} = \sum_{j=1}^{\ell}\frac{1}{c+j-1}\cdot\frac{(c+j-1)_k}{(c+j)_k}.$$

We thus obtain

$$\frac{d}{dc}\frac{(c)_{\ell+k}}{(c)_k} = \frac{(c)_{\ell}(c+\ell)_k}{(c)_k}\sum_{j=1}^{\ell}\frac{1}{c+j-1}\cdot\frac{(c+j-1)_k}{(c+j)_k}. \qquad (79)$$

Differentiating both sides of (78) with respect to c and using (79), with the aid of (30) and (45), we can get the desired identity (77). □

Theorem 22. *Let $\ell \in \mathbb{N}$, $m \in \mathbb{N}_0$, and $\alpha, \beta \in \mathbb{C}$. Furthermore, let c, $\frac{1+\alpha+\beta-m}{2} \in \mathbb{C}\setminus\mathbb{Z}_0^-$. Then*

$$\sum_{j=1}^{\ell}\frac{1}{c+j-1}{}_4F_3\!\left[\begin{array}{c}\alpha,\,\beta,\,c+\ell,\,c+j-1;\\ \frac{1+\alpha+\beta-m}{2},\,c,\,c+j;\end{array}\frac{1}{2}\right]$$
$$= \frac{2^{\beta-1}\Gamma\!\left(\frac{1+\alpha+\beta-m}{2}\right)}{\Gamma(\beta)(c)_\ell}\sum_{j=0}^{\ell-1}(\alpha)_j\,\mathcal{B}_j(c,\ell)\sum_{r=0}^{m}\binom{m}{r}\frac{\Gamma\!\left(\frac{\beta+j+r}{2}\right)}{\Gamma\!\left(\frac{1+\alpha-m+j+r}{2}\right)}. \qquad (80)$$

Theorem 23. *Let $\ell \in \mathbb{N}$, $m \in \mathbb{N}_0$ with $2\ell \leq m$. Furthermore, let $\alpha \in \mathbb{C}$, and $\beta, c \in \mathbb{C}\setminus\mathbb{Z}_0^-$. Then*

$$\sum_{j=1}^{\ell}\frac{1}{c+j-1}{}_4F_3\!\left[\begin{array}{c}\alpha,\,1-\alpha-m,\,c+\ell,\,c+j-1;\,1\\ \beta,\,c,\,c+j;\,2\end{array}\right] = \frac{2^{-\alpha-m}\Gamma(\beta)}{(c)_\ell\,\Gamma(\beta-\alpha)}$$
$$\times \sum_{j=0}^{\ell-1}(\alpha)_j\,(1-\alpha-m)_j\,\mathcal{B}_j(c,\ell)\sum_{r=0}^{m-2j}\binom{m-2j}{r}\frac{\Gamma\!\left(\frac{\beta-\alpha+r}{2}\right)}{\Gamma\!\left(\frac{\beta+\alpha+r}{2}+j\right)}. \qquad (81)$$

Theorem 24. *Let $\ell \in \mathbb{N}$, $m \in \mathbb{N}_0$, and $\alpha, \beta \in \mathbb{C}$. Furthermore, let $\alpha - \beta + m \in \mathbb{C}\setminus\mathbb{Z}^-$ and $c \in \mathbb{C}\setminus\mathbb{Z}_0^-$. Further let $\Re(\beta) < \frac{2+m-\ell}{2}$. Then*

$$\sum_{j=1}^{\ell}\frac{1}{c+j-1}{}_4F_3\!\left[\begin{array}{c}\alpha,\,\beta,\,c+\ell,\,c+j-1;\\ 1+\alpha-\beta+m,\,c,\,c+j;\end{array}-1\right]$$
$$= \frac{\Gamma(1+\alpha-\beta+m)}{2(c)_\ell\,\Gamma(\alpha)(1-\beta)_m}$$
$$\times \sum_{j=0}^{\ell-1}\mathcal{B}_j(c,\ell)\sum_{r=0}^{m+j}\binom{m+j}{r}\frac{(-1)^r\,\Gamma\!\left(\frac{r+\alpha+j}{2}\right)}{\Gamma\!\left(\frac{r+\alpha-j}{2}+1-\beta\right)}. \qquad (82)$$

Theorem 25. *Let $\ell \in \mathbb{N}$, $m \in \mathbb{N}_0$, and $\alpha, \beta \in \mathbb{C}$. Furthermore, let c, $\frac{1+\alpha+\beta+m}{2} \in \mathbb{C} \setminus \mathbb{Z}_0^-$. Then*

$$\sum_{j=1}^{\ell} \frac{1}{c+j-1} {}_4F_3 \left[\begin{array}{c} \alpha, \beta, c+\ell, c+j-1; \\ \frac{1+\alpha+\beta+m}{2}, c, c+j; \end{array} \frac{1}{2} \right]$$
$$= \frac{2^{\alpha-1}}{\Gamma(\alpha)(c)_\ell} \frac{\Gamma(\frac{1+\alpha+\beta+m}{2})\Gamma(\frac{1-\alpha+\beta-m}{2})}{\Gamma(\frac{1-\alpha+\beta+m}{2})} \tag{83}$$
$$\times \sum_{j=0}^{\ell-1} (\beta)_j \mathcal{B}_j(c,\ell) \sum_{r=0}^{m} \binom{m}{r} \frac{(-1)^r \Gamma\left(\frac{\alpha+j+r}{2}\right)}{\Gamma(\frac{1+\beta+j+r-m}{2})}.$$

Theorem 26. *Let $\ell \in \mathbb{N}$, $m \in \mathbb{N}_0$. Furthermore, let $\alpha \in \mathbb{C}$, and $\beta, c \in \mathbb{C} \setminus \mathbb{Z}_0^-$. Then*

$$\sum_{j=1}^{\ell} \frac{1}{c+j-1} {}_4F_3 \left[\begin{array}{c} \alpha, 1-\alpha+m, c+\ell, c+j-1; \\ \beta, c, c+j; \end{array} \frac{1}{2} \right]$$
$$= \frac{2^{m-\alpha} \Gamma(\alpha-m) \Gamma(\beta)}{\Gamma(\alpha) \Gamma(\beta-\alpha)(c)_\ell} \sum_{j=0}^{\ell-1} (-1)^j \mathcal{B}_j(c,\ell) \tag{84}$$
$$\times \sum_{r=0}^{m+2j} (-1)^r \binom{m+2j}{r} \frac{\Gamma(\frac{\beta-\alpha+r}{2})}{\Gamma\left(\frac{\beta+\alpha+r}{2} - m - j\right)}.$$

7. Particular Cases

We address the straightforward special instances of Theorems 3, 6, 9, 12, 15 and 18 when $\ell = 1$, which are specified in the following corollaries. The following are the identities from Section 2: $\mathcal{A}_0(c,1) = c$ and $\mathcal{A}_1(c,1) = 1$.

Corollary 1. *Let $m \in \mathbb{N}_0$, and $\alpha, \beta \in \mathbb{C}$. Furthermore, let $\alpha - \beta - m \in \mathbb{C} \setminus \mathbb{Z}^-$ and $c \in \mathbb{C} \setminus \mathbb{Z}_0^-$. Further let $\Re(\beta) < \frac{1-m}{2}$. Then*

$$_3F_2 \left[\begin{array}{c} \alpha, \beta, c+1; \\ 1+\alpha-\beta-m, c; \end{array} -1 \right]$$
$$= \frac{\Gamma(1+\alpha-\beta-m)}{2\Gamma(\alpha)} \left\{ \sum_{r=0}^{m} \binom{m}{r} \frac{\Gamma\left(\frac{r+\alpha}{2}\right)}{\Gamma\left(\frac{r+\alpha}{2}+1-\beta-m\right)} \right. \tag{85}$$
$$\left. - \frac{\beta}{c} \sum_{r=0}^{m-1} \binom{m-1}{r} \frac{\Gamma\left(\frac{r+\alpha+1}{2}\right)}{\Gamma\left(\frac{r+\alpha+1}{2}+1-\beta-m\right)} \right\}.$$

Corollary 2. *Let $m \in \mathbb{N}_0$, and $\alpha, \beta \in \mathbb{C}$. Furthermore, let c, $\frac{1+\alpha+\beta-m}{2} \in \mathbb{C} \setminus \mathbb{Z}_0^-$. Then*

$$_3F_2 \left[\begin{array}{c} \alpha, \beta, c+1; \\ \frac{1+\alpha+\beta-m}{2}, c; \end{array} \frac{1}{2} \right] = \frac{2^{\beta-1} \Gamma\left(\frac{1+\alpha+\beta-m}{2}\right)}{\Gamma(\beta)}$$
$$\times \sum_{r=0}^{m} \binom{m}{r} \left\{ \frac{\Gamma\left(\frac{\beta+r}{2}\right)}{\Gamma\left(\frac{1+\alpha-m+r}{2}\right)} + \frac{\alpha \Gamma\left(\frac{\beta+1+r}{2}\right)}{c \Gamma\left(\frac{2+\alpha-m+r}{2}\right)} \right\}. \tag{86}$$

Corollary 3. Let $\ell, m \in \mathbb{N}_0$. Furthermore, let $\alpha \in \mathbb{C}$, and $\beta, c \in \mathbb{C} \setminus \mathbb{Z}_0^-$. Then

$$\begin{aligned}
{}_3F_2\left[\begin{array}{c} \alpha, 1-\alpha-m, c+1; \\ \beta, c; \end{array} \frac{1}{2}\right] \\
= \frac{2^{-\alpha-m}\Gamma(\beta)}{\Gamma(\beta-\alpha)} \Bigg\{ \sum_{r=0}^{m} \binom{m}{r} \frac{\Gamma\left(\frac{\beta-\alpha+r}{2}\right)}{\Gamma\left(\frac{\beta+\alpha+r}{2}\right)} \\
+ \frac{\alpha(1-\alpha-m)}{c} \sum_{r=0}^{m-2} \binom{m-2}{r} \frac{\Gamma\left(\frac{\beta-\alpha+r}{2}\right)}{\Gamma\left(\frac{\beta+\alpha+r}{2}+1\right)} \Bigg\}.
\end{aligned} \tag{87}$$

Corollary 4. Let $\ell, m \in \mathbb{N}_0$, and $\alpha, \beta \in \mathbb{C}$. Furthermore, let $\alpha - \beta + m \in \mathbb{C} \setminus \mathbb{Z}^-$ and $c \in \mathbb{C} \setminus \mathbb{Z}_0^-$. Further let $\Re(\beta) < \frac{2+m-\ell}{2}$. Then

$$\begin{aligned}
{}_3F_2\left[\begin{array}{c} \alpha, \beta, c+1; \\ 1+\alpha-\beta+m, c; \end{array} -1 \right] \\
= \frac{\Gamma(1+\alpha-\beta+m)}{2\Gamma(\alpha)(1-\beta)_m} \Bigg\{ \sum_{r=0}^{m} \binom{m}{r} \frac{(-1)^r \Gamma\left(\frac{r+\alpha}{2}\right)}{\Gamma\left(\frac{r+\alpha}{2}+1-\beta\right)} \\
+ \frac{1}{c} \sum_{r=0}^{m+1} \binom{m+1}{r} \frac{(-1)^r \Gamma\left(\frac{r+\alpha+1}{2}\right)}{\Gamma\left(\frac{r+\alpha+1}{2}-\beta\right)} \Bigg\}.
\end{aligned} \tag{88}$$

Corollary 5. Let $m \in \mathbb{N}_0$, and $\alpha, \beta \in \mathbb{C}$. Furthermore, let $c, \frac{1+\alpha+\beta+m}{2} \in \mathbb{C} \setminus \mathbb{Z}_0^-$. Then

$$\begin{aligned}
{}_3F_2\left[\begin{array}{c} \alpha, \beta, c+1; \\ \frac{1+\alpha+\beta+m}{2}, c; \end{array} \frac{1}{2}\right] = \frac{2^{\alpha-1}}{\Gamma(\alpha)} \frac{\Gamma\left(\frac{1+\alpha+\beta+m}{2}\right)\Gamma\left(\frac{1-\alpha+\beta-m}{2}\right)}{\Gamma\left(\frac{1-\alpha+\beta+m}{2}\right)} \\
\times \sum_{r=0}^{m} \binom{m}{r}(-1)^r \Bigg\{ \frac{\Gamma\left(\frac{\alpha+r}{2}\right)}{\Gamma\left(\frac{1+\beta+r-m}{2}\right)} + \frac{\beta \Gamma\left(\frac{1+\alpha+r}{2}\right)}{c\Gamma\left(\frac{2+\beta+r-m}{2}\right)} \Bigg\}.
\end{aligned} \tag{89}$$

Corollary 6. Let $m \in \mathbb{N}_0$. Furthermore, let $\alpha \in \mathbb{C}$, and $\beta, c \in \mathbb{C} \setminus \mathbb{Z}_0^-$. Then

$$\begin{aligned}
{}_3F_2\left[\begin{array}{c} \alpha, 1-\alpha+m, c+1; \\ \beta, c; \end{array} \frac{1}{2}\right] = \frac{2^{m-\alpha}\Gamma(\beta)\Gamma(\alpha-m)}{\Gamma(\alpha)\Gamma(\beta-\alpha)} \\
\times \Bigg\{ \sum_{r=0}^{m}(-1)^r \binom{m}{r} \frac{\Gamma\left(\frac{\beta-\alpha+r}{2}\right)}{\Gamma\left(\frac{\beta+\alpha+r}{2}-m\right)} \\
- \frac{1}{c}\sum_{r=0}^{m+2}(-1)^r \binom{m+2}{r} \frac{\Gamma\left(\frac{\beta-\alpha+r}{2}\right)}{\Gamma\left(\frac{\beta+\alpha+r}{2}-m-1\right)} \Bigg\}.
\end{aligned} \tag{90}$$

Additionally, the following corollary demonstrates the special case of Theorem 4 where $\ell = 1 = \rho$.

Corollary 7. Let $m \in \mathbb{N}_0$, and $\alpha, \beta \in \mathbb{C}$. Furthermore, let $\alpha - \beta - m \in \mathbb{C} \setminus \mathbb{Z}^-$ and $c, d \in \mathbb{C} \setminus \mathbb{Z}_0^-$. Further let $\Re(\beta) < -\frac{m}{2}$. Then

$$
{}_4F_3 \left[\begin{array}{c} \alpha, \beta, c+1, d+1; \\ 1+\alpha-\beta-m, c, d; \end{array} -1 \right] = \frac{\Gamma(1+\alpha-\beta-m)}{2cd\,\Gamma(\alpha)}
$$
$$
\times \left\{ cd \sum_{r=0}^{m} \binom{m}{r} \frac{\Gamma\left(\frac{r+\alpha}{2}\right)}{\Gamma\left(\frac{r+\alpha}{2}+1-\beta-m\right)} \right.
$$
$$
- \beta d \sum_{r=0}^{m-1} \binom{m-1}{r} \frac{\Gamma\left(\frac{r+1+\alpha}{2}\right)}{\Gamma\left(\frac{r+1+\alpha}{2}+1-\beta-m\right)}
$$
$$
- \beta(c+1) \sum_{r=0}^{m-1} \binom{m-1}{r} \frac{\Gamma\left(\frac{r+1+\alpha}{2}\right)}{\Gamma\left(\frac{r+1+\alpha}{2}+1-\beta-m\right)}
$$
$$
\left. + \beta(\beta+1) \sum_{r=0}^{m-2} \binom{m-2}{r} \frac{\Gamma\left(\frac{r+2+\alpha}{2}\right)}{\Gamma\left(\frac{r+2+\alpha}{2}+1-\beta-m\right)} \right\}.
$$
(91)

8. Concluding Remarks and Posing Problems

Beginning with Gauss's celebrated summation formula for ${}_2F_1(1)$ (5), an astoundingly huge number of summation formulae for ${}_pF_q$ ($p, q \in \mathbb{N}_0$), with a variety of arguments, have been given (see, e.g., [9]). Following this trend, we established families of generalized summation formulas for ${}_{t+2}F_{t+1}$ ($t \in \mathbb{N}$) with its arguments -1 and $1/2$ in Sections 4 and 5. We did so by introducing two sequences of new numbers in Definition 1 and their derivatives in Definition 2, as well as by selecting the six generalized summation formulas (15)–(20) above. Furthermore, in Section 6, we demonstrated two identities related to finite sums of ${}_4F_3$ by differentiating both sides of two formulae given here with respect to a particular parameter, among many others. Further, in Section 7, we provided simple specific identities for a few selected formulae in Sections 4 and 5.

In this study, the sequences of new numbers

$$
\{\mathcal{A}_j(\alpha,\ell)\}_{j=0}^{\ell} \quad \text{and} \quad \{\mathcal{B}_j(\alpha,\ell)\}_{j=0}^{\ell}
$$

in Section 2 were helpful in establishing certain generalized summation formulas for ${}_pF_q$ with specific arguments. Further, it is expected that the newly introduced numbers would be used substantially in other fields of study.

We conclude this paper by posing some problems:
(i) Give more detailed accounts of omitted proofs of Theorems in Sections 4 and 5.
(ii) Try to give more general formulas than those in Theorems 5, 8 and 11 as in the shape of the left-handed member of (10).
(iii) Try to establish generalized summation formulas for ${}_pF_q$ based on certain known ones in the literature, by using the similar technique in this paper, with a particular aid of the sequences of newly introduced numbers in Section 2.
(iv) Try to directly prove Equation (33) and Equation (35) from Definitions 1 and 2.

Author Contributions: Writing—original draft, J.C., M.I.Q., A.H.B., J.M.; Writing—review and editing, J.C., M.I.Q., A.H.B., J.M. All authors have read and agreed to the published version of the manuscript.

Funding: The Junesang Choi was supported by the Basic Science Research Program through the National Research Foundation of Korea (NRF) funded by the Ministry of Education (NRF-2020R1I1A1A01052440).

Institutional Review Board Statement: Not applicable.

Informed Consent Statement: Not applicable.

Data Availability Statement: Not applicable.

Acknowledgments: The authors are very grateful to the anonymous referees for their constructive and encouraging comments which improved this paper. They checked each formula in this article numerically using Mathematica.

Conflicts of Interest: The authors have no conflict of interest.

References

1. Srivastava, H.M.; Choi, J. *Zeta and q-Zeta Functions and Associated Series and Integrals*; Elsevier Science Publishers: Amsterdam, The Netherlands; London, UK; New York, NY, USA, 2012.
2. Andrews, G.E.; Askey, R.; Roy, R. *Special Functions*; Encyclopedia of Mathematics and its Applications 71; Cambridge University Press: Cambridge, UK, 1999.
3. Bailey, W.N. *Generalized Hypergeometric Series*; Cambridge University Press: London, UK, 1935.
4. Carlson, B.C. *Special Functions of Applied Mathematics*; Academic Press: New York, NY, USA; San Francisco, CA, USA; London, UK, 1977.
5. Erdélyi, A.; Magnus, W.; Oberhettinger, F.; Tricomi, F.G. *Higher Transcendental Functions*; McGraw-Hill Book Company: New York, NY, USA; Toronto, ON, Canada; London, UK, 1953; Volume I.
6. Rainville, E.D. *Special Functions*; Macmillan Company: New York, 1960; Reprinted by Chelsea Publishing Company, Bronx: New York, NY, USA, 1971.
7. Slater, L.J. *Generalized Hypergeometric Functions*; Cambridge at the University Press: London, UK; New York, NY, USA, 1966.
8. Srivastava, H.M.; Manocha, H.L. *A Treatise on Generating Functions*; Halsted Press (Ellis Horwood Limited): Chichester, UK; John Wiley and Sons: New York, NY, USA; Brisbane, Australia; Toronto, ON, Canada, 1984.
9. Prudnikov, A.P.; Brychkov, Y.A.; Marichev, O.I. *Integrals and Series, More Special Functions*; Nauka: Moscow, Russia, 1986, (In Russian); Gould, G.G., Translator; Gordon and Breach Science Publishers: New York, NY, USA; Philadelphia, PA, USA; London, UK; Paris, France; Montreux, Switzerland; Tokyo, Japan; Melbourne, Australia, 1990; Volume 3.
10. Gauss, C.F. Disquisitiones Generales Circa Seriem Infinitam $1 + \frac{\alpha \cdot \beta}{1 \cdot \gamma} x + \frac{\alpha(\alpha+1) \cdot \beta(\beta+1)}{1 \cdot 2 \cdot \gamma(\gamma+1)} x^2 + \frac{\alpha(\alpha+1)(\alpha+2) \cdot \beta(\beta+1)(\beta+2)}{1 \cdot 2 \cdot 3 \cdot \gamma(\gamma+1)(\gamma+2)} x^3 + \cdots$. *Comment. Soc. Regiae Sci. Gottingensis Recent.* **1813**, *2*, 1–46.
11. Whittaker, E.T.; Watson, G.N. *A Course of Modern Analysis: An Introduction to the General Theory of Infinite Processes and of Analytic Functions; With an Account of the Principal Transcendental Functions*, 4th ed.; Cambridge University Press: Cambridge, UK; London, UK; New York, NY, USA, 1963.
12. Choi, J. Contiguous extensions of Dixon's theorem on the sum of a $_3F_2$. *J. Inequ. App.* **2010**, *2010*, 589618. [CrossRef]
13. Ebisu, A. On a strange evaluation of the hypergeometric series by Gosper. *Ramanujan J.* **2013**, *32*, 101–108. [CrossRef]
14. Joshi, C.M.; McDonald, J.B. Some finite summation theorems and an asymptotic expansion for the generalized hypergeometric series. *J. Math. Anal. Appl.* **1972**, *40*, 278–285. [CrossRef]
15. Kim, Y.S.; Rakha, M.A.; Rathie, A.K. Extensions of certain classical summation theorems for the series $_2F_1$, $_3F_2$, and $_4F_3$ with applications in Ramanujan's summations. *Int. J. Math. Math. Sci.* **2010**, *2010*, 309503. [CrossRef]
16. Kim, Y.S.; Rathie, A.K. Applications of Generalized Kummer's summation theorem for the series $_2F_1$. *Bull. Korean Math. Soc.* **2009**, *46*, 1201–1211. [CrossRef]
17. Kim, Y.S.; Rathie, A.K. Applications of generalized Gauss's second summationtheorem for the series $_2F_1$. *Math. Commun.* **2011**, *16*, 481–489.
18. Kim, Y.S.; Rathie, A.K. A new proof of Saalschütz's theorem for the series $_3F_2(1)$ and its contiguous results with applications. *Commun. Korean Math. Soc.* **2012**, *27*, 129–135. [CrossRef]
19. Kim, Y.S.; Rathie, A.K.; Paris, R.B. An extension of Saalschütz's summation theorem for the series $_{r+3}F_{r+2}$. *Integral Transforms Spec. Funct.* **2013**, *24*, 916–921. [CrossRef]
20. Koepf, W. *Hypergeometric Summation: An Algorithmic Approach to Summation and Special Function Identities*; Friedr. Vieweg & Sohn Verlagsgesellschaft mbH: Braunschweig/Wiesbaden, Germany, 1998.
21. Krattenthaler, C.; Rao, K.S. Automatic generation of hypergeometric identities by the beta integral method. *J. Comput. Appl. Math.* **2003**, *160*, 159–173. [CrossRef]
22. Lavoie, J.L.; Grondin, F.; Rathie, A.K. Generalizations of Watson's theorem on the sum of a $_3F_2$. *Indian J. Math.* **1992**, *34*, 23–32.
23. Lavoie, J.L.; Grondin, F.; Arora, A.K.R.K. Generalizations of Dixon's theorem on the sum of a $_3F_2$. *Math. Comp.* **1994**, *62*, 267–276. [CrossRef]
24. Lavoie, J.L.; Grondin, F.; Rathie, A.K. Generalizations of Whipple's theorem on the sum of a $_3F_2$. *J. Comput. Appl. Math.* **1996**, *72*, 293–300. [CrossRef]
25. Zeilberger, D. A fast algorithm for proving terminating hypergeometric identities. *Discrete Math.* **1990**, *80*, 207–211. [CrossRef]
26. Choi, J.; Quine, J.R. E.W. Barnes' approach of the multiple Gamma functions. *J. Korean Math. Soc.* **1992**, *29*, 127–140.
27. Choi, J.; Srivastava, H.M. A note on a multiplication formula for the multiple Gamma function Γ_n. *Ital. J. Pure Appl. Math.* **2008**, *23*, 179–188.
28. Carlson, B.C. Some extensions of Lardner's relations between $_0F_3$ and Bessel functions. *SIAM J. Math. Anal.* **1970**, *1*, 232–242. [CrossRef]

29. Choi, J.; Rathie, A.K. Generalizations of two summation formulas for the generalized hypergeometric function of higher order due to Exton. *Commun. Korean Math. Soc.* **2010**, *25*, 385–389. [CrossRef]
30. Exton, H. Some new summation formulae for the generalized hypergeometric function of higher order. *J. Comput. Appl. Math.* **1997**, *79*, 183–187. [CrossRef]
31. Henrici, P. A triple product theorem for hypergeometric series. *SIAM J. Math. Anal.* **1987**, *18*, 1513–1518. [CrossRef]
32. Karlsson, P.W.; Srivastava, H.M. A note on Henrici's triple product theorem. *Proc. Am. Math. Soc.* **1990**, *110*, 85–88. [CrossRef]
33. Lardner, T.J. Relations between $_0F_3$ and Bessel functions. *SIAM Rev.* **1969**, *11*, 69–72. [CrossRef]
34. Osler, T.J. An identity for simplifying certain generalized hypergeometric functions. *Math. Comp.* **1975**, *29*, 888–893. [CrossRef]
35. Srivastava, H.M. A note on certain identities involving generalized hypergeometric series. *Indag. Math.* **1979**, *82*, 191–201. [CrossRef]
36. Srivastava, H.M. A certain family of sub-exponential series. *Int. J. Math. Ed. Sci. Tech.* **1994**, *25*, 211–216. [CrossRef]
37. Tremblay, R.; Fugère, B.J. Products of two restricted hypergeometric functions. *J. Math. Anal. Appl.* **1996**, *198*, 844–852. [CrossRef]
38. Gottschalk, J.E.; Maslen, E.N. Reduction formulae for the generalized hypergeometric functions of one variable. *J. Phys. A Math. Gen.* **1988**, *21*, 1983–1998. [CrossRef]
39. Karlsson, P.W. Hypergeometric functions with integral parameter differences. *J. Math. Phys.* **1971**, *12*, 270–271. [CrossRef]
40. Karlsson, P.W. Reduction of hypergoemetric functions with integral parameter differences. *Indag. Math.* **1974**, *77*, 195–198. [CrossRef]
41. Karlsson, P.W. Some reduction formulae for double power series and Kampé de Fériet functions. *Indaga. Math.* **1984**, *87*, 31–36. [CrossRef]
42. Miller, A.R. A summation formula for Clausen's series $_3F_2(1)$ with an application to Goursat's function $_2F_2(x)$. *J. Phys. A Math. Gen.* **2005**, *38*, 3541–3545. [CrossRef]
43. Minton, B.M. Generalized hypergeometric function of unit argument. *J. Math. Phys.* **1970**, *11*, 1375–1376. [CrossRef]
44. Panda, R. A note on certain reducible cases of the generalized hypergeometric function. *Indag. Math.* **1976**, *79*, 41–45. [CrossRef]
45. Srivastava, H.M. Generalized hypergeometric functions with integral parameter differences. *Nederl. Akad. Wetensch. Indag. Math.* **1973**, *76*, 38–40. [CrossRef]
46. Miller, A.R.; Paris, R.B. Certain transformations and summations for generalized hypergeometric series with integral parameter differences. *Integral Transforms Spec. Funct.* **2011**, *22*, 67–77. [CrossRef]
47. Miller, A.R.; Srivastava, H.M. Karlsson-Minton summation theorems for the generalized hypergeometric series of unit argument. *Integral Transforms Spec. Funct.* **2010**, *21*, 603–612. [CrossRef]
48. Kummer, E.E. Über die hypergeometrische Reihe $1 + \frac{\alpha \cdot \beta}{1 \cdot \gamma} x + \frac{\alpha(\alpha+1)\beta(\beta+1)}{1 \cdot 2 \cdot \gamma \cdot (\gamma+1)} x^2 + \cdots$. *J. Reine Angew. Math.* **1836**, *15*, 127–172.
49. Choi, J.; Rathie, A.K.; Srivastava, H.M. A Generalization of a formula due to Kummer. *Integral Transforms Spec. Funct.* **2011**, *22*, 851–859. [CrossRef]
50. Choi, J. Certain applications of generalized Kummer's summation formulas for $_2F_1$. *Symmetry* **2021**, *13*, 1538. [CrossRef]
51. Choi, J.; Rathie, A.K.; Malani, S. Kummer's theorem and its contiguous identities. *Taiwan. J. Math.* **2007**, *11*, 1521–1527. [CrossRef]
52. Qureshi, M.I.; Baboo, M.S. Some unified and generalized Kummer's first summation theorems with applications in Laplace transform technique. *Asia Pac. J. Math.* **2016**, *3*, 10–23.
53. Rakha, M.A.; Rathie, A.K. Generalizations of classical summation theorems for the series $_2F_1$ and $_3F_2$ with applications. *Integral Transforms Spec. Func.* **2011**, *22*, 823–840. [CrossRef]
54. Vidunas, R. A generalization of Kummer's identity. *Rocky Mt. J. Math.* **2002**, *32*, 919–936. [CrossRef]
55. Mittag–Leffler, G.M. Sur la nouvelle fonction $E_\alpha(x)$. *Comptes R. Acad. Sci. Paris* **1903**, *137*, 554–558.
56. Wiman, A. Über den fundamentalsatz in der theorie der funktionen $E_\alpha(x)$. *Acta Math.* **1905**, *29*, 191–201. [CrossRef]
57. Wiman, A. Über die nullsteliun der funktionen $E_\alpha(x)$. *Acta Math.* **1905**, *29*, 217–234. [CrossRef]
58. Djrbashian, M.M. *Harmonic Analysis and Boundary Value Problems in the Complex Domain*; Birkhauser Verlay: Basel, Switzerland; Boston, MA, USA; Berlin, Germany, 1996.
59. Djrbashian, M.M. *Integral Transforms and Representations of Functions in the Complex Domain*; Nauka: Moscow, Russia, 1966. (In Russian)
60. Fox, C. The asymptotic expansion of generalized hypergeometric functions. *Proc. Lond. Math. Soc.* **1928**, *27*, 389–400. [CrossRef]
61. Wright, E.M. The asymptotic expansion of the generalized hypergeometric function. *J. Lond. Math. Soc.* **1935**, *10*, 286–293. [CrossRef]
62. Wright, E.M. The asymptotic expansion of integral functions defined by Taylor series. *Philos. Trans. R. Soc. Lond. A* **1940**, *238*, 423–451. [CrossRef]
63. Wright, E.M. The asymptotic expansion of the generalized hypergeometric function II. *Proc. Lond. Math. Soc.* **1940**, *46*, 389–408. [CrossRef]
64. Srivastava, H.M.; Karlsson, P.W. *Multiple Gaussian Hypergeometric Series*; Halsted Press (Ellis Horwood Limited): Chichester, UK; John Wiley and Sons: New York, NY, USA; Brisbane, Australia; Toronto, ON, Canada, 1985.
65. Mainardi, F.; Gorenflo, R. Time-fractional derivatives in relaxation processes: A tutorial survey. *Fract. Calc. Appl. Anal.* **2007**, *10*, 269–308.
66. Abdalla, M.; Akel, M.; Choi, J. Certain matrix Riemann–Liouville fractional integrals associated with functions involving generalized Bessel matrix polynomials. *Symmetry* **2021**, *13*, 622. [CrossRef]

67. Khan, N.U.; Aman, M.; Usman, T.; Choi, J. Legendre-Gould Hopper-based Sheffer polynomials and operational methods. *Symmetry* **2020**, *12*, 2051. [CrossRef]
68. Khan, N.U.; Usman, T.; Choi, J. A new class of generalized polynomials involving Laguerre and Euler polynomials. *Hacet. J. Math. Stat.* **2021**, *50*, 1–13. [CrossRef]
69. Nahid, T.; Alam, P.; Choi, J. Truncated-exponential-based Appell-type Changhee polynomials. *Symmetry* **2020**, *12*, 1588. [CrossRef]
70. Usman, T.; Saif, M.; Choi, J. Certain identities associated with (p,q)-binomial coefficients and (p,q)-Stirling polynomials of the second kind. *Symmetry* **2020**, *12*, 1436. [CrossRef]
71. Yasmin, G.; Islahi, H.; Choi, J. q-generalized tangent based hybrid polynomials. *Symmetry* **2021**, *13*, 791. [CrossRef]
72. Choi, J. Notes on formal manipulations of double series. *Commun. Korean Math. Soc.* **2003**, *18*, 781–789. [CrossRef]
73. Alzer, H.; Choi, J. Four parametric linear Euler sums. *J. Math. Anal. Appl.* **2020**, *484*, 123661. [CrossRef]

 *fractal and fractional*

Article

Hilfer–Hadamard Fractional Boundary Value Problems with Nonlocal Mixed Boundary Conditions

Bashir Ahmad [1,†] and Sotiris K. Ntouyas [2,*,†]

1. Nonlinear Analysis and Applied Mathematics (NAAM)-Research Group, Department of Mathematics, Faculty of Science, King Abdulaziz University, P.O. Box 80203, Jeddah 21589, Saudi Arabia; bashirahmad_qau@yahoo.com
2. Department of Mathematics, University of Ioannina, 451 10 Ioannina, Greece
* Correspondence: sntouyas@uoi.gr
† These authors contributed equally to this work.

Abstract: This paper is concerned with the existence and uniqueness of solutions for a Hilfer–Hadamard fractional differential equation, supplemented with mixed nonlocal (multi-point, fractional integral multi-order and fractional derivative multi-order) boundary conditions. The existence of a unique solution is obtained via Banach contraction mapping principle, while the existence results are established by applying the fixed point theorems due to Krasnoselskiĭ and Schaefer and Leray–Schauder nonlinear alternatives. We demonstrate the application of the main results by presenting numerical examples. We also derive the existence results for the cases of convex and non-convex multifunctions involved in the multi-valued analogue of the problem at hand.

Keywords: Hilfer–Hadamard fractional derivative; Riemann–Liouville fractional derivative; Caputo fractional derivative; fractional differential equations; inclusions; nonlocal boundary conditions; existence and uniqueness; fixed point

1. Introduction

Fractional calculus is regarded as the generalization of the integer-order integration and differentiation in the sense that it deals with derivative and integral operators of an arbitrary real or complex order. This branch of mathematical analysis gained much importance during the last few decades owing to its widespread applications in a variety of disciplines, such as mechanical engineering, bioengineering, biology, physics, chemistry, economics, viscoelasticity, acoustics, optics, robotics, control theory, electronics, etc. The main reason for the popularity of fractional calculus is that mathematical models based on fractional-order operators are considered to be more realistic than the ones relying on classical calculus as such operators are nonlocal in nature and can trace the history of the phenomena under consideration. For the theoretical development of the subject, we refer the reader to the monographs [1–9] and the references therein. For some recent applications of fractional calculus concerning structural mechanics and, more specifically, nonlocal elasticity, see [10–12].

Fractional-order boundary value problems constitute an important and interesting area of research. It reflects from the literature on the topic that a good deal of work on fractional differential equations involve either Caputo or Riemann–Liouville fractional derivatives. However, these derivatives are found to be inappropriate in the study of some engineering problems. In order to tackle such inaccuracies, some new fractional-order derivative operators such as Hadamard, Erdelyi–Kober, Katugampola, etc., were proposed. In [13], Hilfer introduced a new derivative, which is known as the Hilfer fractional derivative and can generalize both Riemann–Liouville and Caputo derivatives. For some applications of this derivative, we refer the interested reader to the investigations [14,15]. For some recent results on initial and boundary value problems involving Hilfer fractional derivative, for instance, see [16–22] and the references cited therein.

The fractional derivative presented by Hadamard in 1892 [23] differs from the well known Caputo derivative in two significant ways: (i) its kernel involves a logarithmic function with an arbitrary exponent and (ii) the Hadamard derivative of a constant is not zero. One can find applications of the Hadamard derivative and integral operators in the paper [24] and the monograph [2]. The Hadamard calculus can be obtained by changing $d/dt \to t\,d/dt$, $(t-s)^{(\cdot)} \to (\log_e t - \log_e s)^{(\cdot)}$ and $ds \to (1/s)ds$ in Riemann–Liouville and Caputo fractional derivatives. Later, the modification of Hilfer fractional derivative resulted in the concept of the Hilfer–Hadamard derivative.

Existence results for Hilfer–Hadamard fractional differential equations of order in $(0,1]$ were studied by several researchers, for instance, see [25–27]. To the best of our knowledge, only a few results are available in the literature concerning boundary value problems for Hilfer–Hadamard fractional differential equations of order in $(1,2]$. Recently, in [28], the authors applied the tools of the fixed-point theory to study the existence and uniqueness of solutions for a boundary value problem of Hilfer–Hadamard fractional differential equations with nonlocal integro-multi-point boundary conditions:

$$\begin{cases} {}^{HH}D_1^{\alpha,\beta}x(t) = f(t,x(t)), \quad t \in [1,T], \\ x(1) = 0, \quad \sum_{i=1}^{m}\theta_i x(\xi_i) = \lambda\, {}^{H}I_1^{\delta}x(\eta), \end{cases}$$

where ${}^{HH}D_1^{\alpha,\beta}$ denote the Hilfer–Hadamard fractional derivative operator of order $\alpha \in (1,2]$ and type $\beta \in [0,1]$, $\theta_i, \lambda \in \mathbb{R}$ and $i=1,2,\ldots,m$, are given constants, $f:[1,T]\times \mathbb{R} \to \mathbb{R}$ is a given continuous function and ${}^{H}I^{\delta}$ is the Hadamard fractional integral of order $\delta > 0$ and $\eta, \xi_i \in (1,T), i=1,2,\ldots,m$.

In [29], the existence of solutions for the following system of sequential fractional differential equations involving Hilfer–Hadamard type differential operators ${}_H\mathfrak{D}^{\cdot,\cdot}$ of different orders was discussed:

$$\begin{cases} ({}_H\mathfrak{D}_{1^+}^{\alpha_1,\beta_1} + \lambda_1{}_H\mathfrak{D}_{1^+}^{\alpha_1-1,\beta_1})u(t) = f(t,u(t),v(t)), \quad 1 < \alpha_1 \le 2, \ t \in [1,e], \\ ({}_H\mathfrak{D}_{1^+}^{\alpha_2,\beta_2} + \lambda_2{}_H\mathfrak{D}_{1^+}^{\alpha_2-1,\beta_2})v(t) = g(t,u(t),v(t)), \quad 1 < \alpha_2 \le 2, \ t \in [1,e], \\ u(1) = 0, \ u(e) = A_1, \ A_1 \in \mathbb{R}_+, \\ v(1) = 0, \ v(e) = A_2, \ A_2 \in \mathbb{R}_+, \end{cases}$$

where $\lambda_1, \lambda_2 \in \mathbb{R}_+$ and $f,g:[1,e]\times \mathbb{R}^2 \to \mathbb{R}$ are given continuous functions.

Motivated by the aforementioned work, our goal in this paper is to enrich the literature on boundary value problems of Hilfer–Hadamard fractional differential equations of order in $(1,2]$. In precise terms, we introduce and study a nonlocal mixed Hilfer–Hadamard boundary value problem of the following form:

$$\begin{cases} {}^{HH}D_1^{\alpha,\beta}x(t) = f(t,x(t)), \quad t \in [1,T], \\ x(1) = 0, \quad x(T) = \sum_{j=1}^{m}\eta_j x(\xi_j) + \sum_{i=1}^{n}\zeta_i\, {}^{H}I_1^{\phi_i}x(\theta_i) + \sum_{k=1}^{r}\lambda_k\, {}_HD_1^{\omega_k}x(\mu_k), \end{cases} \quad (1)$$

where ${}^{HH}D_1^{\alpha,\beta}$ denotes the Hilfer–Hadamard fractional derivative operator of order $\alpha \in (1,2]$ and type $\beta \in [0,1]$ and $\eta_j, \zeta_i, \lambda_k \in \mathbb{R}$ are given constants, $f:[1,T]\times \mathbb{R} \to \mathbb{R}$ is a given continuous function, ${}^{H}I^{\phi_i}$ is the Hadamard fractional integral operator of order $\phi_i > 0$ and $\xi_j, \theta_i, \mu_k \in (1,T)$, $j=1,2,\ldots,m$, $i=1,2,\ldots,n$, $k=1,2,\ldots,r$. We also study the multi-valued analogue of the problem (1).

Concerning the significance of problem (1), we recall that the Hilfer fractional derivative interpolates between the Riemann–Liouville and Caputo derivatives [13]. Analogously, the Hilfer–Hadamard type fractional derivative covers the cases of the Riemann–Liouville–

Hadamard and Caputo–Hadamard fractional derivatives. Thus, the present study will be useful for improving the works related to glass forming materials [14], Turbulent Flow Model [30], etc. Furthermore, several results involving the Caputo–Hadamard fractional derivative [31–34] can be extended to the framework of Hilfer–Hadamard fractional derivative.

It is well known that the nonlocal condition is more appropriate than the local condition (initial and/or boundary) with respect to describing certain features of applied mathematics and physics correctly (see the survey paper [35]). More specifically, the boundary conditions arising in the study of boundary value problems of nonlocal elasticity are always nonlocal in nature. This is due to the fact that the long-range interactions within nonlocal solids give rise to nonlocal traction (force) boundary conditions.

Here, we remark that there are only two articles [28,29] in the literature (to the best of our knowledge) concerning boundary value problems for Hilfer–Hadamard fractional differential equations of the order in $(1,2]$. Much of the known studies in the literature deals with initial value problems of Hilfer–Hadamard fractional differential equations of the order in $(0,1]$. The two classes of problems are entirely different. The methodology employed to study the Hilfer–Hadamard fractional differential equations of the order in $(0,1]$ is different from the one applied to such equations of the order in $(1,2]$. Thus, our main objective in this paper is to enrich the new research area on Hilfer–Hadamard fractional differential equations of the order in $(1,2]$. Moreover, the mixed boundary conditions introduced in the present study are of a more general type and include multi-point, fractional integral multi-order and fractional derivative multi-order contributions.

One can notice that the boundary conditions considered in problem (1) reduce to several special cases such as (i) nonlocal multi-point boundary conditions if we choose all $\zeta_i = 0, i = 1, 2, \ldots, n$ and $\lambda_k = 0, k = 1, 2, \ldots, r$; (ii) nonlocal Hadamard fractional integral boundary conditions when all $\eta_j = 0, j = 1, 2, \ldots, m$ and $\lambda_k = 0, k = 1, 2, \ldots, r$; and (iii) nonlocal Hadamard fractional boundary conditions if we take all $\eta_j = 0, j = 1, 2, \ldots, m$ and $\zeta_i = 0, i = 0, k = 1, 2, \ldots, n$. Likewise, we can consider the combination of nonlocal multipoint and Hadamard fractional integral conditions when we fix all $\lambda_k = 0$, $k = 1, 2, \ldots, r$ and so on. Thus, the results presented in this paper are significant as they specialize to the ones associated with several interesting boundary conditions. Another novelty in the present work is concerned with the derivation of the existence results for the Hilfer–Hadamard fractional differential inclusions of the order in $(1,2]$ supplemented with the mixed boundary conditions. Thus, the investigation of single-valued and multi-valued nonlocal nonlinear Hilfer–Hadamard fractional boundary value problems of the order in $(1,2]$ enhances the scope of the literature on the topic.

The remaining part of this manuscript is arranged as follows. Section 2 contains some basic notions and known results of fractional differential calculus. In Section 3, we first prove an auxiliary result that plays a key role in transforming the given problem into a fixed point problem. Then, based on Banach's contraction mapping principle, we establish the existence of a unique solution for the problem (1). By using the fixed point theorems due to Krasnoselskiĭ and Schaefer and nonlinear alternative of Leray–Schauder type, we prove some existence results for problem (1). Examples illustrating the applicability of the main results are also presented in this section. The existence results for the multi-valued analogue of the problem (1) are obtained in Section 4. Some interesting observations are presented in the last section of the paper.

2. Preliminaries

In this section, we recall some basic concepts.

Definition 1. (*Hadamard fractional integral* [2]). *Let* $f : [a, \infty) \to \mathbb{R}$. *Then, the Hadamard fractional integral of order* $\alpha > 0$ *is defined as follows:*

$$^H I_a^\alpha f(t) = \frac{1}{\Gamma(\alpha)} \int_a^t \left(\log \frac{t}{z}\right)^{\alpha-1} \frac{f(z)}{z} dz, \quad t > a, \tag{2}$$

provided that the integral exists, where $\log(.) = \log_e(.)$.

Definition 2. *(Hadamard fractional derivative [2]). For a function $f : [a, \infty) \to \mathbb{R}$, the Hadamard fractional derivative of order $\alpha > 0$ is defined as follows:*

$$_H D_a^\alpha f(t) = \delta^n \left({}^H I_a^{n-\alpha} f \right)(t), \quad n = [\alpha] + 1, \tag{3}$$

where $\delta^n = (t \frac{d}{dt})^n$ and $[\alpha]$ denote the integer part of the real number α.

Lemma 1. *[2] If $\alpha > 0, \beta > 0$ and $0 < a < b < \infty$, then*

(i) $\left({}^H I_{a^+}^\alpha \left(\log \frac{t}{a} \right)^{\beta-1} \right)(x) = \frac{\Gamma(\beta)}{\Gamma(\beta+\alpha)} \left(\log \frac{x}{a} \right)^{\beta+\alpha-1}$;

(ii) $\left({}_H D_{a^+}^\alpha \left(\log \frac{t}{a} \right)^{\beta-1} \right)(x) = \frac{\Gamma(\beta)}{\Gamma(\beta-\alpha)} \left(\log \frac{x}{a} \right)^{\beta-\alpha-1}$.

In particular, if $\beta = 1$, then the following is the case:

$$({}_H D_{a^+}^\alpha)(1) = \frac{1}{\Gamma(1-\alpha)} \left(\log \frac{x}{a} \right)^{-\alpha} \neq 0, \ 0 < \alpha < 1.$$

Definition 3. *(Hilfer–Hadamard fractional derivative [15]). Let $f \in L^1(a,b)$ and $n-1 < \alpha < n$, $0 \le \beta \le 1$. We define the Hilfer–Hadamard fractional derivative of order α and type β for f as follows:*

$$\begin{aligned}
({}^{HH} D_a^{\alpha,\beta} f)(t) &= \left({}^H I_a^{\beta(n-\alpha)} \delta^n \, {}^H I_a^{(n-\alpha)(1-\beta)} f \right)(t) \\
&= \left({}^H I_a^{\beta(n-\alpha)} \delta^n \, {}^H I_a^{(n-\gamma)} f \right)(t) \\
&= \left({}^H I_a^{\beta(n-\alpha)} \, {}_H D_a^\gamma f \right)(t), \quad \gamma = \alpha + n\beta - \alpha\beta,
\end{aligned}$$

where ${}^H I_a^{(.)}$ and ${}_H D_a^{(.)}$ are defined by (2) and (3), respectively.

Here, we remark that the Hilfer–Hadamard fractional derivative reduces to the Hadamard fractional derivative for $\beta = 0$ and corresponds to the Caputo–Hadamard derivative for $\beta = 1$ given in the following equation:

$$_H^C D_a^\alpha f(t) = \left({}^H I_a^{n-\alpha} \delta^n f \right)(t), \quad n = [\alpha] + 1.$$

Next, we recall the following known theorem that will be used in the sequel.

Theorem 1. *([5]). Let $\alpha > 0, 0 \le \beta \le 1, \gamma = \alpha + n\beta - \alpha\beta, n = [\alpha] + 1$ and $0 < a < b < \infty$. If $f \in L^1(a,b)$ and $({}^H I_a^{n-\gamma} f)(t) \in AC_\delta^n[a,b]$, then the following is the case:*

$$\begin{aligned}
{}^H I_a^\alpha ({}^{HH} D_a^{\alpha,\beta} f)(t) &= {}^H I_a^\gamma ({}^{HH} D_a^\gamma f)(t) \\
&= f(t) - \sum_{j=0}^{n-1} \frac{(\delta^{(n-j-1)}({}^H I_a^{n-\gamma} f))(a)}{\Gamma(\gamma-j)} \left(\log \frac{t}{a} \right)^{\gamma-j-1}.
\end{aligned}$$

Observe that $\Gamma(\gamma - j)$ exists for all $j = 1, 2, \ldots, n-1$ for $\gamma \in [\alpha, n]$.

3. Main Results

This section is concerned with the existence and uniqueness of solutions for the nonlinear Hilfer–Hadamard fractional boundary value problem (1). First of all, we prove an auxiliary lemma dealing with the linear variant of the boundary value problem (1), which will be used to transform the problem at hand into an equivalent fixed point problem. In the case $n = [\alpha] + 1 = 2$, we have $\gamma = \alpha + (2-\alpha)\beta$.

Lemma 2. *Let $h \in C([1,T], \mathbb{R})$ and that*

$$\Lambda = (\log T)^{\gamma-1} - \sum_{j=1}^{m} \eta_j (\log \xi_j)^{\gamma-1} - \sum_{i=1}^{n} \zeta_i \frac{\Gamma(\gamma)}{\Gamma(\gamma+\phi_i)} (\log \theta_i)^{\gamma+\phi_i-1}$$
$$- \sum_{k=1}^{r} \lambda_k \frac{\Gamma(\gamma)}{\Gamma(\gamma-\omega_k)} (\log \mu_k)^{\gamma-\omega_k-1} \neq 0. \quad (4)$$

Then, x is a solution of the following linear Hilfer–Hadamard fractional boundary value problem:

$$\begin{cases} {}^{HH}D_1^{\alpha,\beta} x(t) = h(t,), \quad t \in [1,T], \\ x(1) = 0, \quad x(T) = \sum_{j=1}^{m} \eta_j x(\xi_j) + \sum_{i=1}^{n} \zeta_i \, {}^H I_1^{\phi_i} x(\theta_i) + \sum_{k=1}^{r} \lambda_k \, {}_H D_1^{\omega_k} x(\mu_k), \end{cases} \quad (5)$$

if and only if it satisfies the integral equation:

$$x(t) = {}^H I_1^\alpha h(t) + \frac{(\log t)^{\gamma-1}}{\Lambda} \left\{ \sum_{j=1}^{m} {}^H I^\alpha h(\xi_j) + \sum_{i=1}^{n} \zeta_i \, {}^H I^{\alpha+\phi_i} h(\theta_i) \right.$$
$$\left. + \sum_{k=1}^{r} \lambda_k \, {}^H I^{\alpha-\omega_k} h(\mu_k) - {}^H I^\alpha h(T) \right\}, \quad t \in [1,T]. \quad (6)$$

Proof. Applying the Hadamard fractional integral operator of order α from 1 to t on both sides of Hilfer–Hadamard fractional differential equation in (5) and using Theorem 1, we find that

$$x(t) - \frac{\delta({}_H I_{1+}^{2-\gamma} x)(1)}{\Gamma(\gamma)} (\log t)^{\gamma-1} - \frac{({}_H I_{1+}^{2-\gamma} x)(1)}{\Gamma(\gamma-1)} (\log t)^{\gamma-2} = {}^H I_1^\alpha h(t), \quad (7)$$

which can be rewritten as follows:

$$x(t) = c_0 (\log t)^{\gamma-1} + c_1 (\log t)^{\gamma-2} + \frac{1}{\Gamma(\alpha)} \int_1^t \frac{h(s)}{s} \left(\log \frac{t}{s}\right)^{\alpha-1} ds, \quad (8)$$

where c_0 and c_1 are arbitrary constants. Using the first boundary condition ($x(1) = 0$) in (8) yields $c_1 = 0$, since $\gamma \in [\alpha, 2]$. In consequence, (8) takes the following form:

$$x(t) = c_0 (\log t)^{\gamma-1} + \frac{1}{\Gamma(\alpha)} \int_1^t \left(\log \frac{t}{s}\right)^{\alpha-1} \frac{h(s)}{s} ds. \quad (9)$$

Now, inserting (9) into the second boundary condition:

$$x(T) = \sum_{j=1}^{m} \eta_j x(\xi_j) + \sum_{i=1}^{n} \zeta_i \, {}^H I_1^{\phi_i} x(\theta_i) + \sum_{k=1}^{r} \lambda_k \, {}_H D_1^{\omega_k} x(\mu_k),$$

and using notation (4), we obtain the following:

$$c_0 = \frac{1}{\Lambda} \left\{ \sum_{j=1}^{m} {}^H I^\alpha h(\xi_j) + \sum_{i=1}^{n} \zeta_i \, {}^H I^{\alpha+\phi_i} h(\theta_i) + \sum_{k=1}^{r} \lambda_k \, {}^H I^{\alpha-\omega_k} h(\mu_k) - {}^H I^\alpha h(T) \right\}.$$

Substituting the value of c_0 in (9) results in Equation (6) as desired. By direct computation, one can obtain the converse of the lemma. The proof is completed. □

Let $X = C([1,T], \mathbb{R})$ be the Banach space endowed with the norm

$$\|x\| := \max_{t \in [1,T]} |x(t)|.$$

In view of Lemma 2 and Definition 1, we introduce an operator $\mathcal{F}: X \to X$ associated with the problem (1) as follows:

$$\begin{aligned}\mathcal{F}(x)(t) &= \frac{1}{\Gamma(\alpha)}\int_1^t \left(\log\frac{t}{z}\right)^{\alpha-1}\frac{f(z,x(z))}{z}dz \\ &+ \frac{(\log t)^{\gamma-1}}{\Lambda}\Bigg\{\sum_{j=1}^m \frac{\eta_j}{\Gamma(\alpha)}\int_1^{\xi_j}\left(\log\frac{\xi_j}{z}\right)^{\alpha-1}\frac{f(z,x(z))}{z}dz \\ &+ \sum_{i=1}^n \frac{\zeta_i}{\Gamma(\alpha+\phi_i)}\int_1^{\theta_i}\left(\log\frac{\theta_i}{z}\right)^{\alpha+\phi_i-1}\frac{f(z,x(z))}{z}dz \\ &+ \sum_{k=1}^r \frac{\lambda_k}{\Gamma(\alpha-\omega_k)}\int_1^{\mu_k}\left(\log\frac{\mu_k}{z}\right)^{\alpha-\omega_k-1}\frac{f(z,x(z))}{z}dz \\ &- \frac{1}{\Gamma(\alpha)}\int_1^T\left(\log\frac{T}{z}\right)^{\alpha-1}\frac{f(z,x(z))}{z}dz\Bigg\},\ t\in[1,T]. \end{aligned} \quad (10)$$

In the sequel, we use the following notation:

$$\begin{aligned}\Omega &= \frac{(\log T)^\alpha}{\Gamma(\alpha+1)} + \frac{(\log T)^{\gamma-1}}{|\Lambda|}\Bigg\{\sum_{i=j}^m \frac{|\eta_j|(\log\xi_j)^\alpha}{\Gamma(\alpha+1)} + \sum_{i=1}^n \frac{|\zeta_i|(\log\theta_i)^{\alpha+\phi_i}}{\Gamma(\alpha+\phi_i+1)} \\ &+ \sum_{k=1}^r \frac{|\lambda_k|(\log\mu_k)^{\alpha-\omega_k}}{\Gamma(\alpha-\omega_k+1)} + \frac{(\log T)^\alpha}{\Gamma(\alpha+1)}\Bigg\}.\end{aligned} \quad (11)$$

3.1. Uniqueness Result

Here, by applying Banach's contraction mapping principle [36], we prove the existence of a unique solution for problem (1).

Theorem 2. *Suppose that the following condition holds:*
(H_1) *There exists a constant $l > 0$ such that for all $t \in [1,T]$ and $u_i \in \mathbb{R}$, $i=1,2,$.*

$$|f(t,u_1) - f(t,u_2)| \le l|u_1 - u_2|.$$

Then, the nonlinear Hilfer–Hadamard fractional boundary value problem (1) has a unique solution on $[1,T]$ if $l\Omega < 1$, where Ω is defined by (11).

Proof. We will verify that the operator $\mathcal{F}$ defined by (10) satisfies the hypotheses of Banach's contraction mapping principle. Fixing $N = \max_{t\in[1,T]}|f(t,0)| < \infty$ and using the assumption (H_1), we obtain the following:

$$|f(t,x(t))| \le l|x(t)| + |f(t,0)| \le l\|x\| + N. \quad (12)$$

The proof is divided into two steps.
Step I: We show that $\mathcal{F}(B_r) \subset B_r$, where $B_r = \{x \in X : \|x\| < r\}$ with $r \ge N\Omega/(1-l\Omega)$. Let $x \in B_r$. Then, we have the following:

$$
\begin{aligned}
|\mathcal{F}(x)(t)| &\leq \frac{1}{\Gamma(\alpha)}\int_1^t \left(\log \frac{t}{z}\right)^{\alpha-1}\frac{|f(z,x(z))|}{z}dz \\
&\quad + \frac{(\log t)^{\gamma-1}}{\Lambda}\bigg\{\sum_{j=1}^m \frac{|\eta_j|}{\Gamma(\alpha)}\int_1^{\xi_j}\left(\log\frac{\xi_j}{z}\right)^{\alpha-1}\frac{|f(z,x(z))|}{z}dz \\
&\quad + \sum_{i=1}^n \frac{|\zeta_i|}{\Gamma(\alpha+\phi_i)}\int_1^{\theta_i}\left(\log\frac{\theta_i}{z}\right)^{\alpha+\phi_i-1}\frac{|f(z,x(z))|}{z}dz \\
&\quad + \sum_{k=1}^r \frac{|\lambda_k|}{\Gamma(\alpha-\omega_k)}\int_1^{\mu_k}\left(\log\frac{\mu_k}{z}\right)^{\alpha-\omega_k-1}\frac{|f(z,x(z))|}{z}dz \\
&\quad + \frac{1}{\Gamma(\alpha)}\int_1^T\left(\log\frac{T}{z}\right)^{\alpha-1}\frac{|f(z,x(z))|}{z}dz\bigg\} \\
&\leq \frac{(\log T)^\alpha}{\Gamma(\alpha+1)}(l\|x\|+N) + \frac{(\log T)^{\gamma-1}}{|\Lambda|}\bigg\{\sum_{i=j}^m \frac{|\eta_j|(\log \xi_j)^\alpha}{\Gamma(\alpha+1)}+\sum_{i=1}^n \frac{|\zeta_i|(\log\theta_i)^{\alpha+\phi_i}}{\Gamma(\alpha+\phi_i+1)} \\
&\quad + \sum_{k=1}^r \frac{|\lambda_k|(\log\mu_k)^{\alpha-\omega_k}}{\Gamma(\alpha-\omega_k+1)}+\frac{(\log T)^\alpha}{\Gamma(\alpha+1)}\bigg\}(l\|x\|+N) \\
&\leq \bigg[\frac{(\log T)^\alpha}{\Gamma(\alpha+1)}+\frac{(\log T)^{\gamma-1}}{|\Lambda|}\bigg\{\sum_{i=j}^m \frac{|\eta_j|(\log\xi_j)^\alpha}{\Gamma(\alpha+1)}+\sum_{i=1}^n \frac{|\zeta_i|(\log\theta_i)^{\alpha+\phi_i}}{\Gamma(\alpha+\phi_i+1)} \\
&\quad + \sum_{k=1}^r \frac{|\lambda_k|(\log\mu_k)^{\alpha-\omega_k}}{\Gamma(\alpha-\omega_k+1)}+\frac{(\log T)^\alpha}{\Gamma(\alpha+1)}\bigg\}\bigg](lr+N) \\
&= \Omega(lr+N) \leq r.
\end{aligned}
$$

Thus, the following is the case:

$$\|\mathcal{F}(x)\| = \max_{t\in[1,T]}|\mathcal{F}(u)(t)| \leq r,$$

which means that $\mathcal{F}(B_r) \subset B_r$.

Step II: To show that the operator $\mathcal{F}$ is a contraction, let $x_1, x_2 \in X$. Then, for any $t \in [1,T]$, we have the following:

$$
\begin{aligned}
&|\mathcal{F}(x_2)(z)-\mathcal{F}(x_1)(z)| \\
&\leq \frac{1}{\Gamma(\alpha)}\int_1^t\left(\log\frac{t}{z}\right)^{\alpha-1}\frac{|f(z,x_2(z))-f(z,x_1(z))|}{z}dz \\
&\quad + \frac{(\log t)^{\gamma-1}}{|\Lambda|}\bigg\{\sum_{j=1}^m \frac{|\eta_j|}{\Gamma(\alpha)}\int_1^{\xi_j}\left(\log\frac{\xi_j}{z}\right)^{\alpha-1}\frac{|f(z,x_2(z))-f(z,x_1(z))|}{z}dz \\
&\quad + \sum_{i=1}^n \frac{|\zeta_i|}{\Gamma(\alpha+\phi_i)}\int_1^{\theta_i}\left(\log\frac{\theta_i}{z}\right)^{\alpha+\phi_i-1}\frac{|f(z,x_2(z))-f(z,x_1(z))|}{z}dz \\
&\quad + \sum_{k=1}^r \frac{|\lambda_k|}{\Gamma(\alpha-\omega_k)}\int_1^{\mu_k}\left(\log\frac{\mu_k}{z}\right)^{\alpha-\omega_k-1}\frac{|f(z,x_2(z))-f(z,x_1(z))|}{z}dz \\
&\quad + \frac{1}{\Gamma(\alpha)}\int_1^T\left(\log\frac{t}{z}\right)^{\alpha-1}\frac{|f(z,x_2(z))-f(z,x_1(z))|}{z}dz\bigg\} \\
&\leq \bigg[\frac{(\log T)^\alpha}{\Gamma(\alpha+1)}+\frac{(\log T)^{\gamma-1}}{|\Lambda|}\bigg\{\sum_{i=j}^m \frac{|\eta_j|(\log\xi_j)^\alpha}{\Gamma(\alpha+1)}+\sum_{i=1}^n \frac{|\zeta_i|(\log\theta_i)^{\alpha+\phi_i}}{\Gamma(\alpha+\phi_i+1)} \\
&\quad + \sum_{k=1}^r \frac{|\lambda_k|(\log\mu_k)^{\alpha-\omega_k}}{\Gamma(\alpha-\omega_k+1)}+\frac{(\log T)^\alpha}{\Gamma(\alpha+1)}\bigg\}\bigg]l\|x_2-x_1\|.
\end{aligned}
$$

Thus, the following is the case:

$$\|\mathcal{F}(x_2) - \mathcal{F}(x_1)\| = \max_{t\in[1,T]} |\mathcal{F}(x_2)(t) - \mathcal{F}(x_1)(t)| \leq l\Omega \|x_2 - x_1\|,$$

which, in view of $l\Omega < 1$, shows that the operator $\mathcal{F}$ is a contraction. Hence, the operator $\mathcal{F}$ has a unique fixed point by Banach's contraction mapping principle. Therefore, the problem (1) has a unique solution on $[1, T]$. The proof is completed. □

3.2. Existence Results

In this subsection, we present different criteria for the existence of solutions for the problem (1). First, we prove an existence result based on Krasnoselskiĭ's fixed point theorem [37].

Theorem 3. *Let $f : [1, T] \times \mathbb{R} \to \mathbb{R}$ be a continuous function satisfying (H_1). In addition, we assume that the following condition is satisfied:*

(H_2) *There exists a continuous function $\phi \in C([1, T], \mathbb{R}^+)$ such that*

$$|f(t, x)| \leq \phi(t), \text{ for each } (t, u) \in [1, T] \times \mathbb{R}.$$

Then, the nonlinear Hilfer–Hadamard fractional boundary value problem (1) has at least one solution on $[1, T]$, provided that the following condition holds:

$$\frac{(\log T)^{\gamma-1}}{|\Lambda|} \left\{ \sum_{i=j}^{m} \frac{|\eta_j|(\log \xi_j)^{\alpha}}{\Gamma(\alpha+1)} + \sum_{i=1}^{n} \frac{|\zeta_i|(\log \theta_i)^{\alpha+\phi_i}}{\Gamma(\alpha+\phi_i+1)} \right. $$
$$\left. + \sum_{k=1}^{r} \frac{|\lambda_k|(\log \mu_k)^{\alpha-\omega_k}}{\Gamma(\alpha-\omega_k+1)} + \frac{(\log T)^{\alpha}}{\Gamma(\alpha+1)} \right\} l < 1. \tag{13}$$

Proof. By assumption (H_2), we can fix $\rho \geq \Omega \|\phi\|$ and consider a closed ball $B_\rho = \{x \in C([1,T],\mathbb{R}) : \|x\| \leq \rho\}$, where $\|\phi\| = \sup_{t\in[1,T]} |\phi(t)|$ and Ω is given by (11). We verify the hypotheses of Krasnoselskiĭ's fixed point theorem [37] by splitting the operator $\mathcal{F}$ defined by (10) on B_ρ to $C([1,T],\mathbb{R})$ as $\mathcal{F} = \mathcal{F}_1 + \mathcal{F}_2$, where $\mathcal{F}_1$ and $\mathcal{F}_2$ are defined by the following:

$$(\mathcal{F}_1 x)(t) = \frac{1}{\Gamma(\alpha)} \int_1^t \left(\log \frac{t}{z}\right)^{\alpha-1} \frac{f(z, x(z))}{z} dz, \; t \in [1, T],$$

$$(\mathcal{F}_2 x)(t) = \frac{(\log t)^{\gamma-1}}{\Lambda} \left\{ \sum_{j=1}^{m} \frac{\eta_j}{\Gamma(\alpha)} \int_1^{\xi_j} \left(\log \frac{\xi_j}{z}\right)^{\alpha-1} \frac{f(z, x(z))}{z} dz \right.$$
$$+ \sum_{i=1}^{n} \frac{\zeta_i}{\Gamma(\alpha+\phi_i)} \int_1^{\theta_i} \left(\log \frac{\theta_i}{z}\right)^{\alpha+\phi_i-1} \frac{f(z, x(z))}{z} dz$$
$$+ \sum_{k=1}^{r} \frac{\lambda_k}{\Gamma(\alpha-\omega_k)} \int_1^{\mu_k} \left(\log \frac{\mu_k}{z}\right)^{\alpha-\omega_k-1} \frac{f(z, x(z))}{z} dz$$
$$\left. - \frac{1}{\Gamma(\alpha)} \int_1^{T} \left(\log \frac{t}{z}\right)^{\alpha-1} \frac{f(z, x(z))}{z} dz \right\}, \; t \in [1, T].$$

For any $x, y \in B_\rho$, we have the following:

$$\begin{aligned}
|(\mathcal{F}_1 x)(t) + (\mathcal{F}_2 y)(t)| &\leq \frac{1}{\Gamma(\alpha)} \int_1^t \left(\log \frac{t}{z}\right)^{\alpha-1} \frac{|f(z,x(z))|}{z} dz \\
&\quad + \frac{(\log t)^{\gamma-1}}{|\Lambda|} \left\{ \sum_{j=1}^m \frac{|\eta_j|}{\Gamma(\alpha)} \int_1^{\xi_j} \left(\log \frac{\xi_j}{z}\right)^{\alpha-1} \frac{|f(z,x(z))|}{z} dz \right. \\
&\quad + \sum_{i=1}^n \frac{|\zeta_i|}{\Gamma(\alpha+\phi_i)} \int_1^{\theta_i} \left(\log \frac{\theta_i}{z}\right)^{\alpha+\phi_i-1} \frac{|f(z,x(z))|}{z} dz \\
&\quad + \sum_{k=1}^r \frac{|\lambda_k|}{\Gamma(\alpha-\omega_k)} \int_1^{\mu_k} \left(\log \frac{\mu_k}{z}\right)^{\alpha-\omega_k-1} \frac{|f(z,x(z))|}{z} dz \\
&\quad \left. + \frac{1}{\Gamma(\alpha)} \int_1^T \left(\log \frac{t}{z}\right)^{\alpha-1} \frac{|f(z,x(z))|}{z} dz \right\} \\
&\leq \left[\frac{(\log T)^\alpha}{\Gamma(\alpha+1)} + \frac{(\log T)^{\gamma-1}}{|\Lambda|} \left\{ \sum_{i=j}^m \frac{|\eta_j|(\log \xi_j)^\alpha}{\Gamma(\alpha+1)} + \sum_{i=1}^n \frac{|\zeta_i|(\log \theta_i)^{\alpha+\phi_i}}{\Gamma(\alpha+\phi_i+1)} \right. \right. \\
&\quad \left. \left. + \sum_{k=1}^r \frac{|\lambda_k|(\log \mu_k)^{\alpha-\omega_k}}{\Gamma(\alpha-\omega_k+1)} + \frac{(\log T)^\alpha}{\Gamma(\alpha+1)} \right\} \right] \|\phi\| \\
&= \Omega \|\phi\| \leq \rho.
\end{aligned}$$

Hence, $\|\mathcal{F}_1 x + \mathcal{F}_2 y\| \leq \rho$, which shows that $\mathcal{F}_1 x + \mathcal{F}_2 y \in B_\rho$. By condition (13), it is easy to prove that the operator $\mathcal{F}_2$ is a contraction mapping. The operator $\mathcal{F}_1$ is continuous by the continuity of f. Moreover, $\mathcal{F}_1$ is uniformly bounded on B_ρ, since

$$\|\mathcal{F}_1 x\| \leq \frac{(\log T)^\alpha}{\Gamma(\alpha+1)} \|\phi\|.$$

Finally, we prove the compactness of the operator $\mathcal{F}_1$. For $t_1, t_2 \in [1,T], t_1 < t_2$, we have the following case:

$$\begin{aligned}
&|\mathcal{F}_1 x(t_2) - \mathcal{F}_1 x(t_1)| \\
&\leq \frac{1}{\Gamma(\alpha)} \int_1^{t_1} \left[\left(\log \frac{t_2}{z}\right)^{\alpha-1} - \left(\log \frac{t_1}{z}\right)^{\alpha-1} \right] \frac{|f(z,x(z))|}{z} dz \\
&\quad + \frac{1}{\Gamma(\alpha)} \int_{t_1}^{t_2} \left(\log \frac{t_2}{z}\right)^{\alpha-1} \frac{|f(z,x(z))|}{z} dz \\
&\leq \frac{\|\phi\|}{\Gamma(\alpha+1)} \left[2(\log t_2 - \log t_1)^\alpha + |(\log t_2)^\alpha - (\log t_1)^\alpha| \right],
\end{aligned}$$

which tends to zero independently of $x \in B_\rho$, as $t_1 \to t_2$. Thus, $\mathcal{F}_1$ is equicontinuous. By the application of the Arzelá–Ascoli theorem, we deduce that operator $\mathcal{F}_1$ is compact on B_ρ. Thus, the hypotheses of Krasnoselskiĭ's fixed point theorem [37] hold true. In consequence, there exists at least one solution for the nonlinear Hilfer–Hadamard fractional boundary value problem (1) on $[1,T]$, which completes the proof. □

Our next existence result is based on Schaefer's fixed point theorem [38].

Theorem 4. *Let $f : [1,T] \times \mathbb{R} \to \mathbb{R}$ be a continuous function satisfying the following assumption:*
(H3) *There exists a real constant $M > 0$ such that for all $t \in [1,T], u \in \mathbb{R}$,*

$$|f(t,u)| \leq M.$$

Then, there exists at least one solution for the nonlinear Hilfer–Hadamard fractional boundary value problem (1) on $[1,T]$.

Proof. We will prove that the operator $\mathcal{F}$, defined by (10), has a fixed point by using Schaefer's fixed point theorem [38]. The proof is given in two steps.

Step I. We show that the operator $\mathcal{F} : X \to X$ is completely continuous.

Let us first establish that $\mathcal{F}$ is continuous. Let $\{x_n\}$ be a sequence such that $x_n \to x$ in X. Then, for each $t \in [1, T]$, we have the following:

$$|\mathcal{F}(x_n)(t) - \mathcal{F}(x)(t)|$$
$$\leq \frac{1}{\Gamma(\alpha)} \int_1^t \left(\log \frac{t}{z}\right)^{\alpha-1} \frac{|f(z, x_n(z)) - f(z, x(z))|}{z} dz$$
$$+ \frac{(\log t)^{\gamma-1}}{|\Lambda|} \left\{ \sum_{j=1}^m \frac{|\eta_j|}{\Gamma(\alpha)} \int_1^{\xi_j} \left(\log \frac{\xi_j}{z}\right)^{\alpha-1} \frac{|f(x, x_n(z)) - f(z, x(z))|}{z} dz \right.$$
$$+ \sum_{i=1}^n \frac{|\zeta_i|}{\Gamma(\alpha + \phi_i)} \int_1^{\theta_i} \left(\log \frac{\theta_i}{z}\right)^{\alpha+\phi_i-1} \frac{|f(z, x_n(z)) - f(z, x(z))|}{z} dz$$
$$+ \sum_{k=1}^r \frac{|\lambda_k|}{\Gamma(\alpha - \omega_k)} \int_1^{\mu_k} \left(\log \frac{\mu_k}{z}\right)^{\alpha-\omega_k-1} \frac{|f(z, x_n(z)) - f(z, x(z))|}{z} dz$$
$$+ \left. \frac{1}{\Gamma(\alpha)} \int_1^T \left(\log \frac{t}{z}\right)^{\alpha-1} \frac{|f(z, x_n(z)) - f(z, x(z))|}{z} dz \right\}.$$

Taking into account the fact that f is continuous, that is, $|f(s, x_n(s)) - f(s, x(s))| \to 0$ as $x_n \to x$, we obtain from the foregoing inequality that the following is the case:

$$\|\mathcal{F}(x_n) - \mathcal{F}(x)\| \to 0 \text{ as } x_n \to x.$$

Hence, $\mathcal{F}$ is continuous.

Now we show that the operator $\mathcal{F}$ which maps bounded sets into bounded sets in X. For $R > 0$, let $B_R = \{x \in X : \|x\| \leq R\}$. Then, for $t \in [1, T]$, we have the following case:

$$|\mathcal{F}(x)(t)| \leq \frac{1}{\Gamma(\alpha)} \int_1^t \left(\log \frac{t}{z}\right)^{\alpha-1} \frac{|f(z, x(z))|}{z} dz$$
$$+ \frac{(\log t)^{\gamma-1}}{|\Lambda|} \left\{ \sum_{j=1}^m \frac{|\eta_j|}{\Gamma(\alpha)} \int_1^{\xi_j} \left(\log \frac{\xi_j}{z}\right)^{\alpha-1} \frac{|f(z, x(z))|}{z} dz \right.$$
$$+ \sum_{i=1}^n \frac{|\zeta_i|}{\Gamma(\alpha + \phi_i)} \int_1^{\theta_i} \left(\log \frac{\theta_i}{z}\right)^{\alpha+\phi_i-1} \frac{|f(z, x(z))|}{z} dz$$
$$+ \sum_{k=1}^r \frac{|\lambda_k|}{\Gamma(\alpha - \omega_k)} \int_1^{\mu_k} \left(\log \frac{\mu_k}{z}\right)^{\alpha-\omega_k-1} \frac{|f(z, x(z))|}{z} dz$$
$$+ \left. \frac{1}{\Gamma(\alpha)} \int_1^T \left(\log \frac{t}{z}\right)^{\alpha-1} \frac{|f(z, x(z))|}{z} dz \right\}$$
$$\leq \frac{(\log T)^\alpha}{\Gamma(\alpha+1)} M + \frac{(\log T)^{\gamma-1}}{|\Lambda|} \left\{ \sum_{i=j}^m \frac{|\eta_j|(\log \xi_j)^\alpha}{\Gamma(\alpha+1)} + \sum_{i=1}^n \frac{|\zeta_i|(\log \theta_i)^{\alpha+\phi_i}}{\Gamma(\alpha + \phi_i + 1)} \right.$$
$$+ \left. \sum_{k=1}^r \frac{|\lambda_k|(\log \mu_k)^{\alpha-\omega_k}}{\Gamma(\alpha - \omega_k + 1)} + \frac{(\log T)^\alpha}{\Gamma(\alpha+1)} \right\} M,$$

which, after taking the norm for $t \in [1, T]$, results in the following inequality:

$$\|\mathcal{F}(x)\| \leq \frac{(\log T)^{\alpha}}{\Gamma(\alpha+1)} M + \frac{(\log T)^{\gamma-1}}{|\Lambda|} \left\{ \sum_{i=j}^{m} \frac{|\eta_j|(\log \xi_j)^{\alpha}}{\Gamma(\alpha+1)} + \sum_{i=1}^{n} \frac{|\zeta_i|(\log \theta_i)^{\alpha+\phi_i}}{\Gamma(\alpha+\phi_i+1)} \right.$$
$$\left. + \sum_{k=1}^{r} \frac{|\lambda_k|(\log \mu_k)^{\alpha-\omega_k}}{\Gamma(\alpha-\omega_k+1)} + \frac{(\log T)^{\alpha}}{\Gamma(\alpha+1)} \right\} M.$$

Next, we show that bounded sets are mapped into equicontinuous sets by $\mathcal{F}$. For $t_1, t_2 \in [1, T]$, $t_1 < t_2$ and $u \in B_R$, we obtain the following:

$$|\mathcal{F}(x)(t_2) - \mathcal{F}(x)(t_1)|$$
$$\leq \frac{1}{\Gamma(\alpha)} \int_1^{t_1} \left[\left(\log \frac{t_2}{z}\right)^{\alpha-1} - \left(\log \frac{t_1}{z}\right)^{\alpha-1} \right] \frac{|f(z, x(z))|}{z} dz$$
$$+ \frac{1}{\Gamma(\alpha)} \int_{t_1}^{t_2} \left(\log \frac{t_2}{z}\right)^{\alpha-1} \frac{|f(z, x(z))|}{z} dz$$
$$+ \frac{|(\log t_2)^{\gamma-1} - (\log t_1)^{\gamma-1}|}{|\Lambda|} \left\{ \sum_{j=1}^{m} \frac{|\eta_j|}{\Gamma(\alpha)} \int_1^{\xi_j} \left(\log \frac{\xi_j}{z}\right)^{\alpha-1} \frac{|f(z, x(z))|}{z} dz \right.$$
$$+ \sum_{i=1}^{n} \frac{|\zeta_i|}{\Gamma(\alpha+\phi_i)} \int_1^{\theta_i} \left(\log \frac{\theta_i}{z}\right)^{\alpha+\phi_i-1} \frac{|f(z, x(z))|}{z} dz$$
$$+ \sum_{k=1}^{r} \frac{|\lambda_k|}{\Gamma(\alpha-\omega_k)} \int_1^{\mu_k} \left(\log \frac{\mu_k}{z}\right)^{\alpha-\omega_k-1} \frac{|f(z, x(z))|}{z} dz$$
$$\left. + \frac{1}{\Gamma(\alpha)} \int_1^{T} \left(\log \frac{t}{z}\right)^{\alpha-1} \frac{|f(z, x(z))|}{z} dz \right\}$$
$$\leq \frac{M}{\Gamma(\alpha+1)} \left[2(\log t_2 - \log t_1)^{\alpha} + |(\log t_2)^{\alpha} - (\log t_1)^{\alpha}| \right]$$
$$+ \frac{|(\log t_2)^{\gamma-1} - (\log t_1)^{\gamma-1}|}{|\Lambda|} \left\{ \sum_{i=j}^{m} \frac{|\eta_j|(\log \xi_j)^{\alpha}}{\Gamma(\alpha+1)} + \sum_{i=1}^{n} \frac{|\zeta_i|(\log \theta_i)^{\alpha+\phi_i}}{\Gamma(\alpha+\phi_i+1)} \right.$$
$$\left. + \sum_{k=1}^{r} \frac{|\lambda_k|(\log \mu_k)^{\alpha-\omega_k}}{\Gamma(\alpha-\omega_k+1)} + \frac{(\log T)^{\alpha}}{\Gamma(\alpha+1)} \right\} M,$$

which tends to zero independently of $x \in B_R$, as $t_1 \to t_2$. Thus, the operator $\mathcal{F}: X \to X$ is completely continuous by applying the Arzelá–Ascoli theorem.

Step II.: We show that the set $\mathcal{E} = \{x \in X \mid x = \nu \mathcal{F}(x), 0 \leq \nu \leq 1\}$ is bounded. Let $x \in \mathcal{E}$, then $x = \nu \mathcal{F}(x)$. For any $t \in [1, T]$, we have $x(t) = \nu \mathcal{F}(x)(t)$. Then, in view of the hypothesis (H_3), as in Step I, we obtain

$$|x(t)| \leq \frac{(\log T)^{\alpha}}{\Gamma(\alpha+1)} M + \frac{(\log T)^{\gamma-1}}{|\Lambda|} \left\{ \sum_{i=j}^{m} \frac{|\eta_j|(\log \xi_j)^{\alpha}}{\Gamma(\alpha+1)} + \sum_{i=1}^{n} \frac{|\zeta_i|(\log \theta_i)^{\alpha+\phi_i}}{\Gamma(\alpha+\phi_i+1)} \right.$$
$$\left. + \sum_{k=1}^{r} \frac{|\lambda_k|(\log \mu_k)^{\alpha-\omega_k}}{\Gamma(\alpha-\omega_k+1)} + \frac{(\log T)^{\alpha}}{\Gamma(\alpha+1)} \right\} M.$$

Thus, the following is the case:

$$\|x\| \leq \frac{(\log T)^\alpha}{\Gamma(\alpha+1)}M + \frac{(\log T)^{\gamma-1}}{|\Lambda|}\left\{\sum_{i=j}^{m}\frac{|\eta_j|(\log \zeta_j)^\alpha}{\Gamma(\alpha+1)} + \sum_{i=1}^{n}\frac{|\zeta_i|(\log \theta_i)^{\alpha+\phi_i}}{\Gamma(\alpha+\phi_i+1)}\right.$$

$$\left. + \sum_{k=1}^{r}\frac{|\lambda_k|(\log \mu_k)^{\alpha-\omega_k}}{\Gamma(\alpha-\omega_k+1)} + \frac{(\log T)^\alpha}{\Gamma(\alpha+1)}\right\}M,$$

which shows that the set $\mathcal{E}$ is bounded. Thus, it follows by Schaefer's fixed point theorem [38] that the operator $\mathcal{F}$ has at least one fixed point. Therefore, there exists at least one solution for the nonlinear Hilfer–Hadamard fractional boundary value problem (1) on $[1, T]$. This completes the proof. □

We apply the Leray–Schauder nonlinear alternative [39] to prove our last existence result.

Theorem 5. *Let $f \in C([1, T] \times \mathbb{R}, \mathbb{R})$. In addition, it is assumed that the following conditions are satisfied:*

(H_4) *There exist $p \in C([1, T], \mathbb{R}^+)$ and a continuous nondecreasing function $\psi : \mathbb{R}^+ \to \mathbb{R}^+$ such that $|f(t, u)| \leq p(t)\psi(\|u\|)$ for each $(t, u) \in [1, T] \times \mathbb{R}$;*

(H_5) *There exists a constant $K > 0$ such that*

$$\frac{K}{\Omega\|p\|\psi(K)} > 1,$$

where Ω is defined by (11).

Then, the nonlinear Hilfer–Hadamard fractional boundary value problem (1) has at least one solution on $[1, T]$.

Proof. As argued in Theorem 4, one can obtain that the operator $\mathcal{F}$ is completely continuous. Next, we establish that we can find an open set $U \subseteq C([1, T], \mathbb{R})$ with $x \neq \mu \mathcal{F}(x)$ for $\mu \in (0, 1)$ and $x \in \partial U$.

Let $x \in C([1, T], \mathbb{R})$ be such that $x = \mu \mathcal{F}(x)$ for some $0 < \mu < 1$. Then, for each $t \in [1, T]$, we have the following case:

$$|x(t)| \leq \frac{1}{\Gamma(\alpha)}\int_1^t \left(\log \frac{t}{z}\right)^{\alpha-1}\frac{|f(z, x(z))|}{z}dz$$

$$+ \frac{(\log t)^{\gamma-1}}{|\Lambda|}\left\{\sum_{j=1}^{m}\frac{|\eta_j|}{\Gamma(\alpha)}\int_1^{\zeta_j}\left(\log \frac{\zeta_j}{z}\right)^{\alpha-1}\frac{|f(z, x(z))|}{z}dz\right.$$

$$+ \sum_{i=1}^{n}\frac{|\zeta_i|}{\Gamma(\alpha+\phi_i)}\int_1^{\theta_i}\left(\log \frac{\theta_i}{z}\right)^{\alpha+\phi_i-1}\frac{|f(z, x(z))|}{z}dz$$

$$+ \sum_{k=1}^{r}\frac{|\lambda_k|}{\Gamma(\alpha-\omega_k)}\int_1^{\mu_k}\left(\log \frac{\mu_k}{z}\right)^{\alpha-\omega_k-1}\frac{|f(z, x(z))|}{z}dz$$

$$\left. + \frac{1}{\Gamma(\alpha)}\int_1^T\left(\log \frac{t}{z}\right)^{\alpha-1}\frac{|f(z, x(z))|}{z}dz\right\}$$

$$\leq \left[\frac{(\log T)^\alpha}{\Gamma(\alpha+1)} + \frac{(\log T)^{\gamma-1}}{|\Lambda|}\left\{\sum_{i=j}^{m}\frac{|\eta_j|(\log \zeta_j)^\alpha}{\Gamma(\alpha+1)} + \sum_{i=1}^{n}\frac{|\zeta_i|(\log \theta_i)^{\alpha+\phi_i}}{\Gamma(\alpha+\phi_i+1)}\right.\right.$$

$$\left.\left. + \sum_{k=1}^{r}\frac{|\lambda_k|(\log \mu_k)^{\alpha-\omega_k}}{\Gamma(\alpha-\omega_k+1)} + \frac{(\log T)^\alpha}{\Gamma(\alpha+1)}\right\}\right]\|p\|\psi(\|x\|).$$

Consequently, we obtain
$$\frac{\|x\|}{\Omega \|p\| \psi(\|x\|)} \leq 1.$$

In view of (H_5), there is no solution x such that $\|x\| \neq K$. Let us set the following:
$$U = \{x \in C([1,T], \mathbb{R}) : \|x\| < K\}.$$

The operator $\mathcal{F} : \overline{U} \to C([1,T], \mathbb{R})$ is continuous and completely continuous. Note that there is no $u \in \partial U$ such that $x = \mu \mathcal{F}(x)$ for some $\mu \in (0,1)$, by the choice of U. Thus, it follows by the Leray–Schauder nonlinear alternative [39] that $\mathcal{F}$ has a fixed point $x \in \overline{U}$ which is a solution of the nonlinear Hilfer–Hadamard fractional boundary value problem (1). This ends the proof. □

3.3. Examples

Consider the following Hilfer–Hadamard fractional boundary value problem:
$$\begin{cases} {}^{HH}D_1^{\alpha,\beta}x(t) = f(t,x(t)), \quad t \in [1,T], \\ x(1) = 0, \quad x(T) = \sum_{j=1}^{m} \eta_j x(\xi_j) + \sum_{i=1}^{n} \zeta_i \, {}^{H}I_1^{\phi_i} x(\theta_i) + \sum_{k=1}^{r} \lambda_k \, {}^{H}D_1^{\omega_k} x(\mu_k), \end{cases} \quad (14)$$

with $\alpha = 5/3, \beta = 3/4, T = 5, m = 4, n = 3, r = 2, \eta_1 = 1/15, \eta_2 = 1/10, \eta_3 = 2/15, \eta_4 = 1/6, \xi_1 = 5/4, \xi_2 = 3/2, \xi_3 = 7/4, \xi_4 = 2, \zeta_i = 1/18, \zeta_2 = 1/9, \zeta_3 = 1/6, \phi_1 = 1/2, \phi_2 = 1, \phi_3 = 3/2, \theta_1 = 5/2, \theta_2 = 3, \theta_3 = 7/2, \lambda_1 = 1/28, \lambda_2 = 1/14, \omega_1 = 1/4, \omega_2 = 2/3, \mu_1 = 4, \mu_2 = 9/2$ and $f(t, x(t))$ to be fixed later. Using the given data, it is found that $\gamma = 23/12, \Lambda \approx 0.662923$ (Λ is given by (4)), $\Omega \approx 39.388095$ (Ω is given by (11)).

(a). For illustrating Theorem 2, we take
$$f(t,x) = \frac{ae^{-(t-1)^2}}{15} \arctan x + \frac{\sin t + 1}{\sqrt{t^3 + 1}}, \quad t \in [1,5], \quad (15)$$

where a is a positive real constant. Obviously, the nonlinear function $f(t,x)$ satisfies the assumption (H_1) with $l = a/15$ and the condition $l\Omega < 1$ holds for $a < 0.3808257$. Thus, the hypothesis of Theorem 2 is satisfied; hence, the problem (14) with $f(t,x)$ given by (15) has a unique solution on $[1,5]$.

(b). In order to illustrate Theorem 3, we take the following function:
$$f(t,x) = \frac{|x|}{[(t-1)^2 + 100](1 + |x|)} + \cos t, \quad t \in [1,5]. \quad (16)$$

It is easy to verify that the nonlinear function $f(t,x)$ satisfies the assumption (H_1) with $l = 1/100$ and that
$$\frac{(\log T)^{\gamma-1}}{|\Lambda|} \left\{ \sum_{i=j}^{m} \frac{|\eta_j|(\log \xi_j)^{\alpha}}{\Gamma(\alpha+1)} + \sum_{i=1}^{n} \frac{|\zeta_i|(\log \theta_i)^{\alpha+\phi_i}}{\Gamma(\alpha+\phi_i+1)} \right.$$
$$\left. + \sum_{k=1}^{r} \frac{|\lambda_k|(\log \mu_k)^{\alpha-\omega_k}}{\Gamma(\alpha-\omega_k+1)} + \frac{(\log T)^{\alpha}}{\Gamma(\alpha+1)} \right\} l \approx 0.379190 < 1.$$

Thus, the assumptions of Theorem 3 hold true. Therefore, by the conclusion of Theorem 3, there exists at least one solution for the problem (14) with $f(t,x)$ given by (16) on $[1,5]$.

(c). Now, we illustrate Theorem 4 with the aid of the following nonlinear function:
$$f(t,x) = \frac{e^{1-t}}{t^2+4} \cos^4\left(\frac{3+2|x|}{1+|x|^2}\right) + \frac{1}{10}\sqrt{(t^2+11)}, \quad t \in [1,5]. \quad (17)$$

It is easy to obtain that $|f(t,x)| \leq 4/5$. Thus, by the conclusion of Theorem 4, the problem (14) with $f(t,x)$ given by (17) has at least one solutions on $[1,5]$.

(d). Finally, we demonstrate the application of Theorem 5 by considering the nonlinear function:

$$f(t,x) = \frac{2}{[(t-1)^2 + 100]}\left(|x|\cos(5+2|x|) + \frac{1}{2}\right), \ t \in [1,5]. \tag{18}$$

Note that $|f(t,x)| \leq p(t)\psi(\|x\|)$, where $p(t) = 2[(t-1)^2 + 100]^{-1}$ ($\|p\| = 1/50$) and $\psi(\|x\|) = \|x\| + 1/2$. By the condition (H_5), we find that $K > 1.85584$. Thus, all the assumptions of Theorem 5 hold true; hence, its conclusion ensures the existence of at least one solution for the problem (14) with $f(t,x)$ given by (18) on $[1,5]$.

4. Multi-Valued Case

This section is devoted to the study of the multi-valued case of the boundary value problem (1) given as follows:

$$\begin{cases} {}^{HH}D_1^{\alpha,\beta}x(t) \in F(t,x(t)), \quad t \in [1,T], \\ x(1) = 0, \quad x(T) = \sum_{j=1}^m \eta_j x(\xi_j) + \sum_{i=1}^n \zeta_i \, {}^HI_1^{\phi_i}x(\theta_i) + \sum_{k=1}^r \lambda_k \, {}_HD_1^{\omega_k}x(\mu_k), \end{cases} \tag{19}$$

where the symbols are the same as defined in problem (1) and $F : J \times \mathbb{R} \to \mathcal{P}(\mathbb{R})$ is a multi-valued map. By $\mathcal{P}(\mathbb{R})$, we denote the family of all nonempty subsets of $\mathbb{R}$.

Next, we define

$$\mathcal{P}_p = \{Y \in \mathcal{P}(X) : Y \neq \emptyset \text{ and has the property } p\},$$

where $(X, \|\cdot\|)$ is a normed space. Thus, $\mathcal{P}_{cl}, \mathcal{P}_b, \mathcal{P}_{cp}$ and $\mathcal{P}_{cp,c}$ respectively, denote the classes of all closed, bounded, compact and compact and convex sets in X.

Define the set of selections of F for each $\omega \in C([1,T],\mathbb{R})$ as

$$S_{F,\omega} := \{z \in L^1([1,T],\mathbb{R}) : z(t) \in F(t,\omega(t)) \text{ for a.e. } t \in [1,T]\}.$$

4.1. Existence Results for the Problem (19)

Let us first define the solution for Hilfer–Hadamard inclusions fractional boundary value problem (19).

Definition 4. *A function $x \in C([1,T],\mathbb{R})$ is called a solution of the multi-valued problem (19) if we can find a function $v \in L^1([1,T],\mathbb{R})$ with $v(t) \in F(t,x)$ almost everywhere on $[1,T]$ such that*

$$\begin{aligned}x(t) =& \frac{1}{\Gamma(\alpha)}\int_1^t \left(\log\frac{t}{z}\right)^{\alpha-1}\frac{v(z)}{z}dz + \frac{(\log t)^{\gamma-1}}{\Lambda}\left\{\sum_{j=1}^m \frac{\eta_j}{\Gamma(\alpha)}\int_1^{\xi_j}\left(\log\frac{\xi_j}{z}\right)^{\alpha-1}\frac{v(z)}{z}dz \right. \\ &+ \sum_{i=1}^n \frac{\zeta_i}{\Gamma(\alpha+\phi_i)}\int_1^{\theta_i}\left(\log\frac{\theta_i}{z}\right)^{\alpha+\phi_i-1}\frac{v(z)}{z}dz \\ &+ \sum_{k=1}^r \frac{\lambda_k}{\Gamma(\alpha-\omega_k)}\int_1^{\mu_k}\left(\log\frac{\mu_k}{z}\right)^{\alpha-\omega_k-1}\frac{v(z)}{z}dz \\ &\left. - \frac{1}{\Gamma(\alpha)}\int_1^T\left(\log\frac{T}{z}\right)^{\alpha-1}\frac{v(z)}{z}dz\right\},\end{aligned}$$

where Λ is given by (4).

4.1.1. Case 1: Convex-Valued Multifunctions

In the first theorem, dealing with convex-valued multifunctions, we assume that the multifunction F is L^1-Carathéodory and apply the nonlinear alternative for Kakutani maps [39] and closed graph operator theorem [40] to prove it.

Theorem 6. *Assume that the following conditions are satisfied:*
(A_1) *The multifunction $F : [1, T] \times \mathbb{R} \to \mathcal{P}_{cp,c}(\mathbb{R})$ is L^1-Carathéodory;*
(A_2) *There exist a nondecreasing function $\chi \in C([0, \infty)(0, \infty))$ and a continuous function $q : [1, T] \to \mathbb{R}^+$ such that*
$$\|F(t, \omega)\|_\mathcal{P} := \sup\{|z| : z \in F(t, \omega)\} \le q(t)\chi(\|\omega\|) \text{ for each } (t, \omega) \in [1, T] \times \mathbb{R};$$
(A_3) *There exists $M > 0$ satisfying the following inequality:*

$$\frac{M}{\chi(M)\|q\|\Omega} > 1,$$

where Ω is given by (11).
Then, there exists at least one solution for the inclusions problem (19) on $[1, T]$.

Proof. Associated with the Hilfer–Hadamard fractional inclusions boundary value problem (19), we introduce a multi-valued operator, $N : C([1, T], \mathbb{R}) \to \mathcal{P}(C([1, T], \mathbb{R}))$, as follows:

$$N(x) = \left\{ h \in C([1, T], \mathbb{R}) : \\ h(t) = \begin{cases} \frac{1}{\Gamma(\alpha)} \int_1^t \left(\log \frac{t}{z}\right)^{\alpha-1} \frac{v(z)}{z} dz \\ + \frac{(\log t)^{\gamma-1}}{\Lambda} \left\{ \sum_{j=1}^m \frac{\eta_j}{\Gamma(\alpha)} \int_1^{\zeta_j} \left(\log \frac{\zeta_j}{z}\right)^{\alpha-1} \frac{v(z)}{z} dz \right. \\ + \sum_{i=1}^n \frac{\zeta_i}{\Gamma(\alpha + \phi_i)} \int_1^{\theta_i} \left(\log \frac{\theta_i}{z}\right)^{\alpha+\phi_i-1} \frac{v(z)}{z} dz \\ + \sum_{k=1}^r \frac{\lambda_k}{\Gamma(\alpha - \omega_k)} \int_1^{\mu_k} \left(\log \frac{\mu_k}{z}\right)^{\alpha-\omega_k-1} \frac{v(z)}{z} dz \\ \left. - \frac{1}{\Gamma(\alpha)} \int_1^T \left(\log \frac{T}{z}\right)^{\alpha-1} \frac{v(z)}{z} dz \right\}, \ v \in S_{F,x}. \end{cases} \right.$$

In what follows, we will prove in several steps that the operator N satisfies the hypotheses of the Leray–Schauder nonlinear alternative for Kakutani maps [39].

Step 1. *N is bounded on bounded sets of $C([1, T], \mathbb{R})$.*

For a fixed $r > 0$, let $B_r = \{x \in C[1, T], \mathbb{R}) : \|x\| \le r\}$ be a bounded set in $C([1, T], \mathbb{R})$. For each $h \in N(x)$ and $x \in B_r$, there exists $v \in S_{F,x}$ such that

$$\begin{aligned} h(t) &= \frac{1}{\Gamma(\alpha)} \int_1^t \left(\log \frac{t}{z}\right)^{\alpha-1} \frac{v(z)}{z} dz + \frac{(\log t)^{\gamma-1}}{\Lambda} \left\{ \sum_{j=1}^m \frac{\eta_j}{\Gamma(\alpha)} \int_1^{\zeta_j} \left(\log \frac{\zeta_j}{z}\right)^{\alpha-1} \frac{v(z)}{z} dz \right. \\ &+ \sum_{i=1}^n \frac{\zeta_i}{\Gamma(\alpha+\phi_i)} \int_1^{\theta_i} \left(\log \frac{\theta_i}{z}\right)^{\alpha+\phi_i-1} \frac{v(z)}{z} dz \\ &+ \sum_{k=1}^r \frac{\lambda_k}{\Gamma(\alpha-\omega_k)} \int_1^{\mu_k} \left(\log \frac{\mu_k}{z}\right)^{\alpha-\omega_k-1} \frac{v(z)}{z} dz \\ &\left. - \frac{1}{\Gamma(\alpha)} \int_1^T \left(\log \frac{T}{z}\right)^{\alpha-1} \frac{v(z)}{z} dz \right\}, \ t \in [1, T]. \end{aligned}$$

For $t \in [1, T]$, using the assumption (A_2), we obtain the following inequality:

$$|h(t)| \leq \frac{1}{\Gamma(\alpha)} \int_1^t \left(\log \frac{t}{z}\right)^{\alpha-1} \frac{|v(z)|}{z} dz$$

$$+ \frac{(\log t)^{\gamma-1}}{\Lambda} \left\{ \sum_{j=1}^m \frac{|\eta_j|}{\Gamma(\alpha)} \int_1^{\xi_j} \left(\log \frac{\xi_j}{z}\right)^{\alpha-1} \frac{|v(z)|}{z} dz \right.$$

$$+ \sum_{i=1}^n \frac{|\zeta_i|}{\Gamma(\alpha+\phi_i)} \int_1^{\theta_i} \left(\log \frac{\theta_i}{z}\right)^{\alpha+\phi_i-1} \frac{|v(z)|}{z} dz$$

$$+ \sum_{k=1}^r \frac{|\lambda_k|}{\Gamma(\alpha-\omega_k)} \int_1^{\mu_k} \left(\log \frac{\mu_k}{z}\right)^{\alpha-\omega_k-1} \frac{|v(z)|}{z} dz$$

$$+ \frac{1}{\Gamma(\alpha)} \int_1^T \left(\log \frac{t}{z}\right)^{\alpha-1} \frac{|v(z)|}{z} dz \right\}$$

$$\leq \left[\frac{(\log T)^\alpha}{\Gamma(\alpha+1)} + \frac{(\log T)^{\gamma-1}}{|\Lambda|} \left\{ \sum_{i=j}^m \frac{|\eta_j|(\log \xi_j)^\alpha}{\Gamma(\alpha+1)} + \sum_{i=1}^n \frac{|\zeta_i|(\log \theta_i)^{\alpha+\phi_i}}{\Gamma(\alpha+\phi_i+1)} \right. \right.$$

$$+ \sum_{k=1}^r \frac{|\lambda_k|(\log \mu_k)^{\alpha-\omega_k}}{\Gamma(\alpha-\omega_k+1)} + \frac{(\log T)^\alpha}{\Gamma(\alpha+1)} \right\} \right] \|p\| \chi(\|x\|),$$

which yields

$$\|h\| \leq \|p\| \chi(r) \Omega.$$

Step 2. *Bounded sets are mapped by N into equicontinuous sets of $C([1, T], \mathbb{R})$.* Let $x \in B_r$ and $h \in N(x)$. Then, there exists $v \in S_{F,x}$ such that

$$h(t) = \frac{1}{\Gamma(\alpha)} \int_1^t \left(\log \frac{t}{z}\right)^{\alpha-1} \frac{v(z)}{z} dz + \frac{(\log t)^{\gamma-1}}{\Lambda} \left\{ \sum_{j=1}^m \frac{\eta_j}{\Gamma(\alpha)} \int_1^{\xi_j} \left(\log \frac{\xi_j}{z}\right)^{\alpha-1} \frac{v(z)}{z} dz \right.$$

$$+ \sum_{i=1}^n \frac{\zeta_i}{\Gamma(\alpha+\phi_i)} \int_1^{\theta_i} \left(\log \frac{\theta_i}{z}\right)^{\alpha+\phi_i-1} \frac{v(z)}{z} dz$$

$$+ \sum_{k=1}^r \frac{\lambda_k}{\Gamma(\alpha-\omega_k)} \int_1^{\mu_k} \left(\log \frac{\mu_k}{z}\right)^{\alpha-\omega_k-1} \frac{v(z)}{z} dz$$

$$- \frac{1}{\Gamma(\alpha)} \int_1^T \left(\log \frac{T}{z}\right)^{\alpha-1} \frac{v(z)}{z} dz \right\}, \quad t \in [1, T].$$

Let $t_1, t_2 \in [1, T]$, $t_1 < t_2$. Then, we obtain

$$|\mathcal{F}(x)(t_2) - \mathcal{F}(x)(t_1)|$$

$$\leq \frac{1}{\Gamma(\alpha)} \int_1^{t_1} \left[\left(\log \frac{t_2}{z}\right)^{\alpha-1} - \left(\log \frac{t_1}{z}\right)^{\alpha-1} \right] \frac{|f(z, x(z))|}{z} dz$$

$$+ \frac{1}{\Gamma(\alpha)} \int_{t_1}^{t_2} \left(\log \frac{t_2}{z}\right)^{\alpha-1} \frac{|v(z)|}{z} dz$$

$$+ \frac{|(\log t_2)^{\gamma-1} - (\log t_1)^{\gamma-1}|}{|\Lambda|} \left\{ \sum_{j=1}^m \frac{|\eta_j|}{\Gamma(\alpha)} \int_1^{\xi_j} \left(\log \frac{\xi_j}{z}\right)^{\alpha-1} \frac{|v(z)|}{z} dz \right.$$

$$+ \sum_{i=1}^n \frac{|\zeta_i|}{\Gamma(\alpha+\phi_i)} \int_1^{\theta_i} \left(\log \frac{\theta_i}{z}\right)^{\alpha+\phi_i-1} \frac{|v(z)|}{z} dz$$

$$+ \sum_{k=1}^r \frac{|\lambda_k|}{\Gamma(\alpha-\omega_k)} \int_1^{\mu_k} \left(\log \frac{\mu_k}{z}\right)^{\alpha-\omega_k-1} \frac{|v(z)|}{z} dz$$

$$+ \frac{1}{\Gamma(\alpha)} \int_1^T \left(\log \frac{t}{z}\right)^{\alpha-1} \frac{|v(z)|}{z} dz \right\}$$

$$\leq \frac{\|p\| \chi(r)}{\Gamma(\alpha+1)} \left[2(\log t_2 - \log t_1)^\alpha + |(\log t_2)^\alpha - (\log t_1)^\alpha| \right]$$

$$+ \frac{|(\log t_2)^{\gamma-1} - (\log t_1)^{\gamma-1}|}{|\Lambda|} \left\{ \sum_{i=j}^m \frac{|\eta_j|(\log \xi_j)^\alpha}{\Gamma(\alpha+1)} + \sum_{i=1}^n \frac{|\zeta_i|(\log \theta_i)^{\alpha+\phi_i}}{\Gamma(\alpha+\phi_i+1)} \right.$$

$$+ \sum_{k=1}^r \frac{|\lambda_k|(\log \mu_k)^{\alpha-\omega_k}}{\Gamma(\alpha-\omega_k+1)} + \frac{(\log T)^\alpha}{\Gamma(\alpha+1)} \right\} \|p\| \chi(r) \to 0,$$

as $t_1 \to t_2$ independently of $x \in B_r$. Hence, $N : C([1,T], \mathbb{R}) \to \mathcal{P}(C[1,T], \mathbb{R}))$ is completely continuous by Arzelá–Ascoli theorem.

Step 3. *For each $x \in C([1,T], \mathbb{R})$, $N(x)$ is convex.*

For $h_1, h_2 \in N(x)$, there exist $v_1, v_2 \in S_{F,x}$ such that, for each $t \in [1,T]$, we have

$$\begin{aligned}
h_\varpi(t) &= \frac{1}{\Gamma(\alpha)} \int_1^t \left(\log \frac{t}{z}\right)^{\alpha-1} \frac{v_\varpi(z)}{z} dz \\
&+ \frac{(\log t)^{\gamma-1}}{\Lambda} \Bigg\{ \sum_{j=1}^m \frac{\eta_j}{\Gamma(\alpha)} \int_1^{\xi_j} \left(\log \frac{\xi_j}{z}\right)^{\alpha-1} \frac{v_\varpi(z)}{z} dz \\
&+ \sum_{i=1}^n \frac{\zeta_i}{\Gamma(\alpha+\phi_i)} \int_1^{\theta_i} \left(\log \frac{\theta_i}{z}\right)^{\alpha+\phi_i-1} \frac{v_\varpi(z)}{z} dz \\
&+ \sum_{k=1}^r \frac{\lambda_k}{\Gamma(\alpha-\omega_k)} \int_1^{\mu_k} \left(\log \frac{\mu_k}{z}\right)^{\alpha-\omega_k-1} \frac{v_\varpi(z)}{z} dz \\
&- \frac{1}{\Gamma(\alpha)} \int_1^T \left(\log \frac{T}{z}\right)^{\alpha-1} \frac{v_\varpi(z)}{z} dz \Bigg\}, \quad \varpi = 1,2.
\end{aligned}$$

Let $0 \leq \sigma \leq 1$. Then, for each $t \in [1,T]$, we have the following:

$$\begin{aligned}
&[\sigma h_1 + (1-\sigma)h_2](t) \\
&= \frac{1}{\Gamma(\alpha)} \int_1^t \left(\log \frac{t}{z}\right)^{\alpha-1} \frac{[\sigma v_1(z) + (1-\sigma)v_2(z)]}{z} dz \\
&+ \frac{(\log t)^{\gamma-1}}{\Lambda} \Bigg\{ \sum_{j=1}^m \frac{\eta_j}{\Gamma(\alpha)} \int_1^{\xi_j} \left(\log \frac{\xi_j}{z}\right)^{\alpha-1} \frac{[\sigma v_1(z) + (1-\sigma)v_2(z)]}{z} dz \\
&+ \sum_{i=1}^n \frac{\zeta_i}{\Gamma(\alpha+\phi_i)} \int_1^{\sigma_i} \left(\log \frac{\sigma_i}{z}\right)^{\alpha+\phi_i-1} \frac{[\sigma v_1(z) + (1-\sigma)v_2(z)]}{z} dz \\
&+ \sum_{k=1}^r \frac{\lambda_k}{\Gamma(\alpha-\omega_k)} \int_1^{\mu_k} \left(\log \frac{\mu_k}{z}\right)^{\alpha-\omega_k-1} \frac{[\sigma v_1(z) + (1-\sigma)v_2(z)]}{z} dz \\
&- \frac{1}{\Gamma(\alpha)} \int_1^T \left(\log \frac{T}{z}\right)^{\alpha-1} \frac{[\sigma v_1(z) + (1-\sigma)v_2(z)]}{z} dz \Bigg\}.
\end{aligned}$$

Since $S_{F,x}$ is convex (F has convex values), $\sigma h_1 + (1-\sigma)h_2 \in N(x)$. In consequence, N is convex-valued.

Next, it will be shown that the operator N is upper semicontinuous. By using the fact that a completely continuous operator, which has a closed graph, is upper semicontinuous ([41] (Proposition 1.2)), it is enough to prove that the operator N has a closed graph. This will be established in the following step.

Step 4. *The graph of N is closed.*

Let $x_n \to x_*$, $h_n \in N(x_n)$ and $h_n \to h_*$. Then, we show that $h_* \in N(x_*)$. Observe that $h_n \in N(x_n)$ implies that there exists $v_n \in S_{F,x_n}$ such that, for each $t \in [1,T]$, we have

$$h_n(t) = \frac{1}{\Gamma(\alpha)} \int_1^t \left(\log \frac{t}{z}\right)^{\alpha-1} \frac{v_n(z)}{z} dz$$
$$+ \frac{(\log t)^{\gamma-1}}{\Lambda} \left\{ \sum_{j=1}^m \frac{\eta_j}{\Gamma(\alpha)} \int_1^{\xi_j} \left(\log \frac{\xi_j}{z}\right)^{\alpha-1} \frac{v_n(z)}{z} dz \right.$$
$$+ \sum_{i=1}^n \frac{\zeta_i}{\Gamma(\alpha+\phi_i)} \int_1^{\theta_i} \left(\log \frac{\theta_i}{z}\right)^{\alpha+\phi_i-1} \frac{v_n(z)}{z} dz$$
$$+ \sum_{k=1}^r \frac{\lambda_k}{\Gamma(\alpha-\omega_k)} \int_1^{\mu_k} \left(\log \frac{\mu_k}{z}\right)^{\alpha-\omega_k-1} \frac{v_n(z)}{z} dz$$
$$\left. - \frac{1}{\Gamma(\alpha)} \int_1^T \left(\log \frac{T}{z}\right)^{\alpha-1} \frac{v_n(z)}{z} dz \right\}.$$

For each $t \in [1, T]$, we must have $v_* \in S_{F,x_*}$ and the following expression:

$$h_*(t) = \frac{1}{\Gamma(\alpha)} \int_1^t \left(\log \frac{t}{z}\right)^{\alpha-1} \frac{v_*(z)}{z} dz$$
$$+ \frac{(\log t)^{\gamma-1}}{\Lambda} \left\{ \sum_{j=1}^m \frac{\eta_j}{\Gamma(\alpha)} \int_1^{\xi_j} \left(\log \frac{\xi_j}{z}\right)^{\alpha-1} \frac{v_*(z)}{z} dz \right.$$
$$+ \sum_{i=1}^n \frac{\zeta_i}{\Gamma(\alpha+\phi_i)} \int_1^{\theta_i} \left(\log \frac{\theta_i}{z}\right)^{\alpha+\phi_i-1} \frac{v_*(z)}{z} dz$$
$$+ \sum_{k=1}^r \frac{\lambda_k}{\Gamma(\alpha-\omega_k)} \int_1^{\mu_k} \left(\log \frac{\mu_k}{z}\right)^{\alpha-\omega_k-1} \frac{v_*(z)}{z} dz$$
$$\left. - \frac{1}{\Gamma(\alpha)} \int_1^T \left(\log \frac{T}{z}\right)^{\alpha-1} \frac{v_*(z)}{z} dz \right\}.$$

Consider the continuous linear operator $\Phi : L^1([1, T], \mathbb{R}) \to C([1, T], \mathbb{R})$ defined as follows:

$$v \to \Phi(v)(t) = \frac{1}{\Gamma(\alpha)} \int_1^t \left(\log \frac{t}{z}\right)^{\alpha-1} \frac{v(z)}{z} dz$$
$$+ \frac{(\log t)^{\gamma-1}}{\Lambda} \left\{ \sum_{j=1}^m \frac{\eta_j}{\Gamma(\alpha)} \int_1^{\xi_j} \left(\log \frac{\xi_j}{z}\right)^{\alpha-1} \frac{v(z)}{z} dz \right.$$
$$+ \sum_{i=1}^n \frac{\zeta_i}{\Gamma(\alpha+\phi_i)} \int_1^{\theta_i} \left(\log \frac{\theta_i}{z}\right)^{\alpha+\phi_i-1} \frac{v(z)}{z} dz$$
$$+ \sum_{k=1}^r \frac{\lambda_k}{\Gamma(\alpha-\omega_k)} \int_1^{\mu_k} \left(\log \frac{\mu_k}{z}\right)^{\alpha-\omega_k-1} \frac{v(z)}{z} dz$$
$$\left. - \frac{1}{\Gamma(\alpha)} \int_1^T \left(\log \frac{T}{z}\right)^{\alpha-1} \frac{v(z)}{z} dz \right\}.$$

Clearly $\|h_n - h_*\| \to 0$ as $n \to \infty$, and consequently, by the closed graph operator theorem [40], $\Phi \circ S_{F,x}$ is a closed graph operator. Moreover, we have $h_n \in \Phi(S_{F,x_n})$ and the following:

$$h_*(t) = \frac{1}{\Gamma(\alpha)} \int_1^t \left(\log \frac{t}{z}\right)^{\alpha-1} \frac{v_*(z)}{z} dz$$
$$+ \frac{(\log t)^{\gamma-1}}{\Lambda} \left\{ \sum_{j=1}^m \frac{\eta_j}{\Gamma(\alpha)} \int_1^{\xi_j} \left(\log \frac{\xi_j}{z}\right)^{\alpha-1} \frac{v_*(z)}{z} dz \right.$$
$$+ \sum_{i=1}^n \frac{\zeta_i}{\Gamma(\alpha + \phi_i)} \int_1^{\theta_i} \left(\log \frac{\theta_i}{z}\right)^{\alpha+\phi_i-1} \frac{v_*(z)}{z} dz$$
$$+ \sum_{k=1}^r \frac{\lambda_k}{\Gamma(\alpha - \omega_k)} \int_1^{\mu_k} \left(\log \frac{\mu_k}{z}\right)^{\alpha-\omega_k-1} \frac{v_*(z)}{z} dz$$
$$\left. - \frac{1}{\Gamma(\alpha)} \int_1^T \left(\log \frac{T}{z}\right)^{\alpha-1} \frac{v_*(z)}{z} dz \right\},$$

for some $v_* \in S_{F,x_*}$. Thus, N has a closed graph, which implies that the operator N is upper semicontinuous.

Step 5. There exists an open set $U \subseteq C([1,T],\mathbb{R})$ such that, for any $k \in (0,1)$ and all $x \in \partial U$, $x \notin kN(x)$.

Let $x \in kN(x)$, $k \in (0,1)$. Then, there exists $v \in L^1([1,T],\mathbb{R})$ with $v \in S_{F,x}$ such that, for $t \in [1,T]$, we have the following case.

$$x(t) = k \frac{1}{\Gamma(\alpha)} \int_1^t \left(\log \frac{t}{z}\right)^{\alpha-1} \frac{v(z)}{z} dz$$
$$+ k \frac{(\log t)^{\gamma-1}}{\Lambda} \left\{ \sum_{j=1}^m \frac{\eta_j}{\Gamma(\alpha)} \int_1^{\xi_j} \left(\log \frac{\xi_j}{z}\right)^{\alpha-1} \frac{v(z)}{z} dz \right.$$
$$+ \sum_{i=1}^n \frac{\zeta_i}{\Gamma(\alpha + \phi_i)} \int_1^{\theta_i} \left(\log \frac{\theta_i}{z}\right)^{\alpha+\phi_i-1} \frac{v(z)}{z} dz$$
$$+ \sum_{k=1}^r \frac{\lambda_k}{\Gamma(\alpha - \omega_k)} \int_1^{\mu_k} \left(\log \frac{\mu_k}{z}\right)^{\alpha-\omega_k-1} \frac{v(z)}{z} dz$$
$$\left. - \frac{1}{\Gamma(\alpha)} \int_1^T \left(\log \frac{T}{z}\right)^{\alpha-1} \frac{v(z)}{z} dz \right\}.$$

Following the computation as in Step 2, for each $t \in [1,T]$, we have the following inequality:

$$|x(t)| \leq \left[\frac{(\log T)^\alpha}{\Gamma(\alpha+1)} + \frac{(\log T)^{\gamma-1}}{|\Lambda|} \left\{ \sum_{i=j}^m \frac{|\eta_j|(\log \xi_j)^\alpha}{\Gamma(\alpha+1)} + \sum_{i=1}^n \frac{|\zeta_i|(\log \theta_i)^{\alpha+\phi_i}}{\Gamma(\alpha+\phi_i+1)} \right. \right.$$
$$\left. \left. + \sum_{k=1}^r \frac{|\lambda_k|(\log \mu_k)^{\alpha-\omega_k}}{\Gamma(\alpha-\omega_k+1)} + \frac{(\log T)^\alpha}{\Gamma(\alpha+1)} \right\} \right] \|p\| \chi(\|x\|)$$
$$= \|p\| \chi(\|x\|) \Omega.$$

In consequence, we obtain

$$\frac{\|x\|}{\chi(\|x\|)\|p\|\Omega} \leq 1.$$

By assumption (A_3), we can find M such that $\|x\| \neq M$. Let us set the following:

$$U = \{x \in C([1,T], \mathbb{R}) : \|x\| < M\}.$$

Notice that $N : \overline{U} \to \mathcal{P}(C([1,T], \mathbb{R}))$ is a compact multi-valued map with convex closed values, which is upper semicontinuous; moreover, from the choice of U, there is no $x \in \partial U$ such that $x \in kN(x)$ for some $k \in (0,1)$. Hence, we deduce by the Leray–Schauder nonlinear alternative for Kakutani maps [39] that N has a fixed point $x \in \overline{U}$. This implies the existence of at least one solution for the inclusions problem (19) on $[1,T]$. The proof is complete. □

4.1.2. Case 2: Nonconvex Valued Multifunctions

Here, we prove an existence result for the Hilfer–Hadamard fractional inclusions boundary value problem (19) with a non-convex valued multi-valued map via a fixed point theorem for multivalued maps due to Covitz and Nadler [42].

Definition 5. ([43]) *Let (X,d) be a metric space induced from the normed space $(X; \|\cdot\|)$ and $H_d : \mathcal{P}(X) \times \mathcal{P}(X) \to \mathbb{R} \cup \{\infty\}$ be defined as follows:*

$$H_d(A,B) = \max\{\sup_{a \in A} d(a,B), \sup_{b \in B} d(A,b)\},$$

where $d(A,b) = \inf_{a \in A} d(a,b)$ and $d(a,B) = \inf_{b \in B} d(a,b)$.

Theorem 7. *Let the following conditions hold:*

(B_1) $F : [1,T] \times \mathbb{R} \to \mathcal{P}_{cp}(\mathbb{R})$ *is such that $F(\cdot, x) : [1,T] \to \mathcal{P}_{cp}(\mathbb{R})$ is measurable for each $x \in \mathbb{R}$;*

(B_2) $H_d(F(t,x), F(t,\bar{x})) \leq \varrho(t)|x - \bar{x}|$ *for almost all $t \in [1,T]$ and $x, \bar{x} \in \mathbb{R}$ with $\varrho \in C([1,T], \mathbb{R}^+)$ and $d(0, F(t,0)) \leq \varrho(t)$ for almost all $t \in [1,T]$.*

Then, the Hilfer–Hadamard inclusions fractional boundary value problem (19) has at least one solution on $[1,T]$ if

$$\Omega \|\varrho\| < 1,$$

where Ω is given by (11).

Proof. We verify that the operator $N : C([1,T], \mathbb{R}) \to \mathcal{P}(C([1,T], \mathbb{R}))$, defined at the beginning of the proof of Theorem 6, satisfies the hypotheses of the fixed point theorem for multivalued maps due to Covitz and Nadler [42].

Step I. N is nonempty and closed for every $v \in S_{F,x}$.

It follows by the measurable selection theorem ([44], (Theorem III.6)) that the set-valued map $F(\cdot, x(\cdot))$ is measurable; hence, it admits a measurable selection $v : [1,T] \to \mathbb{R}$. In view of the assumption (B_2), we obtain $|v(t)| \leq \varrho(t)(1 + |x(t)|)$, that is, $v \in L^1([1,T], \mathbb{R})$; hence, F is integrably bounded. In consequence, we deduce that $S_{F,x} \neq \emptyset$.

Now, we show that $N(x) \in \mathcal{P}_{cl}(C([1,T], \mathbb{R}))$ for each $x \in C([1,T], \mathbb{R})$. For that, let $\{u_n\}_{n \geq 0} \in N(x)$ with $u_n \to u$ $(n \to \infty)$ in $C([1,T], \mathbb{R})$. Then, $u \in C([1,T], \mathbb{R})$, and we can find $v_n \in S_{F,x_n}$ satisfying the following equation:

$$u_n(t) = \frac{1}{\Gamma(\alpha)} \int_1^t \left(\log \frac{t}{z}\right)^{\alpha-1} \frac{v_n(z)}{z} dz$$
$$+ \frac{(\log t)^{\gamma-1}}{\Lambda} \left\{ \sum_{j=1}^m \frac{\eta_j}{\Gamma(\alpha)} \int_1^{\zeta_j} \left(\log \frac{\zeta_j}{z}\right)^{\alpha-1} \frac{v_n(z)}{z} dz \right.$$
$$+ \sum_{i=1}^n \frac{\zeta_i}{\Gamma(\alpha+\phi_i)} \int_1^{\theta_i} \left(\log \frac{\theta_i}{z}\right)^{\alpha+\phi_i-1} \frac{v_n(z)}{z} dz$$
$$+ \sum_{k=1}^r \frac{\lambda_k}{\Gamma(\alpha-\omega_k)} \int_1^{\mu_k} \left(\log \frac{\mu_k}{z}\right)^{\alpha-\omega_k-1} \frac{v_n(z)}{z} dz$$
$$\left. - \frac{1}{\Gamma(\alpha)} \int_1^T \left(\log \frac{T}{z}\right)^{\alpha-1} \frac{v_n(z)}{z} dz \right\},$$

for each $t \in [1, T]$. Then, we can obtain a sub-sequence (if necessary) v_n converging to v in $L^1([1, T], \mathbb{R})$, as F has compact values. Thus, $v \in S_{F,x}$ and for each $t \in [1, T]$, we have the following:

$$u_n(t) \to v(t) = \frac{1}{\Gamma(\alpha)} \int_1^t \left(\log \frac{t}{z}\right)^{\alpha-1} \frac{v(z)}{z} dz$$
$$+ \frac{(\log t)^{\gamma-1}}{\Lambda} \left\{ \sum_{j=1}^m \frac{\eta_j}{\Gamma(\alpha)} \int_1^{\zeta_j} \left(\log \frac{\zeta_j}{z}\right)^{\alpha-1} \frac{v(z)}{z} dz \right.$$
$$+ \sum_{i=1}^n \frac{\zeta_i}{\Gamma(\alpha+\phi_i)} \int_1^{\theta_i} \left(\log \frac{\theta_i}{z}\right)^{\alpha+\phi_i-1} \frac{v(z)}{z} dz$$
$$+ \sum_{k=1}^r \frac{\lambda_k}{\Gamma(\alpha-\omega_k)} \int_1^{\mu_k} \left(\log \frac{\mu_k}{z}\right)^{\alpha-\omega_k-1} \frac{v(z)}{z} dz$$
$$\left. - \frac{1}{\Gamma(\alpha)} \int_1^T \left(\log \frac{T}{z}\right)^{\alpha-1} \frac{v(z)}{z} dz \right\}.$$

Thus, $u \in N(x)$.

Step II. In this step, it will be shown that there exists $0 < m_0 < 1$ ($m_0 = \Omega \|\varrho\|$) such that the following is the case.

$$H_d(N(x), N(\bar{x})) \leq m_0 \|x - \bar{x}\| \text{ for each } x, \bar{x} \in C([1, T], \mathbb{R}).$$

Let $x, \bar{x} \in C([1, T], \mathbb{R})$ and $h_1 \in N(x)$. Then, there exists $v_1(t) \in F(t, x(t))$ such that, for each $t \in [1, T]$, we have

$$h_1(t) = \frac{1}{\Gamma(\alpha)} \int_1^t \left(\log \frac{t}{z}\right)^{\alpha-1} \frac{v_1(z)}{z} dz$$
$$+ \frac{(\log t)^{\gamma-1}}{\Lambda} \left\{ \sum_{j=1}^m \frac{\eta_j}{\Gamma(\alpha)} \int_1^{\zeta_j} \left(\log \frac{\zeta_j}{z}\right)^{\alpha-1} \frac{v_1(z)}{z} dz \right.$$
$$+ \sum_{i=1}^n \frac{\zeta_i}{\Gamma(\alpha+\phi_i)} \int_1^{\theta_i} \left(\log \frac{\theta_i}{z}\right)^{\alpha+\phi_i-1} \frac{v_1(z)}{z} dz$$
$$+ \sum_{k=1}^r \frac{\lambda_k}{\Gamma(\alpha-\omega_k)} \int_1^{\mu_k} \left(\log \frac{\mu_k}{z}\right)^{\alpha-\omega_k-1} \frac{v_1(z)}{z} dz$$
$$\left. - \frac{1}{\Gamma(\alpha)} \int_1^T \left(\log \frac{T}{z}\right)^{\alpha-1} \frac{v_1(z)}{z} dz \right\}.$$

By (B_2), we have
$$H_d(F(t,x), F(t,\bar{x})) \leq \varrho(t)|x(t) - \bar{x}(t)|.$$

Thus, there exists $\vartheta(t) \in F(t, \bar{x}(t))$ such that
$$|v_1(t) - \vartheta| \leq \varrho(t)|x(t) - \bar{x}(t)|, \quad t \in [1, T].$$

Let us define $\mathcal{V} : [1, T] \to \mathcal{P}(\mathbb{R})$ by
$$\mathcal{V}(t) = \{\vartheta \in \mathbb{R} : |v_1(t) - \vartheta| \leq \varrho(t)|x(t) - \bar{x}(t)|\}.$$

Then, there exists a function $v_2(t)$ that is a measurable selection of $\mathcal{V}$, since the multivalued operator $\mathcal{V}(t) \cap F(t, \bar{x}(t))$ is measurable (Proposition III.4 [44]). Hence, $v_2(t) \in F(t, \bar{x}(t))$ and for each $t \in [1, T]$, we have $|v_1(t) - v_2(t)| \leq \varrho(t)|x(t) - \bar{x}(t)|$. Thus, for each $t \in [1, T]$, we have

$$\begin{aligned}
h_2(t) &= \frac{1}{\Gamma(\alpha)} \int_1^t \left(\log \frac{t}{z}\right)^{\alpha-1} \frac{v_2(z)}{z} dz \\
&+ \frac{(\log t)^{\gamma-1}}{\Lambda} \Bigg\{ \sum_{j=1}^m \frac{\eta_j}{\Gamma(\alpha)} \int_1^{\xi_j} \left(\log \frac{\xi_j}{z}\right)^{\alpha-1} \frac{v_2(z)}{z} dz \\
&+ \sum_{i=1}^n \frac{\zeta_i}{\Gamma(\alpha+\phi_i)} \int_1^{\theta_i} \left(\log \frac{\theta_i}{z}\right)^{\alpha+\phi_i-1} \frac{v_2(z)}{z} dz \\
&+ \sum_{k=1}^r \frac{\lambda_k}{\Gamma(\alpha-\omega_k)} \int_1^{\mu_k} \left(\log \frac{\mu_k}{z}\right)^{\alpha-\omega_k-1} \frac{v_2(z)}{z} dz \\
&- \frac{1}{\Gamma(\alpha)} \int_1^T \left(\log \frac{T}{z}\right)^{\alpha-1} \frac{v_2(z)}{z} dz \Bigg\}.
\end{aligned}$$

Hence, we have the following:

$$\begin{aligned}
|h_1(t) - h_2(t)| &\leq \frac{1}{\Gamma(\alpha)} \int_1^t \left(\log \frac{t}{z}\right)^{\alpha-1} \frac{|v_2(z) - v_1(z)|}{z} dz \\
&+ \frac{(\log t)^{\gamma-1}}{\Lambda} \Bigg\{ \sum_{j=1}^m \frac{\eta_j}{\Gamma(\alpha)} \int_1^{\xi_j} \left(\log \frac{\xi_j}{z}\right)^{\alpha-1} \frac{|v_2(z) - v_1(z)|}{z} dz \\
&+ \sum_{i=1}^n \frac{\zeta_i}{\Gamma(\alpha+\phi_i)} \int_1^{\theta_i} \left(\log \frac{\theta_i}{z}\right)^{\alpha+\phi_i-1} \frac{|v_2(z) - v_1(z)|}{z} dz \\
&+ \sum_{k=1}^r \frac{\lambda_k}{\Gamma(\alpha-\omega_k)} \int_1^{\mu_k} \left(\log \frac{\mu_k}{z}\right)^{\alpha-\omega_k-1} \frac{|v_2(z) - v_1(z)|}{z} dz \\
&- \frac{1}{\Gamma(\alpha)} \int_1^T \left(\log \frac{T}{z}\right)^{\alpha-1} \frac{|v_2(z) - v_1(z)|}{z} dz \Bigg\} \\
&\leq \Bigg[\frac{(\log T)^\alpha}{\Gamma(\alpha+1)} + \frac{(\log T)^{\gamma-1}}{|\Lambda|} \Bigg\{ \sum_{i=j}^m \frac{|\eta_j|(\log \xi_j)^\alpha}{\Gamma(\alpha+1)} + \sum_{i=1}^n \frac{|\zeta_i|(\log \theta_i)^{\alpha+\phi_i}}{\Gamma(\alpha+\phi_i+1)} \\
&+ \sum_{k=1}^r \frac{|\lambda_k|(\log \mu_k)^{\alpha-\omega_k}}{\Gamma(\alpha-\omega_k+1)} + \frac{(\log T)^\alpha}{\Gamma(\alpha+1)} \Bigg\} \Bigg] \|\varrho\| \|x - \bar{x}\|,
\end{aligned}$$

which yields
$$\|h_1 - h_2\| \leq \Omega \|\varrho\| \|x - \bar{x}\|.$$

On switching the roles of x and $\bar{x}$, we obtain the following case:
$$H_d(N(x), N(\bar{x})) \le \Omega \|\varrho\| \|x - \bar{x}\|,$$
which shows that N is a contraction. Consequently, the conclusion of the fixed point theorem for multivalued maps due to Covitz and Nadler [42] applies; hence, the operator N has a fixed point x which corresponds to a solution of the Hilfer–Hadamard inclusions fractional boundary value problem (19). The proof is complete. □

4.2. Examples

Let us consider the Hilfer–Hadamard fractional inclusions boundary value problem:
$$\begin{cases} {}^{HH}D_1^{\alpha,\beta}x(t) \in F(t,x(t)), & t \in [1,T], \\ x(1) = 0, \quad x(T) = \sum_{j=1}^{m} \eta_j x(\zeta_j) + \sum_{i=1}^{n} \zeta_i \, {}^H I_1^{\phi_i} x(\theta_i) + \sum_{k=1}^{r} \lambda_k \, {}_H D_1^{\omega_k} x(\mu_k), \end{cases} \quad (20)$$

with the values of the parameters taken in the problem (14), while the multi-valued map $F : J \times \mathbb{R} \to \mathcal{P}(\mathbb{R})$ will be defined below.

We first illustrate Theorem 6 by taking the following multi-valued map:
$$F(t, x(t)) = \left[\frac{e^{-(1-t)^2}}{\sqrt{(1-t)^4 + 75}} \left(\frac{|x|^2}{(1+|x|)} + \frac{1}{5} \right), \frac{1}{100} \left(x + \frac{\cos x}{9} \right) \right]. \quad (21)$$

It is easy to check that $\|F(t,x)\|_\mathcal{P} \le q(t) \chi(\|x\|)$, where $q(t)$ and $\chi(\|x\|)$ are given as follows:
$$q(t) = \frac{e^{-(1-t)^2}}{\sqrt{(1-t)^4 + 75}}, \quad \chi(\|x\|) = \|x\| + \frac{1}{5}.$$

Using the values $\|q\| = 1/75$, $\chi(\|x\|) = \|x\| + \frac{1}{5}$ and $\Omega \approx 39.388095$ (see Section 3.3) in the condition (A_3), that is, $\dfrac{M}{\chi(M) \|q\| \Omega} > 1$, we find that $M > 0.221207$. Thus, all the assumptions of Theorem 6 are satisfied; hence, its conclusion implies that problem (20) with $F(t, x(t))$ given by (21) has a solution on $[1,5]$.

Next, we demonstrate the application of Theorem 7 by choosing the following multi-valued map:
$$F(t, x(t)) = \left[\frac{1 + \arctan x}{(20 + t)^2}, \left(\frac{\sqrt{t^2 + 11}}{360} \right) \left(\frac{1 + 5|x|}{1 + 4|x|} \right) \right]. \quad (22)$$

Clearly, F is measurable for all $x \in \mathbb{R}$, satisfying the following inequality:
$$H_d(F(t,x), F(t,\bar{x})) \le \left(\frac{\sqrt{t^2 + 11}}{360} \right) |x - \bar{x}|, \ x, \bar{x} \in \mathbb{R}, \ t \in [1,5].$$

Fixing $\varrho(t) = \sqrt{t^2 + 11}/360$, we have $\|\varrho\| = 1/60$ and $d(0, \mathcal{F}(t,0)) \le \varrho(t), t \in [1,5]$. Furthermore, $\|\varrho\| \Omega \approx 0.656468 < 1$. Clearly, the hypothesis of Theorem 7 holds true; consequently, we deduce by its conclusion that there exists a solution for the problem (20) with $F(t,x(t))$ given by (22) on $[1,5]$.

5. Conclusions

In this paper, we have presented the existence and uniqueness criteria for the solutions of a Hilfer–Hadamard fractional differential equation complemented with mixed nonlocal (multi-point, fractional integral multi-order and fractional derivative multi-order) boundary conditions. Firstly, we have converted the given nonlinear problem into a fixed point problem. Once the fixed point operator is available, we can make use of the Banach contraction mapping principle to obtain the uniqueness result. The first two existence results are proved by applying the fixed point theorems due to Schaefer and Krasnoselskiĭ, while the third existence result is based on Leray–Schauder nonlinear alternative. Next, we present the existence results for the corresponding inclusions problem. The first result for the inclusions problem deals with the convex-valued multivalued map and is obtained by applying the Leray–Schauder alternative for multivalued maps, while the non-convex valued multivalued map case relies on the Covitz–Nadler fixed point theorem for contractive multivalued maps. All the results obtained for single and multivalued maps are well illustrated by numerical examples. It is imperative to mention that our results are new in the given configuration and enrich the literature on boundary value problems involving Hilfer–Hadamard fractional differential equations and inclusions of order in $(1,2]$.

Author Contributions: Conceptualization, B.A. and S.K.N.; methodology, B.A. and S.K.N.; validation, B.A. and S.K.N.; formal analysis, B.A. and S.K.N.; writing—original draft preparation, B.A. and S.K.N. All authors have read and agreed to the published version of the manuscript.

Funding: This project was funded by the Deanship of Scientific Research (DSR) at King Abdulaziz University, Jeddah, under grant no. (RG-43-130-41).

Institutional Review Board Statement: Not applicable.

Informed Consent Statement: Not applicable.

Data Availability Statement: Not applicable.

Acknowledgments: This project was funded by the Deanship of Scientific Research (DSR) at King Abdulaziz University, Jeddah, under grant no. (RG-43-130-41). The authors, therefore, acknowledge with thanks DSR technical and financial support. The authors also thank the reviewers for their constructive remarks on our work.

Conflicts of Interest: The authors declare no conflict of interest.

References

1. Diethelm, K. *The Analysis of Fractional Differential Equations*; Lecture Notes in Mathematics; Springer: New York, NY, USA, 2010.
2. Kilbas, A.A.; Srivastava, H.M.; Trujillo, J.J. *Theory and Applications of the Fractional Differential Equations*; North-Holland Mathematics Studies; Elsevier: Amsterdam, The Netherlands, 2006; Volume 204.
3. Lakshmikantham, V.; Leela, S.; Devi, J.V. *Theory of Fractional Dynamic Systems*; Cambridge Scientific Publishers: Cambridge, UK, 2009.
4. Miller, K.S.; Ross, B. *An Introduction to the Fractional Calculus and Differential Equations*; John Wiley: NewYork, NY, USA, 1993.
5. Samko, S.G.; Kilbas, A.A.; Marichev, O.I. *Fractional Integrals and Derivatives*; Gordon and Breach Science: Yverdon, Switzerland, 1993.
6. Podlubny, I. *Fractional Differential Equations*; Academic Press: New York, NY, USA, 1999.
7. Ahmad, B.; Alsaedi, A.; Ntouyas, S.K.; Tariboon, J. *Hadamard-Type Fractional Differential Equations, Inclusions and Inequalities*; Springer: Cham, Switzerland, 2017.
8. Zhou, Y. *Basic Theory of Fractional Differential Equations*; World Scientific: Singapore, 2014.
9. Ahmad, B.; Ntouyas, S.K. *Nonlocal Nonlinear Fractional-Order Boundary Value Problems*; World Scientific: Singapore, 2021.
10. Patnaik, S.; Sidhardh, S.; Semperlotti, F. Towards a unified approach to nonlocal elasticity via fractional-order mechanics. *Int. J. Mech. Sci.* **2021**, *189*, 105992. [CrossRef]
11. Patnaik, S.; Sidhardh, S.; Semperlotti, F. Geometrically nonlinear analysis of nonlocal plates using fractional calculus. *Int. J. Mech. Sci.* **2020**, *179*, 105710. [CrossRef]
12. Stempin, P.; Sumelka, W. Space-fractional Euler-Bernoulli beam model-Theory and identification for silver nanobeam bending. *Int. J. Mech. Sci.* **2020**, *186*, 105902. [CrossRef]
13. Hilfer, R. (Ed.) *Applications of Fractional Calculus in Physics*; World Scientific: Singapore, 2000.
14. Hilfer, R. Experimental evidence for fractional time evolution in glass forming materials. *J. Chem. Phys.* **2002**, *284* , 399–408. [CrossRef]
15. Hilfer, R.; Luchko, Y.; Tomovski, Z. Operational method for the solution of fractional differential equations with generalized Riemann-Liouville fractional derivatives. *Frac. Calc. Appl. Anal.* **2009**, *12*, 299–318.
16. Furati, K.M.; Kassim, N.D.; Tatar, N.E. Existence and uniqueness for a problem involving Hilfer fractional derivative. *Comput. Math. Appl.* **2012**, *64*, 1616–1626. [CrossRef]
17. Gu, H.; Trujillo, J.J. Existence of mild solution for evolution equation with Hilfer fractional derivative. *Appl. Math. Comput.* **2015**, *257*, 344–354. [CrossRef]
18. Wang, J.; Zhang, Y. Nonlocal initial value problems for differential equations with Hilfer fractional derivative. *Appl. Math. Comput.* **2015**, *266*, 850–859. [CrossRef]

19. Sousa, J.V.d.; de Oliveira, E.C. On the ψ-Hilfer fractional derivative. *Commun. Nonlinear Sci. Numer. Simul.* **2018**, *60*, 72–91. [CrossRef]
20. Asawasamrit, S.; Kijjathanakorn, A.; Ntouyas, S.K.; Tariboon, J. Nonlocal boundary value problems for Hilfer fractional differential equations. *Bull. Korean Math. Soc.* **2018**, *55*, 1639–1657.
21. Wongcharoen, A.; Ahmad, B.; Ntouyas, S.K.; Tariboon, J. Three-point boundary value problems for Langevin equation with Hilfer fractional derivative. *Adv. Math. Phys.* **2020**, *2020*, 9606428. [CrossRef]
22. Ntouyas, S.K. A survey on existence results for boundary value problems of Hilfer fractional differential equations and inclusions. *Foundations* **2021**, *1*, 63–98. [CrossRef]
23. Hadamard, J. Essai sur l'etude des fonctions donnees par leur developpment de Taylor. *J. Mat. Pure Appl. Ser.* **1892**, *8*, 101–186.
24. Butzer, P.L.; Kilbas, A.A.; Trujillo, J.J. Mellin transform analysis and integration by parts for Hadamard-type fractional integrals. *J. Math. Anal. Appl.* **2002**, *270*, 1–15. [CrossRef]
25. Qassim, M.D.; Furati, K.M.; Tatar, N.-E. On a differential equation involving Hilfer-Hadamard fractional derivative. *Abstr. Appl. Anal.* **2012**, *2012*, 391062. [CrossRef]
26. Vivek, D.; Shahb, K.; Kanagarajan, K. Dynamical analysis of Hilfer-Hadamard type fractional pantograph equations via successive approximation. *J. Taibah Univ. Sci.* **2019**, *13*, 225–230. [CrossRef]
27. Bachira, F.S.; Abbas, S.; Benbachir, M.; Benchohra, M. Hilfer-Hadamard fractional differential equations; Existence and Attractivity, *Adv. Theory Nonl. Anal. Appl.* **2021**, *5*, 49–57.
28. Promsakon, C.; Ntouyas, S.K.; Tariboon, J. Hilfer-Hardamard nonlocal integro-multi-point fractional boundary value problems. *J. Funct. Spaces*, submitted for publication.
29. Saengthong, W.; Thailert, E.; Ntouyas, S.K. Existence and uniqueness of solutions for system of Hilfer-Hadamard sequential fractional differential equations with two point boundary conditions. *Adv. Differ. Equ.* **2019**, *2019*, 525. [CrossRef]
30. Wang, G.; Ren, X.; Zhang, L.; Ahmad, B. Explicit iteration and unique positive solution for a Caputo-Hadamard fractional turbulent flow model. *IEEE Access* **2019**, *7*, 109833–109839. [CrossRef]
31. Graef, J.R.; Grace, S.R.; Tunc, E. Asymptotic behavior of solutions of nonlinear fractional equations with Caputo-type Hadamard derivatives. *Fract. Calc. Appl. Anal.* **2017**, *20*, 71–87. [CrossRef]
32. Abbas, S.; Benchohra, M.; Hamidi, N.; Henderson, J. Caputo-Hadamard fractional differential equations in Banach spaces. *Fract. Calc. Appl. Anal.* **2018**, *21*, 1027–1045. [CrossRef]
33. Ma, L. Comparison theorems for Caputo-Hadamard fractional differential equations. *Fractals* **2019**, *27*, 1950036. [CrossRef]
34. Li, C.; Li, Z.; Wang, Z. Mathematical analysis and the local discontinuous Galerkin method for Caputo-Hadamard fractional partial differential equation. *J. Sci. Comput.* **2020**, *85*, 41. [CrossRef]
35. Ntouyas, S.K. Nonlocal Initial and Boundary Value Problems: A survey. In *Handbook on Differential Equations: Ordinary Differential Equations*; Canada, A., Drabek, P., Fonda, A., Eds.; Elsevier Science B.V.: Amsterdam, The Netherlands, 2005; pp. 459–555.
36. Deimling, K. *Nonlinear Functional Analysis*; Springer: New York, NY, USA, 1985.
37. Krasnoselskiĭ, M.A. Two remarks on the method of successive approximations. *Uspekhi Mat. Nauk* **1955**, *10*, 123–127.
38. Smart, D.R. *Fixed Point Theory*; Cambridge University Press: Cambridge, UK, 1974.
39. Granas, A.; Dugundji, J. *Fixed Point Theory*; Springer: New York, NY, USA, 2003.
40. Lasota, A.; Opial, Z. An application of the Kakutani-Ky Fan theorem in the theory of ordinary differential equations, *Bull. Acad. Pol. Sci. Ser. Sci. Math. Astronom. Phys.* **1965**, *13*, 781–786.
41. Deimling, K. *Multivalued Differential Equations*; De Gruyter: Berlin, Germany, 1992.
42. Covitz, H.; Nadler, S.B., Jr. Multivalued contraction mappings in generalized metric spaces. *Israel J. Math.* **1970**, *8*, 5–11. [CrossRef]
43. Kisielewicz, M. *Differential Inclusions and Optimal Control*; Kluwer: Dordrecht, The Netherlands, 1991.
44. Castaing, C.; Valadier, M. *Convex Analysis and Measurable Multifunctions*; Lecture Notes in Mathematics 580; Springer: Berlin/Heidelberg, Germany; New York, NY, USA, 1977.

fractal and fractional

Article

Certain Recurrence Relations of Two Parametric Mittag-Leffler Function and Their Application in Fractional Calculus

Dheerandra Shanker Sachan [1], Shailesh Jaloree [2] and Junesang Choi [3,*]

1. Department of Mathematics, St.Mary's PG College, Vidisha 464001, India; sachan.dheerandra17@gmail.com
2. Department of Mathematics, Samrat Ashok Technological Institute, Vidisha 464001, India; shailesh_jaloree@rediffmail.com
3. Department of Mathematics, Dongguk University, Gyeongju 38066, Korea
* Correspondence: junesangchoi@gmail.com; Tel.: +82-010-6525-2262

Abstract: The purpose of this paper is to develop some new recurrence relations for the two parametric Mittag-Leffler function. Then, we consider some applications of those recurrence relations. Firstly, we express many of the two parametric Mittag-Leffler functions in terms of elementary functions by combining suitable pairings of certain specific instances of those recurrence relations. Secondly, by applying Riemann–Liouville fractional integral and differential operators to one of those recurrence relations, we establish four new relations among the Fox–Wright functions, certain particular cases of which exhibit four relations among the generalized hypergeometric functions. Finally, we raise several relevant issues for further research.

Keywords: gamma function; Beta function; Mittag-Leffler function; Generalized Mittag-Leffler functions; generalized hypergeometric function; Fox–Wright function; recurrence relations; Riemann–Liouville fractional calculus operators

MSC: 26A33; 32D15; 33B10; 33B15; 33C20; 33E12; 65Q30

1. Introduction and Preliminaries

Magnus Gösta Mittag-Leffler (1846–1927), a Swedish mathematician, invented the function $E_\mu(z)$ (1) in conjunction with the summation technique for divergent series, which is eponymously referred to as the Mittag-Leffler (M-L) function and represented by the following convergent power series across the whole complex plane (see [1–4]):

$$E_\mu(z) = \sum_{n=0}^{\infty} \frac{z^n}{\Gamma(\mu n + 1)} \quad (\Re(\mu) > 0,\ z \in \mathbb{C}), \qquad (1)$$

where $\Gamma(\cdot)$ is the familiar Gamma function (see, e.g., ([5], [Section 1.1])). Let $\mathbb{C}$, $\mathbb{R}$, $\mathbb{R}^+$, $\mathbb{Z}$, and $\mathbb{N}$ represent the sets of complex numbers, real numbers, positive real numbers, integers, and positive integers, respectively, in this and subsequent sections. Further, for some $\ell \in \mathbb{Z}$, let $\mathbb{Z}_{\leq \ell}$ denote the set of integers less than or equal to ℓ. The function (1) is an entire function of order $1/\Re(\mu)$ and type 1 (see, e.g., ([6], [Section 4.1])). Numerous mathematicians have investigated the properties of this entire function (see, e.g., [7–16]). This function (1) reduces to a number of elementary and special functions such as (see, e.g., ([6], [Section 3.2]))

$$E_1(\pm z) = e^{\pm z}, \quad E_2(-z^2) = \cos z, \quad E_2(z^2) = \cosh z, \qquad (2)$$

and

$$E_{\frac{1}{2}}\left(\pm z^{\frac{1}{2}}\right) = e^z \left[1 + \mathrm{erf}\left(\pm z^{\frac{1}{2}}\right)\right] = e^z \mathrm{erfc}\left(\mp z^{\frac{1}{2}}\right), \qquad (3)$$

where erf (erfc) represents the error function (complementary error function)

$$\operatorname{erf}(z) := \frac{2}{\sqrt{\pi}} \int_0^z e^{-u^2}\, du, \quad \operatorname{erfc} := 1 - \operatorname{erf}(z) \quad (z \in \mathbb{C}), \qquad (4)$$

and $z^{\frac{1}{2}}$ denotes the principal branch of the associated multi-valued function. Numerous generalizations of the function (1) have been developed, including (5) and (7). In the 1960s, the Mittag-Leffler function began to be classified as a particular instance of the broader class of Fox H-function, which may have an arbitrary number of parameters in their Mellin–Barnes contour integral representation. Among the special instances of the H-function is the Fox–Wright function (8) (see, e.g., [10,17–19]). This function (1) and its numerous generalizations are significant because they are involved in a large number of applied problems (see, e.g., [20–27]). The Mittag-Leffler function (1) and its numerous extensions have been used, particulary, in conjunction with fractional calculus such as: The resolvant for a particular case of Volterra's equation (see ([28], [Theorem 4.2])) is explicitly expressed in terms of the Mittag-Leffler function (1); The two parametric Mittag-Leffler function $E_{\mu,\varrho}(z)$ (5) and its slight extension are used as solutions of the linear integral equation of the Abel–Volterra type (see ([29], [Theorem 5.3])); Kilbas and Saigo ([30], [Theorems 6 and 7]) employed an extension of the Mittag-Leffler function as solutions of the Abel–Volterra integral equations (see also [31,32]). The monograph ([33], [pages 132, 143, 144, 206]) demonstrated in detail that the Mittag-Leffler function (1) and its modest modification are solutions to a certain class of fractional differential equations; the monograph ([34], [p. 21]) referred to the two parametric Mittag-Leffler function (5) and (1), as well as their fascinating Laplace transform formula. Choi et al. [35] proposed an extension of the Prabhakar function (7) that is used to determine a number of its features and formulae, including higher-order differential equations, many integral transformations, and several fractional derivative and integral formulas. Due to the breadth of applications in connection with fractional calculus, the Mittag-Leffler function has been dubbed the "Queen Function of the Fractional Calculus" by certain academics in the past (see [36]; see also [17]). Haubold et al. [24] provided a concise overview of the Mittag-Leffler function, extended Mittag-Leffler functions, Mittag-Leffler type functions, and their intriguing and useful features.

The two parametric Mittag-Leffler function $E_{\mu,\varrho}(z)$ is defined by (see, e.g., [8,9], ([6], [Chapter 4]))

$$E_{\mu,\varrho}(z) = \sum_{n=0}^{\infty} \frac{z^n}{\Gamma(\mu n + \varrho)} \quad (\Re(\mu) > 0,\ \varrho, z \in \mathbb{C}). \qquad (5)$$

The function (5) is an entire function of order $1/\Re(\mu)$ and type 1 with the argument z under the constraints $\mu, \varrho \in \mathbb{R}^+$ (see, e.g., [37–39]), whose fundamental properties, such as asymptotic behavior and zero distribution, have been explored by a number of researchers (see, e.g., [8–11,40,41]). The function (5) reduces to some special functions such as (see, e.g., ([6], [Equation 4.3.13]))

$$E_{1/2,1}(z) = e^{z^2}(1 + \operatorname{erf} z) = e^{z^2} \operatorname{erfc}(-z). \qquad (6)$$

The interested reader may refer to ([6], [Chapter 4]) for further basic properties and many other relations, representations and applications of the two parametric Mittag-Leffler function (5).

The three parametric Mittag-Leffler function (also known as the Prabhakar function [42]) $E_{\mu,\varrho}^{\xi}(z)$ is defined by (see [43]; see also ([6], [Chapter 5]), [42])

$$E_{\mu,\varrho}^{\xi}(z) = \sum_{n=0}^{\infty} \frac{(\xi)_n z^n}{n!\, \Gamma(\mu n + \varrho)} \quad (\Re(\mu) > 0,\ \Re(\varrho) > 0,\ \xi, z \in \mathbb{C}), \qquad (7)$$

which is also an entire function of order $1/\Re(\mu)$ and type 1. Giusti et al. [42] offered an excellent survey paper in which they covered important findings and applications of the Prabhakar function (7), together with major historical events leading to its discovery and subsequent development, its capacity of introducing an upgraded scheme for fractional calculus, an overview of the advances made in applying this new general framework to physics and renewal processes, a collection of results on its numerical evaluation, and as many as 159 references.

The Mittag-Leffler function (1), its slight generalization (5), the Prabhakar function (7), and a number of other parameterized extensions are found to be particular instances of the following Fox–Wright function defined by (see [44–46]; see also [47], ([48], [p. 21]))

$$
{}_p\Psi_q\left[\begin{matrix}(\alpha_1, A_1), \ldots, (\alpha_p, A_p); \\ (\beta_1, B_1), \ldots, (\beta_q, B_q);\end{matrix} z\right] = \sum_{k=0}^{\infty} \frac{\prod_{j=1}^{p}\Gamma(\alpha_j + A_j k)}{\prod_{j=1}^{q}\Gamma(\beta_j + B_j k)} \frac{z^k}{k!}, \tag{8}
$$

where the coefficients $A_j \in \mathbb{R}$ $(j = 1, \ldots, p)$ and $B_j \in \mathbb{R}$ $(j = 1, \ldots, q)$; $\alpha_j \in \mathbb{C}$ $(j = 1, \ldots, p)$ and $\beta_j \in \mathbb{C}$ $(j = 1, \ldots, q)$. The convergence constraints of (8) are given as follows (see ([49], [Theorem 1.5])): Let

$$
\Omega := \sum_{j=1}^{q} B_j - \sum_{j=1}^{p} A_j,
$$

$$
\omega := \prod_{j=1}^{p} |A_j|^{-A_j} \prod_{j=1}^{q} |B_j|^{B_j}, \tag{9}
$$

$$
\nu := \sum_{j=1}^{q} \beta_j - \sum_{j=1}^{p} \alpha_j + \frac{p-q}{2}.
$$

Then
(i) If $\Omega > -1$, then the series (8) is absolutely convergent for all $z \in \mathbb{C}$;
(ii) If $\Omega = -1$, the series (8) is absolutely convergent for $|z| < \omega$;
(iii) If $\Omega = -1$ and $\Re(\nu) > \frac{1}{2}$, the series (8) is absolutely convergent for $|z| = \omega$.

The Fox–Wright function ${}_p\Psi_q$ is an extension of generalized hypergeometric function ${}_pF_q$ and a particular case of the H-function (see, e.g., ([19], [Equation (1.140)]), [50]):

$$
{}_p\Psi_q\left[\begin{matrix}(\alpha_1, 1), \ldots, (\alpha_p, 1); \\ (\beta_1, 1), \ldots, (\beta_q, 1);\end{matrix} z\right] = \frac{\prod_{j=1}^{p}\Gamma(\alpha_j)}{\prod_{j=1}^{q}\Gamma(\beta_j)} {}_pF_q\left[\begin{matrix}\alpha_1, \ldots, \alpha_p; \\ \beta_1, \ldots, \beta_q;\end{matrix} z\right] \tag{10}
$$

and

$$
{}_p\Psi_q\left[\begin{matrix}(\alpha_1, A_1), \ldots, (\alpha_p, A_p); \\ (\beta_1, B_1), \ldots, (\beta_q, B_q);\end{matrix} z\right]
$$
$$
= H^{1,p}_{p,q+1}\left[-z \,\middle|\, \begin{matrix}(1-\alpha_1, A_1), \ldots, (1-\alpha_p, A_p) \\ (0,1), (1-\beta_1, B_1), \ldots, (1-\beta_q, B_q)\end{matrix}\right]. \tag{11}
$$

There have been many introductions and investigations of fractional integrals and derivatives. We recall the left-sided and right-sided Riemann–Liouville fractional integrals $I_{a+}^\alpha f$ and $I_{b-}^\alpha f$ of order $\alpha \in \mathbb{C}$ defined as (see, e.g., [33,34,49])

$$(I_{a+}^\alpha f)(x) = \frac{1}{\Gamma(\alpha)} \int_a^x (x-\tau)^{\alpha-1} f(\tau) \, d\tau \quad (x > a, \, \Re(\alpha) > 0), \tag{12}$$

and

$$(I_{b-}^\alpha f)(x) = \frac{1}{\Gamma(\alpha)} \int_x^b (\tau-x)^{\alpha-1} f(\tau) \, d\tau \quad (b > x, \, \Re(\alpha) > 0), \tag{13}$$

respectively. The Riemann–Liouville fractional derivatives $D_{a+}^\alpha f$ and $D_{b-}^\alpha f$ of order $\alpha \in \mathbb{C}$ ($\Re(\alpha) \geq 0$) are defined by

$$(D_{a+}^\alpha f)(x) = \left(\frac{d}{dx}\right)^n (I_{a+}^{n-\alpha} f)(x) \quad (x > a) \tag{14}$$

and

$$(D_{b-}^\alpha f)(x) = \left(-\frac{d}{dx}\right)^n (I_{b-}^{n-\alpha} f)(x) \quad (x < b), \tag{15}$$

where $n = [\Re(\alpha)] + 1$. Here and elsewhere, $[x]$ denotes the largest integer less than or equal to $x \in \mathbb{R}$.

We recall some of the recurrence relations for the two parametric Mittag-Leffler function (5) and the three parametric Mittag-Leffler function (7).

1.1. Two Parametric Mittag-Leffler Function:

$$E_{\mu,\varrho}(z) = z \, E_{\mu,\mu+\varrho}(z) + \frac{1}{\Gamma(\varrho)} \tag{16}$$

(see, e.g., ([10], [Equation (23)]), ([51], [Equation (5)])).

Further, for all $\mu, \varrho \in \mathbb{R}^+$,

$$E_{\mu,\varrho}(z) = z^2 \, E_{\mu,\varrho+2\mu}(z) + \frac{1}{\Gamma(\varrho)} + \frac{z}{\Gamma(\varrho+\mu)}, \tag{17}$$

$$E_{\mu,\varrho}(z) = z^3 \, E_{\mu,\varrho+3\mu}(z) + \frac{1}{\Gamma(\varrho)} + \frac{z}{\Gamma(\varrho+\mu)} + \frac{z^2}{\Gamma(\varrho+2\mu)}, \tag{18}$$

$$E_{\mu,\varrho}(z) = z^4 \, E_{\mu,\varrho+4\mu}(z) + \frac{1}{\Gamma(\varrho)} + \frac{z}{\Gamma(\varrho+\mu)} + \frac{z^2}{\Gamma(\varrho+2\mu)} + \frac{z^3}{\Gamma(\varrho+3\mu)} \tag{19}$$

(see, e.g., ([6], [Lemma 4.1]), [51]).

Generally, for $\Re(\mu) > 0$, $\Re(\varrho) > 0$, and $\ell \in \mathbb{N}$,

$$E_{\mu,\varrho}(z) = z^\ell \, E_{\mu,\varrho+\ell\mu}(z) + \sum_{k=0}^{\ell-1} \frac{z^k}{\Gamma(\varrho+k\mu)} \tag{20}$$

(see [52]; see also ([24], [Theorem 5.2])).

Gupta and Debnath (see ([51], [Equation (30)])) presented the following interesting differential recurrence relation for the two parametric Mittag-Leffler function (5):

$$E_{\mu,n+1}(z) = n(n+2) \, E_{\mu,n+3}(z) + \mu z[\mu + 2(n+1)] \, E'_{\mu,n+3}(z) \\ + z^2 \, E''_{\mu,n+3}(z) + E_{\mu,n+2}(z) \quad (n \in \mathbb{N}), \tag{21}$$

where
$$E_{\mu,\varrho}^{(\ell)}(z) = \frac{d^\ell}{dz^\ell} E_{\mu,\varrho}(z) \quad (\ell \in \mathbb{N}_0). \tag{22}$$

The following recurrence relation reveals that, under the restrictions, computation of $E_{\mu,\varrho}(z)$ for the case $\mu > 1$ may be reduced to the case $0 < \mu \leq 1$ (see, e.g., ([53], [Equation (6)]), ([10], [Chapter XVIII]), ([54], [p. 24]), ([55], [Equation (2.2)]), [56]):

$$E_{\mu,\varrho}(z) = \frac{1}{2m+1} \sum_{k=-m}^{m} E_{\mu/(2m+1),\varrho}\left(z^{1/(2m+1)} e^{2\pi i k/(2m+1)}\right) \tag{23}$$

$$(\mu \in \mathbb{R}^+,\ \varrho \in \mathbb{R},\ z \in \mathbb{C},\ m = [(\mu-1)/2]+1).$$

1.2. Three Parametric Mittag-Leffler Function and Its Various Extensions

Giusti et al. pointed out the following interesting recurrence relations for the Prabhakar function (7) (see ([42], [Equations (4.2) and (4.3)])):

$$E_{\mu,\varrho}^{\zeta+1}(z) = \frac{E_{\mu,\varrho-1}^{\zeta}(z) + (1 - \varrho + \mu\zeta) E_{\mu,\varrho}^{\zeta}(z)}{\mu\zeta} \tag{24}$$

(see [43]) and

$$E_{\mu,\varrho}^{\zeta+1}(z) = \frac{E_{\mu,\varrho-\mu-1}^{\zeta}(z) + (1 - \varrho + \mu) E_{\mu,\varrho-\mu}^{\zeta}(z)}{\mu\zeta z} \quad (z \in \mathbb{C} \setminus \{0\}) \tag{25}$$

(see [57]).

Shukla and Prajapati ([58], [Theorem 1]) offered an intriguing differential recurrence relation for an extended Mittag-Leffler function, which can be made by replacing $(\zeta)_n$ in (7) with $(\zeta)_{qn}$ under the restriction $q \in (0,1) \cup \mathbb{N}$ (see [59]), which was further generalized and investigated by Srivastava and Tomovski [50] who substituted a $k \in \mathbb{C}$ for the $q \in (0,1) \cup \mathbb{N}$ under constraints $\Re(\mu) > \max\{0, \Re(k) - 1\}$ and $\Re(k) > 0$.

Salim ([60], [Theorem 2.2]) presented two interesting differential recurrence relation for an extended Mittag-Leffler function, which can be derived by replacing $n!$ in (7) with a Pochhammer symbol, say $(\eta)_n$ under the constraint $\Re(\eta) > 0$.

Kurulay and Bayram ([61], [Theorems 4]) established an intriguing differential recurrence relation for the Prabhakar function (7).

Dhakar and Sharma ([62], [Theorem 2.1]) provided an interesting differential recurrence relation for the k-Mittag-Leffler function (see [63]), which may be obtained by substituting $(\zeta)_{n,k}$ and $\Gamma_k(\mu n + \varrho)$ for $(\zeta)_n$ and $\Gamma(\mu n + \varrho)$ in (7), respectively. Here the k-Pochhammer symbol is defined as follows (see [64]):

$$(\gamma)_{n,k} := \begin{cases} \dfrac{\Gamma_k(\gamma + nk)}{\Gamma_k(\gamma)} & (n \in \mathbb{N};\ k \in \mathbb{R}^+;\ \gamma \in \mathbb{C} \setminus \{0\}), \\ \gamma(\gamma + k) \cdots (\gamma + (n-1)k) & (n \in \mathbb{N};\ \gamma \in \mathbb{C}), \end{cases} \tag{26}$$

where Γ_k is the k-gamma function defined by

$$\Gamma_k(z) = \int_0^\infty e^{-\frac{t^k}{k}} t^{z-1} dt \quad \left(\Re(z) > 0;\ k \in \mathbb{R}^+\right). \tag{27}$$

Sharma and Jain ([65], [Theorem 1]) gave an interesting recurrence relation for the q-analog (or extension) of the Prabhakar function (7).

Gehlot ([66], [Theorem 2.1]) provided an intriguing differential recurrence relation for the following *p-k* Mittag-Leffler function (see [67]):

$$_pE_{k,\mu,\varrho}^{\zeta,q}(z) = \sum_{n=0}^{\infty} \frac{_p(\zeta)_{nq,k}\, z^n}{n!\, _p\Gamma_k(\mu n + \varrho)} \tag{28}$$

$(k, p \in \mathbb{R}^+,\ q \in (0,1) \cup \mathbb{N},\ \min\{\Re(\mu), \Re(\varrho), \Re(\zeta)\} > 0)$.

Here the *p-k* Pochhammer symbol $_p(\alpha)_{n,k}$ and the *p-k* Gamma function $_p\Gamma_k$ are defined by

$$_p(\alpha)_{n,k} = \left(\frac{\alpha p}{k}\right)\left(\frac{\alpha p}{k} + p\right) \cdots \left(\frac{\alpha p}{k} + (n-1)p\right) \tag{29}$$

$(k, p \in \mathbb{R}^+,\ n \in \mathbb{N},\ \Re(\alpha) > 0)$

and $_p(\alpha)_{0,k} := 1$;

$$_p\Gamma_k(z) = \int_0^\infty e^{-\frac{t^k}{p}}\, t^{z-1}\, dt \quad (\Re(z) > 0;\ p, k \in \mathbb{R}^+). \tag{30}$$

Further Gehlot ([67], [Theorem 2.3]) presented an interesting recurrence relation for the *p-k* Mittag-Leffler function.

Choi et al. ([35], [Theorem 3.1]) established a differential recurrence relation for the extended Mittag-Leffler function, which may be obtained by replacing $(\zeta)_n$ in (7) with the generalized Pochhammer symbol $(\zeta; p)_n$. Here the generalized Pochhammer symbol $(\zeta; p)_\nu$ $(\zeta, \nu \in \mathbb{C})$ is defined by

$$(\zeta; p)_\nu := \begin{cases} \frac{\Gamma_p(\zeta+\nu)}{\Gamma(\zeta)} & (\Re(p) > 0), \\ (\zeta)_\nu & (p = 0), \end{cases} \tag{31}$$

where $\Gamma_p(z)$ is the generalized gamma function given as follows:

$$\Gamma_p(z) := \begin{cases} \int_0^\infty t^{z-1} e^{-t - \frac{p}{t}}\, dt & (\Re(p) > 0,\ z \in \mathbb{C}), \\ \Gamma(z) & (p = 0,\ \Re(z) > 0). \end{cases} \tag{32}$$

The aim of this article is to explore some new recurrence relations for the two parametric Mittag-Leffler function. Then, we discuss several applications of such recurrence relations. To begin, we express a number of the two parametric Mittag-Leffler functions in terms of elementary functions by combining appropriate pairings of particular instances of those recurrence relations. Second, we establish four new relations among the Fox–Wright functions by applying Riemann–Liouville fractional integral and differential operators to one of those recurrence relations. Certain particular cases of the Fox–Wright function relations exhibit four relations among the generalized hypergeometric functions. Finally, we propose several pertinent research questions.

Further, for our purpose, we recall the classical Beta function (see, e.g., ([5], [p. 8]))

$$B(\alpha, \beta) = \begin{cases} \int_0^1 \tau^{\alpha-1}(1-\tau)^{\beta-1}\, d\tau & (\Re(\alpha) > 0,\ \Re(\beta) > 0) \\ \frac{\Gamma(\alpha)\Gamma(\beta)}{\Gamma(\alpha+\beta)} & (\alpha, \beta \in \mathbb{C} \setminus \mathbb{Z}_{\leq 0}). \end{cases} \tag{33}$$

The following formula is one of a number of definite integrals that may be expressed in terms of the Beta function (see, e.g., ([5], [p. 9, Equation (49)])):

$$\int_a^b (\tau-a)^{\alpha-1}(b-\tau)^{\beta-1}\, d\tau = (b-a)^{\alpha+\beta-1}\, B(\alpha, \beta) \tag{34}$$

$(b > a, \Re(\alpha) > 0, \Re(\beta) > 0).$

2. Recurrence Relations

This section explores some new recurrence relations for the two parametric Mittag-Leffler function (5).

Theorem 1. *Let $\mu, \varrho, z \in \mathbb{C}$ with $\Re(\mu) > 0$. Then*

$$E_{\mu,\varrho}(z) = \varrho(\varrho+1)E_{\mu,\varrho+2}(z) - \varrho(\varrho+1)zE_{\mu,\mu+\varrho+2}(z) + zE_{\mu,\mu+\varrho}(z) \quad (\Re(\varrho) > 0); \tag{35}$$

$$E_{\mu,\varrho}(z) = z^3 E_{\mu,3\mu+\varrho}(z) - z^2(\mu+\varrho)E_{\mu,2\mu+\varrho+1}(z) + z(\mu+\varrho)E_{\mu,\mu+\varrho+1}(z) + \frac{z^2}{\Gamma(2\mu+\varrho)} + \frac{1}{\Gamma(\varrho)} \quad (\Re(\varrho) > 0); \tag{36}$$

$$(\varrho-1)E_{\mu,\varrho}(z) = z^3 E_{\mu,3\mu+\varrho-1}(z) - zE_{\mu,\mu+\varrho-1}(z) + z(\varrho-1)E_{\mu,\mu+\varrho}(z) + \frac{z^2}{\Gamma(2\mu+\varrho-1)} + \frac{z}{\Gamma(\mu+\varrho-1)} + \frac{1}{\Gamma(\varrho-1)} \quad (\Re(\varrho) > 1); \tag{37}$$

$$(\varrho-1)E_{\mu,\varrho}(z) = z(\varrho-1)E_{\mu,\mu+\varrho}(z) - z^2 E_{\mu,2\mu+\varrho-1}(z) + \frac{1}{z}E_{\mu,\varrho-\mu-1}(z) - \frac{z}{\Gamma(\mu+\varrho-1)} - \frac{1}{z\Gamma(\varrho-\mu-1)} \quad (\Re(\varrho) > 1, \Re(\varrho-\mu) > 1); \tag{38}$$

$$(\varrho-1)E_{\mu,\varrho}(z) = (\varrho-1)z^2 E_{\mu,2\mu+\varrho}(z) - z^2 E_{\mu,2\mu+\varrho-1}(z) + \frac{1}{z}E_{\mu,\varrho-\mu-1}(z) - \frac{1}{z\Gamma(\varrho-\mu-1)} - \frac{\mu z}{\Gamma(\mu+\varrho)} \quad (\Re(\varrho) > 0, \Re(\varrho-\mu) > 1); \tag{39}$$

$$(\varrho-2)(\varrho-1)E_{\mu,\varrho}(z) = z(\varrho-1)(\varrho-2)E_{\mu,\mu+\varrho}(z) + z^3 E_{\mu,3\mu+\varrho-2}(z) - zE_{\mu,\mu+\varrho-2}(z) + \frac{z^2}{\Gamma(2\mu+\varrho-2)} + \frac{z}{\Gamma(\mu+\varrho-2)} + \frac{1}{\Gamma(\varrho-2)} \quad (\Re(\varrho) > 2); \tag{40}$$

$$(\mu+\varrho-2)(\mu+\varrho-1)E_{\mu,\varrho}(z) = (\mu+\varrho-2)(\mu+\varrho-1)z^2 E_{\mu,2\mu+\varrho}(z) + z^3 E_{\mu,3\mu+\varrho-2}(z) - z^2 E_{\mu,2\mu+\varrho-2}(z) + \frac{z^2}{\Gamma(2\mu+\varrho-2)} + \frac{z}{\Gamma(\mu+\varrho-2)} + \frac{(\mu+\varrho-2)(\mu+\varrho-1)}{\Gamma(\varrho)} \tag{41}$$

$$(\Re(\varrho) > 0, \Re(\varrho+\mu) > 2).$$

Proof. We establish only (35). Let $\mathcal{R}_1$ be the right-handed member of (35). By using (5), we obtain

$$\mathcal{R}_1 = \varrho(\varrho+1)\sum_{k=0}^{\infty} \frac{z^k}{\Gamma(\mu k+\varrho+2)} - \varrho(\varrho+1)\sum_{k=0}^{\infty} \frac{z^{k+1}}{\Gamma(\mu k+\mu+\varrho+2)} + \sum_{k=0}^{\infty} \frac{z^{k+1}}{\Gamma(\mu k+\mu+\varrho)}. \tag{42}$$

Setting $k + 1 = k'$ on the second and third summations in (42) and dropping the prime on k, we obtain

$$\begin{aligned}\mathcal{R}_1 &= \varrho(\varrho+1)\sum_{k=0}^{\infty}\frac{z^k}{\Gamma(\mu k+\varrho+2)} - \varrho(\varrho+1)\sum_{k=1}^{\infty}\frac{z^k}{\Gamma(\mu k+\varrho+2)} + \sum_{k=1}^{\infty}\frac{z^k}{\Gamma(\mu k+\varrho)} \\ &= \frac{\varrho(\varrho+1)}{\Gamma(\varrho+2)} + \sum_{k=1}^{\infty}\frac{z^k}{\Gamma(\mu k+\varrho)} = \frac{1}{\Gamma(\varrho)} + \sum_{k=1}^{\infty}\frac{z^k}{\Gamma(\mu k+\varrho)} \\ &= \sum_{k=0}^{\infty}\frac{z^k}{\Gamma(\mu k+\varrho)} = E_{\mu,\varrho}(z).\end{aligned} \quad (43)$$

For the third equality of (43), the following fundamental relation for the Gamma function

$$\Gamma(z+1) = z\Gamma(z) \quad (44)$$

is used. The proof of (35) is complete.

Likewise, the remaining relations (36)–(41) may be established. We omit specifics. □

Taking $\varrho = 1$ in (35), (36), and (41), we obtain some relations between the Mittag-Leffler function (1) and the two parametric Mittag-Leffler function (5) in the following corollary.

Corollary 1. *Let $\mu, z \in \mathbb{C}$ with $\Re(\mu) > 0$. Then*

$$E_\mu(z) = 2E_{\mu,3}(z) - 2zE_{\mu,\mu+3}(z) + zE_{\mu,\mu+1}(z); \quad (45)$$

$$\begin{aligned}E_\mu(z) = &z^3 E_{\mu,3\mu+1}(z) - z^2(\mu+1)E_{\mu,2\mu+2}(z) \\ &+ z(\mu+1)E_{\mu,\mu+2}(z) + \frac{z^2}{\Gamma(2\mu+1)} + 1;\end{aligned} \quad (46)$$

$$\begin{aligned}\mu(\mu-1)E_\mu(z) = &\mu(\mu-1)z^2 E_{\mu,2\mu+1}(z) + z^3 E_{\mu,3\mu-1}(z) \\ &- z^2 E_{\mu,2\mu-1}(z) + \frac{z^2}{\Gamma(2\mu-1)} + \frac{z}{\Gamma(\mu-1)} + \mu(\mu-1).\end{aligned} \quad (47)$$

Similar to (2), the following corollary provides certain interesting expressions of several elementary functions in terms of the two parametric functions (5).

Corollary 2. *Let $z \in \mathbb{C}$. Then*

$$e^z = E_1(z) = zE_{1,2}(z) + 2E_{1,3}(z) - 2zE_{1,4}(z); \quad (48)$$

$$e^z = E_1(z) = 2zE_{1,3}(z) + \left(z^3 - 2z^2\right)E_{1,4}(z) + \frac{z^2}{2} + 1; \quad (49)$$

$$\cos z = E_2(-z^2) = \left(2 - z^2\right)E_{2,3}(-z^2) + 2z^2 E_{2,5}(-z^2); \quad (50)$$

$$\begin{aligned}\cos z = E_2(-z^2) = &-3z^2 E_{2,4}(-z^2) - 3z^4 E_{2,6}(-z^2) \\ &- z^6 E_{2,7}(-z^2) + \frac{z^4}{24} + 1;\end{aligned} \quad (51)$$

$$\begin{aligned}\cos z = E_2(-z^2) = &-\frac{z^4}{2}E_{2,3}(-z^2) + \left(z^4 - \frac{z^6}{2}\right)E_{2,5}(-z^2) \\ &+ \frac{z^4}{4} - \frac{z^2}{2} + 1;\end{aligned} \quad (52)$$

$$\sin z = z\, E_{2,2}(-z^2) = \left(6z - z^3\right) E_{2,4}(-z^2) + 6z^3 E_{2,6}(-z^2); \tag{53}$$

$$\sin z = z\, E_{2,2}(-z^2) = -4z^3 E_{2,5}(-z^2) - 4z^5 E_{2,7}(-z^2) \\ -z^7 E_{2,8}(-z^2) + \frac{z^5}{120} + z; \tag{54}$$

$$\sin z = z\, E_{2,2}(-z^2) = z^3 E_{2,3}(-z^2) - z^3 E_{2,4}(-z^2) \\ - z^7 E_{2,7}(-z^2) + \frac{z^5}{24} - \frac{z^3}{2} + z; \tag{55}$$

$$\sin z = z\, E_{2,2}(-z^2) = -\frac{z^5}{6} E_{2,4}(-z^2) + \left(z^5 - \frac{z^7}{6}\right) E_{2,6}(-z^2) \\ + \frac{z^5}{36} - \frac{z^3}{6} + z; \tag{56}$$

$$\cosh z = E_2(z^2) = \left(2 + z^2\right) E_{2,3}(z^2) - 2z^2 E_{2,5}(z^2); \tag{57}$$

$$\cosh z = E_2(z^2) = 3z^2 E_{2,4}(z^2) - 3z^4 E_{2,6}(z^2) \\ + z^6 E_{2,7}(z^2) + \frac{z^4}{24} + 1; \tag{58}$$

$$\cosh z = E_2(z^2) = -\frac{z^4}{2} E_{2,3}(z^2) + \left(z^4 + \frac{z^6}{2}\right) E_{2,5}(z^2) \\ + \frac{z^4}{4} + \frac{z^2}{2} + 1; \tag{59}$$

$$\sinh z = z\, E_{2,2}(z^2) = \left(6z + z^3\right) E_{2,4}(z^2) - 6z^3 E_{2,6}(z^2); \tag{60}$$

$$\sinh z = z\, E_{2,2}(z^2) = 4z^3 E_{2,5}(z^2) - 4z^5 E_{2,7}(z^2) \\ + z^7 E_{2,8}(z^2) + \frac{z^5}{120} + z; \tag{61}$$

$$\sinh z = z\, E_{2,2}(z^2) = -z^3 E_{2,3}(z^2) + z^3 E_{2,4}(z^2) + \frac{z^5}{24} \\ + z^7 E_{2,7}(z^2) + \frac{z^3}{2} + z; \tag{62}$$

$$\sinh z = z\, E_{2,2}(z^2) = -\frac{z^5}{6} E_{2,4}(z^2) + \left(z^5 + \frac{z^7}{6}\right) E_{2,6}(z^2) \\ + \frac{z^5}{36} + \frac{z^3}{6} + z. \tag{63}$$

Proof. Setting $\mu = 1$ in (45) and (46), respectively, yields (48) and (49).

Putting $\mu = 2$ and replacing z by $-z^2$ in (45), (46), and (47), respectively, gives (50), (51), and (52).

Taking $\mu = 2, \varrho = 2$ and replacing z by $-z^2$ in (35), (36), (37), and (41), respectively, produces (53), (54), (55), and (56).

Setting $\mu = 2$ and replacing z by z^2 in (45), (46), and (47), respectively, offers (57), (58), and (59).

Putting $\mu = 2, \varrho = 2$ and replacing z by z^2 in (35), (36), (37), and (41), respectively, affords (60), (61), (62), and (63). □

By combining appropriate pairings of the identities in Corollary 2, such as (2), we may express several of the two parametric Mittag-Leffler functions in terms of elementary functions, as stated in the following corollary.

Corollary 3. *The following formulas hold.*

$$E_{2,3}(-z^2) = -\frac{1}{z^2}(\cos z - 1) \quad (z \in \mathbb{C} \setminus \{0\}); \tag{64}$$

$$E_{2,5}(-z^2) = \frac{1}{z^4}\left(\cos z + \frac{z^2}{2} - 1\right) \quad (z \in \mathbb{C} \setminus \{0\}); \tag{65}$$

$$E_{2,4}(-z^2) = \frac{1}{z^2}\left(1 - \frac{\sin z}{z}\right) \quad (z \in \mathbb{C} \setminus \{0\}); \tag{66}$$

$$E_{2,6}(-z^2) = \frac{1}{z^4}\left(-1 + \frac{z^2}{6} + \frac{\sin z}{z}\right) \quad (z \in \mathbb{C} \setminus \{0\}); \tag{67}$$

$$E_{2,3}(z^2) = \frac{1}{z^2}(\cosh z - 1) \quad (z \in \mathbb{C} \setminus \{0\}); \tag{68}$$

$$E_{2,5}(z^2) = \frac{1}{z^4}(\cosh z - 1) - \frac{1}{2z^2} \quad (z \in \mathbb{C} \setminus \{0\}); \tag{69}$$

$$E_{2,4}(z^2) = \frac{1}{z^3}\sinh z - \frac{1}{z^2} \quad (z \in \mathbb{C} \setminus \{0\}); \tag{70}$$

$$E_{2,6}(z^2) = \frac{1}{z^5}\sinh z - \frac{1}{z^4} - \frac{1}{6z^2} \quad (z \in \mathbb{C} \setminus \{0\}). \tag{71}$$

Proof. From (50) and (52), we can obtain (64) and (65).

From (53) and (56), we can derive (66) and (67).

From (57) and (59), we can obtain (68) and (69).

From (60) and (63), we can find (70) and (71). □

3. Certain Relations among the Fox–Wright Functions

In this section, by applying the Riemann–Liouville fractional integrals and derivatives to the recurrence relation (35), we obtain four relations among the Fox–Wright functions $_p\Psi_q$, as stated in the following theorems. We also consider some particular cases of our main results.

Theorem 2. *Let $\mu, \varrho, x \in \mathbb{R}^+$, and $a \in \mathbb{C}$. Then*

$$\begin{aligned}
{}_1\Psi_1\!\left[\begin{matrix}(1,1);\\(\mu+\varrho,\mu);\end{matrix}\,ax^\mu\right] &- ax^\mu {}_1\Psi_1\!\left[\begin{matrix}(1,1);\\(2\mu+\varrho,\mu);\end{matrix}\,ax^\mu\right] \\
&= \varrho(\varrho+1)\,{}_2\Psi_2\!\left[\begin{matrix}(\varrho,\mu),(1,1);\\(\varrho+2,\mu),(\mu+\varrho,\mu);\end{matrix}\,ax^\mu\right] \\
&\quad - a\varrho(\varrho+1)x^\mu\,{}_2\Psi_2\!\left[\begin{matrix}(\mu+\varrho,\mu),(1,1);\\(\mu+\varrho+2,\mu),(2\mu+\varrho,\mu);\end{matrix}\,ax^\mu\right].
\end{aligned} \tag{72}$$

Proof. A particular case of (34) is

$$\int_0^x \tau^{\alpha-1}(x-\tau)^{\beta-1}\,d\tau = x^{\alpha+\beta-1}\,B(\alpha,\beta) \tag{73}$$

$$(x > 0,\ \Re(\alpha) > 0,\ \Re(\beta) > 0).$$

Replacing z by az^μ in (35) and multiplying both sides of the resulting identity by $z^{\varrho-1}$, we obtain

$$\begin{aligned}
z^{\varrho-1}E_{\mu,\varrho}(az^\mu) &= \varrho(\varrho+1)z^{\varrho-1}E_{\mu,\varrho+2}(az^\mu) - a\varrho(\varrho+1)z^{\mu+\varrho-1}E_{\mu,\mu+\varrho+2}(az^\mu) \\
&\quad + az^{\mu+\varrho-1}E_{\mu,\mu+\varrho}(az^\mu).
\end{aligned} \tag{74}$$

Taking the left-sided Riemann–Liouville fractional integral (12) on both sides of (74), we obtain

$$\left(I_{0+}^{\mu}\left[z^{\varrho-1}E_{\mu,\varrho}(az^{\mu})\right]\right)(x) = \varrho(\varrho+1)\left(I_{0+}^{\mu}\left[z^{\varrho-1}E_{\mu,\varrho+2}(az^{\mu})\right]\right)(x)$$
$$- a\varrho(\varrho+1)\left(I_{0+}^{\mu}\left[z^{\mu+\varrho-1}E_{\mu,\mu+\varrho+2}(az^{\mu})\right]\right)(x) \quad (75)$$
$$+ a\left(I_{0+}^{\mu}\left[z^{\mu+\varrho-1}E_{\mu,\mu+\varrho}(az^{\mu})\right]\right)(x).$$

Let $\mathcal{L}_1$ be the left-handed member of (75). Interchanging the integral and summation in (5), which may be verified under restrictions, we obtain

$$\mathcal{L}_1 = \frac{1}{\Gamma(\mu)} \sum_{k=0}^{\infty} \frac{a^k}{\Gamma(\mu k + \varrho)} \int_0^x (x-\tau)^{\mu-1} \tau^{\mu k+\varrho-1}\, d\tau. \quad (76)$$

Using (73) in the integral in (76), we obtain

$$\mathcal{L}_1 = x^{\mu+\varrho-1} \sum_{k=0}^{\infty} \frac{\Gamma(k+1)}{\Gamma(\mu k + \mu + \varrho)} \frac{(ax^{\mu})^k}{k!} = x^{\mu+\varrho-1}\, {}_1\Psi_1\left[\begin{array}{c}(1,1);\\(\mu+\varrho,\mu);\end{array} ax^{\mu}\right]. \quad (77)$$

Now, let $\mathcal{R}_1$ be the right-handed member of (75). Similarly, as in obtaining (77), we derive

$$\mathcal{R}_1 = \varrho(\varrho+1)x^{\mu+\varrho-1}\, {}_2\Psi_2\left[\begin{array}{c}(\varrho,\mu),(1,1);\\(\varrho+2,\mu),(\mu+\varrho,\mu);\end{array} ax^{\mu}\right]$$
$$- a\varrho(\varrho+1)x^{2\mu+\varrho-1}\, {}_2\Psi_2\left[\begin{array}{c}(\mu+\varrho,\mu),(1,1);\\(\mu+\varrho+2,\mu),(2\mu+\varrho,\mu);\end{array} ax^{\mu}\right] \quad (78)$$
$$+ ax^{2\mu+\varrho-1}\, {}_1\Psi_1\left[\begin{array}{c}(1,1);\\(2\mu+\varrho,\mu);\end{array} ax^{\mu}\right].$$

Finally, equating the two identities in (77) and (78), we have (72). □

Theorem 3. *Let $x > 0$, $0 < \mu < 1 - \varrho$, and $a \in \mathbb{C}$. Then*

$$x^{\mu}\, {}_2\Psi_2\left[\begin{array}{c}(1-\mu-\varrho,\mu),(1,1);\\(\varrho,\mu),(1-\varrho,\mu);\end{array} ax^{-\mu}\right]$$
$$= \varrho(\varrho+1)x^{\mu}\, {}_2\Psi_2\left[\begin{array}{c}(1-\mu-\varrho,\mu),(1,1);\\(\varrho+2,\mu),(1-\varrho,\mu);\end{array} ax^{-\mu}\right] \quad (79)$$
$$- a\varrho(\varrho+1)\, {}_2\Psi_2\left[\begin{array}{c}(1-\varrho,\mu),(1,1);\\(\mu+\varrho+2,\mu),(\mu-\varrho+1,\mu);\end{array} ax^{-\mu}\right]$$
$$+ a\, {}_2\Psi_2\left[\begin{array}{c}(1-\varrho,\mu),(1,1);\\(\mu+\varrho,\mu),(\mu-\varrho+1,\mu);\end{array} ax^{-\mu}\right].$$

Proof. The following formula is readily derived:

$$\int_x^{\infty} (\tau-x)^{\alpha-1}\tau^{\beta-1}\, d\tau = x^{\alpha+\beta-1} B(\alpha, 1-\alpha-\beta) \quad (80)$$
$$(x > 0,\ 0 < \Re(\alpha) < 1 - \Re(\beta)).$$

Replacing z by $az^{-\mu}$ in (35) and multiplying both sides of the resulting identity by $z^{\varrho-1}$, we obtain

$$z^{\varrho-1}E_{\mu,\varrho}(az^{-\mu}) = \varrho(\varrho+1)z^{\varrho-1}E_{\mu,\varrho+2}(az^{-\mu}) - a\varrho(\varrho+1)z^{-\mu+\varrho-1}E_{\mu,\mu+\varrho+2}(az^{-\mu})$$
$$+ az^{-\mu+\varrho-1}E_{\mu,\mu+\varrho}(az^{-\mu}). \quad (81)$$

Taking the right-sided Riemann–Liouville fractional integral (13) on both sides of (81), we obtain

$$\left(I_{\infty-}^{\mu}\left[z^{\varrho-1}E_{\mu,\varrho}(az^{-\mu})\right]\right)(x) = \varrho(\varrho+1)\left(I_{\infty-}^{\mu}\left[z^{\varrho-1}E_{\mu,\varrho+2}(az^{-\mu})\right]\right)(x)$$
$$- a\varrho(\varrho+1)\left(I_{\infty-}^{\mu}\left[z^{-\mu+\varrho-1}E_{\mu,\mu+\varrho+2}(az^{-\mu})\right]\right)(x) \quad (82)$$
$$+ a\left(I_{\infty-}^{\mu}\left[z^{-\mu+\varrho-1}E_{\mu,\mu+\varrho}(az^{-\mu})\right]\right)(x).$$

Now, similarly as in the proof of Theorem 2, applying (13) to each term of (82) and using (80), we can obtain the desired identity (79). □

Theorem 4. *Let $\mu, \varrho, x \in \mathbb{R}^+$. Further, let $a \in \mathbb{C}$ and $n = [\mu] + 1 \in \mathbb{N}$. Then*

$$x^{-\mu}{}_1\Psi_1\left[\begin{array}{c}(1,1);\\(\varrho-\mu,\mu);\end{array}ax^{\mu}\right] - a \cdot {}_1\Psi_1\left[\begin{array}{c}(1,1);\\(\varrho,\mu);\end{array}ax^{\mu}\right]$$
$$= \varrho(\varrho+1)x^{-\mu}{}_2\Psi_2\left[\begin{array}{c}(\varrho,\mu),(1,1);\\(\varrho+2,\mu),(\varrho-\mu,\mu);\end{array}ax^{\mu}\right] \quad (83)$$
$$- a\varrho(\varrho+1){}_2\Psi_2\left[\begin{array}{c}(\mu+\varrho,\mu),(1,1);\\(\mu+\varrho+2,\mu),(\varrho,\mu);\end{array}ax^{\mu}\right].$$

Proof. Similar to (75), we have

$$\left(D_{0+}^{\mu}\left[z^{\varrho-1}E_{\mu,\varrho}(az^{\mu})\right]\right)(x) = \varrho(\varrho+1)\left(D_{0+}^{\mu}\left[z^{\varrho-1}E_{\mu,\varrho+2}(az^{\mu})\right]\right)(x)$$
$$- a\varrho(\varrho+1)\left(D_{0+}^{\mu}\left[z^{\mu+\varrho-1}E_{\mu,\mu+\varrho+2}(az^{\mu})\right]\right)(x) \quad (84)$$
$$+ a\left(D_{0+}^{\mu}\left[z^{\mu+\varrho-1}E_{\mu,\mu+\varrho}(az^{\mu})\right]\right)(x).$$

Let $\mathcal{L}$ be the left-handed member of (84). Using (14), (12), and (5), we obtain

$$\mathcal{L} = \left(\frac{d}{dx}\right)^n\left\{\frac{1}{\Gamma(n-\mu)}\int_0^x(x-\tau)^{n-\mu-1}\tau^{\varrho-1}\sum_{k=0}^{\infty}\frac{a^k}{\Gamma(\mu k+\varrho)}\tau^{\mu k}d\tau\right\}$$
$$= \left(\frac{d}{dx}\right)^n\left\{\frac{1}{\Gamma(n-\mu)}\sum_{k=0}^{\infty}\frac{a^k}{\Gamma(\mu k+\varrho)}\int_0^x(x-\tau)^{n-\mu-1}\tau^{\varrho+\mu k-1}d\tau\right\}, \quad (85)$$

under the restrictions of which integral and summation can be interchanged. Using (73) to evaluate the integral in (85) and interchanging differentiation and summation, we obtain

$$\mathcal{L} = \sum_{k=0}^{\infty}\frac{a^k}{\Gamma(n-\mu+\varrho+\mu k)}\left(\frac{d}{dx}\right)^n\left(x^{n-\mu+\varrho+\mu k-1}\right). \quad (86)$$

Employing the following easily derivable formula

$$\left(\frac{d}{dx}\right)^n\left(x^{\lambda}\right) = (-1)^n(-\lambda)_n x^{\lambda-n} = (-1)^n\frac{\Gamma(n-\lambda)}{\Gamma(-\lambda)}x^{\lambda-n} \quad (87)$$

in (86), we find

$$\mathcal{L} = x^{\varrho-\mu-1}\sum_{k=0}^{\infty}\frac{(ax^{\mu})^k}{\Gamma(n-\mu+\varrho+\mu k)}\cdot(-1)^n\frac{\Gamma(1+\mu-\varrho-\mu k)}{\Gamma(1-n+\mu-\varrho-\mu k)}. \quad (88)$$

Using the following well-known formula

$$\Gamma(z)\Gamma(1-z) = \frac{\pi}{\sin(\pi z)} \quad (z \in \mathbb{C} \setminus \mathbb{Z}) \tag{89}$$

in (88), we derive

$$\mathcal{L} = x^{\varrho-\mu-1} \sum_{k=0}^{\infty} \frac{(ax^\mu)^k}{\Gamma(\varrho - \mu + \mu k)}$$

$$= x^{\varrho-\mu-1} {}_1\Psi_1 \left[\begin{matrix} (1,1); \\ (\varrho - \mu, \mu); \end{matrix} ax^\mu \right]. \tag{90}$$

Further, the other three fractional derivatives in (84) can be evaluated as in (90). Finally, all of those evaluations that are used in (84) gives the desired identity (83). □

Theorem 5. *Let $\mu, \varrho, x \in \mathbb{R}^+$ and $0 < n - \mu < 1 - \varrho$. Further, let $a \in \mathbb{C}$ and $n = [\mu] + 1 \in \mathbb{N}$. Then*

$$
{}_2\Psi_2 \left[\begin{matrix} (\mu - \varrho + 1, \mu), (1,1); \\ (\varrho, \mu), (1 - \varrho, \mu); \end{matrix} ax^{-\mu} \right]
$$

$$
= \varrho(\varrho + 1) {}_2\Psi_2 \left[\begin{matrix} (1 + \mu - \varrho, \mu), (1,1); \\ (\varrho + 2, \mu), (1 - \varrho, \mu); \end{matrix} ax^{-\mu} \right]
$$

$$
- a\varrho(\varrho + 1) x^{-\mu} {}_2\Psi_2 \left[\begin{matrix} (1 + 2\mu - \varrho, \mu), (1,1); \\ (\mu + \varrho + 2, \mu), (1 + \mu - \varrho, \mu); \end{matrix} ax^{-\mu} \right] \tag{91}
$$

$$
+ a x^{-\mu} {}_2\Psi_2 \left[\begin{matrix} (1 + 2\mu - \varrho, \mu), (1,1); \\ (\mu + \varrho, \mu), (1 + \mu - \varrho, \mu); \end{matrix} ax^{-\mu} \right].
$$

Proof. Similar to (82), we have

$$
\left(D^\mu_{\infty-} \left[z^{\varrho-1} E_{\mu,\varrho}(az^{-\mu}) \right] \right)(x)
$$

$$
= \varrho(\varrho + 1) \left(D^\mu_{\infty-} \left[z^{\varrho-1} E_{\mu,\varrho+2}(az^{-\mu}) \right] \right)(x)
$$

$$
- a\varrho(\varrho + 1) \left(D^\mu_{\infty-} \left[z^{-\mu+\varrho-1} E_{\mu,\mu+\varrho+2}(az^{-\mu}) \right] \right)(x) \tag{92}
$$

$$
+ a \left(D^\mu_{\infty-} \left[z^{-\mu+\varrho-1} E_{\mu,\mu+\varrho}(az^{-\mu}) \right] \right)(x).
$$

As in the proof of Theorem 4, by using (80), we can evaluate all of the four fractional derivatives in (92). For example,

$$\left(D^\mu_{\infty-} \left[z^{\varrho-1} E_{\mu,\varrho}(az^{-\mu}) \right] \right)(x) = x^{\varrho-\mu-1} {}_2\Psi_2 \left[\begin{matrix} (\mu - \varrho + 1, \mu), (1,1); \\ (\varrho, \mu), (1 - \varrho, \mu); \end{matrix} ax^{-\mu} \right]. \tag{93}$$

Hence, all the evaluations which are set in (92) yields the desired identity (91). □

Since the results presented in Theorems 2–5 are quite general, they may reduce to yield a number of interesting identities. For example, setting $\mu = 1$ in the results, with the aid of (10), we obtain several new relations among the generalized hypergeometric functions ${}_pF_q$, as stated in the following corollary (for current research of summation and reduction formulae for ${}_pF_q$, see [68,69]).

Corollary 4. *Let $a \in \mathbb{C}$. Then*

$$
{}_1F_1 \left[\begin{matrix} 1; \\ \varrho + 1; \end{matrix} ax \right] - \frac{ax}{\varrho + 1} {}_1F_1 \left[\begin{matrix} 1; \\ \varrho + 2; \end{matrix} ax \right]
$$

$$
= {}_2F_2 \left[\begin{matrix} \varrho, 1; \\ \varrho + 2, \varrho + 1; \end{matrix} ax \right] - \frac{\varrho ax}{(\varrho + 2)(\varrho + 1)} {}_2F_2 \left[\begin{matrix} \varrho + 1, 1; \\ \varrho + 3, \varrho + 2; \end{matrix} ax \right] \tag{94}
$$

$$(x \in \mathbb{C},\ \varrho \in \mathbb{C} \setminus \mathbb{Z}_{\leq -1});$$

$$x\,{}_2F_2\!\left[\begin{matrix}-\varrho, 1;\\ \varrho+2, 1-\varrho;\end{matrix}\ \frac{a}{x}\right] - x\,{}_2F_2\!\left[\begin{matrix}-\varrho, 1;\\ \varrho, 1-\varrho;\end{matrix}\ \frac{a}{x}\right]$$
$$= \frac{a\varrho}{(\varrho-1)(\varrho+2)}\,{}_2F_2\!\left[\begin{matrix}1-\varrho, 1;\\ \varrho+3, 2-\varrho;\end{matrix}\ \frac{a}{x}\right] - \frac{a}{\varrho-1}\,{}_2F_2\!\left[\begin{matrix}1-\varrho, 1;\\ \varrho+1, 2-\varrho;\end{matrix}\ \frac{a}{x}\right] \tag{95}$$

$$(x \in \mathbb{C} \setminus \{0\},\ \varrho \in \mathbb{C} \setminus \mathbb{Z});$$

$${}_1F_1\!\left[\begin{matrix}1;\\ \varrho-1;\end{matrix}\ ax\right] - \frac{ax}{\varrho-1}\,{}_1F_1\!\left[\begin{matrix}1;\\ \varrho;\end{matrix}\ ax\right]$$
$$= {}_2F_2\!\left[\begin{matrix}\varrho, 1;\\ \varrho+2, \varrho-1;\end{matrix}\ ax\right] - \frac{\varrho\,ax}{(\varrho+2)(\varrho-1)}\,{}_2F_2\!\left[\begin{matrix}\varrho+1, 1;\\ \varrho+3, \varrho;\end{matrix}\ ax\right]; \tag{96}$$

$$(x \in \mathbb{C},\ \varrho \in \mathbb{C} \setminus \mathbb{Z}_{\leq 1});$$

$${}_2F_2\!\left[\begin{matrix}2-\varrho, 1;\\ \varrho+2, 1-\varrho;\end{matrix}\ \frac{a}{x}\right] - {}_2F_2\!\left[\begin{matrix}2-\varrho, 1;\\ \varrho, 1-\varrho;\end{matrix}\ \frac{a}{x}\right]$$
$$= \frac{a(\varrho-2)}{x(\varrho-1)(\varrho+2)}\,{}_2F_2\!\left[\begin{matrix}3-\varrho, 1;\\ \varrho+3, 2-\varrho;\end{matrix}\ \frac{a}{x}\right] - \frac{a(\varrho-2)}{x\varrho(\varrho-1)}\,{}_2F_2\!\left[\begin{matrix}3-\varrho, 1;\\ \varrho+1, 2-\varrho;\end{matrix}\ \frac{a}{x}\right] \tag{97}$$

$$(x \in \mathbb{C} \setminus \{0\},\ \varrho \in \mathbb{C} \setminus \mathbb{Z}).$$

It is worth noting that the restriction on each identity in Corollary 4 may be widened via analytic continuation. Further, each side of the identities (94) and (96) is easily checked to become 1.

4. Concluding Remarks and Posing Problems

We reviewed the birth of the Mittag-Leffler function and its several extensions (among numerous ones) together with their diverse applications to a variety of research areas, particularly, fractional calculus. We recalled many known recurrence (or differential recurrence) relations for the two parametric Mittag-Leffler function (5) and the three parametric Mittag-Leffler function (7). Then, we established a number of recurrence relations for the two parametric Mittag-Leffler function (5). Further, by using appropriate pairings of those recurrence relations presented in this paper, we demonstrated that certain particular cases of the two parametric Mittag-Leffler function (5) can be expressed in terms of elementary functions. Further, by applying the Riemann–Liouville fractional integral and derivative operators to one of the recurrence relations for the two parametric Mittag-Leffler function (5), we derived four new relations among the Fox–Wright functions. Finally, we provided four relations among the generalized hypergeometric functions $_pF_q$ as a particular case of those Fox–Wright function relations.

As in Section 3, by using the other formulas (36)–(41), we may derive some relations among the Fox–Wright functions $_p\Psi_q$ together with their corresponding relations among the generalized hypergeometric functions $_pF_q$ as particular cases. For example, as with Theorem 2, using (36), we obtain

$$
\begin{aligned}
{}_1\Psi_1 &\left[\begin{array}{c}(1,1);\\(\mu+\varrho,\mu);\end{array} ax^\mu\right] - a^3 x^{3\mu} {}_1\Psi_1\left[\begin{array}{c}(1,1);\\(4\mu+\varrho,\mu);\end{array} ax^\mu\right] \\
&= (\mu+\varrho)\, a\, x^\mu\, {}_2\Psi_2\left[\begin{array}{c}(\mu+\varrho,\mu),\,(1,1);\\(\mu+\varrho+1,\mu),\,(2\mu+\varrho,\mu);\end{array} ax^\mu\right] \\
&\quad - (\mu+\varrho)\, a^2 x^{2\mu}\, {}_2\Psi_2\left[\begin{array}{c}(2\mu+\varrho,\mu),\,(1,1);\\(2\mu+\varrho+1,\mu),\,(3\mu+\varrho,\mu);\end{array} ax^\mu\right] \\
&\quad + \frac{a^2 x^{2\mu}}{\Gamma(3\mu+\varrho)} + \frac{1}{\Gamma(\mu+\varrho)}
\end{aligned}
\qquad (98)
$$

$$(\mu,\,\varrho,\,x \in \mathbb{R}^+,\, a \in \mathbb{C})$$

and

$$
\begin{aligned}
{}_1F_1&\left[\begin{array}{c}1;\\1+\varrho;\end{array} ax\right] - \frac{a^3 x^3}{(\varrho+1)(\varrho+2)(\varrho+3)} {}_1F_1\left[\begin{array}{c}1;\\\varrho+4;\end{array} ax\right] \\
&= \frac{ax}{\varrho+1}\, {}_2F_2\left[\begin{array}{c}\varrho+1,\,1;\\\varrho+2,\,\varrho+2;\end{array} ax\right] - \frac{a^2 x^2}{(\varrho+2)^2}\, {}_2F_2\left[\begin{array}{c}\varrho+2,\,1;\\\varrho+3,\,\varrho+3;\end{array} ax\right] \\
&\quad + \frac{a^2 x^2}{(\varrho+1)(\varrho+2)} + 1 \quad (\varrho \in \mathbb{C} \setminus \mathbb{Z}_{\leq -1},\, a,\, x \in \mathbb{C}).
\end{aligned}
\qquad (99)
$$

Here we pose the following problems:

(i) Based on the other recurrence relations (36)–(41), by using Riemann–Liouville fractional integral and derivative operators, try to present certain relations among the Fox–Wright functions $_p\Psi_q$;
(ii) Demonstrate some particular cases of the identities given in (i);
(iii) Based on the recurrence relations (35)–(41), by using the other fractional integral and derivative operators, try to present certain relations among some special functions;
(iv) Consider certain particular cases of identities that will be derived in (iii).

Author Contributions: Writing—original draft, D.S.S., S.J. and J.C.; Writing—review and editing, D.S.S., S.J. and J.C. All authors have read and agreed to the published version of the manuscript.

Funding: The third-named author was supported by the Basic Science Research Program through the National Research Foundation of Korea (NRF) funded by the Ministry of Education (NRF-2020R1I1A1A01052440).

Institutional Review Board Statement: Not applicable.

Informed Consent Statement: Not applicable.

Data Availability Statement: Not applicable.

Acknowledgments: The authors are very thankful of the anonymous referees' constructive and encouraging comments, which helped to improve this article.

Conflicts of Interest: The authors have no conflict of interest.

References

1. Mittag-Leffler, M.G. Sur l'intégrale de Laplace-Abel. *C. R. Acad. Sci. Paris (Ser. II)* **1902**, *136*, 937–939.
2. Mittag-Leffler, M.G. Sur la nouvelle fonction $E_\alpha(x)$. *C. R. Acad. Sci. Paris (Ser. II)* **1903**, *137*, 554–558.
3. Mittag-Leffler, M.G. Sopra la funzione $E_\alpha(x)$. *Rend. Accad. Lincei* **1904**, *13*, 3–5.
4. Buhl, A. *Séries Analytiques, Sommabilité*, Number 7 in Mémorial des Sciences Mathématiques; Gauthier-Villars: Paris, France, 1925.
5. Srivastava, H.M.; Choi, J. *Zeta and q-Zeta Functions and Associated Series and Integrals*; Elsevier Science Publishers: Amsterdam, The Netherlands; London, UK; New York, NY, USA, 2012.
6. Gorenflo, R.; Kilbas, A.A.; Mainardi, F.; Rogosin, S. *Mittag-Leffler Functions, Related Topics and Applications*, 2nd ed.; Springer: Berlin, Germany, 2020.

7. Abramowitz, M.; Stegun, I.A. (Eds.) *Handbook of Mathematical Functions with Formulas, Graphs, and Mathematical Tables*; Applied Mathematics Series 55, Ninth Printing; National Bureau of Standards: Washington, DC, USA, 1972; Reprint of the 1972 Edition, Dover Publications, Inc.: New York, NY, USA, 1992.
8. Djrbashian, M.M. *Harmonic Analysis and Boundary Value Problems in the Complex Domain*; Birkhauser Verlay: Basel, Switzerland; Boston, MA, USA; Berlin, Germany, 1996.
9. Djrbashian, M.M. *Integral Transforms and Representations of Functions in the Complex Domain*; Nauka: Moscow, Russia, 1966.
10. Erdélyi, A.; Magnus, W.; Oberhettinger, F.; Tricomi, F.G. *Higher Transcendental Functions*; McGraw-Hill Book Company: New York, NY, USA; Toronto, ON, Canada; London, UK, 1955; Volume 3.
11. Gorenflo, R.; Luchko, Y.; Rogosin, S. *Mittag-Leffler Type Functions: Notes on Growth Properties and Distribution of Zeros*; Preprint No. A04-97; Serie A Mathematik; Freie Universität Berlin: Berlin, Germany, 1997.
12. Gorenflo, R.; Mainardi, F. *Fractional Oscillations and Mittag-Leffler Type Functions*; Preprint No. A14-96; Freie Universität Berlin, Serie A Mathematik; Berlin, Germany, 1996.
13. Gorenflo, R.; Mainardi, F. The Mittag-Leffler type functions in the Riemann-Liouville fractional calculus. In *Boundary Value Problems, Special Functions and Fractional Calculus*; Kilbas, A.A., Ed.; Proceedings of International Conference (Minsk, 1996); Belarusian State University: Minsk, Belarus, 1996; pp. 215–225.
14. Mittag-Leffler, G. Sur la représentation analytique d'une branche uniforme d'une fonction monogène (cinquième note). *Acta Math.* **1905**, *29*, 101–181. [CrossRef]
15. Wiman, A. Über den fundamentalsatz in der theorie der funktionen $E_\alpha(x)$. *Acta Math.* **1905**, *29*, 191–201. [CrossRef]
16. Wiman, A. Über die nullsteliun der funktionen $E_\alpha(x)$. *Acta Math.* **1905**, *29*, 217–234. [CrossRef]
17. Mainardi, F. Why the Mittag-Leffler Function can be considered the queen function of the fractional calculus? *Entropy* **2020**, *22*, 1359. [CrossRef]
18. Mathai, A.M.; Saxena, R.K. *The H-Function with Applications in Statistics and Other Disciplines*; Halsted Press (John Wiley & Sons): New York, NY, USA; London, UK; Sydney, Australia; Toronto, ON, Canada, 1978.
19. Mathai, A.M.; Saxena, R.K.; Haubold, H.J. *The H-Function: Theory and Applications*; Springer: New York, NY, USA, 2010.
20. Bagley, R.L. On the fractional order initial value problem and its engineering applications. In *Fractional Calculus and Its Applications*; Proceedings of International Conference (Tokyo, 1989); College of Engineering, Nihon University: Tokyo, Japan, 1990; pp. 12–20.
21. Beyer, Y.; Kempfle, S. Definition of physically consistent damping laws with fractional derivatives. *Z. Angew. Math. Mech.* **1995**, *75*, 623–635. [CrossRef]
22. Caputo, M.; Mainardi, F. Linear models of dissipation in anelastic solids. *Riv. Nuovo Cimento* **1971**, *1*, 161–198. [CrossRef]
23. Goldsmith, P.L. The calculation of true practicle size distributions from the sizes observed in a thin slice. *Br. J. Appl. Phys.* **1967**, *18*, 813–830. [CrossRef]
24. Haubold, H.J.; Mathai, A.M.; Saxena, R.K. Mittag-Leffler functions and their applications. *J. Appl. Math.* **2011**, *2011*, 298628. [CrossRef]
25. Nonnenmacher, T.F.; Glockle, W.G. A fractional model for mechanical stress relaxation. *Philos. Mag. Lett.* **1991**, *64*, 89–93. [CrossRef]
26. Schneider, W.R.; Wyss, W. Fractional diffusion and wave equations. *J. Math. Phys.* **1989**, *30*, 134–144. [CrossRef]
27. Humbert, P.; Agrawal, R.P. Sur la fonction de Mittag-Leffler et quelques-unes de ses généralisations. *Bull. Sci. Math.* **1953**, *77*, 180–185.
28. Hille, E.; Tamarkin, J.D. On the theory of linear integral equations. *Ann. Math.* **1930**, *31*, 479–528. doi: 10.2307/1968241. [CrossRef]
29. Kilbas, A.A.; Saigo, M. On solution of integral equations of Abel-Volterra type. *Differ. Integral Equ.* **1995**, *8*, 993–1011.
30. Kilbas, A.A.; Saigo, M. On Mittag-Leffler type function, fractional calculus operators and solutions of integral equations. *Integral Transform. Spec. Funct.* **1996**, *4*, 355–370. [CrossRef]
31. Kilbas, A.A.; Saigo, M. Fractional integrals and derivatives of Mittag-Leffler type function (Russian). *Doklady Akad. Nauk Belarusi* **1995**, *39*, 22–26.
32. Kilbas, A.A.; Saigo, M. Solution of Abel type integral equations of second kind and differential equations of fractional order (Russian). *Doklady Akad. Nauk Belarusi* **1995**, *39*, 29–34.
33. Miller, K.S.; Ross, B. *An Introduction to the Fractional Calculus and Fractional Differential Equations*; Wiley: New York, NY, USA, 1993.
34. Samko, S.G.; Kilbas, A.A.; Marichev, O.I. *Fractional Integrals and Derivatives: Theory and Applications*; Gordon and Breach Science Publishers: Yverdon, Switzerland; Amsterdam, The Netherlands; Reading, UK; Tokyo, Japan; Paris, France; Berlin, Germany; Langhorne, PA, USA; Victoria, Australia, 1993.
35. Choi, J.; Parmar, R.K.; Chopra, P. Extended Mittag-Leffler function and associated fractional calculus operators. *Georgian Math. J.* **2019**, *27*, 11. [CrossRef]
36. Mainardi, F.; Gorenflo, R. Time-fractional derivatives in relaxation processes: A tutorial survey. *Fract. Calc. Appl. Anal.* **2007**, *10*, 269–308.
37. Agarwal, R.P. A propos d'une note de M. Pierre Humbert. *CR Acad. Sci. Paris* **1953**, *236*, 2031–2032.
38. Fry, C.G.; Hughes, H.K. Asymptotic developments of certain integral functions. *Duke Math. J.* **1942**, *9*, 791–802. [CrossRef]
39. Humbert, P. Quelques resultatsrelatifs a la fonction de Mittag-Leffler. *CR Acad. Sci. Paris* **1953**, *236*, 1467–1468.
40. Schneider, W.R. Completely monotone generalized Mittag-Leffler functions. *Expo. Math.* **1996**, *14*, 3–24.
41. Sedletskii, A.M. Asymptotic formulas for zero of a function of Mittag-Leffler type (Russian). *Anal. Math.* **1994**, *20*, 117–132.

42. Giusti, A.; Colombaro, I.; Garra, R.; Garrappa, R.; Polito, F.; Popolizio, M.; Mainardi, F. A practical guide to Prabhakar fractional calculus. *Fract. Calc. Appl. Anal.* **2020**, *23*, 9–54. [CrossRef]
43. Prabhakar, T.R. A singular integral equation with a generalized Mittag–Leffler function in the kernel. *Yokohama Math. J.* **1971**, *19*, 7–15.
44. Wright, E.M. The asymptotic expansion of the generalized hypergeometric function. *Proc. Lond. Math. Soc.* **1940**, *46*, 389–408. [CrossRef]
45. Wright, E.M. The asymptotic expansion of integral functions defined by Taylor series. *Philos. Trans. R. Soc. Lond. A* **1940**, *238*, 423–451. [CrossRef]
46. Wright, E.M. The asymptotic expansion of integral functions defined by Taylor series (second paper). *Philos. Trans. R. Soc. Lond. A* **1941**, *239*, 217–232. [CrossRef]
47. Fox, C. The asymptotic expansion of generalized hypergeometric functions. *Proc. Lond. Math. Soc.* **1928**, *27*, 389–400. [CrossRef]
48. Srivastava, H.M.; Karlsson, P.W. *Multiple Gaussian Hypergeometric Series*; Halsted Press (Ellis Horwood Limited): Chichester, UK; John Wiley and Sons: New York, NY, USA; Chichester, UK; Brisbane, Australia; Toronto, ON, Canada, 1985.
49. Kilbas, A.A.; Srivastava, H.M.; Trujillo, J.J. *Theory and Applications of Fractional Differential Equations*; North-Holland Mathematical Studies, Elsevier (North-Holland) Science Publishers: Amsterdam, The Netherlands; London, UK; New York, NY, USA, 2006; Volume 204.
50. Srivastava, H.M.; Tomovski, Ž. Fractional calculus with an integral operator containing a generalized Mittag-Leffler function in the kernel. *Appl. Math. Comput.* **2009**, *211*, 198–210. [CrossRef]
51. Gupta, I.S.; Debnath, L. Some properties of the Mittag-Leffler functions. *Integral Transform. Spec. Funct.* **2007**, *18*, 329–336. [CrossRef]
52. Saxena, R.K. Certain properties of generalized Mittag-Leffler function. In Proceedings of the 3rd Annual Conference of the Society for Special Functions and Their Applications, Chennai, India, 22 June–3 July 2002; pp. 77–81.
53. Hilfer, R.; Seybold, H.J. Computation of the generalized Mittag-Leffler function and its inverse in the complex plane. *Integral Transform. Spec. Funct.* **2006**, *17*, 637–652. [CrossRef]
54. Podlubny, I. *Fractional Differential Equations*; Academic Press: San Diego, CA, USA; Boston, MA, USA; New York, NY, USA; London, UK; Sydney, Australia; Tokyo, Japan; Toronto, ON, Canada, 1999.
55. Seybold, H.; Hilfer, R. Numerical Algorithm for calculating the generalized Mittag-Leffler function. *SIAM J. Numer. Anal.* **2008**, *47*, 69–88. [CrossRef]
56. Berberan-Santos, M.N. Properties of the Mittag-Leffler relaxation function. *J. Math. Chem.* **2005**, *38*, 629–635. [CrossRef]
57. Garra, R.; Garrappa, R. The Prabhakar or three parameter Mittag-Leffler function: Theory and application. *Commun. Nonlinear Sci. Numer. Simul.* **2018**, *56*, 314–329. [CrossRef]
58. Shukla, A.K.; Prajapati, J.C. On a recurrence relation of generalized Mittag-Leffler function. *Surv. Math. Appl.* **2009**, *4* 133–138.
59. Shukla, A.K.; Prajapati, J.C. On a generalization of Mittag-Leffler function and its properties. *J. Math. Anal. Appl.* **2007**, *336*, 797–811. [CrossRef]
60. Salim, T.O. Some properties relating to the generalized Mittag-Leffler function. *Adv. Appl. Math. Anal.* **2009**, *4*, 21–30.
61. Kurulay, M.; Bayram, M. Some properties of the Mittag-Leffler functions and their relation with the Wright functions. *Adv. Diff. Equ.* **2012**, *2012*, 181. [CrossRef]
62. Dhakar, V.S.; Sharma, K. On a recurrence relation of K-Mittag-Leffler function. *Commun. Korean Math. Soc.* **2013**, *28*, 851–856. [CrossRef]
63. Dorrego, G.A.; Cerutti, R.A. The k-Mittag-Leffler function. *Int. J. Contemp. Math. Sci.* **2012**, *7*, 705–716.
64. Díaz, R.; Pariguan, E. On hypergeometric functions and k-Pochhammer symbol. *Divulg. Mat.* **2007**, *15*, 179–192.
65. Sharma, S.; Jain, R. On some recurrence relations of generalized q-Mittag Leffler function. *Math. Aeterna* **2016**, *6*, 791–795.
66. Gehlot, K.S. Recurrence relation and integral representation of p-k Mittag-Leffler function. *Palest. J. Math.* **2021**, *10*, 290–298.
67. Gehlot, K.S. The p-k Mittag-Leffler function. *Palest. J. Math.* **2018**, *7*, 628–632.
68. Choi, J. Certain applications of generalized Kummer's summation formulas for $_2F_1$. *Symmetry* **2021**, *13*, 1538. [CrossRef]
69. Choi, J.; Qureshi, M.I.; Bhat, A.H.; Majid, J. Reduction formulas for generalized hypergeometric series associated with new sequences and applications. *Fractal Fract.* **2021**, *5*, 150. [CrossRef]

fractal and fractional

Article

Fejér–Hadamard Type Inequalities for (α, h-m)-p-Convex Functions via Extended Generalized Fractional Integrals

Ghulam Farid [1], Muhammad Yussouf [2] and Kamsing Nonlaopon [3,*]

[1] Department of Mathematics, COMSATS University Islamabad, Attock Campus, Attock 43600, Pakistan; ghlmfarid@ciit-attock.edu.pk
[2] Department of Mathematics, Government Graduate Talim-ul-Islam College, Chenab Nagar 35460, Pakistan; mohsin2013.uos@gmail.com
[3] Department of Mathematics, Faculty of Science, Khon Kaen University, Khon Kaen 40002, Thailand
* Correspondence: nkamsi@kku.ac.th; Tel.: +668-6642-1582

Abstract: Integral operators of a fractional order containing the Mittag-Leffler function are important generalizations of classical Riemann–Liouville integrals. The inequalities that are extensively studied for fractional integral operators are the Hadamard type inequalities. The aim of this paper is to find new versions of the Fejér–Hadamard (weighted version of the Hadamard inequality) type inequalities for (α, h-m)-p-convex functions via extended generalized fractional integrals containing Mittag-Leffler functions. These inequalities hold simultaneously for different types of well-known convexities as well as for different kinds of fractional integrals. Hence, the presented results provide more generalized forms of the Hadamard type inequalities as compared to the inequalities that already exist in the literature.

Keywords: (α, h-m)-p-convex function; Fejér–Hadamard inequality; Mittag-Leffler function; extended generalized fractional integrals

MSC: 26A51; 26A33; 33E12

1. Introduction

Convexity in alliance with integral inequalities is an attractive area of research. Researchers define novel types of convexities for the need of hour. Convex functions and mathematical inequalities play a vital role in the progress of diverse fields of pure and applied sciences. A large number of inequalities have been established for convex functions, see [1–10] and references therein. On the other hand, fractional integral and derivative operators are important tools to generalize the classical concepts and methods that are based on ordinary integration and derivative. Fractional integral and derivative operators lead in the study of fractional differential equations [11], fractional initial and boundary value problems [12], and fractional dynamical systems [13].

The Mittag-Leffler function appears in the solutions of fractional differential equations, the same as how likely the exponential function appears in solving differential equations. Fractional integral operators containing Mittag-Leffler function are also developed and applied to study well-known real world problems, see [14–19] and references therein.

In recent years, fractional integral operators, as well as new classes of functions closely related to convex functions, play a significant role in extensions and generalizations of classical inequalities. The most celebrated inequality which is investigated for fractional integrals is the well-known Hadamard inequality. Sarikaya, in [6,20], proved two versions of the Hadamard inequality for convex functions by using Riemann–Liouville fractional integral operators. After that, plenty of such versions have been published by defining new kinds of functions related to convex functions via fractional integrals related to Riemann–Liouville integrals. Farid, in [21], gave the Hadamard and the Fejér–Hadamard inequalities for convex functions by using the fractional integral operators containing Mittag-Leffler functions. For other known and new classes of functions, the Hadamard and the Fejér–Hadamard fractional inequalities can be found in [22–30] and references therein.

Inspired by a huge number of findings in the credit of the Hadamard and the Fejér–Hadamard inequalities for fractional integrals, we are motivated in establishing the generalized forms of such type of inequalities by utilizing the class of so called $(\alpha, h\text{-}m)\text{-}p$-convex functions and the fractional integral operators involving Mittag-Leffler functions. Next, we give definitions and important notions which will be utilized in proving the results of this paper.

Definition 1 ([31]). *A function $f : [a,b] \to \mathbb{R}$ is said to be convex, if the following inequality holds:*

$$f(ta + (1-t)b) \leq tf(a) + (1-t)f(b), \quad \forall t \in [0,1]. \tag{1}$$

The Fejér–Hadamard inequality for convex functions is stated in the forthcoming theorem, while the Hadamard inequality can be deduced for $\hbar \equiv 1$.

Theorem 1 ([32]). *Let $f : [a,b] \to \mathbb{R}$ be a convex function with $a < b$. Then, the following inequality holds:*

$$f\left(\frac{a+b}{2}\right)\int_a^b \hbar(x)dx \leq \int_a^b f(x)\hbar(x)dx \leq \frac{f(a)+f(b)}{2}\int_a^b \hbar(x)dx,$$

where $\hbar : [a,b] \to \mathbb{R}$ is non-negative, integrable and symmetric function about $\frac{a+b}{2}$.

In [30], Jia et al. defined the class of $(\alpha, h\text{-}m)\text{-}p$-convex functions given as follows:

Definition 2. *Let $J \subseteq \mathbb{R}$ be an interval containing $(0,1)$. Let $I \subset (0,\infty)$ be a real interval and $p \in \mathbb{R} \setminus \{0\}$. Moreover, let $h : J \to \mathbb{R}$ be a non-negative function. A function $f : I \to \mathbb{R}$ is said to be $(\alpha, h\text{-}m)\text{-}p$-convex, if*

$$f\left((ta^p + m(1-t)b^p)^{\frac{1}{p}}\right) \leq h(t^\alpha)f(a) + mh(1-t^\alpha)f(b), \quad \forall t \in (0,1), (\alpha, m) \in [0,1]^2, \tag{2}$$

provided $(ta^p + m(1-t)b^p)^{\frac{1}{p}} \in I$.

The inequality (2), provides the definition of $(h\text{-}m)\text{-}p$-convex functions for $\alpha = 1$; $(\alpha, m)\text{-}p$-convex functions for $h(t) = t$; $(\alpha, h)\text{-}p$-convex functions for $m = 1$; $(s,m)\text{-}p$-convex functions for $h(t) = t^s, \alpha = 1$; $(\alpha, h\text{-}m)$-convex functions for $p = 1$; $(h\text{-}m)$-convex functions for $\alpha = p = 1$; $(h\text{-}p)$-convex functions for $\alpha = m = 1$; h-convex function for $\alpha = p = m = 1$; s-convex functions for $h(t) = t^s, \alpha = p = m = 1$; m-convex functions for $h(t) = t, \alpha = p = 1$; p-convex functions for $h(t) = t, \alpha = m = 1$; convex functions for $h(t) = t, \alpha = p = m = 1$.

Next, we give the definition of integral operators involving Mittag-Leffler functions as follows:

Definition 3 ([33]). *Let $f,g : [a,b] \to \mathbb{R}$, $0 < a < b$ be the functions such that f be positive and $f \in L_1[a,b]$ and g be a differentiable and strictly increasing. Moreover, let $\frac{\varphi}{x}$ be an increasing function on $[a,\infty)$ and $\mu, \psi, \phi, \rho, c \in \mathbb{C}$, $\Re(\psi), \Re(\phi) > 0$, $\Re(c) > \Re(\rho) > 0$ with $q \geq 0, \sigma, r > 0$ and $0 < k \leq r + \sigma$. Then, for $x \in [a,b]$, the integral operators are defined by:*

$$({_g}F^{\varphi,\rho,r,k,c}_{\sigma,\psi,\phi,\mu,a^+}f)(x;q) = \int_a^x \frac{\varphi(g(x)-g(t))}{g(x)-g(t)} E^{\rho,r,k,c}_{\sigma,\psi,\phi}(\mu(g(x)-g(t))^\sigma;q)f(t)d(g(t)), \tag{3}$$

110

$$\left({}_g F^{\varphi,\rho,r,k,c}_{\sigma,\psi,\phi,\mu,b^-} f\right)(x;q) = \int_x^b \frac{\varphi(g(t)-g(x))}{g(t)-g(x)} E^{\rho,r,k,c}_{\sigma,\psi,\phi}(\mu(g(t)-g(x))^\sigma;q) f(t) d(g(t)). \quad (4)$$

Here

$$E^{\rho,r,k,c}_{\sigma,\psi,\phi}(t;q) = \sum_{n=0}^\infty \frac{\beta_q(\rho+nk, c-\rho)(c)_{nk} t^n}{\beta(\rho,c-\rho)\Gamma(\sigma n+\psi)(\phi)_{nr}}, \quad (5)$$

is the generalized Mittag-Leffler function and $(c)_{nk}$ is the generalized Pochhammer symbol defined as follows:

$$(c)_{nk} = \frac{\Gamma(c+nk)}{\Gamma(c)},$$

$\Gamma(.)$ is gamma function defined as follows:

$$\Gamma(\psi) = \int_0^\infty e^{-t} t^{\psi-1} dt, \quad \psi > 0,$$

β_q is the extension of beta function defined as follows:

$$\beta_q(x,y) = \int_0^1 t^{x-1}(1-t)^{y-1} e^{-\frac{q}{t(1-t)}} dt.$$

The following definition of extracted generalized fractional integral operators from Definition 3 is very useful to obtain the Fejér–Hadamard type inequalities.

Definition 4 ([27]). *Let $f, g : [a,b] \to \mathbb{R}$, $0 < a < b$ be the functions such that f be positive and $f \in L_1[a,b]$ and g be a differentiable and strictly increasing. Moreover, let $\mu, \psi, \phi, \rho, c \in \mathbb{C}$, $\Re(\psi), \Re(\phi) > 0, \Re(c) > \Re(\rho) > 0$ with $q \geq 0, \sigma, r > 0$ and $0 < k \leq r + \sigma$. Then, for $x \in [a,b]$, the integral operators are defined by:*

$$\left({}_g Y^{\rho,r,k,c}_{\sigma,\psi,\phi,\mu,a^+} f\right)(x;q) = \int_a^x (g(x)-g(t))^{\psi-1} E^{\rho,r,k,c}_{\sigma,\psi,\phi}(\mu(g(x)-g(t))^\sigma;q) f(t) d(g(t)), \quad (6)$$

$$\left({}_g Y^{\rho,r,k,c}_{\sigma,\psi,\phi,\mu,b^-} f\right)(x;q) = \int_x^b (g(t)-g(x))^{\psi-1} E^{\rho,r,k,c}_{\sigma,\psi,\phi}(\mu(g(t)-g(x))^\sigma;q) f(t) d(g(t)). \quad (7)$$

Remark 1. *The integral operators (6) and (7) provide the following fractional integral operators:*
(i) For $g(x) = x$, we recover the fractional integral operators defined by Andrić et al. in [34].
(ii) For $g(x) = x$ and $q = 0$, we recover the fractional integral operators defined by Salim-Faraj in [17].
(iii) For $g(x) = x$ and $\phi = r = 1$, we recover the fractional integral operators defined by Rahman et al. in [16].
(iv) For $g(x) = x$, $q = 0$ and $\phi = r = 1$, we recover the fractional integral operators defined by Srivastava-Tomovski in [19].
(v) For $g(x) = x$, $q = 0$ and $\phi = r = k = 1$, we recover the fractional integral operators defined by Prabhakar in [15].
(vi) For $g(x) = x$ and $\mu = q = 0$, we recover the Riemann–Liouville fractional integrals.

In the next section, we give two versions of Fejér–Hadamard inequalities for $(\alpha, h-m)-p$-convex functions via generalized fractional integral operators. Moreover, we give the Fejér–Hadamard inequalities for different classes of convex functions which are deducible from $(\alpha, h-m)-p$-convex functions.

In the whole paper, we use the following notations frequently:

$$\left(\mathcal{F}^{a^+}_{b,\sigma,\psi}\right)(\mu,f) = \left({}_g Y^{\rho,r,k,c}_{\sigma,\psi,\phi,\mu,a^+} f\right)(b;q), \quad \left(\mathcal{F}^{b^-}_{a,\sigma,\psi}\right)(\mu,f) = \left({}_g Y^{\rho,r,k,c}_{\sigma,\psi,\phi,\mu,b^-} f\right)(a;q).$$

2. Fejér–Hadamard Type Inequalities

Theorem 2. *Let $f, g, \hbar : [a,b] \to \mathbb{R}$, $0 < a < mb$, $m \in (0,1]$, Range (g), Range $(\hbar) \subset [a,b]$ be the functions such that f is positive and $f \in L_1[a,b]$, g is differentiable and strictly increasing. If f is $(\alpha, h\text{-}m)\text{-}p$-convex on $[a,b]$ and $\hbar\left((g^p(a) + mg^p(b) - mg(y))^{\frac{1}{p}}\right) = \hbar((g(y))^{\frac{1}{p}})$, then for (6) and (7), we have the following inequalities:*

(i) *If $p > 0$, then*

$$f\left(\left(\frac{g^p(a) + mg^p(b)}{2}\right)^{\frac{1}{p}}\right)\left(\mathcal{F}^{g^{-1}(g^p(a))^+}_{g^{-1}(mg^p(b)),\sigma,\psi}\right)(\mu', \hbar \circ \zeta) \qquad (8)$$

$$\leq h\left(\frac{1}{2^\alpha}\right)\left(\mathcal{F}^{g^{-1}(g^p(a))^+}_{g^{-1}(mg^p(b)),\sigma,\psi}\right)(\mu', f\hbar \circ \zeta) + m^{\psi+1}h\left(\frac{2^\alpha - 1}{2^\alpha}\right)\left(\mathcal{F}^{g^{-1}(g^p(b))^-}_{g^{-1}\left(\frac{g^p(a)}{m}\right),\sigma,\psi}\right)(m^\sigma\mu', f\hbar \circ \zeta)$$

$$\leq \left(h\left(\frac{1}{2^\alpha}\right)f(g(a)) + mh\left(\frac{2^\alpha - 1}{2^\alpha}\right)f(g(b))\right)\int_0^1 t^{\psi-1}E^{\rho,r,k,c}_{\sigma,\psi,\phi}(\mu t^\sigma; q)$$

$$\times \hbar\left((tg^p(a) + m(1-t)g^p(b))^{\frac{1}{p}}\right)h(t^\alpha)dt + m\left(h\left(\frac{1}{2^\alpha}\right)f(g(b)) + mh\left(\frac{2^\alpha - 1}{2^\alpha}\right)f\left(\frac{g(a)}{m^2}\right)\right)$$

$$\times \int_0^1 t^{\psi-1}E^{\rho,r,k,c}_{\sigma,\psi,\phi}(\mu t^\sigma; q)\hbar\left((tg^p(a) + m(1-t)g^p(b))^{\frac{1}{p}}\right)h(1-t^\alpha)dt,$$

$\zeta(z) = g^{\frac{1}{p}}(z), z \in [a^p, mb^p], f\hbar \circ \zeta = (f \circ \zeta)(\hbar \circ \zeta), \mu' = \frac{\mu}{(mg^p(b) - g^p(a))^\sigma}.$

(ii) *If $p < 0$, then*

$$f\left(\left(\frac{g^p(a) + mg^p(b)}{2}\right)^{\frac{1}{p}}\right)\left(\mathcal{F}^{g^{-1}(g^p(a))^-}_{g^{-1}(mg^p(b)),\sigma,\psi}\right)(\mu', \hbar \circ \zeta)$$

$$\leq h\left(\frac{1}{2^\alpha}\right)\left(\mathcal{F}^{g^{-1}(g^p(a))^-}_{g^{-1}(mg^p(b)),\sigma,\psi}\right)(\mu', f\hbar \circ \zeta) + m^{\psi+1}h\left(\frac{2^\alpha - 1}{2^\alpha}\right)\left(\mathcal{F}^{g^{-1}(g^p(b))^+}_{g^{-1}\left(\frac{g^p(a)}{m}\right),\sigma,\psi}\right)(m^\sigma\mu', f\hbar \circ \zeta)$$

$$\leq \left(h\left(\frac{1}{2^\alpha}\right)f(g(a)) + mh\left(\frac{2^\alpha - 1}{2^\alpha}\right)f(g(b))\right)\int_0^1 t^{\psi-1}E^{\rho,r,k,c}_{\sigma,\psi,\phi}(\mu t^\sigma; q)$$

$$\times \hbar\left((tg^p(a) + m(1-t)g^p(b))^{\frac{1}{p}}\right)h(t^\alpha)dt + m\left(h\left(\frac{1}{2^\alpha}\right)f(g(b)) + mh\left(\frac{2^\alpha - 1}{2^\alpha}\right)f\left(\frac{g(a)}{m^2}\right)\right)$$

$$\times \int_0^1 t^{\psi-1}E^{\rho,r,k,c}_{\sigma,\psi,\phi}(\mu t^\sigma; q)\hbar\left((tg^p(a) + m(1-t)g^p(b))^{\frac{1}{p}}\right)h(1-t^\alpha)dt,$$

$\zeta(z) = g^{\frac{1}{p}}(z), z \in [mb^p, a^p], f\hbar \circ \zeta = (f \circ \zeta)(\hbar \circ \zeta), \mu' = \frac{\mu}{(g^p(a) - mg^p(b))^\sigma}.$

Proof. (i) As we know, f is $(\alpha, h\text{-}m)\text{-}p$-convex function; therefore, we can write

$$f\left(\left(\frac{g^p(x) + mg^p(y)}{2}\right)^{\frac{1}{p}}\right) \leq h\left(\frac{1}{2^\alpha}\right)f(g(x)) + mh\left(\frac{2^\alpha - 1}{2^\alpha}\right)f(g(y)). \qquad (9)$$

Putting $g(x) = (tg^p(a) + m(1-t)g^p(b))^{\frac{1}{p}}$ and $g(y) = \left(tg^p(b) + (1-t)\frac{g^p(a)}{m}\right)^{\frac{1}{p}}$ in (9), we obtain

$$f\left(\left(\frac{g^p(a) + mg^p(b)}{2}\right)^{\frac{1}{p}}\right)$$

$$\leq h\left(\frac{1}{2^\alpha}\right)f\left((tg^p(a) + m(1-t)g^p(b))^{\frac{1}{p}}\right) + mh\left(\frac{2^\alpha - 1}{2^\alpha}\right)f\left(\left(tg^p(b) + (1-t)\frac{g^p(a)}{m}\right)^{\frac{1}{p}}\right). \qquad (10)$$

Multiplying (10) by $t^{\psi-1}E^{\rho,r,k,c}_{\sigma,\psi,\phi}(\mu t^\sigma;q)\hbar\left((tg^p(a)+m(1-t)g^p(b))^{\frac{1}{p}}\right)$ and integrating, we have

$$f\left(\left(\frac{g^p(a)+mg^p(b)}{2}\right)^{\frac{1}{p}}\right)\int_0^1 t^{\psi-1}E^{\rho,r,k,c}_{\sigma,\psi,\phi}(\mu t^\sigma;q)\hbar\left((tg^p(a)+m(1-t)g^p(b))^{\frac{1}{p}}\right)dt$$

$$\leq h\left(\frac{1}{2^\alpha}\right)\int_0^1 t^{\psi-1}E^{\rho,r,k,c}_{\sigma,\psi,\phi}(\mu t^\sigma;q)f\left((tg^p(a)+m(1-t)g^p(b))^{\frac{1}{p}}\right)$$

$$\times \hbar\left((tg^p(a)+m(1-t)g^p(b))^{\frac{1}{p}}\right)dt + mh\left(\frac{2^\alpha-1}{2^\alpha}\right)\int_0^1 t^{\psi-1}E^{\rho,r,k,c}_{\sigma,\psi,\phi}(\mu t^\sigma;q)$$

$$f\left(\left(tg^p(b)+(1-t)\frac{g^p(a)}{m}\right)^{\frac{1}{p}}\right)\hbar\left((tg^p(a)+m(1-t)g^p(b))^{\frac{1}{p}}\right)dt.$$

We use substitution $tg^p(a)+m(1-t)g^p(b)=g(x)$ in the integral appearing on the left hand side. In the integrals appearing on the right hand side, we use substitution $g(y)=\left(tg^p(b)+(1-t)\frac{g^p(a)}{m}\right)$ that is $tg^p(a)+m(1-t)g^p(b)=g^p(a)+mg^p(b)-mg(y)$. By utilizing the condition $\hbar\left((g^p(a)+mg^p(b)-mg(y))^{\frac{1}{p}}\right)=\hbar((g(y))^{\frac{1}{p}})$ and the Equations (6) and (7), the first inequality of (8) can be achieved.

From the right hand side of (10) for $(\alpha,h\text{-}m)\text{-}p$-convex function f, we can write

$$h\left(\frac{1}{2^\alpha}\right)f\left((tg^p(a)+m(1-t)g^p(b))^{\frac{1}{p}}\right)+mh\left(\frac{2^\alpha-1}{2^\alpha}\right)f\left(\left(tg^p(b)+(1-t)\frac{g^p(a)}{m}\right)^{\frac{1}{p}}\right) \quad (11)$$

$$\leq \left(h\left(\frac{1}{2^\alpha}\right)f(g(a))+mh\left(\frac{2^\alpha-1}{2^\alpha}\right)f(g(b))\right)h(t^\alpha)$$

$$+m\left(h\left(\frac{1}{2^\alpha}\right)f(g(b))+mh\left(\frac{2^\alpha-1}{2^\alpha}\right)f\left(\frac{g(a)}{m^2}\right)\right)h(1-t^\alpha).$$

Now, multiplying (11) by $t^{\psi-1}E^{\rho,r,k,c}_{\sigma,\psi,\phi}(\mu t^\sigma;q)\hbar\left((tg^p(a)+m(1-t)g^p(b))^{\frac{1}{p}}\right)$ and integrating, we have

$$h\left(\frac{1}{2^\alpha}\right)\int_0^1 t^{\psi-1}E^{\rho,r,k,c}_{\sigma,\psi,\phi}(\mu t^\sigma;q)f\left((tg^p(a)+m(1-t)g^p(b))^{\frac{1}{p}}\right)\hbar\left((tg^p(a)+m(1-t)g^p(b))^{\frac{1}{p}}\right)dt$$

$$+mh\left(\frac{2^\alpha-1}{2^\alpha}\right)\int_0^1 t^{\psi-1}E^{\rho,r,k,c}_{\sigma,\psi,\phi}(\mu t^\sigma;q)f\left(\left(tg^p(b)+(1-t)\frac{g^p(a)}{m}\right)^{\frac{1}{p}}\right)\hbar\left((tg^p(a)+m(1-t)g^p(b))^{\frac{1}{p}}\right)dt$$

$$\leq \left(h\left(\frac{1}{2^\alpha}\right)f(g(a))+mh\left(\frac{2^\alpha-1}{2^\alpha}\right)f(g(b))\right)\int_0^1 t^{\psi-1}E^{\rho,r,k,c}_{\sigma,\psi,\phi}(\mu t^\sigma;q)\hbar\left((tg^p(a)+m(1-t)g^p(b))^{\frac{1}{p}}\right)h(t^\alpha)dt$$

$$+m\left(h\left(\frac{1}{2^\alpha}\right)f(g(b))+mh\left(\frac{2^\alpha-1}{2^\alpha}\right)f\left(\frac{g(a)}{m^2}\right)\right)\int_0^1 t^{\psi-1}E^{\rho,r,k,c}_{\sigma,\psi,\phi}(\mu t^\sigma;q)$$

$$\times \hbar\left((tg^p(a)+m(1-t)g^p(b))^{\frac{1}{p}}\right)h(1-t^\alpha)dt.$$

Again, setting $tg^p(a)+m(1-t)g^p(b)=g(x)$ in the first term of left hand side. While using $g(y)=\left(tg^p(b)+(1-t)\frac{g^p(a)}{m}\right)$, that is $tg^p(a)+m(1-t)g^p(b)=g^p(a)+mg^p(b)-mg(y)$ in the second term of left hand side and utilizing condition $\hbar\left((g^p(a)+mg^p(b)-mg(y))^{\frac{1}{p}}\right)=\hbar((g(y))^{\frac{1}{p}})$. Then, using the Equations (6) and (7), the second inequality of (8) is achieved.

(ii) Proof is similar to the proof of (i). □

Remark 2. (i) For $g = I$ and $\mu = q = 0$ in Theorem 2 (i), Theorem 3.1 [30] is achieved.
(ii) For $\alpha = m = p = 1$ and $h(t) = t$ in Theorem 2 (i), Theorem 7 [27] is achieved.
(iii) For $\alpha = m = 1$, $p = -1$ and $h(t) = t$ in Theorem 2 (ii), Theorem 2.5 [26] is achieved.
(iv) For $\alpha = m = 1$, $p = -1$, $\mu = q = 0$, $g = I$, $\hbar(x) = 1$ and $h(t) = t$ in Theorem 2 (ii), Theorem 4 [24] is achieved.
(v) For $\alpha = m = 1$, $p = -1$, $\mu = q = 0$, $g = I$, $\psi = 1$ and $h(t) = t$ in Theorem 2 (ii), Theorem 8 [3] is achieved.
(vi) For $\alpha = m = 1$, $p = -1$, $\mu = q = 0$, $g = I$, $\psi = 1$, $\hbar(x) = 1$ and $h(t) = t$ in Theorem 2 (ii), Theorem 2.4 [5] is achieved.
(vii) For $\alpha = m = 1$, $p > 0$ and $h(t) = t$ in Theorem 2 (i), Theorem 4 (i) [29] is achieved.
(viii) For $\alpha = m = 1$, $p < 0$ and $h(t) = t$ in Theorem 2 (ii), Theorem 4 (ii) [29] is achieved.
(ix) For $p = -1$ in Theorem 2 (ii), Theorem 4 [25] is achieved.
(x) For $\alpha = m = 1$, $p = 1$, $h(t) = t$, $\mu = q = 0$, $g = I$ and $\psi = 1$ in Theorem 2 (i), classical Fejér–Hadamard inequality [32] is achieved.
(xi) For $\alpha = m = 1$, $p = 1$, $h(t) = t$, $\mu = q = 0$, $g = I$, $\hbar(x) = 1$ and $\psi = 1$ in Theorem 2 (i), classical Hadamard inequality [35,36] is achieved.
(xii) For $p = 1$ in Theorem 2 (i), the result for $(\alpha, h\text{-}m)$-convex function is achieved.
(xiii) For $\alpha = 1$, $p = 1$, in Theorem 2 (i), Theorem 2.2 [28] is achieved. Further, if $h(t) = t$, then Theorem 2.1 [28] is achieved.
(xiv) For $p = 1$ and $h(t) = t$ in Theorem 2 (i), the result for (α, m)-convex function is achieved.
(xv) For $\alpha = 1$ and $p = 1$ in Theorem 2 (i), the result for $(h\text{-}m)$-convex function is achieved.

Theorem 3. *Under the suppositions of Theorem 2, we have the following inequalities:*
(i) *If $p > 0$, then*

$$f\left(\left(\frac{g^p(a)+mg^p(b)}{2}\right)^{\frac{1}{p}}\right)\left(\mathcal{F}^{g^{-1}\left(\frac{g^p(a)+mg^p(b)}{2}\right)^+}_{g^{-1}(mg^p(b)),\sigma,\psi}\right)(2^\sigma \mu', \hbar \circ \zeta) \leq h\left(\frac{1}{2^\alpha}\right)$$

$$\times \left(\mathcal{F}^{g^{-1}\left(\frac{g^p(a)+mg^p(b)}{2}\right)^+}_{g^{-1}(mg^p(b)),\sigma,\psi}\right)(2^\sigma \mu', f\hbar \circ \zeta) + m^{\psi+1} h\left(\frac{2^\alpha-1}{2^\alpha}\right)\left(\mathcal{F}^{g^{-1}\left(\frac{g^p(a)+mg^p(b)}{2m}\right)^-}_{g^{-1}\left(\frac{g^p(a)}{m}\right),\sigma,\psi}\right)((2m)^\sigma \mu', f\hbar \circ \zeta)$$

$$\leq \left(h\left(\frac{1}{2^\alpha}\right)f(g(a))+mh\left(\frac{2^\alpha-1}{2^\alpha}\right)f(g(b))\right)\int_0^1 t^{\psi-1} E^{\rho,r,k,c}_{\sigma,\psi,\phi}(\mu t^\sigma; q) \quad (12)$$

$$\times \hbar\left(\left(\frac{t}{2}g^p(a)+m\left(1-\frac{t}{2}\right)g^p(b)\right)^{\frac{1}{p}}\right)h\left(\left(\frac{t}{2}\right)^\alpha\right)dt + m\left(h\left(\frac{1}{2^\alpha}\right)f(g(b))+mh\left(\frac{2^\alpha-1}{2^\alpha}\right)f\left(\frac{g(a)}{m^2}\right)\right)$$

$$\times \int_0^1 t^{\psi-1} E^{\rho,r,k,c}_{\sigma,\psi,\phi}(\mu t^\sigma; q) \hbar\left(\left(\frac{t}{2}g^p(a)+m\left(1-\frac{t}{2}\right)g^p(b)\right)^{\frac{1}{p}}\right)h\left(1-\left(\frac{t}{2}\right)^\alpha\right)dt,$$

$\zeta(z) = g^{\frac{1}{p}}(z)$, $z \in [a^p, mb^p]$, $f\hbar \circ \zeta = (f \circ \zeta)(\hbar \circ \zeta)$, $\mu' = \frac{\mu}{(mg^p(b)-g^p(a))^\sigma}$.

(ii) *If $p < 0$, then*

$$f\left(\left(\frac{g^p(a)+mg^p(b)}{2}\right)^{\frac{1}{p}}\right)\left(\mathcal{F}^{g^{-1}\left(\frac{g^p(a)+mg^p(b)}{2}\right)^-}_{g^{-1}(mg^p(b)),\sigma,\psi}\right)(2^\sigma \mu', \hbar \circ \zeta) \leq h\left(\frac{1}{2^\alpha}\right)$$

$$\times \left(\mathcal{F}^{g^{-1}\left(\frac{g^p(a)+mg^p(b)}{2}\right)^-}_{g^{-1}(mg^p(b)),\sigma,\psi}\right)(2^\sigma \mu', f\hbar \circ \zeta) + m^{\psi+1} h\left(\frac{2^\alpha-1}{2^\alpha}\right)\left(\mathcal{F}^{g^{-1}\left(\frac{g^p(a)+mg^p(b)}{2m}\right)^+}_{g^{-1}\left(\frac{g^p(a)}{m}\right),\sigma,\psi}\right)((2m)^\sigma \mu', f\hbar \circ \zeta)$$

$$\leq \left(h\left(\frac{1}{2^\alpha}\right)f(g(a))+mh\left(\frac{2^\alpha-1}{2^\alpha}\right)f(g(b))\right)\int_0^1 t^{\psi-1} E^{\rho,r,k,c}_{\sigma,\psi,\phi}(\mu t^\sigma; q)$$

$$\times \hbar\left(\left(\frac{t}{2}g^p(a)+m\left(1-\frac{t}{2}\right)g^p(b)\right)^{\frac{1}{p}}\right)h\left(\left(\frac{t}{2}\right)^\alpha\right)dt + m\left(h\left(\frac{1}{2^\alpha}\right)f(g(b))+mh\left(\frac{2^\alpha-1}{2^\alpha}\right)f\left(\frac{g(a)}{m^2}\right)\right)$$

$$\times \int_0^1 t^{\psi-1} E^{\rho,r,k,c}_{\sigma,\psi,\phi}(\mu t^\sigma; q) \hbar\left(\left(\frac{t}{2}g^p(a)+m\left(1-\frac{t}{2}\right)g^p(b)\right)^{\frac{1}{p}}\right)h\left(1-\left(\frac{t}{2}\right)^\alpha\right)dt,$$

$\zeta(z) = g^{\frac{1}{p}}(z)$, $z \in [mb^p, a^p]$, $f\hbar \circ \zeta = (f \circ \zeta)(\hbar \circ \zeta)$, $\mu' = \frac{\mu}{(g^p(a) - mg^p(b))^\sigma}$.

Proof. (i) Putting $g(x) = \left(\frac{t}{2}g^p(a) + m(1-\frac{t}{2})g^p(b)\right)^{\frac{1}{p}}$ and $g(y) = \left(\frac{t}{2}g^p(b) + (1-\frac{t}{2})\frac{g^p(a)}{m}\right)^{\frac{1}{p}}$ in (9), we obtain

$$f\left(\left(\frac{g^p(a) + mg^p(b)}{2}\right)^{\frac{1}{p}}\right) \leq h\left(\frac{1}{2^\alpha}\right) f\left(\left(\frac{t}{2}g^p(a) + m(1-\frac{t}{2})g^p(b)\right)^{\frac{1}{p}}\right) + mh\left(\frac{2^\alpha - 1}{2^\alpha}\right) f\left(\left(\frac{t}{2}g^p(b) + (1-\frac{t}{2})\frac{g^p(a)}{m}\right)^{\frac{1}{p}}\right). \tag{13}$$

Multiplying (13) by $t^{\psi - 1} E^{\rho,r,k,c}_{\sigma,\psi,\phi}(\mu t^\sigma; q) \hbar\left(\left(\frac{t}{2}g^p(a) + m(1-\frac{t}{2})g^p(b)\right)^{\frac{1}{p}}\right)$ and integrating, we have

$$f\left(\left(\frac{g^p(a) + mg^p(b)}{2}\right)^{\frac{1}{p}}\right) \int_0^1 t^{\psi - 1} E^{\rho,r,k,c}_{\sigma,\psi,\phi}(\mu t^\sigma; q) \hbar\left(\left(\frac{t}{2}g^p(a) + m\left(1 - \frac{t}{2}\right)g^p(b)\right)^{\frac{1}{p}}\right) dt$$

$$\leq h\left(\frac{1}{2^\alpha}\right) \int_0^1 t^{\psi - 1} E^{\rho,r,k,c}_{\sigma,\psi,\phi}(\mu t^\sigma; q) f\left(\left(\frac{t}{2}g^p(a) + m\left(1 - \frac{t}{2}\right)g^p(b)\right)^{\frac{1}{p}}\right)$$

$$\times \hbar\left(\left(\frac{t}{2}g^p(a) + m\left(1 - \frac{t}{2}\right)g^p(b)\right)^{\frac{1}{p}}\right) dt + mh\left(\frac{2^\alpha - 1}{2^\alpha}\right) \int_0^1 t^{\psi - 1} E^{\rho,r,k,c}_{\sigma,\psi,\phi}(\mu t^\sigma; q)$$

$$f\left(\left(\frac{t}{2}g^p(b) + \left(1 - \frac{t}{2}\right)\frac{g^p(a)}{m}\right)^{\frac{1}{p}}\right) \hbar\left(\left(\frac{t}{2}g^p(a) + m\left(1 - \frac{t}{2}\right)g^p(b)\right)^{\frac{1}{p}}\right) dt.$$

We use the substitution $\frac{t}{2}g^p(a) + m(1 - \frac{t}{2})g^p(b) = g(x)$ in integral appearing in the left hand side. While in the integral appearing in the first term of the right hand side, we use $g(y) = \left(\frac{t}{2}g^p(b) + (1 - \frac{t}{2})\frac{g^p(a)}{m}\right)$ that is $\frac{t}{2}g^p(a) + m(1 - \frac{t}{2})g^p(b) = g^p(a) + mg^p(b) - mg(y)$ in the last term of right hand side. Then, by utilizing the given condition $\hbar\left((g^p(a) + mg^p(b) - mg(y))^{\frac{1}{p}}\right) = \hbar((g(y))^{\frac{1}{p}})$ and the Equations (6) and (7), the first inequality of (12) can be achieved.

From the right hand side of (13) for $(\alpha, h\text{-}m)\text{-}p$-convex function f, we can write

$$h\left(\frac{1}{2^\alpha}\right) f\left(\left(\frac{t}{2}g^p(a) + m\left(1 - \frac{t}{2}\right)g^p(b)\right)^{\frac{1}{p}}\right) + mh\left(\frac{2^\alpha - 1}{2^\alpha}\right) f\left(\left(\frac{t}{2}g^p(b) + \left(1 - \frac{t}{2}\right)\frac{g^p(a)}{m}\right)^{\frac{1}{p}}\right) \tag{14}$$

$$\leq \left(h\left(\frac{1}{2^\alpha}\right) f(g(a)) + mh\left(\frac{2^\alpha - 1}{2^\alpha}\right) f(g(b))\right) h\left(\left(\frac{t}{2}\right)^\alpha\right)$$

$$+ m\left(h\left(\frac{1}{2^\alpha}\right) f(g(b)) + mh\left(\frac{2^\alpha - 1}{2^\alpha}\right) f\left(\frac{g(a)}{m^2}\right)\right) h\left(1 - \left(\frac{t}{2}\right)^\alpha\right).$$

Now, multiplying (14) by $t^{\psi-1}E_{\sigma,\psi,\phi}^{\rho,r,k,c}(\mu t^\sigma;q)\hbar\left(\left(\frac{t}{2}g^p(a)+m\left(1-\frac{t}{2}\right)g^p(b)\right)^{\frac{1}{p}}\right)$ and integrating over $[0,1]$, we have

$$h\left(\frac{1}{2^\alpha}\right)\int_0^1 t^{\psi-1}E_{\sigma,\psi,\phi}^{\rho,r,k,c}(\mu t^\sigma;q)f\left(\left(\frac{t}{2}g^p(a)+m\left(1-\frac{t}{2}\right)g^p(b)\right)^{\frac{1}{p}}\right)$$

$$\times \hbar\left(\left(\frac{t}{2}g^p(a)+m\left(1-\frac{t}{2}\right)g^p(b)\right)^{\frac{1}{p}}\right)dt + mh\left(\frac{2^\alpha-1}{2^\alpha}\right)\int_0^1 t^{\psi-1}E_{\sigma,\psi,\phi}^{\rho,r,k,c}(\mu t^\sigma;q)$$

$$f\left(\left(\frac{t}{2}g^p(b)+\left(1-\frac{t}{2}\right)\frac{g^p(a)}{m}\right)^{\frac{1}{p}}\right)\hbar\left(\left(\frac{t}{2}g^p(a)+m\left(1-\frac{t}{2}\right)g^p(b)\right)^{\frac{1}{p}}\right)dt$$

$$\leq \left(h\left(\frac{1}{2^\alpha}\right)f(g(a))+mh\left(\frac{2^\alpha-1}{2^\alpha}\right)f(g(b))\right)\int_0^1 t^{\psi-1}E_{\sigma,\psi,\phi}^{\rho,r,k,c}(\mu t^\sigma;q)$$

$$\times \hbar\left(\left(\frac{t}{2}g^p(a)+m\left(1-\frac{t}{2}\right)g^p(b)\right)^{\frac{1}{p}}\right)h\left(\left(\frac{t}{2}\right)^\alpha\right)dt + m\left(h\left(\frac{1}{2^\alpha}\right)f(g(b))+mh\left(\frac{2^\alpha-1}{2^\alpha}\right)\right.$$

$$\left.f\left(\frac{g(a)}{m^2}\right)\right)\int_0^1 t^{\psi-1}E_{\sigma,\psi,\phi}^{\rho,r,k,c}(\mu t^\sigma;q)\hbar\left(\left(\frac{t}{2}g^p(a)+m\left(1-\frac{t}{2}\right)g^p(b)\right)^{\frac{1}{p}}\right)h\left(1-\left(\frac{t}{2}\right)^\alpha\right)dt.$$

Again, we set $\frac{t}{2}g^p(a)+m(1-\frac{t}{2})g^p(b)=g(x)$ in integral appearing in the first term of the left hand side. While we use $g(y) = \left(\frac{t}{2}g^p(b)+(1-\frac{t}{2})\frac{g^p(a)}{m}\right)$ that is $\frac{t}{2}g^p(a)+m(1-\frac{t}{2})g^p(b)=g^p(a)+mg^p(b)-mg(y)$ in integral appearing in the second term of the left hand side. By utilizing the condition $\hbar\left((g^p(a)+mg^p(b)-mg(y))^{\frac{1}{p}}\right)=\hbar((g(y))^{\frac{1}{p}})$ and the Equations (6) and (7), the second inequality of (12) can be achieved.
(ii) Proof is similar to the proof of (i). □

Remark 3. (i) For $g=I$ and $\mu=q=0$ in Theorem 3 (i), Theorem 3.3 [30] is achieved.
(ii) For $\alpha=m=1$, $p=-1$ and $h(t)=t$ in Theorem 3 (ii), Theorem 3.3 [26] is achieved.
(iii) For $\alpha=m=1$, $p=-1$ and $h(t)=t$ in Theorem 3 (ii), Theorem 8 [3] is achieved.
(iv) For $\alpha=m=1$, $p=-1$, $\mu=q=0$, $g=I$ and $h(t)=t$ in Theorem 3 (ii), Corollary 2.10 [26] is achieved.
(v) For $\alpha=m=1$, $p>0$ and $h(t)=t$ in Theorem 3 (i), Theorem 6 (i) [29] is achieved.
(vi) For $\alpha=m=1$, $p<0$ and $h(t)=t$ in Theorem 3 (ii), Theorem 6 (ii) [29] is achieved.
(vii) For $p=-1$ in Theorem 3 (ii), Theorem 6 [25] is achieved.
(viii) For $p=1$ in Theorem 3 (i), the result for $(\alpha,h\text{-}m)$-convex function is achieved.
(ix) For $p=1$ and $h(t)=t$ in Theorem 3 (i), the result for (α,m)-convex function is achieved.
(x) For $\alpha=1$ and $p=1$ in Theorem 3 (i), the result for $(h\text{-}m)$-convex function is achieved.

In the following, we list the results for $(h\text{-}m)\text{-}p$-convex, $(\alpha,m)\text{-}p$-convex, $(\alpha,h)\text{-}p$-convex and $(s,m)\text{-}p$-convex functions.

2.1. Results for (h-m)-p-Convex Functions

Theorem 4. *From Theorem 2 for $\alpha = 1$, we have the following inequalities for (h-m)-p-convex functions:*

(i) If $p > 0$, then

$$\frac{1}{h\left(\frac{1}{2}\right)} f\left(\left(\frac{g^p(a)+mg^p(b)}{2}\right)^{\frac{1}{p}}\right) \left(\mathcal{F}_{g^{-1}(mg^p(b)),\sigma,\psi}^{g^{-1}(g^p(a))^+}\right)(\mu',\hbar\circ\zeta)$$

$$\leq \left(\mathcal{F}_{g^{-1}(mg^p(b)),\sigma,\psi}^{g^{-1}(g^p(a))^+}\right)(\mu',f\hbar\circ\zeta) + m^{\psi+1}\left(\mathcal{F}_{g^{-1}\left(\frac{g^p(a)}{m}\right),\sigma,\psi}^{g^{-1}(g^p(b))^-}\right)(m^\sigma\mu',f\hbar\circ\zeta)$$

$$\leq (f(g(a))+mf(g(b)))\int_0^1 t^{\psi-1}E_{\sigma,\psi,\phi}^{\rho,r,k,c}(\mu t^\sigma;q)\hbar\left((tg^p(a)+m(1-t)g^p(b))^{\frac{1}{p}}\right)h(t)dt$$

$$+ m\left(f(g(b))+mf\left(\frac{g(a)}{m^2}\right)\right)\int_0^1 t^{\psi-1}E_{\sigma,\psi,\phi}^{\rho,r,k,c}(\mu t^\sigma;q)\hbar\left((tg^p(a)+m(1-t)g^p(b))^{\frac{1}{p}}\right)h(1-t)dt,$$

$\zeta(z) = g^{\frac{1}{p}}(z)$, $z \in [a^p, mb^p]$, $f\hbar\circ\zeta = (f\circ\zeta)(\hbar\circ\zeta)$, $\mu' = \frac{\mu}{(mg^p(b)-g^p(a))^\sigma}$.

(ii) If $p < 0$, then

$$\frac{1}{h\left(\frac{1}{2}\right)} f\left(\left(\frac{g^p(a)+mg^p(b)}{2}\right)^{\frac{1}{p}}\right) \left(\mathcal{F}_{g^{-1}(mg^p(b)),\sigma,\psi}^{g^{-1}(g^p(a))^-}\right)(\mu',\hbar\circ\zeta)$$

$$\leq \left(\mathcal{F}_{g^{-1}(mg^p(b)),\sigma,\psi}^{g^{-1}(g^p(a))^-}\right)(\mu',f\hbar\circ\zeta) + m^{\psi+1}\left(\mathcal{F}_{g^{-1}\left(\frac{g^p(a)}{m}\right),\sigma,\psi}^{g^{-1}(g^p(b))^+}\right)(m^\sigma\mu',f\hbar\circ\zeta)$$

$$\leq (f(g(a))+mf(g(b)))\int_0^1 t^{\psi-1}E_{\sigma,\psi,\phi}^{\rho,r,k,c}(\mu t^\sigma;q)\hbar\left((tg^p(a)+m(1-t)g^p(b))^{\frac{1}{p}}\right)h(t)dt$$

$$+ m\left(f(g(b))+mf\left(\frac{g(a)}{m^2}\right)\right)\int_0^1 t^{\psi-1}E_{\sigma,\psi,\phi}^{\rho,r,k,c}(\mu t^\sigma;q)\hbar\left((tg^p(a)+m(1-t)g^p(b))^{\frac{1}{p}}\right)h(1-t)dt,$$

$\zeta(z) = g^{\frac{1}{p}}(z)$, $z \in [mb^p, a^p]$, $f\hbar\circ\zeta = (f\circ\zeta)(\hbar\circ\zeta)$, $\mu' = \frac{\mu}{(g^p(a)-mg^p(b))^\sigma}$.

Theorem 5. *From Theorem 3, for $\alpha = 1$, we have the following inequalities for (h-m)-p-convex functions:*

(i) If $p > 0$, then

$$\frac{1}{h\left(\frac{1}{2}\right)} f\left(\left(\frac{g^p(a)+mg^p(b)}{2}\right)^{\frac{1}{p}}\right) \left(\mathcal{F}_{g^{-1}(mg^p(b)),\sigma,\psi}^{g^{-1}\left(\frac{g^p(a)+mg^p(b)}{2}\right)^+}\right)(2^\sigma\mu',\hbar\circ\zeta)$$

$$\leq \left(\mathcal{F}_{g^{-1}(mg^p(b)),\sigma,\psi}^{g^{-1}\left(\frac{g^p(a)+mg^p(b)}{2}\right)^+}\right)(2^\sigma\mu',f\hbar\circ\zeta) + m^{\psi+1}\left(\mathcal{F}_{g^{-1}\left(\frac{g^p(a)}{m}\right),\sigma,\psi}^{g^{-1}\left(\frac{g^p(a)+mg^p(b)}{2m}\right)^-}\right)((2m)^\sigma\mu',f\hbar\circ\zeta)$$

$$\leq (f(g(a))+mf(g(b)))\int_0^1 t^{\psi-1}E_{\sigma,\psi,\phi}^{\rho,r,k,c}(\mu t^\sigma;q)\hbar\left(\left(\frac{t}{2}g^p(a)+m\left(1-\frac{t}{2}\right)g^p(b)\right)^{\frac{1}{p}}\right)h\left(\frac{t}{2}\right)dt$$

$$+ m\left(f(g(b))+mf\left(\frac{g(a)}{m^2}\right)\right)\int_0^1 t^{\psi-1}E_{\sigma,\psi,\phi}^{\rho,r,k,c}(\mu t^\sigma;q)\hbar\left(\left(\frac{t}{2}g^p(a)+m\left(1-\frac{t}{2}\right)g^p(b)\right)^{\frac{1}{p}}\right)h\left(1-\frac{t}{2}\right)dt,$$

$\zeta(z) = g^{\frac{1}{p}}(z)$, $z \in [a^p, mb^p]$, $f\hbar\circ\zeta = (f\circ\zeta)(\hbar\circ\zeta)$, $\mu' = \frac{\mu}{(mg^p(b)-g^p(a))^\sigma}$.

(ii) If $p < 0$, then

$$\frac{1}{h\left(\frac{1}{2}\right)} f\left(\left(\frac{g^p(a)+mg^p(b)}{2}\right)^{\frac{1}{p}}\right)\left(\mathcal{F}_{g^{-1}(mg^p(b)),\sigma,\psi}^{g^{-1}\left(\frac{g^p(a)+mg^p(b)}{2}\right)^{-}}\right)(2^\sigma \mu',\hbar\circ\zeta)$$

$$\leq \left(\mathcal{F}_{g^{-1}(mg^p(b)),\sigma,\psi}^{g^{-1}\left(\frac{g^p(a)+mg^p(b)}{2}\right)^{-}}\right)(2^\sigma \mu',f\hbar\circ\zeta) + m^{\psi+1}\left(\mathcal{F}_{g^{-1}\left(\frac{g^p(a)}{m}\right),\sigma,\psi}^{g^{-1}\left(\frac{g^p(a)+mg^p(b)}{2m}\right)^{+}}\right)((2m)^\sigma \mu',f\hbar\circ\zeta)$$

$$\leq (f(g(a))+mf(g(b)))\int_0^1 t^{\psi-1} E^{\rho,r,k,c}_{\sigma,\psi,\phi}(\mu t^\sigma;q)\hbar\left(\left(\frac{t}{2}g^p(a)+m\left(1-\frac{t}{2}\right)g^p(b)\right)^{\frac{1}{p}}\right)h\left(\frac{t}{2}\right)dt$$

$$+m\left(f(g(b))+mf\left(\frac{g(a)}{m^2}\right)\right)\int_0^1 t^{\psi-1} E^{\rho,r,k,c}_{\sigma,\psi,\phi}(\mu t^\sigma;q)\hbar\left(\left(\frac{t}{2}g^p(a)+m\left(1-\frac{t}{2}\right)g^p(b)\right)^{\frac{1}{p}}\right)h\left(1-\frac{t}{2}\right)dt,$$

$\zeta(z) = g^{\frac{1}{p}}(z)$, $z \in [mb^p, a^p]$, $f\hbar\circ\zeta = (f\circ\zeta)(\hbar\circ\zeta)$, $\mu' = \frac{\mu}{(g^p(a)-mg^p(b))^\sigma}$.

2.2. Results for (α,m)-p-Convex Functions

Theorem 6. *From Theorem 2 for $h(t) = t$, we have the following inequalities for (α,m)-p-convex functions:*
(i) If $p > 0$, then

$$2^\alpha f\left(\left(\frac{g^p(a)+mg^p(b)}{2}\right)^{\frac{1}{p}}\right)\left(\mathcal{F}^{g^{-1}(g^p(a))^{+}}_{g^{-1}(mg^p(b)),\sigma,\psi}\right)(\mu',\hbar\circ\zeta)$$

$$\leq \left(\mathcal{F}^{g^{-1}(g^p(a))^{+}}_{g^{-1}(mg^p(b)),\sigma,\psi}\right)(\mu',f\hbar\circ\zeta) + m^{\psi+1}(2^\alpha-1)\left(\mathcal{F}^{g^{-1}(g^p(b))^{-}}_{g^{-1}\left(\frac{g^p(a)}{m}\right),\sigma,\psi}\right)(m^\sigma \mu',f\hbar\circ\zeta)$$

$$\leq (f(g(a))+m(2^\alpha-1)f(g(b)))\left(\mathcal{F}^{g^{-1}(g^p(a))^{+}}_{g^{-1}(mg^p(b)),\sigma,\psi+\alpha}\right)(\mu',\hbar\circ\zeta) + m(f(g(b))+m(2^\alpha-1)$$

$$f\left(\frac{g(a)}{m^2}\right))\left(\left(\mathcal{F}^{g^{-1}(g^p(a))^{+}}_{g^{-1}(mg^p(b)),\sigma,\psi}\right)(\mu',\hbar\circ\zeta) - \left(\mathcal{F}^{g^{-1}(g^p(a))^{+}}_{g^{-1}(mg^p(b)),\sigma,\psi+\alpha}\right)(\mu',\hbar\circ\zeta)\right),$$

$\zeta(z) = g^{\frac{1}{p}}(z)$, $z \in [a^p, mb^p]$, $f\hbar\circ\zeta = (f\circ\zeta)(\hbar\circ\zeta)$, $\mu' = \frac{\mu}{(mg^p(b)-g^p(a))^\sigma}$.

(ii) If $p < 0$, then

$$2^\alpha f\left(\left(\frac{g^p(a)+mg^p(b)}{2}\right)^{\frac{1}{p}}\right)\left(\mathcal{F}^{g^{-1}(g^p(a))^{-}}_{g^{-1}(mg^p(b)),\sigma,\psi}\right)(\mu',\hbar\circ\zeta)$$

$$\leq \left(\mathcal{F}^{g^{-1}(g^p(a))^{-}}_{g^{-1}(mg^p(b)),\sigma,\psi}\right)(\mu',f\hbar\circ\zeta) + m^{\psi+1}(2^\alpha-1)\left(\mathcal{F}^{g^{-1}(g^p(b))^{+}}_{g^{-1}\left(\frac{g^p(a)}{m}\right),\sigma,\psi}\right)(m^\sigma \mu',f\hbar\circ\zeta)$$

$$\leq (f(g(a))+m(2^\alpha-1)f(g(b)))\left(\mathcal{F}^{g^{-1}(g^p(a))^{-}}_{g^{-1}(mg^p(b)),\sigma,\psi+\alpha}\right)(\mu',\hbar\circ\zeta) + m(f(g(b))+m(2^\alpha-1)$$

$$f\left(\frac{g(a)}{m^2}\right))\left(\left(\mathcal{F}^{g^{-1}(g^p(a))^{-}}_{g^{-1}(mg^p(b)),\sigma,\psi}\right)(\mu',\hbar\circ\zeta) - \left(\mathcal{F}^{g^{-1}(g^p(a))^{-}}_{g^{-1}(mg^p(b)),\sigma,\psi+\alpha}\right)(\mu',\hbar\circ\zeta)\right),$$

$\zeta(z) = g^{\frac{1}{p}}(z)$, $z \in [mb^p, a^p]$, $f\hbar\circ\zeta = (f\circ\zeta)(\hbar\circ\zeta)$, $\mu' = \frac{\mu}{(g^p(a)-mg^p(b))^\sigma}$.

Theorem 7. *From Theorem 3 for $h(t) = t$, we have the following inequalities for (α, m)-p-convex functions:*
(i) *If $p > 0$, then*

$$2^{\alpha} f\left(\left(\frac{g^p(a) + mg^p(b)}{2}\right)^{\frac{1}{p}}\right) \left(\mathcal{F}^{g^{-1}\left(\frac{g^p(a)+mg^p(b)}{2}\right)^+}_{g^{-1}(mg^p(b)),\sigma,\psi}\right)(2^{\sigma}\mu', \hbar \circ \zeta)$$

$$\leq \left(\mathcal{F}^{g^{-1}\left(\frac{g^p(a)+mg^p(b)}{2}\right)^+}_{g^{-1}(mg^p(b)),\sigma,\psi}\right)(2^{\sigma}\mu', f\hbar \circ \zeta) + m^{\psi+1}(2^{\alpha}-1)\left(\mathcal{F}^{g^{-1}\left(\frac{g^p(a)+mg^p(b)}{2m}\right)^-}_{g^{-1}\left(\frac{g^p(a)}{m}\right),\sigma,\psi}\right)((2m)^{\sigma}\mu', f\hbar \circ \zeta)$$

$$\leq \frac{1}{2^{\alpha}}(f(g(a)) + m(2^{\alpha}-1)f(g(b)))\left(\mathcal{F}^{g^{-1}\left(\frac{g^p(a)+mg^p(b)}{2}\right)^+}_{g^{-1}(mg^p(b)),\sigma,\psi+\alpha}\right)(2^{\sigma}\mu', \hbar \circ \zeta) + m(f(g(b)) + m(2^{\alpha}-1)$$

$$f\left(\frac{g(a)}{m^2}\right)\right)\left(\left(\mathcal{F}^{g^{-1}\left(\frac{g^p(a)+mg^p(b)}{2}\right)^+}_{g^{-1}(mg^p(b)),\sigma,\psi}\right)(2^{\sigma}\mu', \hbar \circ \zeta) - \frac{1}{2^{\alpha}}\left(\mathcal{F}^{g^{-1}\left(\frac{g^p(a)+mg^p(b)}{2}\right)^+}_{g^{-1}(mg^p(b)),\sigma,\psi+\alpha}\right)(2^{\sigma}\mu', \hbar \circ \zeta)\right),$$

$\zeta(z) = g^{\frac{1}{p}}(z)$, $z \in [a^p, mb^p]$, $f\hbar \circ \zeta = (f \circ \zeta)(\hbar \circ \zeta)$, $\mu' = \frac{\mu}{(mg^p(b) - g^p(a))^{\sigma}}$.

(ii) *If $p < 0$, then*

$$2^{\alpha} f\left(\left(\frac{g^p(a) + mg^p(b)}{2}\right)^{\frac{1}{p}}\right) \left(\mathcal{F}^{g^{-1}\left(\frac{g^p(a)+mg^p(b)}{2}\right)^-}_{g^{-1}(mg^p(b)),\sigma,\psi}\right)(2^{\sigma}\mu', \hbar \circ \zeta)$$

$$\leq \left(\mathcal{F}^{g^{-1}\left(\frac{g^p(a)+mg^p(b)}{2}\right)^-}_{g^{-1}(mg^p(b)),\sigma,\psi}\right)(2^{\sigma}\mu', f\hbar \circ \zeta) + m^{\psi+1}(2^{\alpha}-1)\left(\mathcal{F}^{g^{-1}\left(\frac{g^p(a)+mg^p(b)}{2m}\right)^+}_{g^{-1}\left(\frac{g^p(a)}{m}\right),\sigma,\psi}\right)((2m)^{\sigma}\mu', f\hbar \circ \zeta)$$

$$\leq \frac{1}{2^{\alpha}}(f(g(a)) + m(2^{\alpha}-1)f(g(b)))\left(\mathcal{F}^{g^{-1}\left(\frac{g^p(a)+mg^p(b)}{2}\right)^-}_{g^{-1}(mg^p(b)),\sigma,\psi+\alpha}\right)(2^{\sigma}\mu', \hbar \circ \zeta) + m(f(g(b)) + m(2^{\alpha}-1)$$

$$f\left(\frac{g(a)}{m^2}\right)\right)\left(\left(\mathcal{F}^{g^{-1}\left(\frac{g^p(a)+mg^p(b)}{2}\right)^-}_{g^{-1}(mg^p(b)),\sigma,\psi}\right)(2^{\sigma}\mu', \hbar \circ \zeta) - \frac{1}{2^{\alpha}}\left(\mathcal{F}^{g^{-1}\left(\frac{g^p(a)+mg^p(b)}{2}\right)^-}_{g^{-1}(mg^p(b)),\sigma,\psi+\alpha}\right)(2^{\sigma}\mu', \hbar \circ \zeta)\right),$$

$\zeta(z) = g^{\frac{1}{p}}(z)$, $z \in [mb^p, a^p]$, $f\hbar \circ \zeta = (f \circ \zeta)(\hbar \circ \zeta)$, $\mu' = \frac{\mu}{(g^p(a) - mg^p(b))^{\sigma}}$.

2.3. Results for (α, h)-p-Convex Functions

Theorem 8. *From Theorem 2 for $m = 1$, we have the following inequalities for (α, h)-p-convex functions:*
(i) *If $p > 0$, then*

$$f\left(\left(\frac{g^p(a) + g^p(b)}{2}\right)^{\frac{1}{p}}\right) \left(\mathcal{F}^{g^{-1}(g^p(a))^+}_{g^{-1}(g^p(b)),\sigma,\psi}\right)(\mu', \hbar \circ \zeta)$$

$$\leq h\left(\frac{1}{2^{\alpha}}\right)\left(\mathcal{F}^{g^{-1}(g^p(a))^+}_{g^{-1}(g^p(b)),\sigma,\psi}\right)(\mu', f\hbar \circ \zeta) + h\left(\frac{2^{\alpha}-1}{2^{\alpha}}\right)\left(\mathcal{F}^{g^{-1}(g^p(b))^-}_{g^{-1}(g^p(a)),\sigma,\psi}\right)(\mu', f\hbar \circ \zeta)$$

$$\leq \left(h\left(\frac{1}{2^{\alpha}}\right)f(g(a)) + h\left(\frac{2^{\alpha}-1}{2^{\alpha}}\right)f(g(b))\right)\int_0^1 t^{\psi-1} E^{\rho,r,k,c}_{\sigma,\psi,\phi}(\mu t^{\sigma}; q)$$

$$\times \hbar\left((tg^p(a) + (1-t)g^p(b))^{\frac{1}{p}}\right)h(t^{\alpha})dt + \left(h\left(\frac{1}{2^{\alpha}}\right)f(g(b)) + h\left(\frac{2^{\alpha}-1}{2^{\alpha}}\right)f(g(a))\right)$$

$$\times \int_0^1 t^{\psi-1} E^{\rho,r,k,c}_{\sigma,\psi,\phi}(\mu t^{\sigma}; q)\hbar\left((tg^p(a) + (1-t)g^p(b))^{\frac{1}{p}}\right)h(1-t^{\alpha})dt,$$

$\zeta(z) = g^{\frac{1}{p}}(z)$, $z \in [a^p, b^p]$, $f\hbar \circ \zeta = (f \circ \zeta)(\hbar \circ \zeta)$, $\mu' = \frac{\mu}{(g^p(b) - g^p(a))^{\sigma}}$.

(ii) If $p < 0$, then

$$f\left(\left(\frac{g^p(a)+g^p(b)}{2}\right)^{\frac{1}{p}}\right)\left(\mathcal{F}^{g^{-1}(g^p(a))^-}_{g^{-1}(g^p(b)),\sigma,\psi}\right)(\mu',\hbar\circ\zeta)$$

$$\leq h\left(\frac{1}{2^\alpha}\right)\left(\mathcal{F}^{g^{-1}(g^p(a))^-}_{g^{-1}(g^p(b)),\sigma,\psi}\right)(\mu',f\hbar\circ\zeta)+h\left(\frac{2^\alpha-1}{2^\alpha}\right)\left(\mathcal{F}^{g^{-1}(g^p(b))^+}_{g^{-1}(g^p(a)),\sigma,\psi}\right)(\mu',f\hbar\circ\zeta)$$

$$\leq \left(h\left(\frac{1}{2^\alpha}\right)f(g(a))+h\left(\frac{2^\alpha-1}{2^\alpha}\right)f(g(b))\right)\int_0^1 t^{\psi-1}E^{\rho,r,k,c}_{\sigma,\psi,\phi}(\mu t^\sigma;q)$$

$$\times \hbar\left((tg^p(a)+(1-t)g^p(b))^{\frac{1}{p}}\right)h(t^\alpha)dt+\left(h\left(\frac{1}{2^\alpha}\right)f(g(b))+h\left(\frac{2^\alpha-1}{2^\alpha}\right)f(g(a))\right)$$

$$\times \int_0^1 t^{\psi-1}E^{\rho,r,k,c}_{\sigma,\psi,\phi}(\mu t^\sigma;q)\hbar\left((tg^p(a)+(1-t)g^p(b))^{\frac{1}{p}}\right)h(1-t^\alpha)dt,$$

$\zeta(z) = g^{\frac{1}{p}}(z), z \in [b^p, a^p], \mu' = \frac{\mu}{(g^p(a)-g^p(b))^\sigma}$.

Theorem 9. *From Theorem 3 for $m = 1$, we have the following inequalities for (α,h)-p-convex functions:*
(i) If $p > 0$, then

$$f\left(\left(\frac{g^p(a)+g^p(b)}{2}\right)^{\frac{1}{p}}\right)\left(\mathcal{F}^{g^{-1}\left(\frac{g^p(a)+g^p(b)}{2}\right)^+}_{g^{-1}(g^p(b)),\sigma,\psi}\right)(2^\sigma\mu',\hbar\circ\zeta)\leq h\left(\frac{1}{2^\alpha}\right)$$

$$\times\left(\mathcal{F}^{g^{-1}\left(\frac{g^p(a)+g^p(b)}{2}\right)^+}_{g^{-1}(g^p(b)),\sigma,\psi}\right)(2^\sigma\mu',f\hbar\circ\zeta)+h\left(\frac{2^\alpha-1}{2^\alpha}\right)\left(\mathcal{F}^{g^{-1}\left(\frac{g^p(a)+g^p(b)}{2}\right)^-}_{g^{-1}(g^p(a)),\sigma,\psi}\right)(2^\sigma\mu',f\hbar\circ\zeta)$$

$$\leq \left(h\left(\frac{1}{2^\alpha}\right)f(g(a))+h\left(\frac{2^\alpha-1}{2^\alpha}\right)f(g(b))\right)\int_0^1 t^{\psi-1}E^{\rho,r,k,c}_{\sigma,\psi,\phi}(\mu t^\sigma;q)$$

$$\times\hbar\left(\left(\frac{t}{2}g^p(a)+\left(1-\frac{t}{2}\right)g^p(b)\right)^{\frac{1}{p}}\right)h\left(\left(\frac{t}{2}\right)^\alpha\right)dt+\left(h\left(\frac{1}{2^\alpha}\right)f(g(b))+h\left(\frac{2^\alpha-1}{2^\alpha}\right)f(g(a))\right)$$

$$\times\int_0^1 t^{\psi-1}E^{\rho,r,k,c}_{\sigma,\psi,\phi}(\mu t^\sigma;q)\hbar\left(\left(\frac{t}{2}g^p(a)+\left(1-\frac{t}{2}\right)g^p(b)\right)^{\frac{1}{p}}\right)h\left(1-\left(\frac{t}{2}\right)^\alpha\right)dt,$$

$\zeta(z) = g^{\frac{1}{p}}(z), z \in [a^p, b^p], f\hbar\circ\zeta = (f\circ\zeta)(\hbar\circ\zeta), \mu' = \frac{\mu}{(g^p(b)-g^p(a))^\sigma}$.

(ii) If $p < 0$, then

$$f\left(\left(\frac{g^p(a)+g^p(b)}{2}\right)^{\frac{1}{p}}\right)\left(\mathcal{F}^{g^{-1}\left(\frac{g^p(a)+g^p(b)}{2}\right)^-}_{g^{-1}(g^p(b)),\sigma,\psi}\right)(2^\sigma\mu',\hbar\circ\zeta)\leq h\left(\frac{1}{2^\alpha}\right)$$

$$\times\left(\mathcal{F}^{g^{-1}\left(\frac{g^p(a)+g^p(b)}{2}\right)^-}_{g^{-1}(g^p(b)),\sigma,\psi}\right)(2^\sigma\mu',f\hbar\circ\zeta)+h\left(\frac{2^\alpha-1}{2^\alpha}\right)\left(\mathcal{F}^{g^{-1}\left(\frac{g^p(a)+g^p(b)}{2}\right)^+}_{g^{-1}(g^p(a)),\sigma,\psi}\right)(2^\sigma\mu',f\hbar\circ\zeta)$$

$$\leq \left(h\left(\frac{1}{2^\alpha}\right)f(g(a))+h\left(\frac{2^\alpha-1}{2^\alpha}\right)f(g(b))\right)\int_0^1 t^{\psi-1}E^{\rho,r,k,c}_{\sigma,\psi,\phi}(\mu t^\sigma;q)$$

$$\times\hbar\left(\left(\frac{t}{2}g^p(a)+\left(1-\frac{t}{2}\right)g^p(b)\right)^{\frac{1}{p}}\right)h\left(\left(\frac{t}{2}\right)^\alpha\right)dt+\left(h\left(\frac{1}{2^\alpha}\right)f(g(b))+h\left(\frac{2^\alpha-1}{2^\alpha}\right)f(g(a))\right)$$

$$\times\int_0^1 t^{\psi-1}E^{\rho,r,k,c}_{\sigma,\psi,\phi}(\mu t^\sigma;q)\hbar\left(\left(\frac{t}{2}g^p(a)+\left(1-\frac{t}{2}\right)g^p(b)\right)^{\frac{1}{p}}\right)h\left(1-\left(\frac{t}{2}\right)^\alpha\right)dt,$$

$\zeta(z) = g^{\frac{1}{p}}(z), z \in [b^p, a^p], f\hbar\circ\zeta = (f\circ\zeta)(\hbar\circ\zeta), \mu' = \frac{\mu}{(g^p(a)-g^p(b))^\sigma}$.

2.4. Results for (s-m)-p-Convex Functions

Theorem 10. *From Theorem 2 for $\alpha = 1$ and $h(t) = t^s$, we have the following inequalities for (s-m)-p-convex functions:*

(i) If $p > 0$, then

$$2^s f\left(\left(\frac{g^p(a)+mg^p(b)}{2}\right)^{\frac{1}{p}}\right)\left(\mathcal{F}^{g^{-1}(g^p(a))^+}_{g^{-1}(mg^p(b)),\sigma,\psi}\right)(\mu',\hbar\circ\zeta)$$

$$\leq \left(\mathcal{F}^{g^{-1}(g^p(a))^+}_{g^{-1}(mg^p(b)),\sigma,\psi}\right)(\mu',f\hbar\circ\zeta) + m^{\psi+1}\left(\mathcal{F}^{g^{-1}(g^p(b))^-}_{g^{-1}\left(\frac{g^p(a)}{m}\right),\sigma,\psi}\right)(m^\sigma\mu',f\hbar\circ\zeta)$$

$$\leq (f(g(a))+mf(g(b)))\left(\mathcal{F}^{g^{-1}(g^p(a))^+}_{g^{-1}(mg^p(b)),\sigma,\psi+s}\right)(\mu',\hbar\circ\zeta)$$

$$+ \int_0^1 t^{\psi-1}(1-t)^s E^{\rho,r,k,c}_{\sigma,\psi,\phi}(\mu t^\sigma;q)\hbar\left((tg^p(a)+m(1-t)g^p(b))^{\frac{1}{p}}\right)dt,$$

$\zeta(z) = g^{\frac{1}{p}}(z),\ z \in [a^p, mb^p],\ f\hbar\circ\zeta = (f\circ\zeta)(\hbar\circ\zeta),\ \mu' = \frac{\mu}{(mg^p(b)-g^p(a))^\sigma}.$

(ii) If $p < 0$, then

$$2^s f\left(\left(\frac{g^p(a)+mg^p(b)}{2}\right)^{\frac{1}{p}}\right)\left(\mathcal{F}^{g^{-1}(g^p(a))^-}_{g^{-1}(mg^p(b)),\sigma,\psi}\right)(\mu',\hbar\circ\zeta)$$

$$\leq \left(\mathcal{F}^{g^{-1}(g^p(a))^+}_{g^{-1}(mg^p(b)),\sigma,\psi}\right)(\mu',f\hbar\circ\zeta) + m^{\psi+1}\left(\mathcal{F}^{g^{-1}(g^p(b))^-}_{g^{-1}\left(\frac{g^p(a)}{m}\right),\sigma,\psi}\right)(m^\sigma\mu',f\hbar\circ\zeta)$$

$$\leq (f(g(a))+mf(g(b)))\left(\mathcal{F}^{g^{-1}(g^p(a))^-}_{g^{-1}(mg^p(b)),\sigma,\psi+s}\right)(\mu',\hbar\circ\zeta)$$

$$+ \int_0^1 t^{\psi-1}(1-t)^s E^{\rho,r,k,c}_{\sigma,\psi,\phi}(\mu t^\sigma;q)\hbar\left((tg^p(a)+m(1-t)g^p(b))^{\frac{1}{p}}\right)dt,$$

$\zeta(z) = g^{\frac{1}{p}}(z),\ z \in [mb^p, a^p],\ f\hbar\circ\zeta = (f\circ\zeta)(\hbar\circ\zeta),\ \mu' = \frac{\mu}{(g^p(a)-mg^p(b))^\sigma}.$

Theorem 11. *From Theorem 3 for $\alpha = 1$ and $h(t) = t^s$, we have the following inequalities for (s-m)-p-convex functions:*

(i) If $p > 0$, then

$$2^s f\left(\left(\frac{g^p(a)+mg^p(b)}{2}\right)^{\frac{1}{p}}\right)\left(\mathcal{F}^{g^{-1}\left(\frac{g^p(a)+mg^p(b)}{2}\right)^+}_{g^{-1}(mg^p(b)),\sigma,\psi}\right)(2^\sigma\mu',\hbar\circ\zeta)$$

$$\leq \left(\mathcal{F}^{g^{-1}\left(\frac{g^p(a)+mg^p(b)}{2}\right)^+}_{g^{-1}(mg^p(b)),\sigma,\psi}\right)(2^\sigma\mu',f\hbar\circ\zeta) + m^{\psi+1}\left(\mathcal{F}^{g^{-1}\left(\frac{g^p(a)+mg^p(b)}{2m}\right)^+}_{g^{-1}\left(\frac{g^p(a)}{m}\right),\sigma,\psi}\right)((2m)^\sigma\mu',f\hbar\circ\zeta)$$

$$\leq \frac{1}{2^s}(f(g(a))+mf(g(b)))\left(\mathcal{F}^{g^{-1}\left(\frac{g^p(a)+mg^p(b)}{2}\right)^+}_{g^{-1}(mg^p(b)),\sigma,\psi+s}\right)(2^\sigma\mu',\hbar\circ\zeta)$$

$$+ m\left(f(g(b))+mf\left(\frac{g(a)}{m^2}\right)\right)\int_0^1 t^{\psi-1}\left(1-\frac{t}{2}\right)^s E^{\rho,r,k,c}_{\sigma,\psi,\phi}(\mu t^\sigma;q)\hbar\left(\left(\frac{t}{2}g^p(a)+m\left(1-\frac{t}{2}\right)g^p(b)\right)^{\frac{1}{p}}\right)dt,$$

$\zeta(z) = g^{\frac{1}{p}}(z),\ z \in [a^p, mb^p],\ f\hbar\circ\zeta = (f\circ\zeta)(\hbar\circ\zeta),\ \mu' = \frac{\mu}{(mg^p(b)-g^p(a))^\sigma}.$

(ii) If $p < 0$, then

$$2^s f\left(\left(\frac{g^p(a)+mg^p(b)}{2}\right)^{\frac{1}{p}}\right)\left(\mathcal{F}^{g^{-1}\left(\frac{g^p(a)+mg^p(b)}{2}\right)^{-}}_{g^{-1}(mg^p(b)),\sigma,\psi}\right)(2^\sigma \mu',\hbar\circ\zeta)$$

$$\leq \left(\mathcal{F}^{g^{-1}\left(\frac{g^p(a)+mg^p(b)}{2}\right)^{-}}_{g^{-1}(mg^p(b)),\sigma,\psi}\right)(2^\sigma \mu',f\hbar\circ\zeta)+m^{\psi+1}\left(\mathcal{F}^{g^{-1}\left(\frac{g^p(a)+mg^p(b)}{2m}\right)^{+}}_{g^{-1}\left(\frac{g^p(a)}{m}\right),\sigma,\psi}\right)((2m)^\sigma \mu',f\hbar\circ\zeta)$$

$$\leq \frac{1}{2^s}(f(g(a))+mf(g(b)))\left(\mathcal{F}^{g^{-1}\left(\frac{g^p(a)+mg^p(b)}{2}\right)^{-}}_{g^{-1}(mg^p(b)),\sigma,\psi+s}\right)(2^\sigma \mu',\hbar\circ\zeta)$$

$$+m\left(f(g(b))+mf\left(\frac{g(a)}{m^2}\right)\right)\int_0^1 t^{\psi-1}\left(1-\frac{t}{2}\right)^s E^{\rho,r,k,c}_{\sigma,\psi,\phi}(\mu t^\sigma;q)\hbar\left(\left(\frac{t}{2}g^p(a)+m\left(1-\frac{t}{2}\right)g^p(b)\right)^{\frac{1}{p}}\right)dt,$$

$\zeta(z) = g^{\frac{1}{p}}(z)$, $z \in [mb^p, a^p]$, $f\hbar\circ\zeta = (f\circ\zeta)(\hbar\circ\zeta)$, $\mu' = \frac{\mu}{(g^p(a)-mg^p(b))^\sigma}$.

Remark 4. *From Theorems 2 and 3, one can deduce results for $(s,m)-p$-Godunova–Levin-convex function of second kind, (p,P)-convex function, Godunova–Levin type harmonic convex function, s-Godunova–Levin type harmonic convex function, $(\alpha,h\text{-}m)$-HA-convex function, (α,h)-HA-convex function, HA-convex function and (α,m)-HA-convex function. Moreover, all the results for operators given in Remark 1 [27] can be obtained.*

3. Conclusions

It is common practice to establish the Hadamard type integral inequalities for new classes of functions related to convex functions. On the other hand, fractional integral operators are used to provide the generalizations of these inequalities. This paper presents the inequalities of Fejér–Hadamard type which simultaneously hold for many kinds of fractional integral operators. The reader can deduce a number of published as well as new Hadamard and Fejér–Hadamard type inequalities from the results of this paper.

Author Contributions: Conceptualization, G.F., M.Y. and K.N.; investigation, G.F., M.Y. and K.N.; methodology, G.F., M.Y. and K.N.; validation, G.F., M.Y. and K.N.; visualization, G.F., M.Y. and K.N.; writing—original draft, G.F. and K.N.; writing—review and editing, G.F. and K.N. All authors have read and agreed to the published version of the manuscript.

Funding: This research received no external funding.

Institutional Review Board Statement: Not applicable.

Informed Consent Statement: Not applicable.

Data Availability Statement: Not applicable.

Acknowledgments: This research has received funding support from the National Science, Research and Innovation Fund (NSRF), Thailand.

Conflicts of Interest: The authors declare no conflict of interest.

References

1. Ahmad, B.; Alsaedi, A.; Kirane, M.; Torebek, B.T. Hermite-Hadamard, Hermite-Hadamard-Fejér, Dragomir-Agarwal and Pachpatte type inequalities for convex functions via new fractional integrals. *J. Comp. Appl. Math.* **2019**, *353*, 120–129. [CrossRef]
2. Butt, S.I.; Bakula, M.K.; Pečarić, J. Steffensen-Grüss inequality. *J. Math. Inequal.* **2021**, *15*, 799–810. [CrossRef]
3. Chen, F.; Wu, S. Hermite-Hadamard type inequalities for harmonically convex functions. *J. Appl. Math.* **2014**, *2014*, 386806. [CrossRef]
4. Fang, Z.B.; Shi, R. On the (p,h)-convex function and some integral inequalities. *J. Inequal. Appl.* **2014**, *2014*, 45. [CrossRef]
5. Kunt, M.; İscan, İ.; Yazi, N.; Gozutok, U. On new inequalities of Hermite-Hadamard-Fejér type inequalities for harmonically convex functions via fractional integrals. *Springer Plus.* **2016**, *5*, 1–19. [CrossRef]
6. Sarikaya, M.Z.; Set, E.; Yaldiz, H.; Basak, N. Hermite-Hadamard inequalities for fractional integrals and related fractional inequalities. *J. Math. Comput. Model.* **2013**, *57*, 2403–2407. [CrossRef]

7. İscan, İ.; Turhan, S.; Maden, S. Hermite-Hadamard and simpson-like type inequalities for differentiable *p*-quasi-convex functions. *Filomat* **2017**, *31*, 5945–5953. [CrossRef]
8. Varošanec, S. On *h*-convexity. *J. Math. Anal. Appl.* **2007**, *326*, 303–311. [CrossRef]
9. İscan, İ. Ostrowski type inequalities for *p*-convex functions. *New Trends Math. Sci.* **2016**, *4*, 140–150. [CrossRef]
10. Mihesan, V.G. *A Generalization of the Convexity*; Seminar on Functional Equations, Approx. and Convex.: Cluj-Napoca, Romania, 1993.
11. Kilbas, A.A.; Srivastava H.M.; Trujillo, J.J. *Theory and Applications of Fractional Differential Equations*; Elsevier: Amsterdam, The Netherlands, 2006.
12. Ahmad, B.; Henderson, J.; Luca, R. *Boundary Value Problems for Fractional Differential Equations and Systems*; World Scientific: Singapore, 2021.
13. Tarasov, V.E. Fractional Dynamical Systems. In *Fractional Dynamics. Nonlinear Physical Science*; Springer: Berlin/Heidelberg, Germany, 2010.
14. Amsalu, H.; Suthar, D.L. Generalized fractional integral operators involving Mittag-Leffler function. *Abstr. Appl. Anal.* **2018**, *2018*, 7034124. [CrossRef]
15. Prabhakar, T.R. A singular integral equation with a generalized Mittag-Leffler function in the kernel. *Yokohama Math. J.* **1971**, *19*, 7–15.
16. Rahman, G.; Baleanu, D.; Qurashi, M.A.; Purohit, S.D.; Mubeen, S.; Arshad, M. The extended Mittag-Leffler function via fractional calculus. *J. Nonlinear Sci. Appl.* **2017**, *10*, 4244–4253. [CrossRef]
17. Salim, T.O.; Faraj, A.W. A generalization of Mittag-Leffler function and integral operator associated with integral calculus. *J. Frac. Calc. Appl.* **2012**, *3*, 1–13.
18. Sachan, D.S.; Jaloree, S.; Choi, J. Certain recurrence relations of two parametric Mittag-Leffler function and their application in fractional calculus. *Fractal Fract.* **2021**, *5*, 215. [CrossRef]
19. Srivastava, H.M.; Tomovski, Z. Fractional calculus with an integral operator containing generalized Mittag-Leffler function in the kernal. *Appl. Math. Comput.* **2009**, *211*, 198–210.
20. Sarikaya, M.Z.; Yildirim, H. On Hermite-Hadamard type inequalities for Riemann-Liouville fractional integrals. *Miskolc Math. Notes* **2017**, *17*, 1049–1059. [CrossRef]
21. Farid, G. A treatment of the Hadamard inequality due to *m*-convexity via generalized fractional integrals. *J. Fract. Calc. Appl.* **2018**, *9*, 8–14.
22. Butt, S.I.; Yousaf, S.; Akdemir, A.O.; Dokuyucu, M.A. New Hadamard-type integral inequalities via a general form of fractional integral operators. *Chaos Solitons Fractals* **2021**, *148*, 111025. [CrossRef]
23. İscan, İ. Hermite-Hadamard-Fejér type inequalities for convex functions via fractional integrals. *Stud. Univ. Babes-Bolyai Math.* **2015**, *60*, 355–366.
24. İscan, İ.; Wu, S. Hermite-Hadamard type inequalities for harmonically convex functions via fractional integrals. *Appl. Math. Comput.* **2014**, *238*, 237–244.
25. Jung, C.Y.; Yussouf, M.; Chu, Y.M.; farid, G.; Kang, S.M. Generalized fractional Hadamard and Fejér-Hadamard inequalities for generalized harmonically convex functions. *J. Math.* **2020**, *2020*, 8245324. [CrossRef]
26. Qiang, X.; Farid, G.; Yussouf, M.; Khan, K.A.; Rehman, A.U. New generalized fractional versions of Hadamard and Fejér inequalities for harmonically convex functions. *J. Inequal. Appl.* **2020**, *2020*, 191. [CrossRef]
27. Rao, Y.; Yussouf, M.; Farid, G.; Pečarić, J.; Tlili, I. Further generalizations of Hadamard and Fejér-Hadamard inequalities and error estimations. *Adv. Diff. Equ.* **2020**, *2020*, 421. [CrossRef]
28. Yang, X.; Farid, G.; Nazeer, W.; Yussouf, M.; Chu, Y.M.; Dong, C. Fractional generalized Hadamard and the Fejér-Hadamard type inequalities for *m*-convex functions. *AIMS Math.* **2020**, *5*, 6325–6340. [CrossRef]
29. Yussouf, M.; Farid, G.; Khan, K.A.; Jung, C.Y. Hadamard and Fejér inequalities for further generalized fractional integrals involving Mittag-Leffler Functions. *J. Math.* **2021**, *2021*, 13. [CrossRef]
30. Jia, W.; Yussouf, M.; Farid, G.; Khan, K.A. Hadamard and Fejér-Hadamard inequalities for (α, h-m)-*p*-convex functions via Riemann-Liouville fractional integrals. *Math. Probl. Eng.* **2021**, *2021*, 12. [CrossRef]
31. Toader, G.H. Some generalizations of the convexity. *Proc. Colloq. Approx. Optim. Cluj-Napoca (Romania)* **1985**, 329–338.
32. Fejér, L. Überdie Fourierreihen II. *Math Naturwiss Anz Ungar Akad Wiss.* **1906**, *24*, 369–390.
33. Farid, G. A unified integral operator and further its consequences. *Open J. Math. Anal.* **2020**, *4*, 1–7. [CrossRef]
34. Andrić, M.; Farid, G.; Pečarić, J. A further extension of Mittag-Leffler function. *Fract. Calc. Appl. Anal.* **2018**, *21*, 1377–1395. [CrossRef]
35. Hadamard, J. Etude sur les proprietes des fonctions entieres e.t en particulier dune fonction consideree par Riemann. *J. Math. Pure Appl.* **1893**, *58*, 171–215.
36. Hermite, C. Sur deux limites d'une intgrale dfinie. *Mathesis* **1883**, *3*, 82.

 *fractal and fractional*

Article

The Mittag–Leffler Functions for a Class of First-Order Fractional Initial Value Problems: Dual Solution via Riemann–Liouville Fractional Derivative

Abdelhalim Ebaid [1,*] and Hind K. Al-Jeaid [2]

1 Computational & Analytical Mathematics and Their Applications Research Group, Department of Mathematics, Faculty of Science, University of Tabuk, Tabuk 71491, Saudi Arabia
2 Department of Mathematical Sciences, Umm Al-Qura University, Makkah 715, Saudi Arabia; hkjeaid@uqu.edu.sa
* Correspondence: aebaid@ut.edu.sa

Abstract: In this paper, a new approach is developed to solve a class of first-order fractional initial value problems. The present class is of practical interest in engineering science. The results are based on the Riemann–Liouville fractional derivative. It is shown that the dual solution can be determined for the considered class. The first solution is obtained by means of the Laplace transform and expressed in terms of the Mittag–Leffler functions. The second solution was determined through a newly developed approach and given in terms of exponential and trigonometric functions. Moreover, the results reduce to the ordinary version as the fractional-order tends to unity. Characteristics of the dual solution are discussed in detail. Furthermore, the advantages of the second solution over the first one is declared. It is revealed that the second solution is real at certain values of the fractional-order. Such values are derived theoretically and accordingly, and the behavior of the real solution is shown through several plots. The present analysis may be introduced for obtaining the solution in a straightforward manner for the first time. The developed approach can be further extended to include higher-order fractional initial value problems of oscillatory types.

Keywords: Mittag–Leffler functions; Riemann–Liouville fractional derivative; initial value problems; Laplace transform; exact solution

1. Introduction

The fractional calculus (FC) is a growing field of research that is usually utilized to investigate the physical phenomena of the memory effect [1–3]. Many scientific models have been analyzed via the FC approach [4–8]. A comprehensive list of FC applications are listed in Refs. [9–14]. For example, the fractional physical model of the projectile motion was discussed by Ebaid [15] and Ebaid et al. [16] utilizing the Caputo fractional derivative (CFD), and their results have been compared with experimental data. In addition, Ahmed et al. [17] implemented the Riemann–Liouville fractional derivative (RLFD) to analyze the same problem. The above models have been formulated in the form of second-order fractional initial value problems (2nd-order FIVPs).

Furthermore, Kumar et al. [18] and Ebaid et al. [19] studied the first-order fractional initial value problems (1st-order FIVPs) describing the absorption of light by the interstellar matter (called Ambartsumian-fractional model) by means of CFD. The exact solution of this model was determined by Ebaid et al. [19] using the Laplace transform (LT). Moreover, the RLFD was used by El-Zahar et al. [20] to provide the solution of the Ambartsumian-fractional model in a closed series form. Recent interesting results and applications of FC can be found in [21–29]. Very recently, El-Dib and Elgazery [30] investigated the nonlinear oscillations utilizing the properties of the RLFD.

The objective of this work is to extend the application of RLFD to a certain class of 1st-order FIVPs of oscillatory nature. Such class is of great importance in the field of engineering. Thus, this paper considers the class of 1st-order FIVPs:

$$^{RL}_{-\infty}D^{\alpha}_t y(t) + \omega^2 y(t) = a\cos(\Omega t), \quad D^{\alpha-1}_t y(0) = A, \quad 0 < \alpha \leq 1, \tag{1}$$

where α is the non-integer order of the RLFD, while a, ω, Ω, and A are constants. Moreover, the present model can be viewed as a forced harmonic-oscillator of the first-order in a fractional form, and it may be of practical interest in engineering science. Although Equation (1) seems simple, obtaining its exact solution is not an easy task due to several factors that will be illustrated. It will also be shown that a dual solution exists. As a solution method, the LT is a basic and effective tool to solve 1st-order FIVPs, even for higher-order FIVPs. The LT will be applied on the current class to construct the first solution in terms of Mittag–Leffler functions. However, a new approach is to be developed in this paper to determine the second solution in which only exponential and trigonometric functions are involved.

Characteristics of these solutions will also be discussed. The advantages of the second solution over the first one will be demonstrated. To our knowledge, the present analysis has not been yet reported in the literature. The rest of the paper is organized as follows. In Section 2, the definition/properties of the RLFD and the Mittag–Leffler functions are introduced. In Section 3, a basic theorem for the particular solution of Equation (1) is introduced. Moreover, it is shown that the present particular solution reduces to the corresponding one in the literature as a special case. In Section 4, the dual solution of 1st-order FIVPs (1) is constructed and analyzed in detail. Section 5 is devoted to studying the characteristics of the established solutions. In addition, the α-values that admit real solutions of the present class are obtained theoretically. Moreover, the behavior of the current solution is discussed. In Section 6, the main conclusions are summarized.

2. Preliminaries

The Riemann–Liouville fractional integral of order α of function $f : [c, d] \to \mathbb{R}$ ($-\infty < c < d < \infty$) is defined as [1–3]

$$_c I^{\alpha}_t f(t) = \frac{1}{\Gamma(\alpha)} \int_c^t \frac{f(\tau)}{(t-\tau)^{1-\alpha}} d\tau, \quad t > c, \; \alpha > 0. \tag{2}$$

The RLFD of order $\alpha \in \mathbb{R}^+_0$ is [1–3]

$$^{RL}_c D^{\alpha}_t f(t) = \frac{1}{\Gamma(n-\alpha)} \frac{d^n}{dt^n} \left(\int_c^t \frac{f(\tau)}{(t-\tau)^{\alpha-n+1}} d\tau \right), \quad n = [\alpha] + 1, \; t > c, \tag{3}$$

where $[\alpha]$ means the integral part of α. For $t \in \mathbb{R}$ and $c \to -\infty$, the RLFD of the functions $e^{i\omega t}$, $\cos(\omega t)$, and $\sin(\omega t)$ are [30,31]

$$\begin{aligned}
^{RL}_{-\infty}D^{\alpha}_t e^{i\omega t} &= (i\omega)^{\alpha} e^{i\omega t}, \\
^{RL}_{-\infty}D^{\alpha}_t \cos(\omega t) &= \omega^{\alpha} \cos\left(\omega t + \frac{\alpha \pi}{2}\right), \\
^{RL}_{-\infty}D^{\alpha}_t \sin(\omega t) &= \omega^{\alpha} \sin\left(\omega t + \frac{\alpha \pi}{2}\right).
\end{aligned} \tag{4}$$

It may be important to mention that the first equation in (4) was implemented by El-Dib and Elgazery [30]. Such implementation is based on the proof introduced by Ortigueira et al. [31]. However, the last two equations in (4) are utilized in [30] without proof, which may be because the authors [30] considered the derivation of these equations an easy task. For this reason, the proof and validity of the last two equations in (4) are provided in the Appendix A. The Laplace transform (LT) of the RLFD, as $c \to 0$, is

$$\mathcal{L}\left[{}_{0}^{RL}D_{t}^{\alpha}y(t)\right] = s^{\alpha}Y(s) - \sum_{i=0}^{n-1} s^{i} D_{t}^{\alpha-i-1} y(0). \tag{5}$$

In view of Equations (4) and (5), we have to distinguish between the properties of the RLFD when $c \to -\infty$ and $c \to 0$, i.e., ${}_{-\infty}^{RL}D_{t}^{\alpha}$ and ${}_{0}^{RL}D_{t}^{\alpha}$, respectively. Thus, the dual solution of Equation (1) is expected.

The Mittag–Leffler function of two parameters is defined by

$$E_{\alpha,\gamma}(z) = \sum_{i=0}^{\infty} \frac{z^{i}}{\Gamma(\alpha i + \gamma)}, \quad (\alpha > 0, \ \gamma > 0). \tag{6}$$

In particular, we have the following properties

$$E_{\alpha,1}(z) = E_{\alpha}(z), \quad E_{1}(z) = e^{z}, \quad E_{2,1}(-z^{2}) = \cos(z), \quad E_{2,2}(-z^{2}) = \frac{\sin z}{z}. \tag{7}$$

The inverse LT of some expressions can be given via the Mittag–Leffler function as

$$\mathcal{L}^{-1}\left(\frac{s^{\alpha-\gamma}}{s^{\alpha} + \omega^{2}}\right) = t^{\gamma-1} E_{\alpha,\gamma}(-\omega^{2} t^{\alpha}), \quad \mathrm{Re}(s) > |\omega^{2}|^{\frac{1}{\alpha}}, \tag{8}$$

which gives the equalities:

$$\mathcal{L}^{-1}\left(\frac{s^{\alpha-1}}{s^{\alpha} + 1}\right) = E_{\alpha}(-t^{\alpha}), \tag{9}$$

$$\mathcal{L}^{-1}\left(\frac{1}{s^{\alpha} + \omega^{2}}\right) = t^{\alpha-1} E_{\alpha,\alpha}(-\omega^{2} t^{\alpha}), \quad \mathrm{Re}(s) > |\omega^{2}|^{\frac{1}{\alpha}}, \tag{10}$$

$$\mathcal{L}^{-1}\left(\frac{s^{-1}}{s^{\alpha} + \omega^{2}}\right) = t^{\alpha} E_{\alpha,\alpha+1}(-\omega^{2} t^{\alpha}), \quad \mathrm{Re}(s) > |\omega^{2}|^{\frac{1}{\alpha}}. \tag{11}$$

3. Analysis

Theorem 1 (The particular solution). *The particular solution $y_p(t)$ of Equation (1) is given by*

$$y_p(t) = \rho_1(\alpha) \cos(\Omega t) + \rho_2(\alpha) \sin(\Omega t), \quad 0 < \alpha \leq 1, \tag{12}$$

where $\rho_1(\alpha)$ and $\rho_2(\alpha)$ are

$$\rho_1(\alpha) = a \left(\frac{\omega^{2} + \Omega^{\alpha} \cos\left(\frac{\pi\alpha}{2}\right)}{\omega^{4} + \Omega^{2\alpha} + 2\omega^{2}\Omega^{\alpha} \cos\left(\frac{\pi\alpha}{2}\right)} \right), \quad \rho_2(\alpha) = a \left(\frac{\Omega^{\alpha} \sin\left(\frac{\pi\alpha}{2}\right)}{\omega^{4} + \Omega^{2\alpha} + 2\omega^{2}\Omega^{\alpha} \cos\left(\frac{\pi\alpha}{2}\right)} \right). \tag{13}$$

Proof. Assume y_p in the form of Equation (12), then

$$\begin{aligned}
{}_{-\infty}^{RL}D_{t}^{\alpha} y_p &= \rho_1(\alpha) \, {}_{-\infty}^{RL}D_{t}^{\alpha} \cos(\Omega t) + \rho_2(\alpha) \, {}_{-\infty}^{RL}D_{t}^{\alpha} \sin(\Omega t), \\
&= \Omega^{\alpha} \cos(\Omega t) \left(\rho_1(\alpha) \cos\left(\frac{\pi\alpha}{2}\right) + \rho_2(\alpha) \sin\left(\frac{\pi\alpha}{2}\right) \right) + \\
&\quad \Omega^{\alpha} \sin(\Omega t) \left(\rho_2(\alpha) \cos\left(\frac{\pi\alpha}{2}\right) - \rho_1(\alpha) \sin\left(\frac{\pi\alpha}{2}\right) \right),
\end{aligned} \tag{14}$$

and hence

$$\begin{aligned}
{}_{-\infty}^{RL}D_{t}^{\alpha} y_p + \omega^{2} y_p &= \left[\left(\Omega^{\alpha} \cos\left(\frac{\pi\alpha}{2}\right) + \omega^{2} \right) \rho_1(\alpha) + \Omega^{\alpha} \sin\left(\frac{\pi\alpha}{2}\right) \rho_2(\alpha) \right] \cos(\Omega t) + \\
&\quad \left[\left(\Omega^{\alpha} \cos\left(\frac{\pi\alpha}{2}\right) + \omega^{2} \right) \rho_2(\alpha) - \Omega^{\alpha} \sin\left(\frac{\pi\alpha}{2}\right) \rho_1(\alpha) \right] \sin(\Omega t).
\end{aligned} \tag{15}$$

Substituting Equation (15) into Equation (1), we obtain the algebraic system:

$$\begin{aligned}\left(\Omega^\alpha \cos\left(\frac{\pi\alpha}{2}\right)+\omega^2\right)\rho_1(\alpha)+\Omega^\alpha \sin\left(\frac{\pi\alpha}{2}\right)\rho_2(\alpha)&=a,\\ \left(\Omega^\alpha \cos\left(\frac{\pi\alpha}{2}\right)+\omega^2\right)\rho_2(\alpha)-\Omega^\alpha \sin\left(\frac{\pi\alpha}{2}\right)\rho_1(\alpha)&=0.\end{aligned} \quad (16)$$

Solving the algebraic system (16) for $\rho_1(\alpha)$ and $\rho_2(\alpha)$, we obtain

$$\rho_1(\alpha)=a\left(\frac{\omega^2+\Omega^\alpha \cos\left(\frac{\pi\alpha}{2}\right)}{\omega^4+\Omega^{2\alpha}+2\omega^2\Omega^\alpha \cos\left(\frac{\pi\alpha}{2}\right)}\right),\ \rho_2(\alpha)=a\left(\frac{\Omega^\alpha \sin\left(\frac{\pi\alpha}{2}\right)}{\omega^4+\Omega^{2\alpha}+2\omega^2\Omega^\alpha \cos\left(\frac{\pi\alpha}{2}\right)}\right). \quad (17)$$

Inserting (17) into (12) and simplifying, we obtain y_p in the form:

$$y_p(t)=a\left(\frac{\omega^2 \cos(\Omega t)+\Omega^\alpha \cos\left(\Omega t-\frac{\pi\alpha}{2}\right)}{\omega^4+\Omega^{2\alpha}+2\omega^2\Omega^\alpha \cos\left(\frac{\pi\alpha}{2}\right)}\right), \quad (18)$$

which completes the proof. □

Lemma 1. *At $a=1$, the particular solution $y_p(t)$ of Equation (1) reduces to*

$$y_p(t)=\frac{\omega^2 \cos(\Omega t)+\Omega^\alpha \cos\left(\Omega t-\frac{\pi\alpha}{2}\right)}{\omega^4+\Omega^{2\alpha}+2\omega^2\Omega^\alpha \cos\left(\frac{\pi\alpha}{2}\right)}. \quad (19)$$

Proof. The proof follows immediately by setting $a=1$ in Equation (18); thus

$$y_p(t)=\frac{\omega^2 \cos(\Omega t)+\Omega^\alpha \cos\left(\Omega t-\frac{\pi\alpha}{2}\right)}{\omega^4+\Omega^{2\alpha}+2\omega^2\Omega^\alpha \cos\left(\frac{\pi\alpha}{2}\right)}, \quad (20)$$

which agrees with the obtained particular integral in Ref. [30] (Equation (7)) using the $\left(D^\alpha+\omega^2\right)^{-1}$ operator. However, our approach is straightforward and easier. □

4. Dual Solution

It is shown in this section that the dual solution of the present class of 1st-order FIVPs can be derived. The first solution is obtained in terms of Mittag–Leffler functions, while the second is provided in terms of exponential and trigonometric functions so that the Mittag–Leffler function can be avoided. Characteristics of these solutions will also be discussed in a subsequent section.

4.1. Solution in Terms of Mittag–Leffler Functions

Applying the LT on Equation (1), yields

$$s^\alpha Y(s)-D_t^{\alpha-1}y(0)+\omega^2 Y(s)=\frac{as}{s^2+\Omega^2}, \quad (21)$$

where $Y(s)$ is the LT of $y(t)$. Solving (21) for $Y(s)$ gives

$$Y(s)=\frac{A}{s^\alpha+\omega^2}+\frac{as}{(s^\alpha+\omega^2)(s^2+\Omega^2)}. \quad (22)$$

The solution $y(t)$ is obtained by applying the inverse LT on $Y(s)$, this gives

$$y(t)=A\mathcal{L}^{-1}\left(\frac{1}{s^\alpha+\omega^2}\right)+\mathcal{L}^{-1}\left[\frac{as}{(s^\alpha+\omega^2)(s^2+\Omega^2)}\right], \quad (23)$$

i.e.,
$$y(t) = At^{\alpha-1}E_{\alpha,\alpha}\left(-\omega^2 t^\alpha\right) + a\mathcal{L}^{-1}\left(\frac{1}{s^\alpha+\omega^2}\right)*\mathcal{L}^{-1}\left(\frac{s}{s^2+\Omega^2}\right), \tag{24}$$

where $(*)$ refers to the convolution operation. Therefore

$$y(t) = At^{\alpha-1}E_{\alpha,\alpha}\left(-\omega^2 t^\alpha\right) + a\int_0^t \tau^{\alpha-1}E_{\alpha,\alpha}\left(-\omega^2\tau^\alpha\right)\cos[\Omega(t-\tau)]d\tau, \tag{25}$$

which can be written as

$$y(t) = At^{\alpha-1}E_{\alpha,\alpha}\left(-\omega^2 t^\alpha\right) + a\cos(\Omega t)\int_0^t \tau^{\alpha-1}E_{\alpha,\alpha}\left(-\omega^2\tau^\alpha\right)\cos(\Omega\tau)d\tau +$$
$$a\sin(\Omega t)\int_0^t \tau^{\alpha-1}E_{\alpha,\alpha}\left(-\omega^2\tau^\alpha\right)\sin(\Omega\tau)d\tau, \tag{26}$$

As $\alpha \to 1$, the solution reduces to

$$y(t) = AE_{1,1}\left(-\omega^2 t\right) + a\cos(\Omega t)\int_0^t E_{1,1}\left(-\omega^2\tau\right)\cos(\Omega\tau)d\tau +$$
$$a\sin(\Omega t)\int_0^t E_{1,1}\left(-\omega^2\tau\right)\sin(\Omega\tau)d\tau, \tag{27}$$

i.e.,

$$y(t) = Ae^{-\omega^2 t} + a\cos(\Omega t)\int_0^t e^{-\omega^2\tau}\cos(\Omega\tau)d\tau + a\sin(\Omega t)\int_0^t e^{-\omega^2\tau}\sin(\Omega\tau)d\tau. \tag{28}$$

Evaluating the involved integrals and simplifying yields

$$y(t) = \left(A - \frac{a\omega^2}{\omega^4+\Omega^2}\right)e^{-\omega^2 t} + \frac{a}{\omega^4+\Omega^2}\left[\omega^2\cos(\Omega t) + \Omega\sin(\Omega t)\right], \tag{29}$$

which agrees with the solution of the ordinary version of the FIVP (1). However, the present solution in fractional form (26) is not analytic at $t=0, \forall \alpha \in (0,1)$ for the existence of term $t^{\alpha-1}$. This phenomena will be avoided in the next section via a new approach to obtaining the exact analytic solution for the FIVP (1) in terms of exponential and trigonometric functions. Equation (26) is non-analytic at $t=0$, which is just a consequence of applying the LT on the RLFD as c tends to zero. This gives the second solution, in terms of exponential and trigonometric functions, an advantage over the first one, in terms of the Mittag–Leffler functions.

4.2. Solution in Terms of Exponential and Trigonometric Functions

The general solution $y(t)$ of the FIVP (1) consists of the complementary solution $y_c(t)$ and the particular solution $y_p(t)$ so that

$$y(t) = y_c(t) + y_p(t), \tag{30}$$

where $y_p(t)$ was already obtained by Theorem 1, while $y_c(t)$ is the solution of the homogeneous part:
$${}^{RL}_{-\infty}D_t^\alpha y_c(t) + \omega^2 y_c(t) = 0. \tag{31}$$

Assume $y_c(t)$ is in the form:
$$y_c(t) = c_1(\alpha)\cos(\delta t) + c_2(\alpha)\sin(\delta t), \tag{32}$$

where $c_1(\alpha)$, $c_2(\alpha)$, and $\delta(\omega)$ are to be determined. Substituting (32) into (31), yields

$$\left[\left(\delta^\alpha \cos\left(\frac{\pi\alpha}{2}\right) + \omega^2\right)c_1(\alpha) + \delta^\alpha \sin\left(\frac{\pi\alpha}{2}\right)c_2(\alpha)\right]\cos(\delta t) +$$
$$\left[\left(\delta^\alpha \cos\left(\frac{\pi\alpha}{2}\right) + \omega^2\right)c_2(\alpha) - \delta^\alpha \sin\left(\frac{\pi\alpha}{2}\right)c_1(\alpha)\right]\sin(\delta t) = 0. \qquad (33)$$

In order to avoid trivial solutions for $c_1(\alpha)$ and $c_2(\alpha)$ in (33), we can set $c_2(\alpha) = ic_1(\alpha)$, without loss of generality. Thus, Equation (33) becomes

$$c_1(\alpha)\left[\delta^\alpha \cos\left(\frac{\pi\alpha}{2} + \omega^2\right) + i\delta^\alpha \sin\left(\frac{\pi\alpha}{2}\right)\right]\cos(\delta t) +$$
$$ic_1(\alpha)\left[\delta^\alpha \cos\left(\frac{\pi\alpha}{2}\right) + \omega^2 + i\delta^\alpha \sin\left(\frac{\pi\alpha}{2}\right)\right]\sin(\delta t) = 0, \qquad (34)$$

which can be reduced to

$$c_1(\alpha)\left[\delta^\alpha\left(\cos\left(\frac{\pi\alpha}{2}\right) + i\sin\left(\frac{\pi\alpha}{2}\right)\right) + \omega^2\right](\cos(\delta t) + i\sin(\delta t)) = 0. \qquad (35)$$

Equation (35) can be further simplified as

$$c_1(\alpha)\left[\delta^\alpha e^{i\frac{\pi\alpha}{2}} + \omega^2\right]e^{i\delta t} = 0. \qquad (36)$$

For a non-trivial complementary solution, we restrict so that $c_1(\alpha) \neq 0$, and hence, Equation (36) becomes

$$\left(\delta e^{i\frac{\pi}{2}}\right)^\alpha + \omega^2 = 0. \qquad (37)$$

Solving this equation for δ, we obtain

$$\delta = -i\left(-\omega^2\right)^{1/\alpha}, \qquad y_c(t) = c_1(\alpha)e^{i\delta t}. \qquad (38)$$

Accordingly,

$$y(t) = c_1(\alpha)e^{i\delta t} + y_p(t), \qquad (39)$$

where $c_1(\alpha)$ can be determined by applying the given initial condition. To do so, we have from Equation (39) that

$$D_t^{\alpha-1}y(t) = c_1(\alpha)D_t^{\alpha-1}e^{i\delta t} + D_t^{\alpha-1}y_p(t) = c_1(\alpha)(i\delta)^{\alpha-1}e^{i\delta t} + D_t^{\alpha-1}y_p(t), \qquad (40)$$

and at $t = 0$, we have

$$D_t^{\alpha-1}y(0) = c_1(\alpha)(i\delta)^{\alpha-1} + D_t^{\alpha-1}y_p(0). \qquad (41)$$

The magnitude $D_t^{\alpha-1}y_p(0)$ is calculated as follows

$$\left[D_t^{\alpha-1}y_p(t)\right]_{t=0} = a\left(\frac{\omega^2 + \Omega^\alpha \cos\left(\frac{\pi\alpha}{2}\right)}{\omega^4 + \Omega^{2\alpha} + 2\omega^2\Omega^\alpha \cos\left(\frac{\pi\alpha}{2}\right)}\right)\left[D_t^{\alpha-1}\cos(\Omega t)\right]_{t=0} +$$
$$a\left(\frac{\Omega^\alpha \sin\left(\frac{\pi\alpha}{2}\right)}{\omega^4 + \Omega^{2\alpha} + 2\omega^2\Omega^\alpha \cos\left(\frac{\pi\alpha}{2}\right)}\right)\left[D_t^{\alpha-1}\sin(\Omega t)\right]_{t=0}. \qquad (42)$$

Thus

$$D_t^{\alpha-1}y_p(0) = a\left(\frac{\omega^2 + \Omega^\alpha \cos\left(\frac{\pi\alpha}{2}\right)}{\omega^4 + \Omega^{2\alpha} + 2\omega^2\Omega^\alpha \cos\left(\frac{\pi\alpha}{2}\right)}\right)\Omega^{\alpha-1}\cos\left(\frac{\pi}{2}(\alpha-1)\right) +$$

$$a\left(\frac{\Omega^\alpha \sin\left(\frac{\pi\alpha}{2}\right)}{\omega^4 + \Omega^{2\alpha} + 2\omega^2\Omega^\alpha \cos\left(\frac{\pi\alpha}{2}\right)}\right)\Omega^{\alpha-1}\sin\left(\frac{\pi}{2}(\alpha-1)\right),$$

$$= \frac{a\Omega^{\alpha-1}}{\omega^4 + \Omega^{2\alpha} + 2\omega^2\Omega^\alpha \cos\left(\frac{\pi\alpha}{2}\right)} \times$$

$$\left[\left(\omega^2 + \Omega^\alpha \cos\left(\frac{\pi\alpha}{2}\right)\right)\cos\left(\frac{\pi}{2}(\alpha-1)\right) + \Omega^\alpha \sin\left(\frac{\pi\alpha}{2}\right)\sin\left(\frac{\pi}{2}(\alpha-1)\right)\right],$$

$$= \frac{a\Omega^{\alpha-1}}{\omega^4 + \Omega^{2\alpha} + 2\omega^2\Omega^\alpha \cos\left(\frac{\pi\alpha}{2}\right)}\left[\omega^2\cos\left(\frac{\pi}{2}(\alpha-1)\right) + \Omega^\alpha\cos\left(\frac{\pi\alpha}{2} - \frac{\pi}{2}(\alpha-1)\right)\right],$$

$$= \frac{a\omega^2\Omega^{\alpha-1}\sin\left(\frac{\pi\alpha}{2}\right)}{\omega^4 + \Omega^{2\alpha} + 2\omega^2\Omega^\alpha \cos\left(\frac{\pi\alpha}{2}\right)}. \tag{43}$$

Substituting (43) into (41) and implementing the given initial condition $D_t^{\alpha-1}y(0) = A$, we obtain

$$c_1(\alpha)(i\delta)^{\alpha-1} + \frac{a\omega^2\Omega^{\alpha-1}\sin\left(\frac{\pi\alpha}{2}\right)}{\omega^4 + \Omega^{2\alpha} + 2\omega^2\Omega^\alpha \cos\left(\frac{\pi\alpha}{2}\right)} = A. \tag{44}$$

Therefore, $c_1(\alpha)$ is given by

$$c_1(\alpha) = A(i\delta)^{1-\alpha} - \frac{a\omega^2\left(\frac{\Omega}{i\delta}\right)^{\alpha-1}\sin\left(\frac{\pi\alpha}{2}\right)}{\omega^4 + \Omega^{2\alpha} + 2\omega^2\Omega^\alpha \cos\left(\frac{\pi\alpha}{2}\right)}. \tag{45}$$

Substituting $\delta = -i(-\omega^2)^{1/\alpha}$ into (54), yields

$$c_1(\alpha) = A(-\omega^2)^{\frac{1}{\alpha}-1} + \frac{a(-\omega^2)^{\frac{1}{\alpha}}\Omega^{\alpha-1}\sin\left(\frac{\pi\alpha}{2}\right)}{\omega^4 + \Omega^{2\alpha} + 2\omega^2\Omega^\alpha \cos\left(\frac{\pi\alpha}{2}\right)}. \tag{46}$$

Hence, the solution takes the final form:

$$y(t) = \left[A(-\omega^2)^{\frac{1}{\alpha}-1} + \frac{a(-\omega^2)^{\frac{1}{\alpha}}\Omega^{\alpha-1}\sin\left(\frac{\pi\alpha}{2}\right)}{\omega^4 + \Omega^{2\alpha} + 2\omega^2\Omega^\alpha \cos\left(\frac{\pi\alpha}{2}\right)}\right]e^{(-\omega^2)^{\frac{1}{\alpha}}t} +$$

$$a\left[\frac{\omega^2\cos(\Omega t) + \Omega^\alpha \cos\left(\Omega t - \frac{\pi\alpha}{2}\right)}{\omega^4 + \Omega^{2\alpha} + 2\omega^2\Omega^\alpha \cos\left(\frac{\pi\alpha}{2}\right)}\right], \tag{47}$$

To check as $\alpha \to 1$, we have

$$y(t) = \left(A - \frac{a\omega^2}{\omega^4 + \Omega^2}\right)e^{-\omega^2 t} + \frac{a}{\omega^4 + \Omega^2}\left[\omega^2\cos(\Omega t) + \Omega\sin(\Omega t)\right], \tag{48}$$

which also agrees with the solution of the ordinary version of the FIVP (1). The advantage of the fractional form (47) is that it is analytic in the whole domain $t \geq 0, \forall \alpha \in (0, 1]$. However, this solution is real at specific/certain values of α. This issue is addressed in detail in the next section.

5. Characteristics of Solutions

For the class of 1st-order FIVP (1), it is observed that the exact solution in Equation (47) depends on whether the quantity $(-\omega^2)^{\frac{1}{\alpha}}$ is real or complex for $0 < \alpha < 1$. Since $\omega \in \mathbb{R}$, then the expression $(\omega^2)^{\frac{1}{\alpha}} \in \mathbb{R}\ \forall\ \alpha \in (0, 1)$. If we write $(-\omega^2)^{\frac{1}{\alpha}} = \epsilon(\omega^2)^{\frac{1}{\alpha}}$, where

$\epsilon = (-1)^{\frac{1}{\alpha}}$, then the solution is real/complex if ϵ is real/complex. The following theorem determines the values of α for the real solutions of the 1st-order FIVP (1).

Theorem 2 (α-values for real solutions). *The solution (47) is real at* $\alpha = \frac{2n-1}{2(k+n-1)}$ *($\epsilon = 1$) and* $\alpha = \frac{2n-1}{2(k+n)-1}$ *($\epsilon = -1$), $\forall\, n, k \in \mathbb{N}^+$.*

Proof. For $0 < \alpha < 1$, the fractional-order α can be assumed as $\alpha = \frac{l_1}{r_1}$ such that $0 < l_1 < r_1$. For odd l_1 and even r_1, the possible values of α belong to the sets $\{\frac{1}{2}, \frac{1}{4}, \frac{1}{6}, \ldots\}$, $\{\frac{3}{4}, \frac{3}{6}, \frac{3}{8}, \ldots\}$, $\{\frac{5}{6}, \frac{5}{8}, \frac{5}{10}, \ldots\}$, $\{\frac{7}{8}, \frac{7}{10}, \frac{7}{12}, \ldots\}$, $\ldots$, which can be written as $\{\frac{1}{2k}\}_{k=1}^\infty$, $\{\frac{3}{2k+2}\}_{k=1}^\infty$, $\{\frac{5}{2k+4}\}_{k=1}^\infty$, $\ldots$, and such sets can be unified as

$$\alpha = \frac{2n-1}{2(k+n-1)}, \quad \forall\, n, k \in \mathbb{N}^+, \tag{49}$$

and in this case, we have $\epsilon = (-1)^{\frac{1}{\alpha}} = (-1)^{\frac{2(k+n-1)}{2n-1}} = 1$.

Furthermore, α can be assumed as $\alpha = \frac{l_1}{l_2}$ for two odd positive integers such that $0 < l_1 < l_2$. In this case, the values of α belong to the sets $\{\frac{1}{3}, \frac{1}{5}, \frac{1}{7}, \ldots\}$, $\{\frac{3}{5}, \frac{3}{7}, \frac{3}{9}, \ldots\}$, $\{\frac{5}{7}, \frac{5}{9}, \frac{5}{11}, \ldots\}$, $\{\frac{7}{9}, \frac{7}{11}, \frac{7}{13}, \ldots\}$, $\ldots$, which can be written as $\{\frac{1}{2k+1}\}_{k=1}^\infty$, $\{\frac{3}{2k+3}\}_{k=1}^\infty$, $\{\frac{5}{2k+5}\}_{k=1}^\infty$, $\ldots$, and such sets can be unified as

$$\alpha = \frac{2n-1}{2k+2n-1}, \quad \forall\, n, k \in \mathbb{N}^+, \tag{50}$$

and we have $\epsilon = -1$. Note that $\epsilon \in \mathbb{C}$ if $\alpha = \frac{r_2}{l_2}$ for any even r_2 and any odd l_2 such that $0 < r_2 < l_2$. □

Numerical Results: Oscillatory Solution

In Figures 1–4, the solution in Equation (47) is plotted at some selected values according to Theorem 2. The periodicity/oscillatory of the solution is clear in these figures. In addition, it can be seen from Figure 4 that our curves for those values of α near to 1 are identical to the ordinary case. Finally, Figures 5 and 6 show the effect of the initial condition A and the parameter Ω on the behavior of the solution.

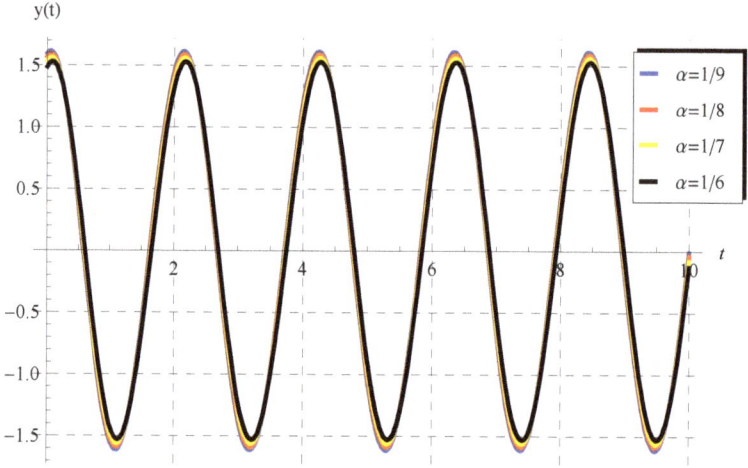

Figure 1. Plots of $y(t)$ in Equation (47) vs. t when $a = 2$, $A = 1$, $\omega = \frac{1}{3}$, and $\Omega = 3$ at different values of $\alpha = \frac{1}{9}, \frac{1}{8}, \frac{1}{7}, \frac{1}{6}$.

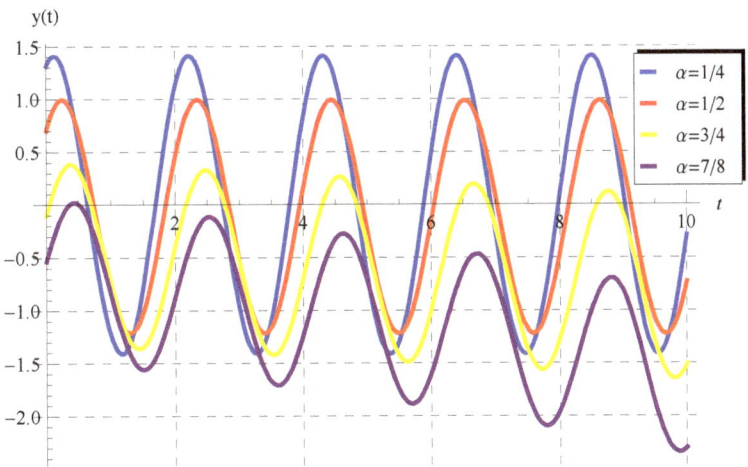

Figure 2. Plots of $y(t)$ in Equation (47) vs. t when $a = 2$, $A = 1$, $\omega = \frac{1}{3}$, and $\Omega = 3$ at different values of $\alpha = \frac{1}{4}, \frac{1}{2}, \frac{3}{4}, \frac{7}{8}$.

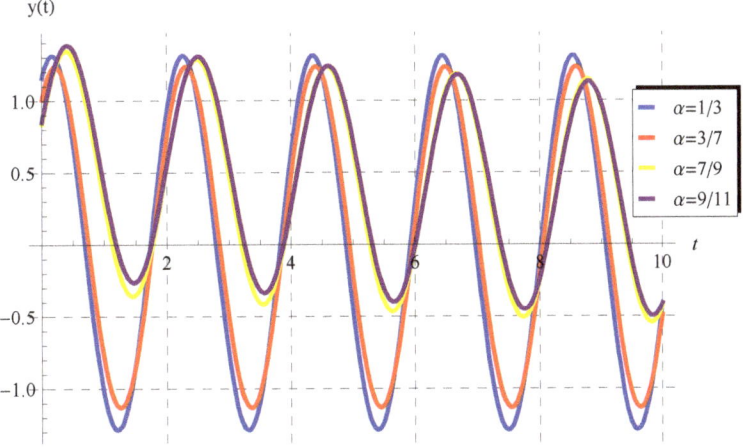

Figure 3. Plots of $y(t)$ in Equation (47) vs. t when $a = 2$, $A = 1$, $\omega = \frac{1}{3}$, and $\Omega = 3$ at different values of $\alpha = \frac{1}{3}, \frac{3}{7}, \frac{7}{9}, \frac{9}{11}$.

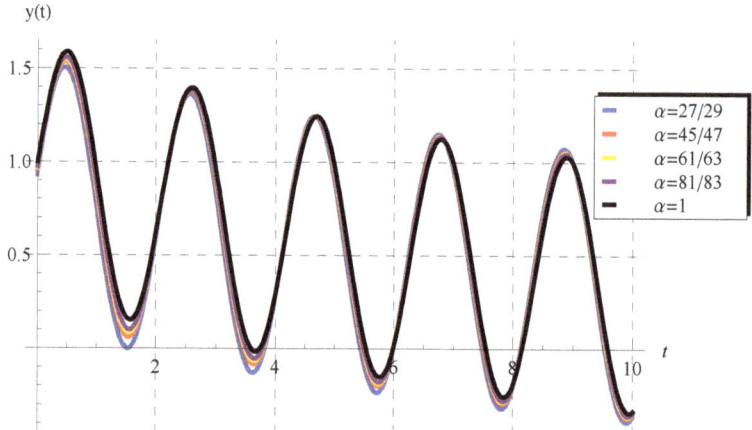

Figure 4. Plots of $y(t)$ in Equation (47) vs. t when $a = 2$, $A = 1$, $\omega = \frac{1}{3}$, and $\Omega = 3$ at different values of $\alpha = \frac{27}{29}, \frac{45}{47}, \frac{61}{63}, \frac{81}{83}, 1$.

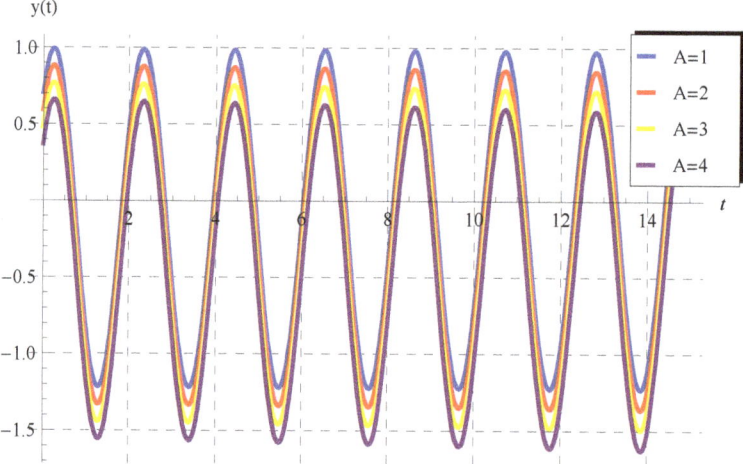

Figure 5. Plots of $y(t)$ in Equation (47) vs. t when $\alpha = \frac{1}{2}$, $a = 2$, $\omega = \frac{1}{3}$, and $\Omega = 3$ at different values of $A = 1, 2, 3, 4$.

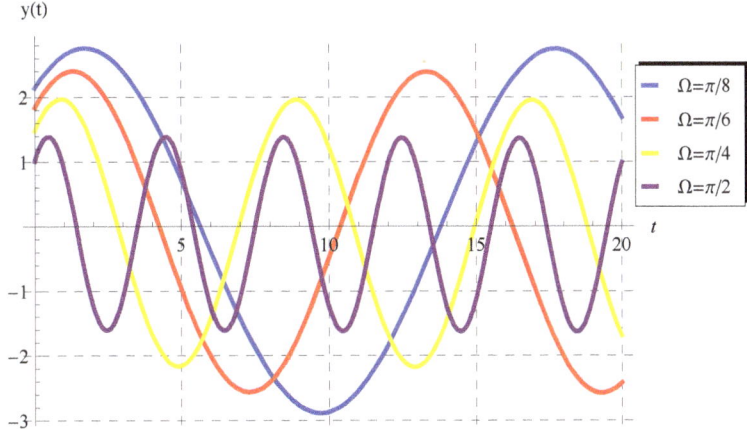

Figure 6. Plots of $y(t)$ in Equation (47) vs. t when $\alpha = \frac{1}{2}$, $A = 1$, $a = 2$, and $\omega = \frac{1}{3}$ at different values of $\Omega = \pi/8, \pi/6, \pi/4, \pi/2$.

6. Conclusions

A class of 1st-order FIVPs was investigated. This class is oscillatory in nature and hence of practical interest in engineering science. A dual solution was determined for the present class. The first solution was obtained through the LT and expressed in terms of the Mittag–Leffler functions. The second solution was derived via a newly developed approach in terms of exponential and trigonometric functions. The advantages of the second solution over the first one were demonstrated. In addition, it was revealed that the second solution is real at certain values of the fractional-order α. Such values of α were derived theoretically. The behavior of the real solution was displayed through several figures. The present analysis may be introduced for the first time to obtain the solution with a straightforward approach. The developed approach can be extended to higher-order FIVPs of oscillatory types. Finally, such approach can be viewed as a corner stone to obtaining periodic solutions for more complex oscillatory problems, such as the forced Duffing oscillator [30].

Author Contributions: Conceptualization, A.E.; methodology, A.E.; software, H.K.A.-J.; validation, H.K.A.-J.; investigation, A.E.; resources, H.K.A.-J. All authors have read and agreed to the published version of the manuscript.

Funding: This research received no external funding.

Informed Consent Statement: Not applicable.

Data Availability Statement: Not applicable.

Conflicts of Interest: The authors declare no conflict of interest.

Appendix A. The Fractional Derivative of Periodic Functions

Theorem A1. *The RLFD (3), as $c \to -\infty$, of the functions $\cos(\omega t)$ and $\sin(\omega t)$ are*

$$_{-\infty}^{RL}D_t^\alpha \cos(\omega t) = \omega^\alpha \cos\left(\omega t + \frac{\alpha \pi}{2}\right), \tag{A1}$$

$$_{-\infty}^{RL}D_t^\alpha \sin(\omega t) = \omega^\alpha \sin\left(\omega t + \frac{\alpha \pi}{2}\right). \tag{A2}$$

Proof. The proof is quite simple. Let us begin with the RLFD (3), as $c \to -\infty$, of the exponential function introduced by Ortigueira et al. [31]:

$$_{-\infty}^{RL}D_t^\alpha e^{i\omega t} = (i\omega)^\alpha e^{i\omega t}. \tag{A3}$$

The identities:

$$e^{i\omega t} = \cos(\omega t) + i\sin(\omega t), \tag{A4}$$

$$i^\alpha = e^{i\frac{\alpha\pi}{2}} = \cos\left(\frac{\alpha\pi}{2}\right) + i\sin\left(\frac{\alpha\pi}{2}\right), \tag{A5}$$

are to be used to derive Equations (A1) and (A2). Substituting (A4) and (A5) into (A3) reads

$$\begin{aligned}
_{-\infty}^{RL}D_t^\alpha \cos(\omega t) + i\, _{-\infty}^{RL}D_t^\alpha \sin(\omega t) &= \omega^\alpha(\cos(\omega t) + i\sin(\omega t))\left[\cos\left(\frac{\alpha\pi}{2}\right) + i\sin\left(\frac{\alpha\pi}{2}\right)\right] \\
&= \omega^\alpha\left[\cos(\omega t)\cos\left(\frac{\alpha\pi}{2}\right) - \sin(\omega t)\sin\left(\frac{\alpha\pi}{2}\right)\right] + \\
&\quad i\omega^\alpha\left[\sin(\omega t)\cos\left(\frac{\alpha\pi}{2}\right) + \cos(\omega t)\sin\left(\frac{\alpha\pi}{2}\right)\right] \\
&= \omega^\alpha \cos\left(\omega t + \frac{\alpha\pi}{2}\right) + i\omega^\alpha \sin\left(\omega t + \frac{\alpha\pi}{2}\right).
\end{aligned} \tag{A6}$$

Comparing the real and imaginary parts of the last equation completes the proof. □

References

1. Miller, K.S.; Ross, B. *An Introduction to the Fractional Calculus and Fractional Differential Equations*; John Wiley & Sons: New York, NY, USA, 1993.
2. Podlubny, I. *Fractional Differential Equations*; Academic Press: San Diego, CA, USA, 1999.
3. Hilfer, R. *Applications of Fractional Calculus in Physics*; World Scientific Publishing Company: Singapore, 2000.
4. Achar, B.N.N.; Hanneken, J.W.; Enck, T.; Clarke, T. Dynamics of the fractional oscillator. *Physica A* **2001**, *297*, 361–367. [CrossRef]
5. Sebaa, N.; Fellah, Z.E.A.; Lauriks, W.; Depollier, C. Application of fractional calculus to ultrasonic wave propagation in human cancellous bone. *Signal Process.* **2006**, *86*, 2668–2677. [CrossRef]
6. Tarasov, V.E. Fractional Heisenberg equation. *Phys. Lett. A* **2008**, *372*, 2984–2988. [CrossRef]
7. Ding, Y.; Yea, H. A fractional-order differential equation model of HIV infection of CD4+T-cells. *Math. Comput. Model.* **2009**, *50*, 386–392. [CrossRef]
8. Wang, S.; Xu, M.; Li, X. Green's function of time fractional diffusion equation and its applications in fractional quantum mechanics. *Nonlinear Anal. Real World Appl.* **2009**, *10*, 1081–1086. [CrossRef]
9. Song, L.; Xu, S.; Yang, J. Dynamical models of happiness with fractional order. *Commun. Nonlinear Sci. Numer. Simul.* **2010**, *15*, 616–628. [CrossRef]
10. Gómez-Aguilara, J.F.; Rosales-García, J.J.; Bernal-Alvarado, J.J. Fractional mechanical oscillators. *Rev. Mex. Física* **2012**, *58*, 348–352.
11. Machado, J.T.; Kiryakova, V.; Mainardi, F. Recent history of fractional calculus. *Commun. Nonlinear Sci. Numer. Simul.* **2011**, *16*, 1140–1153. [CrossRef]
12. Ebaid, A.; El-Sayed, D.M.M.; Aljoufi, M.D. Fractional calculus model for damped mathieu equation: Approximate analytical solution. *Appl. Math. Sci.* **2012**, *6*, 4075–4080.
13. Garcia, J.J.R.; Calderon, M.G.; Ortiz, J.M.; Baleanu, D. Motion of a particle in a resisting medium using fractional calculus approach. *Proc. Rom. Acad. Ser. A* **2013**, *14*, 42–47.
14. Machado, J.T. A fractional approach to the Fermi-Pasta-Ulam problem. *Eur. Phys. J. Spec. Top.* **2013**, *222*, 1795–1803. [CrossRef]
15. Ebaid, A. Analysis of projectile motion in view of the fractional calculus. *Appl. Math. Model.* **2011**, *35*, 1231–1239. [CrossRef]
16. Ebaid, A.; El-Zahar, E.R.; Aljohani, A.F.; Salah, B.; Krid, M.; Machado, J.T. Analysis of the two-dimensional fractional projectile motion in view of the experimental data. *Nonlinear Dyn.* **2019**, *97*, 1711–1720. [CrossRef]
17. Ahmad, B.; Batarfi, H.; Nieto, J.J.; Oscar, O.-Z.; Shammakh, W. Projectile motion via Riemann-Liouville calculus. *Adv. Differ. Equ.* **2015**, *63*, 1–14. [CrossRef]
18. Kumar, D.; Singh, J.; Baleanu, D.; Rathore, S. Analysis of a fractional model of the Ambartsumian equation. *Eur. Phys. J. Plus* **2018**, *133*, 133–259. [CrossRef]
19. Ebaid, A.; Cattani, C.; Juhani1, A.S.A.; El-Zahar, E.R. A novel exact solution for the fractional Ambartsumian equation. *Adv. Differ. Equ.* **2021**, *2021*, 88. [CrossRef]
20. El-Zahar, E.R.; Alotaibi, A.M.; Ebaid, A.; Aljohani, A.F.; Aguilar, J.F.G. The Riemann-Liouville fractional derivative for Ambartsumian equation. *Results Phys.* **2020**, *19*, 103551. [CrossRef]

21. Kaur, D.; Agarwal, P.; Rakshit, M.; Chand, M. Fractional Calculus involving (p, q)-Mathieu Type Series. *Appl. Math. Nonlinear Sci.* **2020**, *5*, 15–34. [CrossRef]
22. Agarwal, P.; Mondal, S.R.; Nisar, K.S. On fractional integration of generalized struve functions of first kind. *Thai J. Math.* **2020**, in press.
23. Agarwal, P.; Singh, R. Modelling of transmission dynamics of Nipah virus (Niv): A fractional order approach. *Phys. A Stat. Mech. Its Appl.* **2020**, *547*, 124243. [CrossRef]
24. Alderremy, A.A.; Saad, K.M.; Agarwal, P.; Aly, S.; Jain, S. Certain new models of the multi space-fractional Gardner equation. *Phys. A Stat. Mech. Its Appl.* **2020**, *545*, 123806. [CrossRef]
25. Feng, Y.; Yang, X.; Liu, J.; Chen, Z. New perspective aimed at local fractional order memristor model on Cantor sets. *Fractals* **2020**, accepted manuscript. [CrossRef]
26. Feng, Y.; Yang, X.; Liu, J. On overall behavior of Maxwell mechanical model by the combined Caputo fractional derivative. *Chin. J. Phys.* **2020**, *66*, 269–276. [CrossRef]
27. Sweilam, N.H.; Al-Mekhlafi, S.M.; Assiri, T.; Atangana, A. Optimal control for cancer treatment mathematical model using Atangana-Baleanu-Caputo fractional derivative. *Adv. Differ. Equ.* **2020**, *334*, 1–21. [CrossRef]
28. Atangana, A.; Qureshi, S. Mathematical Modeling of an Autonomous Nonlinear Dynamical System for Malaria Transmission Using Caputo Derivative. *Fract. Order Anal. Theory Methods Appl.* **2020**, *9*, 225–252. [CrossRef]
29. Elzahar, E.R.; Gaber, A.A.; Aljohani, A.F.; Machado, J.T.; Ebaid, A. Generalized Newtonian fractional model for the vertical motion of a particle. *Appl. Math. Model.* **2020**, *88*, 652–660. [CrossRef]
30. El-Dib, Y.O.; Elgazery, N.S. Effect of Fractional Derivative Properties on the Periodic Solution of the Nonlinear Oscillations. *Fractals* **2020**, *28*, 2050095. [CrossRef]
31. Ortigueira, M.D.; Machado, J.T.; Trujillo, J.J. Fractional derivatives and periodic functions. *Int. J. Dynam. Control* **2015**, *5*, 72–78. [CrossRef]

fractal and fractional

Article

Generalized k-Fractional Integral Operators Associated with Pólya-Szegö and Chebyshev Types Inequalities

Zhiqiang Zhang [1,†], Ghulam Farid [2,†], Sajid Mehmood [3,†], Kamsing Nonlaopon [4,*,†] and Tao Yan [1,†]

1 School of Computer Science, Chengdu University, Chengdu 610000, China; zqzhang@cdu.edu.cn (Z.Z.); yantao@cdu.edu.cn (T.Y.)
2 Department of Mathematics, COMSATS University Islamabad, Attock Campus, Attock 43600, Pakistan; faridphdsms@hotmail.com or ghlmfarid@cuiatk.edu.pk
3 Govt Boys Primary School Sherani, Hazro Attock 43440, Pakistan; smjg227@gmail.com
4 Department of Mathematics, Faculty of Science, Khon Kaen University, Khon Kaen 40002, Thailand
* Correspondence: nkamsi@kku.ac.th
† These authors contributed equally to this work.

Abstract: Inequalities related to derivatives and integrals are generalized and extended via fractional order integral and derivative operators. The present paper aims to define an operator containing Mittag-Leffler function in its kernel that leads to deduce many already existing well-known operators. By using this generalized operator, some well-known inequalities are studied. The results of this paper reproduce Chebyshev and Pólya-Szegö type inequalities for Riemann-Liouville and many other fractional integral operators.

Keywords: Chebyshev inequality; Pólya-Szegö inequality; fractional integral operators; Mittag-Leffler function

1. Introduction

Integral and derivative operators of fractional order are simple and important tools to generalize the classical theories and well-known problems related to integer order derivatives and integrals. Many modern subjects in different fields of mathematics, engineering, and sciences have been introduced due to the applications of fractional derivatives and integrals. These days, fractional integral/derivative operators are very frequently considered by the researchers working on mathematical inequalities to extend the classical literature. One can see the well-known inequalities related to integer order derivatives and integrals have been extended to fractional order derivatives and integrals. These include the inequalities of Chebyshev [1], Hadamard [2], Jensen [3], Pólya-Szegö [4], Petrovic [5], Grüss [6], Ostrowski [7], and many others. Here, we are interested to refer the versions of all these inequalities for Riemann-Liouville fractional integrals studied in [8–13].

It is interesting to compare the integral mean of product of two functions to the product of integral means of these functions. The Chebyshev inequality provides the comparison of integral mean of product of two positive functions of same monotonicity to the product of their integral means. After Chebyshev's inequality, people started to analyze the error bounds of this inequality. For instance, the well-known Grüss inequality gives the error bounds of difference of terms of the Chebyshev inequality (which is well-known as Chebyshev-functional). The well-known Pólya-Szegö inequality gives the estimation of quotient in terms of the Chebyshev inequality for bounded functions. These inequalities have been studied for Riemann-Liouville and other fractional integral operators in [10,14–20].

Next, we give the results, which are necessary to produce the results of this paper. First, we give Chebyshev functional and then the Chebyshev inequality [1] as follows:

$$T(f,g) = \frac{1}{b-a}\int_a^b f(\tau)g(\tau)d\tau - \left(\frac{1}{b-a}\int_a^b f(\tau)d\tau\right)\left(\frac{1}{b-a}\int_a^b g(\tau)d\tau\right), \quad (1)$$

where f and g are two positive and integrable functions over the interval $[a,b]$. If f and g are synchronous on $[a,b]$, then Chebyshev inequality $T(f,g) \geq 0$ is obtained.

The functional (1) has attracted the attention of many researchers due to its application in mathematical analysis. One of the famous inequalities related to functional (1) is the Grüss inequality [6], stated as follows:

$$|T(f,g)| \leq \frac{(U-u)(V-v)}{4},$$

where the positive and integrable functions f and g satisfy

$$u \leq f(\tau) \leq U, \quad v \leq g(\tau) \leq V,$$

for all $\tau \in [a,b]$ and constants $u, U, v, V \in \mathbb{R}$.

Another famous inequality which will be useful to obtain our main results is the Pólya-Szegö inequality [4], stated as follows:

$$\frac{\int_a^b f^2(\tau)d\tau \int_a^b g^2(\tau)d\tau}{\left(\int_a^b f(\tau)g(\tau)d\tau\right)^2} \leq \frac{1}{4}\left(\sqrt{\frac{uv}{UV}} + \sqrt{\frac{UV}{uv}}\right)^2.$$

By using the Pólya-Szegö inequality, Dragomir and Diamond [21] proved the following Grüss type inequality:

$$|T(f,g)| \leq \frac{(U-u)(V-v)}{4(b-a)^2\sqrt{uUvV}}\int_a^b f(\tau)d\tau \int_a^b g(\tau)d\tau,$$

where positive and integrable functions f and g satisfy

$$0 < u \leq f(\tau) \leq U < \infty, \quad 0 < v \leq g(\tau) \leq V < \infty,$$

for all $\tau \in [a,b]$ and constants $u, U, v, V \in \mathbb{R}$.

Inspired by the above-defined inequalities, our aim in this paper is to get some fractional versions of these inequalities. The main objective is to give some new Pólya-Szegö and Chebyshev inequalities for generalized k-fractional integral operator containing Mittag-Leffler function in its kernel. In the upcoming section, we define a new k-fractional integral operator containing Mittag-Leffer function. In Section 3, we will utilize this k-fractional integral operator to obtain the Pólya-Szegö and Chebyshev inequalities. Moreover, the presented results are the generalizations of the results which are already published in [10,14,19].

2. Fractional Integral Operators

Fractional integral operators are very useful in the advancement of mathematical inequalities. A large number of fractional integral inequalities due to different types of fractional integral operators have been established (see [10,14,19,22–29] and references therein). Applications of fractional integral operators in differential equations and other fields can be found in [25,30–35]. The first formulation of fractional integral operator is the Riemann-Liouville fractional integral operator, defined as follows:

Definition 1. Let $f \in L_1[a,b]$. Then Riemann-Liouville fractional integrals of order $\sigma \in \mathbb{C}$, $\Re(\sigma) > 0$ are defined by:

$$\left(\zeta_{a^+}^{\sigma} f\right)(x) = \frac{1}{\Gamma(\sigma)} \int_a^x (x-\tau)^{\sigma-1} f(\tau) d\tau, \quad x > a \tag{2}$$

$$\left(\zeta_{b^-}^{\sigma} f\right)(x) = \frac{1}{\Gamma(\sigma)} \int_x^b (\tau-x)^{\sigma-1} f(\tau) d\tau, \quad x < b, \tag{3}$$

where $\Gamma(.)$ is the gamma function defined as: $\Gamma(\sigma) = \int_0^{\infty} \tau^{\sigma-1} e^{-\tau} d\tau$.

In [36], Andrić et al. introduced the generalized fractional integral operators as follows:

Definition 2. Let $\vartheta, \theta, \sigma, l, \mu, c \in \mathbb{C}$, $\Re(\theta), \Re(\sigma), \Re(l) > 0$, $\Re(c) > \Re(\mu) > 0$ with $p \geq 0$, $r > 0$ and $0 < q \leq r + \Re(\theta)$. Let $f \in L_1[a,b]$ and $x \in [a,b]$. Then the generalized fractional integral operators are defined by:

$$\left(\zeta_{\theta,\sigma,l,\vartheta,a^+}^{\mu,r,q,c} f\right)(x;p) = \int_a^x (x-\tau)^{\sigma-1} E_{\theta,\sigma,l}^{\mu,r,q,c}(\vartheta(x-\tau)^{\theta};p) f(\tau) d\tau, \tag{4}$$

$$\left(\zeta_{\theta,\sigma,l,\vartheta,b^-}^{\mu,r,q,c} f\right)(x;p) = \int_u^b (\tau-x)^{\sigma-1} E_{\theta,\sigma,l}^{\mu,r,q,c}(\vartheta(\tau-x)^{\theta};p) f(\tau) d\tau, \tag{5}$$

where $E_{\theta,\sigma,l}^{\mu,r,q,c}(\tau;p)$ is the generalized Mittag-Leffler function defined by:

$$E_{\theta,\sigma,l}^{\mu,r,q,c}(\tau;p) = \sum_{n=0}^{\infty} \frac{B_p(\mu+nq,c-\mu)}{B(\mu,c-\mu)} \frac{(c)_{nq}}{\Gamma(\theta n + \sigma)} \frac{\tau^n}{(l)_{nr}},$$

$B_p(x,y) = \int_0^1 \tau^{x-1}(1-\tau)^{y-1} e^{-\frac{p}{\tau(1-\tau)}} d\tau$ and $(c)_{nq} = \frac{\Gamma(c+nq)}{\Gamma(c)}$.

In [32], Farid introduced the unified integral operators as follows:

Definition 3. Let $f, \alpha : [a,b] \to \mathbb{R}$, $0 < a < b$ be the functions such that f be a positive and integrable and α be a differentiable and strictly increasing. Also, let $\frac{\varphi}{x}$ be an increasing function on $[a, \infty)$ and $\sigma, l, \mu, c \in \mathbb{C}$, $p, \theta, r \geq 0$ and $0 < q \leq r + \theta$. Then for $x \in [a,b]$ the integral operators are defined by:

$$\left({}_{\alpha}\zeta_{\theta,\sigma,l,\vartheta,a^+}^{\varphi;\mu,r,q,c} f\right)(x;p) = \int_a^x \frac{\varphi(\alpha(x)-\alpha(\tau))}{\alpha(x)-\alpha(\tau)} E_{\theta,\sigma,l}^{\mu,r,q,c}(\vartheta(\alpha(x)-\alpha(\tau))^{\theta};p) f(\tau) d(\alpha(\tau)), \tag{6}$$

$$\left({}_{\alpha}\zeta_{\theta,\sigma,l,\vartheta,b^-}^{\varphi;\mu,r,q,c} f\right)(x;p) = \int_x^b \frac{\varphi(\alpha(\tau)-\alpha(x))}{\alpha(\tau)-\alpha(x)} E_{\theta,\sigma,l}^{\mu,r,q,c}(\vartheta(\alpha(\tau)-\alpha(x))^{\theta};p) f(\tau) d(\alpha(\tau)). \tag{7}$$

The following generalized k-integral operators involving Mittag-Leffler function with some modification is produced, for $\varphi(x) = x^{\frac{\sigma}{k}}$ with $k > 0$, in (6) and (7):

Definition 4. Let $f, \alpha : [a,b] \to \mathbb{R}$, $0 < a < b$ be the functions such that f be a positive and integrable and α be a differentiable and strictly increasing. Also, let $\theta, \sigma, l, \vartheta, \mu, c \in \mathbb{C}$, $p, \theta, r \geq 0$, $0 < q \leq r + \theta$ and $k > 0$. Then for $x \in [a,b]$ the integral operators are defined by:

$$\left({}_{\alpha}^{k}\zeta_{\theta,\sigma,l,\vartheta,a^+}^{\mu,r,q,c} f\right)(x;p) = \int_a^x (\alpha(x)-\alpha(\tau))^{\frac{\sigma}{k}-1} E_{\theta,\sigma,l,k}^{\mu,r,q,c}(\vartheta(\alpha(x)-\alpha(\tau))^{\frac{\theta}{k}};p) f(\tau) d(\alpha(\tau)), \tag{8}$$

$$\left({}_{\alpha}^{k}\zeta_{\theta,\sigma,l,\vartheta,b^-}^{\mu,r,q,c} f\right)(x;p) = \int_x^b (\alpha(\tau)-\alpha(x))^{\frac{\sigma}{k}-1} E_{\theta,\sigma,l,k}^{\mu,r,q,c}(\vartheta(\alpha(\tau)-\alpha(x))^{\frac{\theta}{k}};p) f(\tau) d(\alpha(\tau)), \tag{9}$$

where $E_{\theta,\sigma,l,k}^{\mu,r,q,c}(\tau;p)$ is the Mittag-Leffler function defined by:

$$E_{\theta,\sigma,l,k}^{\mu,r,q,c}(\tau;p) = \sum_{n=0}^{\infty} \frac{B_p(\mu+nq, c-\mu)}{B(\mu, c-\mu)} \frac{(c)_{nq}}{k\Gamma_k(\theta n + \sigma)} \frac{\tau^n}{(l)_{nr}}. \tag{10}$$

Remark 1. *The following integral operators can be deduced from (8) and (9):*

1. *The following integral operator is produced, for $\alpha(x) = x$ in (8):*

$$\left({}^k\zeta_{\theta,\sigma,l,\vartheta,a^+}^{\mu,r,q,c} f\right)(x;p) = \int_a^x (x-\tau)^{\frac{\sigma}{k}-1} E_{\theta,\sigma,l,k}^{\mu,r,q,c}(\vartheta(x-\tau)^{\frac{\theta}{k}};p) f(\tau) d\tau. \tag{11}$$

2. *The following generalized Hadamard integral operator is produced, for $\alpha(x) = \ln x$ in (8):*

$$\left({}^k\zeta_{\theta,\sigma,l,\vartheta,a^+}^{\mu,r,q,c} f\right)(x;p) = \int_a^x \left(\ln\frac{x}{\tau}\right)^{\frac{\sigma}{k}-1} E_{\theta,\sigma,l,k}^{\mu,r,q,c}\left(\vartheta\left(\ln\frac{x}{\tau}\right)^{\frac{\theta}{k}};p\right) f(\tau) \frac{d\tau}{\tau}. \tag{12}$$

3. *The following generalized Katugampola integral operator is produced, for $\alpha(x) = \frac{x^\rho}{\rho}, \rho > 0$ in (8):*

$$\left({}_\rho^k\zeta_{\theta,\sigma,l,\vartheta,a^+}^{\mu,r,q,c} f\right)(x;p) = \int_a^x \left(\frac{x^\rho - \tau^\rho}{\rho}\right)^{\frac{\sigma}{k}-1} E_{\theta,\sigma,l,k}^{\mu,r,q,c}\left(\vartheta\left(\frac{x^\rho - \tau^\rho}{\rho}\right)^{\frac{\theta}{k}};p\right) f(\tau) \tau^{\rho-1} d\tau. \tag{13}$$

4. *The following generalized (k,s)-integral operator is produced, for $\alpha(x) = \frac{x^{s+1}}{s+1}, s \in \mathbb{R} - \{-1\}$ in (8):*

$$\left({}_s^k\zeta_{\theta,\sigma,l,\vartheta,a^+}^{\mu,r,q,c} f\right)(x;p) = \int_a^x \left(\frac{x^{s+1}-\tau^{s+1}}{s+1}\right)^{\frac{\sigma}{k}-1} E_{\theta,\sigma,l,k}^{\mu,r,q,c}\left(\vartheta\left(\frac{x^{s+1}-\tau^{s+1}}{s+1}\right)^{\frac{\theta}{k}};p\right) f(\tau) \tau^s d\tau. \tag{14}$$

5. *The following generalized conformable k-integral operator is produced, for $\alpha(x) = \frac{x^{\lambda+\nu}}{\lambda+\nu}$ in (8):*

$$\left({}_{\lambda,\nu}^k\zeta_{\theta,\sigma,l,\vartheta,a^+}^{\mu,r,q,c} f\right)(x;p) = \int_a^x \left(\frac{x^{\lambda+\nu}-\tau^{\lambda+\nu}}{\lambda+\nu}\right)^{\frac{\sigma}{k}-1} E_{\theta,\sigma,l,k}^{\mu,r,q,c}\left(\vartheta\left(\frac{x^{\lambda+\nu}-\tau^{\lambda+\nu}}{\lambda+\nu}\right)^{\frac{\theta}{k}};p\right) f(\tau) \tau^\lambda d_\nu \tau. \tag{15}$$

Similarly, all above operators can be deduce for generalized k-integral operators (9).

6. *The following generalized conformable (k,s)-integral operators are produced, for $\alpha(x) = \frac{(x-a)^s}{s}, s > 0$ in (8) and $\alpha(x) = \frac{-(b-x)^s}{s}, s > 0$ in (9):*

$$\left({}_s^k\zeta_{\theta,\sigma,l,\vartheta,a^+}^{\mu,r,q,c} f\right)(x;p) = \int_a^x \left(\frac{(x-a)^s-(\tau-a)^s}{s}\right)^{\frac{\sigma}{k}-1} E_{\theta,\sigma,l,k}^{\mu,r,q,c}\left(\vartheta\left(\frac{(x-a)^s-(\tau-a)^s}{s}\right)^{\frac{\theta}{k}};p\right) f(\tau)(\tau-a)^{s-1} d\tau. \tag{16}$$

$$\left({}_s^k\zeta_{\theta,\sigma,l,\theta,b^-}^{\mu,r,q,c} f\right)(x;p) = \int_x^b \left(\frac{(b-x)^s-(b-\tau)^s}{s}\right)^{\frac{\sigma}{k}-1} E_{\theta,\sigma,l,k}^{\mu,r,q,c}\left(\vartheta\left(\frac{(b-x)^s-(b-\tau)^s}{s}\right)^{\frac{\theta}{k}};p\right) f(\tau)(b-\tau)^{s-1} d\tau. \tag{17}$$

Remark 2. *For different choices of parameters involving in the Mittag-Leffler function (10), one can obtain new generalized integral operators.*

Remark 3. *From integral operators (8) and (9), we have the following particular cases:*

1. *The integral operators given in [29] are reproduced, for $k = 1$.*
2. *The integral operators given in [4] and (5) are reproduced, for $k = 1$ and $\alpha(x) = x$.*
3. *The integral operators given in [37] are reproduced, for $k = 1$, $\alpha(x) = x$, and $p = 0$.*
4. *The integral operators given in [38] are reproduced, for $k = 1$, $\alpha(x) = x$, and $r = l = 1$.*
5. *The integral operators given in [39] are reproduced, for $k = 1$, $\alpha(x) = x$, $p = 0$, and $r = l = 1$.*
6. *The integral operators given in [40] are reproduced, for $k = 1$, $\alpha(x) = x$, $p = 0$, and $q = r = l = 1$.*

7. The integral operators given in [26] are reproduced, for $k = 1$, $\alpha(x) = \frac{x^\rho}{\rho}$, $\rho > 0$, and $\vartheta = p = 0$.
8. The integral operators given in [41] are reproduced, for $k = 1$, $\alpha(x) = \ln x$, and $\vartheta = p = 0$.
9. The integral operators given in [42] are reproduced, for $\alpha(x) = \frac{x^{s+1}}{s+1}$ and $\vartheta = p = 0$.
10. The integral operators given in [30] are reproduced, for $k = 1$, $\alpha(x) = \frac{(x)^{\lambda+\nu}}{\lambda+\nu}$ and $\vartheta = p = 0$.
11. The integral operators given in [27] are reproduced, for $\alpha(x) = \frac{(x-a)^s}{s}$, $s > 0$ in (8) and $\alpha(x) = -\frac{(b-x)^s}{s}$, $s > 0$ in (9) with $\vartheta = p = 0$.
12. The integral operators given in [33] are reproduced, for $\alpha(x) = \frac{(x-a)^s}{s}$, $s > 0$ in (8) and $\alpha(x) = -\frac{(b-x)^s}{s}$, $s > 0$ in (9) with $k = 1$ and $\vartheta = p = 0$.
13. The integral operators given in [28] are reproduced, for $\vartheta = p = 0$.
14. The integral operators given in [41] are reproduced, for $\vartheta = p = 0$ and $k = 1$.
15. The integral operators given in [43] are reproduced, for $\vartheta = p = 0$, and $\alpha(x) = x$.
16. The integral operators given in (2) and (3) are reproduced, for $\vartheta = p = 0$, $\alpha(x) = x$, and $k = 1$.

For constant function, from generalized k-fractional integral operator (8), we have

$$\left({}_a^k \zeta_{\theta,\sigma,l,\vartheta,a^+}^{\mu,r,q,c} 1\right)(x;p)$$

$$= \int_a^x (\alpha(x) - \alpha(\tau))^{\frac{\sigma}{k}-1} E_{\theta,\sigma,l,k}^{\mu,r,q,c}(\vartheta(\alpha(x) - \alpha(\tau))^{\frac{\theta}{k}}; p) d(\alpha(\tau))$$

$$= \int_a^x (\alpha(x) - \alpha(\tau))^{\frac{\sigma}{k}-1} \sum_{n=0}^\infty \frac{B_p(\mu+nq, c-\mu)}{B(\mu, c-\mu)} \frac{(c)_{nq}}{k\Gamma_k(\theta n + \sigma)} \frac{\vartheta^n (\alpha(x) - \alpha(\tau))^{\frac{\theta n}{k}}}{(l)_{nr}} d(\alpha(\tau))$$

$$= \sum_{n=0}^\infty \frac{B_p(\mu+nq, c-\mu)}{B(\mu, c-\mu)} \frac{(c)_{nq}}{k\Gamma_k(\theta n + \sigma)} \frac{\vartheta^n}{(l)_{nr}} \int_a^x (\alpha(x) - \alpha(\tau))^{\frac{\theta n}{k} + \frac{\sigma}{k}-1} d(\alpha(\tau))$$

$$= k(\alpha(x) - \alpha(a))^{\frac{\sigma}{k}} \sum_{n=0}^\infty \frac{B_p(\mu+nq, c-\mu)}{B(\mu, c-\mu)} \frac{(c)_{nq}}{k\Gamma_k(\theta n + \sigma)} \frac{\vartheta^n}{(l)_{nr}} \frac{(\alpha(x) - \alpha(a))^{\frac{\theta n}{k}}}{\theta n + \sigma}$$

$$= k(\alpha(x) - \alpha(a))^{\frac{\sigma}{k}} \sum_{n=0}^\infty \frac{B_p(\mu+nq, c-\mu)}{B(\mu, c-\mu)} \frac{(c)_{nq}}{k\Gamma_k(\theta n + \sigma + k)} \frac{\vartheta^n (\alpha(x) - \alpha(a))^{\frac{\theta n}{k}}}{(l)_{nr}}$$

$$= k(\alpha(x) - \alpha(a))^{\frac{\sigma}{k}} E_{\theta,\sigma+k,l,k}^{\mu,r,q,c}(\vartheta(\alpha(x) - \alpha(a))^{\frac{\theta}{k}}; p).$$

Hence

$$\left({}_a^k \zeta_{\theta,\sigma,l,\vartheta,a^+}^{\mu,r,q,c} 1\right)(x;p) = k(\alpha(x) - \alpha(a))^{\frac{\sigma}{k}} E_{\theta,\sigma+k,l,k}^{\mu,r,q,c}(\vartheta(\alpha(x) - \alpha(a))^{\frac{\theta}{k}}; p) := \chi_{a^+}^\sigma(x;p) \quad (18)$$

similarly for constant function, from generalized fractional integral operator (9), we get

$$\left({}_a^k \zeta_{\theta,\sigma,l,\vartheta,b^-}^{\mu,r,q,c} 1\right)(x;p) = k(\alpha(b) - \alpha(x))^{\frac{\sigma}{k}} E_{\theta,\sigma+k,l,k}^{\mu,r,q,c}(\vartheta(\alpha(b) - \alpha(x))^{\frac{\theta}{k}}; p) := \chi_{b^-}^\sigma(x;p). \quad (19)$$

In the all upcoming results, the parameters of the Mittag-Leffler function are considered in $\mathbb{R}$.

3. Pólya-Szegö and Chebyshev Type Inequalities for Generalized k-Fractional Integral Operators

In this section, we obtain Pólya-Szegö and Chebyshev type inequalities for generalized k-fractional integral operators containing Mittag-Leffler function in their kernels. For the reader's convenience we will use a simplified notation:

$$\left({}_a^k \mathbf{Z}_\sigma f\right)(x;p) := \left({}_a^k \zeta_{\theta,\sigma,l,\vartheta,a^+}^{\mu,r,q,c} f\right)(x;p)$$

$$= \int_a^x (\alpha(x) - \alpha(\tau))^{\frac{\sigma}{k}-1} \mathbf{E}_\sigma^k(\vartheta(\alpha(x) - \alpha(\tau))^{\frac{\theta}{k}}; p) f(\tau) d(\alpha(\tau)),$$

where

$$\mathbf{E}_\sigma^k := E_{\theta,\sigma,l,k}^{\mu,r,q,c}(\tau;p)$$
$$= \sum_{n=0}^{\infty} \frac{B_p(\mu+nq, c-\mu)}{B(\mu, c-\mu)} \frac{(c)_{nq}}{k\Gamma_k(\theta n + \sigma)} \frac{\tau^n}{(l)_{nr}}.$$

Theorem 1. *Suppose that:*

- *f and g be two positive and integrable functions on $[0, \infty)$;*
- *$\alpha : [a, b] \to \mathbb{R}$ be an increasing and differentiable function with $\alpha' \in L[a, b]$;*
- *there exist four positive integrable functions ζ_1, ζ_2, η_1 and η_2, such that*

$$0 < \zeta_1(\tau) \leq f(\tau) \leq \zeta_2(\tau), \quad 0 < \eta_1(\tau) \leq g(\tau) \leq \eta_2(\tau) \quad (\tau \in [a, x], x > a). \tag{20}$$

Then for generalized k-fractional integral operator containing Mittag-Leffler function, we have

$$\frac{\left(_a^k\mathbf{Z}_\sigma \eta_1 \eta_2 f^2\right)(x;p) \left(_a^k\mathbf{Z}_\sigma \zeta_1 \zeta_2 g^2\right)(x;p)}{\left[\left(_a^k\mathbf{Z}_\sigma(\zeta_1\eta_1 + \zeta_2\eta_2)fg\right)(x;p)\right]^2} \leq \frac{1}{4}. \tag{21}$$

Proof. From (20) for $\tau \in [a, x]$ with $x > a$, we can write

$$\left(\frac{\zeta_2(\tau)}{\eta_1(\tau)} - \frac{f(\tau)}{g(\tau)}\right)\left(\frac{f(\tau)}{g(\tau)} - \frac{\zeta_1(\tau)}{\eta_2(\tau)}\right) \geq 0,$$

which implies

$$(\zeta_1(\tau)\eta_1(\tau) + \zeta_2(\tau)\eta_2(\tau))f(\tau)g(\tau) \geq \eta_1(\tau)\eta_2(\tau)f^2(\tau) + \zeta_1(\tau)\zeta_2(\tau)g^2(\tau). \tag{22}$$

Multiplying (22) with $(\alpha(x) - \alpha(\tau))^{\frac{\sigma}{k}-1}\mathbf{E}_\sigma^k(\vartheta(\alpha(x) - \alpha(\tau))^\theta; p)\alpha'(\tau)$ on both sides and integrating, we get

$$\int_a^x (\alpha(x) - \alpha(\tau))^{\frac{\sigma}{k}-1}\mathbf{E}_\sigma^k(\vartheta(\alpha(x) - \alpha(\tau))^\theta; p)(\zeta_1(\tau)\eta_1(\tau) + \zeta_2(\tau)\eta_2(\tau))f(\tau)g(\tau)\alpha'(\tau)d\tau$$
$$\geq \int_a^x (\alpha(x) - \alpha(\tau))^{\frac{\sigma}{k}-1}\mathbf{E}_\sigma^k(\vartheta(\alpha(x) - \alpha(\tau))^\theta; p)\eta_1(\tau)\eta_2(\tau)f^2(\tau)\alpha'(\tau)d\tau$$
$$+ \int_a^x (\alpha(x) - \alpha(\tau))^{\frac{\sigma}{k}-1}\mathbf{E}_\sigma^k(\vartheta(\alpha(x) - \alpha(\tau))^\theta; p)\zeta_1(\tau)\zeta_2(\tau)g^2(\tau)\alpha'(\tau)d\tau.$$

Now by using k-fractional integral operator, we get

$$\left(_a^k\mathbf{Z}_\sigma(\zeta_1\eta_1 + \zeta_2\eta_2)fg\right)(x;p) \geq \left(_a^k\mathbf{Z}_\sigma \eta_1\eta_2 f^2\right)(x;p) + \left(_a^k\mathbf{Z}_\sigma \zeta_1\zeta_2 g^2\right)(x;p).$$

By applying AM-GM inequality, we get

$$\left(_a^k\mathbf{Z}_\sigma(\zeta_1\eta_1 + \zeta_2\eta_2)fg\right)(x;p) \geq 2\sqrt{\left(_a^k\mathbf{Z}_\sigma \eta_1\eta_2 f^2\right)(x;p)\left(_a^k\mathbf{Z}_\sigma \zeta_1\zeta_2 g^2\right)(x;p)},$$

which leads to the required inequality (21). □

Corollary 1. *If $\zeta_1 = u, \zeta_2 = U, \eta_1 = v$ and $\eta_2 = V$, then we have*

$$\frac{\left(_a^k\mathbf{Z}_\sigma f^2\right)(x;p)\left(_a^k\mathbf{Z}_\sigma g^2\right)(x;p)}{\left[\left(_a^k\mathbf{Z}_\sigma fg\right)(x;p)\right]^2} \leq \frac{1}{4}\left(\sqrt{\frac{uv}{UV}} + \sqrt{\frac{UV}{uv}}\right)^2.$$

Remark 4. *In Theorem 1, for $\alpha(x) = x$ and $k = 1$, we get [14] (Theorem 1), for $\vartheta = p = 0$ (and $a = 0$), we get [19] (Lemma 3.1), for $\alpha(x) = x, k = 1, \vartheta = p = 0$ (and $a = 0$), we get [10] (Lemma 3.1).*

Theorem 2. *Under the assumptions of Theorem 1 with $\varsigma > 0$, we have*

$$\frac{\left(^k_\alpha \mathbf{Z}_\sigma \zeta_1 \zeta_2\right)(x;p) \left(^k_\alpha \mathbf{Z}_\varsigma \eta_1 \eta_2\right)(x;p) \left(^k_\alpha \mathbf{Z}_\sigma f^2\right)(x;p) \left(^k_\alpha \mathbf{Z}_\varsigma g^2\right)(x;p)}{\left[\left(^k_\alpha \mathbf{Z}_\sigma \zeta_1 f\right)(x;p)\left(^k_\alpha \mathbf{Z}_\varsigma \eta_1 g\right)(x;p) + \left(^k_\alpha \mathbf{Z}_\sigma \zeta_2 f\right)(x;p)\left(^k_\alpha \mathbf{Z}_\varsigma \eta_2 g\right)(x;p)\right]^2} \leq \frac{1}{4}. \quad (23)$$

Proof. From (20), for $\tau, \kappa \in [a,x]$ with $x > a$, we can write

$$\left(\frac{\zeta_1(\tau)}{\eta_2(\kappa)} + \frac{\zeta_2(\tau)}{\eta_1(\kappa)}\right)\frac{f(\tau)}{g(\kappa)} \geq \frac{f^2(\tau)}{g^2(\kappa)} + \frac{\zeta_1(\tau)\zeta_2(\tau)}{\eta_1(\kappa)\eta_2(\kappa)},$$

which imply

$$\zeta_1(\tau)f(\tau)\eta_1(\kappa)g(\kappa) + \zeta_2(\tau)f(\tau)\eta_2(\kappa)g(\kappa) \geq \eta_1(\kappa)\eta_2(\kappa)f^2(\tau) + \zeta_1(\tau)\zeta_2(\tau)g^2(\kappa). \quad (24)$$

Multiplying (24) with $(\alpha(x) - \alpha(\tau))^{\frac{\sigma}{k}-1}(\alpha(x) - \alpha(\kappa))^{\frac{\varsigma}{k}-1}\mathbf{E}^k_\sigma(\vartheta(\alpha(x) - \alpha(\tau))^\theta;p)$ $\mathbf{E}^k_\varsigma(\vartheta(\alpha(x) - \alpha(\kappa))^\theta;p)\alpha'(\tau)\alpha'(\kappa)$ on both sides and integrating, we get

$$\int_a^x \int_a^x (\alpha(x) - \alpha(\tau))^{\frac{\sigma}{k}-1}(\alpha(x) - \alpha(\kappa))^{\frac{\varsigma}{k}-1}\mathbf{E}^k_\sigma(\vartheta(\alpha(x) - \alpha(\tau))^\theta;p)$$
$$\times \mathbf{E}^k_\varsigma(\vartheta(\alpha(x) - \alpha(\kappa))^\theta;p)\zeta_1(\tau)f(\tau)\eta_1(\kappa)g(\kappa)\alpha'(\tau)\alpha'(\kappa)d\tau d\kappa$$
$$+ \int_a^x \int_a^x (\alpha(x) - \alpha(\tau))^{\frac{\sigma}{k}-1}(\alpha(x) - \alpha(\kappa))^{\frac{\varsigma}{k}-1}\mathbf{E}^k_\sigma(\vartheta(\alpha(x) - \alpha(\tau))^\theta;p)$$
$$\times \mathbf{E}^k_\varsigma(\vartheta(\alpha(x) - \alpha(\kappa))^\theta;p)\zeta_2(\tau)f(\tau)\eta_2(\kappa)g(\kappa)\alpha'(\tau)\alpha'(\kappa)d\tau d\kappa$$
$$\geq$$
$$\int_a^x \int_a^x (\alpha(x) - \alpha(\tau))^{\frac{\sigma}{k}-1}(\alpha(x) - \alpha(\kappa))^{\frac{\varsigma}{k}-1}\mathbf{E}^k_\sigma(\vartheta(\alpha(x) - \alpha(\tau))^\theta;p)$$
$$\times \mathbf{E}^k_\varsigma(\vartheta(\alpha(x) - \alpha(\kappa))^\theta;p)\eta_1(\kappa)\eta_2(\kappa)f^2(\tau)\alpha'(\tau)\alpha'(\kappa)d\tau d\kappa$$
$$+ \int_a^x \int_a^x (\alpha(x) - \alpha(\tau))^{\frac{\sigma}{k}-1}(\alpha(x) - \alpha(\kappa))^{\frac{\varsigma}{k}-1}\mathbf{E}^k_\sigma(\vartheta(\alpha(x) - \alpha(\tau))^\theta;p)$$
$$\times \mathbf{E}^k_\varsigma(\vartheta(\alpha(x) - \alpha(\kappa))^\theta;p)\zeta_1(\tau)\zeta_2(\tau)g^2(\kappa)\alpha'(\tau)\alpha'(\kappa)d\tau d\kappa.$$

Now by using k-fractional integral operator, we get

$$\left(^k_\alpha \mathbf{Z}_\sigma \zeta_1 f\right)(x;p)\left(^k_\alpha \mathbf{Z}_\varsigma \eta_1 g\right)(x;p) + \left(^k_\alpha \mathbf{Z}_\sigma \zeta_2 f\right)(x;p)\left(^k_\alpha \mathbf{Z}_\varsigma \eta_2 g(x;p)\right)$$
$$\geq \left(^k_\alpha \mathbf{Z}_\sigma f^2\right)(x;p)\left(^k_\alpha \mathbf{Z}_\varsigma \eta_1 \eta_2\right)(x;p) + \left(^k_\alpha \mathbf{Z}_\sigma \zeta_1 \zeta_2\right)(x;p)\left(^k_\alpha \mathbf{Z}_\varsigma g^2\right)(x;p).$$

By applying AM-GM inequality, we get

$$\left(^k_\alpha \mathbf{Z}_\sigma \zeta_1 f\right)(x;p)\left(^k_\alpha \mathbf{Z}_\varsigma \eta_1 g\right)(x;p) + \left(^k_\alpha \mathbf{Z}_\sigma \zeta_2 f\right)(x;p)\left(^k_\alpha \mathbf{Z}_\varsigma \eta_2 g\right)(x;p)$$
$$\geq 2\sqrt{\left(^k_\alpha \mathbf{Z}_\sigma f^2\right)(x;p)\left(^k_\alpha \mathbf{Z}_\varsigma \eta_1 \eta_2\right)(x;p)\left(^k_\alpha \mathbf{Z}_\sigma \zeta_1 \zeta_2\right)(x;p)\left(^k_\alpha \mathbf{Z}_\varsigma g^2\right)(x;p)},$$

which leads to the required inequality (23). □

Corollary 2. *If $\zeta_1 = u$, $\zeta_2 = U$, $\eta_1 = v$ and $\eta_2 = V$, then we have*

$$\frac{\chi_\sigma(x;p)\,\chi_\varsigma(x;p)\left(^k_\alpha \mathbf{Z}_\sigma f^2\right)(x;p)\left(^k_\alpha \mathbf{Z}_\varsigma g^2\right)(x;p)}{\left[\left(^k_\alpha \mathbf{Z}_\sigma f\right)(x;p)\left(^k_\alpha \mathbf{Z}_\varsigma g\right)(x;p)\right]^2} \leq \frac{1}{4}\left(\sqrt{\frac{uv}{UV}} + \sqrt{\frac{UV}{uv}}\right)^2.$$

Remark 5. *In Theorem 2, for $\alpha(x) = x$ and $k = 1$, we get [14] (Theorem 2), for $\vartheta = p = 0$ (and $a = 0$), we get [19] (Lemma 3.6), for $\alpha(x) = x$, $k = 1$, $\vartheta = p = 0$ (and $a = 0$), we get [10] (Lemma 3.3).*

Theorem 3. *Under the assumptions of Theorem 1 with $\varsigma > 0$, we have*

$$\left({}^k_a\mathbf{Z}_\sigma f^2\right)(x;p)\left({}^k_a\mathbf{Z}_\varsigma g^2\right)(x;p) \leq \left({}^k_a\mathbf{Z}_\sigma(\zeta_2 fg/\eta_1)\right)(x;p)\left({}^k_a\mathbf{Z}_\varsigma(\eta_2 fg/\zeta_1)\right)(x;p). \quad (25)$$

Proof. From (20), for $\tau \in [a,x]$ with $x > a$, we can write

$$\frac{\zeta_2(\tau)f(\tau)g(\tau)}{\eta_1(\tau)} - f^2(\tau) \geq 0 \quad (26)$$

and

$$\frac{\eta_2(\kappa)f(\kappa)g(\kappa)}{\zeta_1(\kappa)} - g^2(\kappa) \geq 0. \quad (27)$$

Multiplying (26) with $(\alpha(x) - \alpha(\tau))^{\frac{\sigma}{k}-1}\mathbf{E}^k_\sigma(\vartheta(\alpha(x) - \alpha(\tau))^\theta;p)\alpha'(\tau)$ and (27) with $(\alpha(x) - \alpha(\kappa))^{\frac{\varsigma}{k}-1}\mathbf{E}^k_\varsigma(\vartheta(\alpha(x) - \alpha(\kappa))^\theta;p)\alpha'(\kappa)$ on both sides and integrating, we get

$$\int_a^x (\alpha(x) - \alpha(\tau))^{\frac{\sigma}{k}-1}\mathbf{E}^k_\sigma(\vartheta(\alpha(x) - \alpha(\tau))^\theta;p)f^2(\tau)\alpha'(\tau)d\tau$$
$$\leq \int_a^x (\alpha(x) - \alpha(\tau))^{\frac{\sigma}{k}-1}\mathbf{E}^k_\sigma(\vartheta(\alpha(x) - \alpha(\tau))^\theta;p)\frac{\zeta_2(\tau)}{\eta_1(\tau)}f(\tau)g(\tau)\alpha'(\tau)d\tau$$

and

$$\int_a^x (\alpha(x) - \alpha(\kappa))^{\frac{\varsigma}{k}-1}\mathbf{E}^k_\varsigma(\vartheta(\alpha(x) - \alpha(\kappa))^\theta;p)g^2(\kappa)\alpha'(\kappa)d\kappa$$
$$\leq \int_a^x (\alpha(x) - \alpha(\kappa))^{\frac{\varsigma}{k}-1}\mathbf{E}^k_\varsigma(\vartheta(\alpha(x) - \alpha(\kappa))^\theta;p)\frac{\eta_2(\kappa)}{\zeta_1(\kappa)}f(\kappa)g(\kappa)\alpha'(\kappa)d\kappa.$$

Now by using k-fractional integral operator, we get

$$\left({}^k_a\mathbf{Z}_\sigma f^2\right)(x;p) \leq \left({}^k_a\mathbf{Z}_\sigma(\zeta_2 fg/\eta_1)\right)(x;p) \quad (28)$$

and

$$\left({}^k_a\mathbf{Z}_\varsigma g^2\right)(x;p) \leq \left({}^k_a\mathbf{Z}_\varsigma(\eta_2 fg/\zeta_1)\right)(x;p). \quad (29)$$

Multiplying (28) with (29), we obtain (25). □

Corollary 3. *If $\zeta_1 = u$, $\zeta_2 = U$, $\eta_1 = v$ and $\eta_2 = V$, then we have*

$$\frac{\left({}^k_a\mathbf{Z}_\sigma f^2\right)(x;p)\left({}^k_a\mathbf{Z}_\varsigma g^2\right)(x;p)}{\left({}^k_a\mathbf{Z}_\sigma fg\right)(x;p)\left({}^k_a\mathbf{Z}_\varsigma fg\right)(x;p)} \leq \frac{UV}{uv}.$$

Remark 6. *In Theorem 3, for $\alpha(x) = x$ and $k = 1$, we get [14] (Theorem 3), for $\alpha(x) = x$, $k = 1$, $\vartheta = p = 0$ (and $a = 0$), we get [10] (Lemma 3.4).*

The Chebyshev type inequalities for generalized k-fractional integral operators are given as follows:

Theorem 4. *Under the assumptions of Theorem 1 with $\varsigma > 0$, we have*

$$\left|\chi_\sigma(x;p)\left({}^k_a\mathbf{Z}_\varsigma fg\right)(x;p) + \chi_\varsigma(x;p)\left({}^k_a\mathbf{Z}_\sigma fg\right)(x;p)\right.$$
$$\left. - \left({}^k_a\mathbf{Z}_\sigma f\right)(x;p)\left({}^k_a\mathbf{Z}_\varsigma g\right)(x;p) - \left({}^k_a\mathbf{Z}_\sigma g\right)(x;p)\left({}^k_a\mathbf{Z}_\varsigma f\right)(x;p)\right|$$
$$\leq |G_{\sigma,\varsigma}(f,\zeta_1,\zeta_2)(x;p) + G_{\varsigma,\sigma}(f,\zeta_1,\zeta_2)(x;p)|^{\frac{1}{2}} \quad (30)$$
$$\times |G_{\sigma,\varsigma}(g,\eta_1,\eta_2)(x;p) + G_{\varsigma,\sigma}(g,\eta_1,\eta_2)(x;p)|^{\frac{1}{2}},$$

where

$$G_{\sigma,\varsigma}(m,n,o)(x;p) = \frac{\chi_\varsigma(x;p)\left[\left(_a^k Z_\sigma(n+o)m\right)(x;p)\right]^2}{4\left(_a^k Z_\sigma no\right)(x;p)} \tag{31}$$
$$-\left(_a^k Z_\sigma m\right)(x;p)\left(_a^k Z_\varsigma m\right)(x;p).$$

Proof. Let f and g be two positive and integrable functions on $[0,\infty)$. For $\tau, \kappa \in [a,x]$ with $x > a$, we define $A(\tau, \kappa)$ as

$$A(\tau,\kappa) = (f(\tau) - f(\kappa))(g(\tau) - g(\kappa)),$$

which imply

$$A(\tau,\kappa) = f(\tau)g(\tau) + f(\kappa)g(\kappa) - f(\tau)g(\kappa) - f(\kappa)g(\tau). \tag{32}$$

Multiplying (32) with $(\alpha(x) - \alpha(\tau))^{\frac{\sigma}{k}-1}(\alpha(x) - \alpha(\kappa))^{\frac{\varsigma}{k}-1}\mathbf{E}^k_\sigma(\vartheta(\alpha(x) - \alpha(\tau))^\theta;p)$ $\mathbf{E}^k_\varsigma(\vartheta(\alpha(x) - \alpha(\kappa))^\theta;p)\alpha'(\tau)\alpha'(\kappa)$ and integrating, we get

$$\int_a^x \int_a^x (\alpha(x) - \alpha(\tau))^{\frac{\sigma}{k}-1}(\alpha(x) - \alpha(\kappa))^{\frac{\varsigma}{k}-1}\mathbf{E}^k_\sigma(\vartheta(\alpha(x) - \alpha(\tau))^\theta;p)$$
$$\times \mathbf{E}^k_\varsigma\left(\vartheta(\alpha(x) - \alpha(\kappa))^\theta;p\right) A(\tau,\kappa)\alpha'(\tau)\alpha'(\kappa)d\tau d\kappa.$$

Now by using k-fractional integral operator, we get

$$\chi_\varsigma(x;p)\left(_a^k Z_\sigma fg\right)(x;p) + \chi_\sigma(x;p)\left(_a^k Z_\varsigma fg\right)(x;p)$$
$$-\left(_a^k Z_\sigma f\right)(x;p)\left(_a^k Z_\varsigma g\right)(x;p) - \left(_a^k Z_\varsigma f\right)(x;p)\left(_a^k Z_\sigma g\right)(x;p). \tag{33}$$

By using Cauchy-Schwartz inequality, we have

$$\left|\int_a^x \int_a^x (\alpha(x) - \alpha(\tau))^{\frac{\sigma}{k}-1}(\alpha(x) - \alpha(\kappa))^{\frac{\varsigma}{k}-1}\mathbf{E}^k_\sigma(\vartheta(\alpha(x) - \alpha(\tau))^\theta;p)\right.$$
$$\left.\times \mathbf{E}^k_\varsigma(\vartheta(\alpha(x) - \alpha(\kappa))^\theta;p) A(\tau,\kappa)\alpha'(\tau)\alpha'(\kappa)d\tau d\kappa\right|$$
$$\leq \left[\int_a^x \int_a^x (\alpha(x) - \alpha(\tau))^{\frac{\sigma}{k}-1}(\alpha(x) - \alpha(\kappa))^{\frac{\varsigma}{k}-1}\mathbf{E}^k_\sigma(\vartheta(\alpha(x) - \alpha(\tau))^\theta;p)\right.$$
$$\times \mathbf{E}^k_\varsigma(\vartheta(\alpha(x) - \alpha(\kappa))^\theta;p)f^2(\tau)\alpha'(\tau)\alpha'(\kappa)d\tau d\kappa$$
$$+ \int_a^x \int_a^x (\alpha(x) - \alpha(\tau))^{\frac{\sigma}{k}-1}(\alpha(x) - \alpha(\kappa))^{\frac{\varsigma}{k}-1}\mathbf{E}^k_\sigma(\vartheta(\alpha(x) - \alpha(\tau))^\theta;p)$$
$$\times \mathbf{E}^k_\varsigma(\vartheta(\alpha(x) - \alpha(\kappa))^\theta;p)f^2(\kappa)\alpha'(\tau)\alpha'(\kappa)d\tau d\kappa$$
$$- 2\int_a^x \int_a^x (\alpha(x) - \alpha(\tau))^{\frac{\sigma}{k}-1}(\alpha(x) - \alpha(\kappa))^{\frac{\varsigma}{k}-1}\mathbf{E}^k_\sigma(\vartheta(\alpha(x) - \alpha(\tau))^\theta;p)$$
$$\left.\times \mathbf{E}^k_\varsigma(\vartheta(\alpha(x) - \alpha(\kappa))^\theta;p)f(\tau)f(\kappa)\alpha'(\tau)\alpha'(\kappa)d\tau d\kappa\right]^{\frac{1}{2}}$$
$$\times \left[\int_a^x \int_a^x (\alpha(x) - \alpha(\tau))^{\frac{\sigma}{k}-1}(\alpha(x) - \alpha(\kappa))^{\frac{\varsigma}{k}-1}\mathbf{E}^k_\sigma(\vartheta(\alpha(x) - \alpha(\tau))^\theta;p)\right.$$
$$\times \mathbf{E}^k_\varsigma(\vartheta(\alpha(x) - \alpha(\kappa))^\theta;p)g^2(\tau)\alpha'(\tau)\alpha'(\kappa)d\tau d\kappa$$
$$+ \int_a^x \int_a^x (\alpha(x) - \alpha(\tau))^{\frac{\sigma}{k}-1}(\alpha(x) - \alpha(\kappa))^{\frac{\varsigma}{k}-1}\mathbf{E}^k_\sigma(\vartheta(\alpha(x) - \alpha(\tau))^\theta;p)$$
$$\times \mathbf{E}^k_\varsigma(\vartheta(\alpha(x) - \alpha(\kappa))^\theta;p)g^2(\kappa)\alpha'(\tau)\alpha'(\kappa)d\tau d\kappa$$

$$-2\int_a^x\int_a^x(\alpha(x)-\alpha(\tau))^{\frac{\varsigma}{k}-1}(\alpha(x)-\alpha(\kappa))^{\frac{\varsigma}{k}-1}\mathbf{E}_\sigma^k(\vartheta(\alpha(x)-\alpha(\tau))^\theta;p)$$
$$\times\mathbf{E}_\varsigma^k(\vartheta(\alpha(x)-\alpha(\kappa))^\theta;p)g(\tau)g(\kappa)\alpha'(\tau)\alpha'(\kappa)d\tau d\kappa\Big]^{\frac{1}{2}}$$
$$\leq\Big[\chi_\varsigma(x;p)\left({}_\alpha^k\mathbf{Z}_\sigma f^2\right)(x;p)+\chi_\sigma(x;p)\left({}_\alpha^k\mathbf{Z}_\varsigma f^2\right)(x;p)-2\left({}_\alpha^k\mathbf{Z}_\sigma f\right)(x;p)\left({}_\alpha^k\mathbf{Z}_\varsigma f\right)(x;p)\Big]^{\frac{1}{2}}$$
$$\times\Big[\chi_\varsigma(x;p)\left({}_\alpha^k\mathbf{Z}_\sigma g^2\right)(x;p)+\chi_\sigma(x;p)\left({}_\alpha^k\mathbf{Z}_\varsigma g^2\right)(x;p)-2\left({}_\alpha^k\mathbf{Z}_\sigma g\right)(x;p)\left({}_\alpha^k\mathbf{Z}_\varsigma g\right)(x;p)\Big]^{\frac{1}{2}}.$$

By taking $\eta_1(t)=\eta_2(t)=g(t)=1$ in Theorem 1, we get the following inequality:

$$\left({}_\alpha^k\mathbf{Z}_\sigma f^2\right)(x;p)\leq\frac{\left[\left({}_\alpha^k\mathbf{Z}_\sigma(\zeta_1+\zeta_2)f\right)(x;p)\right]^2}{4\left({}_\alpha^k\mathbf{Z}_\sigma\zeta_1\zeta_2\right)(x;p)}.$$

This implies

$$\chi_\varsigma(x;p)\left({}_\alpha^k\mathbf{Z}_\sigma f^2\right)(x;p)-\left({}_\alpha^k\mathbf{Z}_\sigma f\right)(x;p)\left({}_\alpha^k\mathbf{Z}_\varsigma f\right)(x;p)$$
$$\leq\frac{\chi_\varsigma(x;p)\left[\left({}_\alpha^k\mathbf{Z}_\sigma(\zeta_1+\zeta_2)f\right)(x;p)\right]^2}{4\left({}_\alpha^k\mathbf{Z}_\sigma\zeta_1\zeta_2\right)(x;p)}-\left({}_\alpha^k\mathbf{Z}_\sigma f\right)(x;p)\left({}_\alpha^k\mathbf{Z}_\varsigma f\right)(x;p) \qquad (34)$$
$$=G_{\sigma,\varsigma}(f,\zeta_1,\zeta_2)(x;p)$$

and

$$\chi_\sigma(x;p)\left({}_\alpha^k\mathbf{Z}_\varsigma f^2\right)(x;p)-\left({}_\alpha^k\mathbf{Z}_\sigma f\right)(x;p)\left({}_\alpha^k\mathbf{Z}_\varsigma f\right)(x;p)$$
$$\leq\frac{\chi_\sigma(x;p)\left[\left({}_\alpha^k\mathbf{Z}_\varsigma(\zeta_1+\zeta_2)f\right)(x;p)\right]^2}{4\left({}_\alpha^k\mathbf{Z}_\varsigma\zeta_1\zeta_2\right)(x;p)}-\left({}_\alpha^k\mathbf{Z}_\sigma f\right)(x;p)\left({}_\alpha^k\mathbf{Z}_\varsigma f\right)(x;p) \qquad (35)$$
$$=G_{\varsigma,\sigma}(f,\zeta_1,\zeta_2)(x;p).$$

Applying the same procedure for $\zeta_1(t)=\zeta_2(t)=f(t)=1$, we get the following inequalities:

$$\chi_\varsigma(x;p)\left({}_\alpha^k\mathbf{Z}_\sigma g^2\right)(x;p)-\left({}_\alpha^k\mathbf{Z}_\sigma g\right)(x;p)\left({}_\alpha^k\mathbf{Z}_\varsigma g\right)(x;p)\leq G_{\sigma,\varsigma}(g,\eta_1,\eta_2)(x;p) \qquad (36)$$

and

$$\chi_\sigma(x;p)\left({}_\alpha^k\mathbf{Z}_\varsigma g^2\right)(x;p)-\left({}_\alpha^k\mathbf{Z}_\sigma g\right)(x;p)\left({}_\alpha^k\mathbf{Z}_\varsigma g\right)(x;p)\leq G_{\varsigma,\sigma}(g,\eta_1,\eta_2)(x;p). \qquad (37)$$

Finally, considering (33) to (37), we arrive at the desired result in (30). □

Theorem 5. *Under the assumptions of Theorem 4, we have*

$$\left|\chi_\sigma(x;p)\left({}_\alpha^k\mathbf{Z}_\sigma fg\right)(x;p)-\left({}_\alpha^k\mathbf{Z}_\sigma f\right)(x;p)\left({}_\alpha^k\mathbf{Z}_\sigma g\right)(x;p)\right|$$
$$\leq|G_{\sigma,\sigma}(f,\zeta_1,\zeta_2)(x;p)\,G_{\sigma,\sigma}(g,\eta_1,\eta_2)(x;p)|^{\frac{1}{2}}, \qquad (38)$$

where

$$G_{\sigma,\sigma}(m,n,o)(x;p)=\frac{\chi_\sigma(x;p)\left[\left({}_\alpha^k\mathbf{Z}_\sigma(n+o)m\right)(x;p)\right]^2}{4\left({}_\alpha^k\mathbf{Z}_\sigma no\right)(x;p)}$$
$$-\left({}_\alpha^k\mathbf{Z}_\sigma m\right)(x;p)\left({}_\alpha^k\mathbf{Z}_\sigma m\right)(x;p).$$

Proof. By taking $\sigma=\varsigma$ in (30), we get the inequality (38). □

Corollary 4. *If* $\zeta_1 = u$, $\zeta_2 = U$, $\eta_1 = v$ *and* $\eta_2 = V$, *then we have*

$$\left| \chi_\sigma(x;p) \left({}^k_a Z_\sigma fg \right)(x;p) - \left({}^k_a Z_\sigma f \right)(x;p) \left({}^k_a Z_\sigma g \right)(x;p) \right|$$
$$\leq \frac{(U-u)(V-v)}{4\sqrt{uUvV}} \left({}^k_a Z_\sigma f \right)(x;p) \left({}^k_a Z_\sigma g \right)(x;p).$$

Remark 7. *In Theorem 4 and Theorem 5, for* $\alpha(x) = x$ *and* $k = 1$*, we get [14] (Theorem 4, Corollary 4), for* $\alpha(x) = x$, $k = 1$, $\vartheta = p = 0$ *(and* $a = 0$*), we get [10] (Theorem 3.6, Theorem 3.7).*

4. Conclusions

We have proved some new Pólya-Szegö and Chebyshev type inequalities for generalized k-fractional integral operators involving Mittag-Leffler function in their kernels. The outcomes of this paper also provide a lot of Pólya-Szegö and Chebyshev type inequalities for several well-known fractional integral operators via parameter substitutions. The classical inequalities and results can be generalized by using the new k-fractional integral operators.

Author Contributions: Conceptualization, Z.Z.; G.F.; S.M.; K.N. and T.Y.; methodology, Z.Z.; G.F.; S.M.; K.N. and T.Y.; validation, G.F.; K.N. and T.Y.; writing—original draft preparation, G.F.; S.M. and K.N.; writing—review and editing, Z.Z.; G.F.; S.M.; K.N. and T.Y. All authors have read and agreed to the published version of the manuscript.

Funding: This research received funding support from the National Science, Research, and Innovation Fund (NSRF), Thailand.

Institutional Review Board Statement: Not applicable.

Informed Consent Statement: Not applicable.

Data Availability Statement: Not applicable.

Acknowledgments: We would like to thank the referees for their valuable comments and helpful advice on our manuscript.

Conflicts of Interest: The authors declare no conflict of interest.

References

1. Chebyshev, P.L. Sur les expressions approximatives des integrales definies par les autres prises entre les mêmes limites. *Proc. Math. Soc. Charkov.* **1882**, *2*, 93–98.
2. Hadamard, J. Etude sur les proprietes des fonctions entieres e.t en particulier dune fonction consideree par Riemann. *J. Math. Pure Appl.* **1893**, *58*, 171–215.
3. Jensen, J.L.W.V. Sur les fonctions convexes et les inégalités entre les valeurs moyennes. *Acta Math.* **1906**, *30*, 175–193. [CrossRef]
4. Pólya, G.; Szegö, G. *Aufgaben und Lehrsatze aus der Analysis*; Springer: Berlin, Germany, 1925.
5. Petrović, M. Sur une fontionnelle. *Publ. Math. Univ. Belgrade* **1932**, *1*, 146–149.
6. Grüss, G. Über das maximum des absolten Betrages von $\frac{1}{b-a}\int_a^b f(x)g(x)dx - \frac{1}{b-a}^2 \int_a^b f(x)dx \int_a^b f(x)dx$. *Math. Z.* **1935**, *39*, 215–226. [CrossRef]
7. Ostrowski, A. Uber die Absolutabweichung einer differentiierbaren funktion von ihrem integralmittelwert. *Comment. Math. Helv.* **1937**, *10*, 226–227. [CrossRef]
8. Belarbi, S.; Dahmani, Z. On some new fractional integral inequalities. *J. Inequal. Appl. Math.* **2009**, *10*, 1–12.
9. Farid, G. Some new Ostrowski type inequalities via fractional integrals. *Int. J. Anal. Appl.* **2017**, *14*, 64–68.
10. Ntouyas, S.K.; Agarwal, P.; Tariboon, J. On Pólya-Szegö and Chebyshev types inequalities involving the Riemann-Liouville fractional integral operators. *J. Math. Inequal.* **2016**, *10*, 491–504. [CrossRef]
11. Sarikaya, M.Z.; Yildirim, H. On Hermite-Hadamard type inequalities for Riemann-Liouville fractional integrals. *Miskolc Math. Notes* **2016**, *17*, 1049–1059. [CrossRef]
12. Sarikaya, M.Z.; Set, E.; Yaldiz, H.; Basak, N. Hermite-Hadamard inequalities for fractional integrals and related fractional inequalities. *J. Math. Comput. Model.* **2013**, *57*, 2403–2407. [CrossRef]
13. Tariboon, J.; Ntouyas, S.K.; Sudsutad, W. Some new Riemann-Liouville fractional integral inequalities. *Int. J. Math. Math. Sci.* **2014**, *2014*, 6. [CrossRef]
14. Andrić, M.; Farid, G.; Mehmood, S.; Pečarić, J., Pólya-Szegö and Chebyshev types inequalities via an extended generalized Mittag-Leffler function. *Math. Ineq. Appl.* **2019**, *22*, 1365–1377. [CrossRef]

15. Butt, S.I.; Akdemir, A.O.; Bhatti, M.Y.; Nadeem, M. New refinements of Chebyshev-Pólya-Szegö-type inequalities via generalized fractional integral operators. *J. Inequal. Appl.* **2020**, *2020*, 157. [CrossRef]
16. Dahmani, Z.; Mechouar, O.; Brahami, S. Certain inequalities related to the Chebyshev's functional involving a type Riemann-Liouville operator. *Bull. Math. Anal. Appl.* **2011**, *3*, 38–44.
17. Deniz, E.; Akdemir, A.O.; Yüksel, E. New extensions of Chebyshev-Pólya-Szegö type inequalities via conformable integrals. *AMIS Math.* **2020**, *5*, 956–965. [CrossRef]
18. Farid, G.; Rehman, A.U.; Mishra, V.N.; Mehmood, S. Fractional integral inequalities of Grüss type via generalized Mittag-Leffler function. *Int. J. Anal. Appl.* **2019**, *17*, 548–558.
19. Rashid, S.; Jarad, F.; Kalsoom, H.; Chu, Y.M. Pólya-Szegö and Chebyshev types inequalities via generalized k-fractional integrals. *Adv. Differ. Equ.* **2020**, *2020*, 125. [CrossRef]
20. Wang, G.; Agarwal, P.; Chand, M. Certain Grüss type inequalities involving the generalized fractional integral operator. *J. Inequal. Appl.* **2014**, *2014*, 1–8. [CrossRef]
21. Dragomir, S.S.; Diamond, N.T. Integral inequalities of Grüss type via Pólya-Szegö and Shisha-Mond results. *East Asian Math. J.* **2003**, *19*, 27–39.
22. Agarwal, P. Some inequalities involving Hadamard-type k-fractional integral operators. *Math. Methods Appl. Sci.* **2017**, *40*, 3882–3891. [CrossRef]
23. Agarwal, P.; Jleli, M.; Tomar, M. Certain Hermite-Hadamard type inequalities via generalized k-fractional integrals. *J. Inequal. Appl.* **2017**, *2017*, 1–10. [CrossRef]
24. Akkurt, A.; Yildirim, M.E.; Yildirim, H. On some integral inequalities for (k,h)-Riemann-Liouville fractional integral. *New Trends Math. Sci.* **2016**, *4*, 138–146. [CrossRef]
25. Andrić, M.; Farid, G.; Pečarić, J. *Analytical Inequalities for Fractional Calculus Operators and the Mittag-Leffler Function*; Element: Zagreb, Croatia, 2021.
26. Chen, H.; Katugampola, U.N. Hermite-Hadamard and Hermite-Hadamard-Fejér type inequalities for generalized fractional integrals. *J. Math. Anal. Appl.* **2017**, *446*, 1274–1291. [CrossRef]
27. Habib, S.; Mubeen, S.; Naeem, M.N. Chebyshev type integral inequalities for generalized k-fractional conformable integrals. *J. Inequal. Spec. Func.* **2018**, *9*, 53–65.
28. Kwun, Y.C.; Farid, G.; Latif, N.; Nazeer, W.; Kang, S.M. Generalized Riemann-Liouville k-fractional integrals associated with Ostrowski type inequalities and error bounds of Hadamard inequalities. *IEEE Access* **2018**, *6*, 64946–64953. [CrossRef]
29. Mehmood, S.; Farid, G.; Khan, K.A.; Yussouf, M. New fractional Hadamard and Fejér-Hadamard inequalities associated with exponentially (h,m)-convex function. *Eng. Appl. Sci. Lett.* **2020**, *3*, 9–18. [CrossRef]
30. Khan, T.U.; Khan, M.A. Generalized conformable fractional operators. *J. Comput. Appl. Math.* **2019**, *346*, 378–389. [CrossRef]
31. Che, Y.; Keir, M.Y.A. Study on the training model of football movement trajectory drop point based on fractional differential equation. *Appl. Math. Nonlinear Sci.* **2021**, 1–6. [CrossRef]
32. Farid, G. A unified integral operator and further its consequences. *Open J. Math. Anal.* **2020**, *4*, 1–7. [CrossRef]
33. Jarad, F.; Ugurlu, E.; Abdeljawad, T.; Baleanu, D. On a new class of fractional operators. *Adv. Differ. Equ.* **2017**, *2017*, 247. [CrossRef]
34. Man, S.; Yang, R.; Educational reform informatisation based on fractional differential equation. *Appl. Math. Nonlinear Sci.* **2021**, 1–10. [CrossRef]
35. Zhao, N.; Yao, F.; Khadidos, A.O.; Muwafak, B.M. The impact of financial repression on manufacturing upgrade based on fractiona Fourier transform and probability. *Appl. Math. Nonlinear Sci.* **2021**, 1–11. [CrossRef]
36. Andrić, M.; Farid, G.; Pečarić, J. A further extension of Mittag-Leffler function. *Fract. Calc. Appl. Anal.* **2018**, *21*, 1377–1395. [CrossRef]
37. Salim, T.O.; Faraj, A.W. A generalization of Mittag-Leffler function and integral operator associated with integral calculus. *J. Frac. Calc. Appl.* **2012**, *3*, 1–13.
38. Rahman, G.; Baleanu, D.; Qurashi, M.A.; Purohit, S.D.; Mubeen, S.; Arshad, M. The extended Mittag-Leffler function via fractional calculus. *J. Nonlinear Sci. Appl.* **2017**, *10*, 4244–4253. [CrossRef]
39. Srivastava, H.M.; Tomovski, Z. Fractional calculus with an integral operator containing generalized Mittag-Leffler function in the kernal. *Appl. Math. Comput.* **2009**, *211*, 198–210.
40. Prabhakar, T.R. A singular integral equation with a generalized Mittag-Leffler function in the kernel. *Yokohama Math. J.* **1971**, *19*, 7–15.
41. Kilbas, A.A.; Srivastava, H.M. Trujillo, J.J. *Theory and Applications of Fractional Differential Equations*; North-Holland Mathematics Studies; Elsevier: New York, NY, USA; London, UK, 2006.
42. Sarikaya, M.Z.; Dahmani, M.; Kiris, M.E.; Ahmad, F. (k,s)-Riemann-Liouville fractional integral and applications. *Hacet. J. Math. Stat.* **2016**, *45*, 77–89. [CrossRef]
43. Mubeen, S.; Habibullah, G.M. k-fractional integrals and applications. *Int. J. Contemp. Math. Sci.* **2012**, *7*, 89–94.

fractal and fractional

Article

Certain Integral and Differential Formulas Involving the Product of Srivastava's Polynomials and Extended Wright Function

Saima Naheed [1], Shahid Mubeen [1], Gauhar Rahman [2], Zareen A. Khan [3] and Kottakkaran Sooppy Nisar [4,*]

[1] Department of Mathematics, University of Sargodha, Sargodha 40100, Pakistan; saima.naheed@uos.edu.pk (S.N.); smjhanda@gmail.com (S.M.)
[2] Department of Mathematics and Statistics, Hazara University, Mansehra 21300, Pakistan; gauhar55uom@gmail.com or drgauhar.rahman@hu.edu.pk
[3] Department of Mathematical Sciences, College of Science, Princess Nourah bint Abdulrahman University, P.O. Box 84428, Riyadh 11671, Saudi Arabia; dr.zareenkhan@ymail.com or zakhan@pnu.edu.sa
[4] Department of Mathematics, College of Arts and Sciences, Prince Sattam Bin Abdulaziz University, Wadi Aldawser 11991, Saudi Arabia
* Correspondence: n.sooppy@psau.edu.sa or ksnisar1@gmail.com

Abstract: Many authors have established various integral and differential formulas involving different special functions in recent years. In continuation, we explore some image formulas associated with the product of Srivastava's polynomials and extended Wright function by using Marichev–Saigo–Maeda fractional integral and differential operators, Lavoie–Trottier and Oberhettinger integral operators. The obtained outcomes are in the form of the Fox–Wright function. It is worth mentioning that some interesting special cases are also discussed.

Keywords: Wright function; Srivastava's polynomials; fractional calculus operators; Lavoie–Trottier integral formula; Oberhettinger integral formula

MSC: 33C20; 33B15; 33C20; 44A20

Citation: Naheed, S.; Mubeen, S.; Rahman, G.; Khan, Z.A.; Nisar, K.S. Certain Integral and Differential Formulas Involving the Product of Srivastava's Polynomials and Extended Wright Function. *Fractal Fract.* **2022**, *6*, 93. https://doi.org/10.3390/fractalfract6020093

Academic Editor: Maja Andrić

Received: 19 November 2021
Accepted: 25 January 2022
Published: 8 February 2022

Publisher's Note: MDPI stays neutral with regard to jurisdictional claims in published maps and institutional affiliations.

Copyright: © 2022 by the authors. Licensee MDPI, Basel, Switzerland. This article is an open access article distributed under the terms and conditions of the Creative Commons Attribution (CC BY) license (https://creativecommons.org/licenses/by/4.0/).

1. Introduction and Preliminaries

Numerous integral formulas including special functions have been anticipated and they are important in solving various problems in science and engineering. Many authors have established certain unified integral formulae associated with special functions [1–5]. A brief study of some important properties of generalized gamma and beta functions defined in the form of Fox–Wright function is presented in [6]. Certain integral formulas involving the generalized Bessel function and Bessel–Maitland function are explored in [7,8]. A new class of integrals associated with hypergeometric function is established by Rakha et al. [9]. Certain fractional calculus operators and their applications are briefly discussed by Samraiz et al. [10]. A brief discussion of generalized Mittag–Leffler function and multivariable Mittag–Leffler function via generalized fractional calculus operators is available in [11,12]. Composition formulas of various fractional calculus operators are studied in [13,14].

Many generalized special functions have been linked to various types of issues in different fields of mathematical sciences. This reason inspired many researchers to explore the field of integrals and associated generalized special functions. Several unified integral formulas derived by many authors involving different type of special functions are discussed in (see, for example, [15–19]). Suthar established the composition formulae for the k-fractional calculus operators associated with k-Wright function [20]. Fractional calculus and integral transforms of general class of polynomials and incomplete Fox–Wright functions is discussed by Jangid et al. [21]. Certain expressions of the Laguerre polynomial and relations of some known functions in terms of generalized Meijer G-functions are explored in [22,23].

Lavoie–Trottier integral formula involving product of Bessel function of the first kind and general class of polynomials are established by Menaria et al. [24]. Suthar et al. [25] explored the certain integral formulae involving product of Srivastava's polynomials and generalized Bessel–Maitland function.

In continuation of the above work, we developed generalized integral formulae involving product of Srivastava's polynomials $S_c^d[r]$ and extended Wright function $R_{\varsigma,\tau}^{\varpi,\varepsilon}(z)$. These formulae are communicated in the form of generalized Fox–Wright function. For our purpose, we start by reviewing some known functions and earlier work. Srivastava [26] established the general class of polynomials $S_c^d[r]$ in the following way.

$$S_c^d[r] = \sum_{k=0}^{[\frac{d}{c}]} \frac{(-c)_{dk}}{k!} G_{c,k} r^k \quad (c = 0, 1, 2, \dots), \tag{1}$$

where d is an arbitrary positive integer and the coefficients $G_{c,k}(c, k \geq 0)$ are arbitrary constants (real or complex). The polynomial family $S_c^d[r]$ have many known polynomials as its special cases. The extended Wright function [27] is defined as follows:

$$R_{\varsigma,\tau}^{\varpi,\varepsilon}(z) = \sum_{n=0}^{\infty} \frac{(\varpi)_n}{(\varepsilon)_n \Gamma(\varsigma n + \tau)} \frac{z^n}{n!}, \tag{2}$$

where $\varsigma > -1$, $\tau, \varpi, \varepsilon \in \mathbb{C}, \varepsilon \neq 0, -1, -2, \dots$ with $z \in \mathbb{C}$. The function $R_{\varsigma,\tau}^{\varpi,\varepsilon}(z)$ is an entire function of order $\frac{1}{1+\varsigma}$.

The two (generalized Wright type) auxiliary functions for any order $\varsigma \in (0,1)$ and for all complex variable $z \neq 0$ are defined as follows:

$$M_{-\varsigma}^{\varpi,\varepsilon} = R_{-\varsigma,1-\varsigma}^{\varpi,\varepsilon}(-z) = \sum_{n=0}^{\infty} \frac{(\varpi)_n}{(\varepsilon)_n \Gamma(1 - \varsigma(n+1))} \frac{(-1)^n z^n}{n!} \tag{3}$$

and

$$F_{-\varsigma}^{\varpi,\varepsilon} = R_{-\varsigma,1-\varsigma}^{\varpi,\varepsilon}(-z) = \sum_{n=0}^{\infty} \frac{(\varpi)_n}{(\varepsilon)_n \Gamma(-\varsigma n)} \frac{(-1)^n z^n}{n!}. \tag{4}$$

For $\varpi = \varepsilon = 0$, we have the normalized Wright function [28] which is given by

$$W_{\varsigma,\tau}(z) = \Gamma(\tau) \sum_{n=0}^{\infty} \frac{z^n}{\Gamma(\tau + n\varsigma) n!}, \tau > -1, \varsigma \in \mathbb{C}. \tag{5}$$

The extended Wright function can also be expressed in the following form

$$R_{\varsigma,\tau}^{\varpi,\varepsilon}(z) = \frac{\Gamma(\varepsilon)}{\Gamma(\varpi)} {}_1\psi_2 \left[\begin{array}{c} (\varpi, 1); \\ (\varepsilon, 1), (\tau, \varsigma); \end{array} z \right], \tag{6}$$

where ${}_u\psi_x$ is the Fox–Wright hypergeometric function defined in [29] as follows: For $a_i, b_j \in \mathbb{C}, A_i, B_j \in \mathbb{R}$ $(A_i, B_j) \neq 0$, where $i = 1, 2, \dots, u; j = 1, 2, \dots, x$ and $(a_i + A_i n), (b_j + B_j n) \in \mathbb{C} \setminus k\mathbb{Z}^-,$

$${}_u\psi_x[(a_1, A_1), \dots, (a_u, A_u); (b_1, B_1), \dots, (b_x, B_x); z] = \sum_{n=0}^{\infty} \frac{\Gamma(a_1 + nA_1) \dots \Gamma(a_u + nA_u) z^n}{\Gamma(b_1 + nB_1) \dots \Gamma(b_x + nB_x) n!}, \tag{7}$$

with convergence condition

$$1 + \sum_{j=1}^{u} B_j - \sum_{i=1}^{v} A_i > 0. \tag{8}$$

The extended Wright function has relationship with some other special functions which are discussed below.

Relation with Mittag–Leffler function:

$$R^{1,\varepsilon}_{0,\tau}(z) = \frac{\Gamma(\varepsilon)}{\Gamma(\tau)} E_{1,\varepsilon}(z), \tag{9}$$

for $\varsigma = 0$, $\varpi = 1$, $\tau, \varepsilon \in \mathbb{C}$ and $Re(\varepsilon) > 0$.

Relation with Meijer G-function:

Extended Wright function and Meijer G-function are related as

$$R^{1,\varepsilon}_{0,\tau}(z) = \frac{\Gamma(\varepsilon)}{\Gamma(\varpi)} G^{1\,1}_{1\,3}\left[z \,\bigg|\, \begin{array}{c} 1-\varpi \\ 0, 1-\tau, 1-\varepsilon \end{array}\right], \tag{10}$$

for $\varsigma = 1$.

Relation with Fox H-function:

From the definition of Fox H-function and extended Wright function, we obtain

$$R^{\varpi,\varepsilon}_{\varsigma,\tau}(-z) = \frac{\Gamma(\varepsilon)}{\Gamma(\varpi)} H^{1\,1}_{1\,3}\left[z \,\bigg|\, \begin{array}{c} (1-\varpi, 1) \\ (0,1),(1-\tau,\varsigma),(1-\varepsilon,1) \end{array}\right]. \tag{11}$$

The Marichev–Saigo–Maeda fractional integral operators [30] for $\xi, \acute{\xi}, \eta, \acute{\eta}, \lambda \in \mathbb{C}$ and $x > 0$ are defined as follows:

$$I^{\xi,\acute{\xi},\eta,\acute{\eta},\lambda}_{0+} f(t) = \frac{x^{-\xi}}{\Gamma(\lambda)} \int_0^x (x-t)^{\lambda-1} t^{-\acute{\xi}} F_3\left(\xi, \acute{\xi}, \eta, \acute{\eta}; \lambda; 1-\frac{t}{x}; 1-\frac{x}{t}\right) f(t) dt \tag{12}$$

and

$$I^{\xi,\acute{\xi},\eta,\acute{\eta},\lambda}_{-} f(t) = \frac{x^{-\acute{\xi}}}{\Gamma(\lambda)} \int_x^\infty (t-x)^{\lambda-1} t^{-\xi} F_3\left(\xi, \acute{\xi}, \eta, \acute{\eta}; \lambda; 1-\frac{x}{t}; 1-\frac{t}{x}\right) f(t) dt, \tag{13}$$

where F_3 is Appell function.

The Marichev–Saigo–Maeda fractional differential operators [31] for $\xi, \acute{\xi}, \eta, \acute{\eta}, \lambda \in \mathbb{C}$ and $x > 0$ are defined as follows:

$$D^{\xi,\acute{\xi},\eta,\acute{\eta},\lambda}_{0+} f(t) = \left(\frac{d}{dt}\right)^m (I^{-\acute{\xi},-\xi,-\acute{\eta}+m,-\eta,-\lambda+m}_{0+} f)(t) \tag{14}$$

and

$$D^{\xi,\acute{\xi},\eta,\acute{\eta},\lambda}_{-} f(t) = \left(-\frac{d}{dt}\right)^m (I^{-\acute{\xi},-\xi,-\acute{\eta},-\eta+m,-\lambda+m}_{-} f)(t). \tag{15}$$

The Saigo fractional integral operators [32] are defined as:

For $w \in \mathbb{R}^+$, $\epsilon, \varrho, \chi \in \mathbb{C}$ with $Re(\epsilon) > 0$,

$$(I^{\epsilon,\varrho,\chi}_{0+} f)(w) = \frac{w^{-\epsilon-\varrho}}{\Gamma(\epsilon)} \int_0^w (w-t)^{\epsilon-1} \times {}_2F_1\left(\epsilon+\varrho, -\chi; \epsilon; \left(1-\frac{t}{w}\right)\right) f(t) dt \tag{16}$$

and

$$(I^{\epsilon,\varrho,\chi}_{-} f)(w) = \frac{1}{\Gamma(\epsilon)} \int_w^\infty (t-w)^{\epsilon-1} t^{-\epsilon-\varrho} \times {}_2F_1\left(\epsilon+\varrho, -\chi; \epsilon; \left(1-\frac{w}{t}\right)\right) f(t) dt. \tag{17}$$

Additionally, the Saigo fractional differential operators [33], for $w > 0$ and $\epsilon, \varrho, \chi \in \mathbb{C}$, $Re(\epsilon) > 0$ are given by

$$(D_{0+}^{\epsilon,\varrho,\chi}f)(w) = (\tfrac{d}{dw})^n(I_{0+}^{-\epsilon+n,-\varrho-n,\epsilon+\chi-n}f)w, \quad n = [Re(\epsilon)+1]$$
$$= (\tfrac{d}{dw})^n \frac{w^{\frac{\epsilon+\varrho}{k}}}{k\Gamma_k(-\epsilon+n)} \int_0^w (w-t)^{\frac{-\epsilon}{k}+n-1}$$
$$\times {}_2F_1((-\epsilon-\varrho,-\chi-\epsilon+n;-\epsilon+n;(1-\tfrac{t}{w}))f(t)dt \qquad (18)$$

and

$$(D_{-}^{\epsilon,\varrho,\chi}f)(w) = (\tfrac{d}{dw})^n(I_{-}^{-\epsilon+n,-\varrho-n,\epsilon+\chi}f)w, \quad n = [Re(\epsilon)+1]$$
$$= (\tfrac{d}{dw})^n \frac{1}{k\Gamma_k(-\epsilon+n)} \int_w^\infty (t-w)^{\frac{-\epsilon-n}{k}-1} t^{\frac{\epsilon+\varrho}{k}}$$
$$\times {}_2F_1((-\epsilon-\varrho,-\chi-\epsilon+n;-\epsilon+n;(1-\tfrac{w}{t}))f(t)dt, \qquad (19)$$

where $[Re(\epsilon)]$ is the integral part of $Re(\epsilon)$ and ${}_2F_1(\epsilon,\varrho,\chi;w)$ is the hypergeometric function.

The left-sided and right-sided Riemann–Liouville fractional integral operators [33] are defined as follows:

$$I_{a+}^\zeta f(y) = \frac{1}{\Gamma(\zeta)} \int_a^y (y-t)^{\zeta-1} f(t)dt \qquad (20)$$

and

$$I_{b-}^\zeta f(y) = \frac{1}{\Gamma(\zeta)} \int_y^b (t-y)^{\zeta-1} f(t)dt, \qquad (21)$$

where $Re(\zeta) > 0$.

The left and right-sided Riemann–Liouville fractional differential operators are given by

$$(D_{a+}^\zeta y)(x) = \left(\frac{d}{dx}\right)^n \frac{1}{\Gamma(n-\zeta)} \int_a^x (x-t)^{\zeta-n+1} y(t)dt, \qquad (22)$$

where $n = [Re(\zeta)]+1, x > a$ and

$$(D_{b-}^\zeta y)(x) = \left(-\frac{d}{dx}\right)^n \frac{1}{\Gamma(p-\zeta)} \int_z^\infty (t-x)^{\zeta-n+1} y(t)dt, \qquad (23)$$

where $n = [Re(\zeta)]+1$ and $x < b$.

Further, we will recall the Lavoie–Trottier integral formula [34] which is given by

$$\int_0^1 y^{\epsilon-1}(1-y)^{2\omega-1}\left(1-\frac{y}{3}\right)^{2\epsilon-1}\left(1-\frac{y}{4}\right)^{\omega-1} dy = \left(\frac{2}{3}\right)^{\epsilon-1} \frac{\Gamma(\gamma)\Gamma(\omega)}{\Gamma(\epsilon+\omega)}, \qquad (24)$$

for $Re(\gamma), Re(\delta) > 0$. Additionally, we evoke Oberhettinger's integral formula [35] given as follows:

$$\int_0^\infty y^{\phi-1}(y+b+\sqrt{y^2+2by})^{-\chi} dy = 2\chi b^{-\chi}\left(\frac{b}{2}\right)^\phi \frac{\Gamma(2\phi)\Gamma(\chi-\phi)}{\Gamma(1+\chi+\phi)}, \qquad (25)$$

where $0 < Re(\phi) < Re(\chi)$.

2. Image Formulas for Marichev–Saigo–Maeda Integral Operators Involving the Product of Srivastava's Polynomials and Extended Wright Function

In this section, we establish the image formulas by applying Marichev–Saigo–Maeda fractional integral operators (12) and (13) to the product of Srivastava's Polynomials (1) and extended Wright function (2).

The following formulas for a power function, under operators (12) and (13) are given in [31] are helpful to prove our main results.

$$[I_{0^+}^{\zeta,\acute{\zeta},\eta,\acute{\eta},\lambda}t^{\nu-1}]y = \Gamma\left[\begin{array}{c} \nu,\ \nu+\lambda-\zeta-\acute{\zeta}-\eta,\ \nu-\acute{\zeta}-\zeta+\acute{\eta} \\ \nu+\acute{\eta},\ \nu+\lambda-\zeta-\acute{\zeta},\ \nu+\lambda-\acute{\zeta}-\eta \end{array}\right]y^{\nu-\zeta-\acute{\zeta}+\lambda-1}, \quad (26)$$

where $Re(\lambda) > 0$ and $Re(\nu) > max\{0, Re(\zeta + \acute{\zeta} + \eta - \lambda), Re(\acute{\zeta} - \acute{\eta})\}$.

Additionally,

$$[I_{0^-}^{\zeta,\acute{\zeta},\eta,\acute{\eta},\lambda}t^{\nu-1}]y = \Gamma\left[\begin{array}{c} 1+\zeta+\acute{\zeta}-\lambda-\nu,\ 1+\zeta+\acute{\eta}-\lambda-\nu,\ 1-\eta-\nu \\ 1-\nu,\ 1+\zeta+\acute{\zeta}+\acute{\eta}-\lambda-\nu,\ 1+\zeta-\eta-\nu \end{array}\right]$$
$$\times y^{\nu+\lambda-\zeta-\acute{\zeta}-1}, \quad (27)$$

where $Re(\lambda) > 0$ and $Re(\nu) < 1 + min\{Re(-\eta), Re(\zeta + \acute{\eta} - \lambda), Re(\zeta + \acute{\zeta} - \lambda)\}$.

We use the notation
$$\Gamma\left[\begin{array}{c} a,b,c \\ d,e,f \end{array}\right] = \frac{\Gamma(a)\Gamma(b)\Gamma(c)}{\Gamma(d)\Gamma(e)\Gamma(f)}.$$

Theorem 1. *For* $\zeta, \acute{\zeta}, \eta, \acute{\eta}, \lambda, \omega \in \mathbb{C}, x > 0$ *such that* $Re(\lambda) > 0, Re(\varsigma) > -1$, $Re(\nu + \varsigma\mu + 2\tau\mu) > max\{0, Re(\zeta + \acute{\zeta} + \eta - \lambda), Re(\acute{\zeta} - \acute{\eta})\}$, *then prove the following formula*

$$[I_{0^+}^{\zeta,\acute{\zeta},\eta,\acute{\eta},\lambda}t^{\nu-1}S_c^d(\sigma t^\alpha)R_{\varsigma,\tau}^{\omega,\varepsilon}(\delta t^\mu)](x)$$
$$= x^{\nu-\zeta-\acute{\zeta}+\lambda-1}\frac{\Gamma(\varepsilon)}{\Gamma(\omega)}\sum_{k=0}^{[\frac{d}{c}]}\frac{(-c)_{dk}}{k!}G_{c,k}(\sigma x^\alpha)^k$$
$$\times {}_4\psi_5\left[\begin{array}{c} (\omega,1), (\nu+\alpha k, \mu), (\nu+\alpha k+\lambda-\zeta-\acute{\zeta}-\eta, \mu), \\ (\nu+\alpha k-\acute{\zeta}+\acute{\eta},\mu); \\ (\varepsilon,1), (\tau,\varsigma), (\nu+\alpha k+\acute{\eta},\mu), (\nu+\alpha k+\lambda-\zeta-\acute{\zeta},\mu), \\ (\nu+\alpha k+\lambda-\acute{\zeta}+\eta,\mu); \end{array} \delta x^\mu\right]. \quad (28)$$

Proof. By using (1) and (2) in the left hand side of (24) and representing it with $\mathfrak{S}$. After some simplification, we obtain

$$\mathfrak{S} = \sum_{k=0}^{[\frac{d}{c}]}\frac{(-c)_{dk}}{k!}G_{c,k}(\sigma)^k \sum_{n=0}^{\infty}\frac{(\omega)_n}{(\varepsilon)_n\Gamma(\varsigma n+\tau)}\frac{(\delta t^\mu)^n}{n!}$$
$$\times [I_{0^+}^{\zeta,\acute{\zeta},\eta,\acute{\eta},\lambda}t^{\nu+\alpha k+\mu n-1}](x). \quad (29)$$

Now, applying the integral Formula (26) to (29), we obtain

$$\mathfrak{S} = \sum_{k=0}^{[\frac{d}{c}]}\frac{(-c)_{dk}}{k!}G_{c,k}\sigma^k \sum_{n=0}^{\infty}\frac{(\omega)_n}{(\varepsilon)_n\Gamma(\varsigma n+\tau)}\frac{(\delta)^n}{n!}$$
$$\times \Gamma\left[\begin{array}{c} \nu+\alpha k+\mu n, \nu+\alpha k+\lambda-\zeta-\acute{\zeta}-\eta+\mu n, \\ \nu+\alpha k-\acute{\zeta}+\acute{\eta}+\mu n \\ \nu+\alpha k+\acute{\eta}+\mu n, \nu+\alpha k+\lambda-\zeta-\acute{\zeta}+\mu n, \\ \nu+\alpha k+\lambda-\acute{\zeta}+\eta+\mu n \end{array}\right]x^{\nu+\alpha k+\mu n-\zeta-\acute{\zeta}+\lambda-1}, \quad (30)$$

which further implies

$$\mathfrak{S} = x^{\nu-\xi-\acute{\xi}+\lambda-1} \frac{\Gamma(\varepsilon)}{\Gamma(\omega)} \sum_{k=0}^{[\frac{d}{c}]} \frac{(-c)_{dk}}{k!} G_{c,k}(\sigma x^\alpha)^k$$

$$\times \sum_{n=0}^{\infty} \Gamma \begin{bmatrix} \omega+n, \nu+\alpha k+\mu n, \nu+\alpha k+\lambda-\xi-\acute{\xi}-\eta+\mu n, \\ \nu+\alpha k-\acute{\xi}+\acute{\eta}+\mu n \\ \varepsilon+n, \tau+\varsigma n, \nu+\alpha k+\acute{\eta}+\mu n, \nu+\alpha k+\lambda-\xi-\acute{\xi}+\mu n, \\ \nu+\alpha k+\lambda-\acute{\xi}+\eta+\mu n \end{bmatrix} \frac{(\delta x^\mu)^n}{n!}. \quad (31)$$

Now, by using definition (7), we arrive at required formula (28). □

Theorem 2. For $\xi, \acute{\xi}, \eta, \acute{\eta}, \lambda, \omega \in \mathbb{C}, x > 0$ such that $Re(\lambda) > 0, Re(\varsigma) > -1$, $Re(1-\beta-\nu-\varsigma\mu-2\tau\mu) < 1 + \min\{Re(-\eta), Re(\xi+\acute{\eta}-\lambda), Re(\xi+\acute{\xi}-\lambda)\}$, then the following formula holds true

$$[I_{0^-}^{\xi,\acute{\xi},\eta,\acute{\eta},\lambda} t^{-\nu-\beta} S_c^d(\sigma t^\alpha) R_{\varsigma,\tau}^{\omega,\varepsilon}(\delta t^{-\mu})](x)$$

$$= x^{\nu+\alpha k-\xi-\acute{\xi}+\lambda-1} \frac{\Gamma(\varepsilon)}{\Gamma(\omega)} \sum_{k=0}^{[\frac{d}{c}]} \frac{(-c)_{dk}}{k!} G_{c,k}(\sigma x^\alpha)^k$$

$$\times {}_4\psi_5 \begin{bmatrix} (\omega,1), (\xi+\acute{\xi}-\lambda+\nu+\beta-\alpha k, \mu), (\xi+\acute{\eta}-\lambda+\nu+\beta-\alpha k, \mu), \\ (\nu+\beta-\eta-\alpha k, \mu); \\ (\varepsilon,1), (\tau,\varsigma), (\nu+\beta-\alpha k, \mu), (\xi+\acute{\xi}+\acute{\eta}-\lambda+\nu+\beta-\alpha k, \mu), \\ (\xi-\eta+\nu+\beta-\alpha k, \mu); \end{bmatrix} \delta x^{-\mu} . \quad (32)$$

Proof. By using (1) and (2) in the left hand side of (32) and representing it by $\mathfrak{V}$. After some simplification, we obtain

$$\mathfrak{V} = \sum_{k=0}^{[\frac{d}{c}]} \frac{(-c)_{dk}}{k!} G_{c,k}(\sigma)^k \sum_{n=0}^{\infty} \frac{(\omega)_n}{(\varepsilon)_n \Gamma(\varsigma n+\tau)} \frac{(\delta)^n}{n!}$$

$$\times [I_{0^+}^{\xi,\acute{\xi},\eta,\acute{\eta},\lambda} t^{1-\nu-\beta+\alpha k-\mu n-1}](x). \quad (33)$$

Now, applying the integral Formula (27) to (33), we obtain

$$\mathfrak{V} = \sum_{k=0}^{[\frac{d}{c}]} \frac{(-c)_{dk}}{k!} G_{c,k} \sigma^k \sum_{n=0}^{\infty} \frac{(\omega)_n}{(\varepsilon)_n \Gamma(\varsigma n+\tau)} \frac{(\delta)^n}{n!}$$

$$\times \Gamma \begin{bmatrix} \xi+\acute{\xi}-\lambda+\nu+\beta-\alpha k+\mu n, \xi+\acute{\eta}-\lambda+\nu+\beta-\alpha k+\mu n, \\ \nu+\beta-\eta-\alpha k+\mu n \\ \nu+\beta-\alpha k+\mu n, \xi+\acute{\xi}+\acute{\eta}-\lambda+\nu+\beta-\alpha k+\mu n, \\ \xi-\eta+\nu+\beta-\alpha k+\mu n \end{bmatrix}$$

$$\times x^{-\nu-\beta+\alpha k-\mu n-\xi-\acute{\xi}+\lambda-1}, \quad (34)$$

which further implies

$$\mathfrak{V} = x^{-\nu-\beta-\zeta-\acute{\zeta}+\lambda-1} \frac{\Gamma(\varepsilon)}{\Gamma(\omega)} \sum_{k=0}^{[\frac{d}{c}]} \frac{(-c)_{dk}}{k!} G_{c,k}(\sigma x^\alpha)^k$$

$$\times \sum_{n=0}^{\infty} \Gamma \begin{bmatrix} \omega + n, \zeta + \acute{\zeta} - \lambda + \nu + \beta - \alpha k + \mu n, \zeta + \acute{\eta} - \lambda + \nu + \beta - \alpha k + \mu n, \\ \nu + \beta - \eta - \alpha k + \mu n \\ \varepsilon + n, \tau + \varsigma n, \nu + \beta - \alpha k + \mu n, \zeta + \acute{\zeta} + \acute{\eta} - \lambda + \nu + \beta - \alpha k + \mu n, \\ \zeta - \eta + \nu + \beta - \alpha k + \mu n \end{bmatrix} \frac{(\delta x^\mu)^n}{n!}. \quad (35)$$

Now, by using definition (7), we arrive at required formula (32). □

Now, we discuss some special cases regarding extended Wright function in the following corollaries.

Corollary 1. *Assume that condition of Theorem 1 is fulfilled then using generalized Wright-type Function (3), the Formula (28) reduces to the following form.*

$$[I_{0+}^{\zeta,\acute{\zeta},\eta,\acute{\eta},\lambda} t^{\nu-1} S_c^d(\sigma t^\alpha) M_{-\varsigma,1-\varsigma}^{\omega,\varepsilon}(-\delta t^\mu)](x)$$

$$= x^{\nu-\zeta-\acute{\zeta}+\lambda-1} \frac{\Gamma(\varepsilon)}{\Gamma(\omega)} \sum_{k=0}^{[\frac{d}{c}]} \frac{(-c)_{dk}}{k!} G_{c,k}(\sigma x^\alpha)^k$$

$$\times {}_4\psi_5 \begin{bmatrix} (\omega,1),(\nu+\alpha k,\mu),(\nu+\alpha k+\lambda-\zeta-\acute{\zeta}-\eta,\mu), \\ (\nu+\alpha k-\acute{\zeta}+\acute{\eta},\mu); \\ (\varepsilon,1),(1-\varsigma,-\varsigma),(\nu+\alpha k+\acute{\eta},\mu),(\nu+\alpha k+\lambda-\zeta-\acute{\zeta},\mu), \\ (\nu+\alpha k+\lambda-\acute{\zeta}+\eta,\mu); \end{bmatrix} - \delta x^\mu. \quad (36)$$

Corollary 2. *Assume that condition of Theorem 2 is fulfilled then using generalized Wright-type Function (3), the Formula (32) reduces to the following form.*

$$[I_{0-}^{\zeta,\acute{\zeta},\eta,\acute{\eta},\lambda} t^{-\nu-\beta} S_c^d(\sigma t^\alpha) M_{-\varsigma,1-\varsigma}^{\omega,\varepsilon}(-\delta t^{-\mu})](x)$$

$$= x^{\nu-\zeta-\acute{\zeta}+\lambda-1} \frac{\Gamma(\varepsilon)}{\Gamma(\omega)} \sum_{k=0}^{[\frac{d}{c}]} \frac{(-c)_{dk}}{k!} G_{c,k}(\sigma x^\alpha)^k$$

$$\times {}_4\psi_5 \begin{bmatrix} (\omega,1),(\zeta+\acute{\zeta}-\lambda+\nu+\beta-\alpha k,\mu),(\zeta+\acute{\eta}-\lambda+\nu+\beta-\alpha k,\mu), \\ (\nu+\beta-\eta-\alpha k,\mu); \\ (\varepsilon,1),(\nu+\beta-\alpha k,\mu),(\zeta+\acute{\zeta}+\acute{\eta}-\lambda+\nu+\beta-\alpha k,\mu), \\ (1-\varsigma,-\varsigma),(\zeta-\eta+\nu+\beta-\alpha k,\mu); \end{bmatrix} - \delta x^{-\mu}. \quad (37)$$

Corollary 3. *Assume that condition of Theorem 1 is fulfilled and for $\varsigma \in (0,1)$ then using generalized Wright-type Function (4), the Formula (28) reduces to the following form.*

$$[I_{0+}^{\zeta,\acute{\zeta},\eta,\acute{\eta},\lambda} t^{\nu-1} S_c^d(\sigma t^\alpha) F_{-\varsigma}^{\omega,\varepsilon}(-\delta t^\mu)](x)$$

$$= x^{\nu-\zeta-\acute{\zeta}+\lambda-1} \frac{\Gamma(\varepsilon)}{\Gamma(\omega)} \sum_{k=0}^{[\frac{d}{c}]} \frac{(-c)_{dk}}{k!} G_{c,k}(\sigma x^\alpha)^k$$

$$\times {}_4\psi_5 \begin{bmatrix} (\omega,1),(\nu+\alpha k,\mu),(\nu+\alpha k+\lambda-\zeta-\acute{\zeta}-\eta,\mu), \\ (\nu+\alpha k-\acute{\zeta}+\acute{\eta},\mu); \\ (\varepsilon,1),(0,-\varsigma),(\nu+\alpha k+\acute{\eta},\mu),(\nu+\alpha k+\lambda-\zeta-\acute{\zeta},\mu), \\ (\nu+\alpha k+\lambda-\acute{\zeta}+\eta,\mu); \end{bmatrix} - \delta x^\mu. \quad (38)$$

Corollary 4. *Assume that condition of Theorem 2 is fulfilled then using generalized Wright-type Function (4), the Formula (32) reduces to the following form.*

$$[I_{0-}^{\zeta,\acute{\zeta},\eta,\acute{\eta},\lambda} t^{-\nu-\beta} S_c^d(\sigma t^\alpha) F_{-\varsigma,0}^{\omega,\varepsilon}(-\delta t^{-\mu})](x)$$

$$= x^{\nu-\zeta-\acute{\zeta}+\lambda-1} \sum_{k=0}^{[\frac{d}{c}]} \frac{(-c)_{dk}}{k!} G_{c,k}(\sigma x^\alpha)^k$$

$$\times {}_4\psi_5 \left[\begin{array}{c} (\omega,1), (\zeta+\acute{\zeta}-\lambda+\nu+\beta-\alpha k,\mu), (\acute{\zeta}+\acute{\eta}-\lambda+\nu+\beta-\alpha k,\mu), \\ (\nu+\beta-\eta-\alpha k,\mu); \\ (\varepsilon,1), (\nu+\beta-\alpha k,\mu), (\zeta+\acute{\zeta}+\acute{\eta}-\lambda+\nu+\beta-\alpha k,\mu), \\ (0,-\varsigma), (\acute{\zeta}-\eta+\nu+\beta-\alpha k,\mu); \end{array} -\delta x^{-\mu} \right]. \quad (39)$$

Corollary 5. *Assume that condition of Theorem 1 is fulfilled then using normalized Wright Function (5), the Formula (28) reduces to the following form.*

$$[I_{0+}^{\zeta,\acute{\zeta},\eta,\acute{\eta},\lambda} t^{\nu-1} S_c^d(\sigma t^\alpha) R_{\varsigma,\tau}(\delta t^\mu)](x)$$

$$= x^{\nu-\zeta-\acute{\zeta}+\lambda-1} \sum_{k=0}^{[\frac{d}{c}]} \frac{(-c)_{dk}}{k!} G_{c,k}(\sigma x^\alpha)^k$$

$$\times {}_3\psi_4 \left[\begin{array}{c} (\nu+\alpha k,\mu), (\nu+\alpha k+\lambda-\zeta-\acute{\zeta}-\eta,\mu), \\ (\nu+\alpha k-\acute{\zeta}+\acute{\eta},\mu); \\ (\tau,\varsigma), (\nu+\alpha k+\acute{\eta},\mu), (\nu+\alpha k+\lambda-\zeta-\acute{\zeta},\mu), \\ (\nu+\alpha k+\lambda-\acute{\zeta}+\eta,\mu); \end{array} \delta x^\mu \right]. \quad (40)$$

Corollary 6. *Assume that condition of Theorem 2 is fulfilled then using normalized Wright Function (5), the Formula (32) reduces to the following form.*

$$[I_{0-}^{\zeta,\acute{\zeta},\eta,\acute{\eta},\lambda} t^{-\nu-\beta} S_c^d(\sigma t^\alpha) R_{\varsigma,\tau}(\delta t^{-\mu})](x)$$

$$= x^{\nu-\zeta-\acute{\zeta}+\lambda-1} \sum_{k=0}^{[\frac{d}{c}]} \frac{(-c)_{dk}}{k!} G_{c,k}(\sigma x^\alpha)^k$$

$$\times {}_3\psi_4 \left[\begin{array}{c} (\zeta+\acute{\zeta}-\lambda+\nu+\beta-\alpha k,\mu), (\acute{\zeta}+\acute{\eta}-\lambda+\nu+\beta-\alpha k,\mu), \\ (\nu+\beta-\eta-\alpha k,\mu); \\ (\nu+\beta-\alpha k,\mu), (\zeta+\acute{\zeta}+\acute{\eta}-\lambda+\nu+\beta-\alpha k,\mu), \\ (\tau,\varsigma), (\acute{\zeta}-\eta+\nu+\beta-\alpha k,\mu); \end{array} \delta x^{-\mu} \right]. \quad (41)$$

Remark 1. *Following are the special cases of results discussed above.*
(i) *The formulas obtained reduce to formulas for Saigo's fractional integral operators (16) and (17) for $\acute{\zeta} = 0$;*
(ii) *By substituting $\eta = -\zeta$ in Saigo's fractional integral operators, we obtain the formulas for Riemann–Liouville fractional integral operators (20) and (21).*

3. Image Formulas for Marichev–Saigo–Maeda Differential Operators Involving the Product of Srivastava's Polynomials and Extended Wright Function

In this section, we establish the image formulas by applying Marichev–Saigo–Maeda fractional differential operators (14) and (15) to the product of Srivastava's Polynomials (1) and extended Wright function (2).

The formulas for a power function, under operators (14) and (15) are given in [36] are helpful to prove our main results.

For $\xi, \acute{\xi}, \eta, \acute{\eta}, \lambda, \nu \in \mathbb{C}$, such that $Re(\xi) > 0$ and $Re(\nu) > max\{0, Re(-\xi + \eta), Re(-\xi - \acute{\xi} - \acute{\eta} + \lambda)\}$, we have

$$[D_{0^+}^{\xi,\acute{\xi},\eta,\acute{\eta},\lambda} t^{\nu-1}]y = \Gamma\left[\begin{array}{c} \nu, \; -\eta + \xi + \nu, \; \xi + \acute{\xi} + \eta - \acute{\lambda} + \nu \\ -\eta + \nu, \; \xi + \acute{\xi} - \lambda + \nu, \; \xi + \acute{\eta} - \lambda + \nu \end{array}\right] y^{\nu + \xi + \acute{\xi} + \lambda - 1} \quad (42)$$

and for $\xi, \acute{\xi}, \eta, \acute{\eta}, \lambda, \nu \in \mathbb{C}$, such that $Re(\xi) > 0$ and $Re(\nu) > max\{0, Re(-\acute{\eta}), Re(\acute{\xi} + \eta - \lambda), Re(\xi + \acute{\xi} - \lambda) + [Re(\lambda)] + 1\}$, we have

$$[D_{0^-}^{\xi,\acute{\xi},\eta,\acute{\eta},\lambda} t^{\nu-1}]y = \Gamma\left[\begin{array}{c} \acute{\xi} + \nu, \; -\xi - \acute{\xi} + \lambda + \nu, \; -\acute{\xi} - \eta + \lambda + \nu \\ \nu, \; -\acute{\xi} + \acute{\eta} + \nu, \; -\xi - \acute{\xi} - \eta + \lambda + \nu \end{array}\right]$$
$$\times y^{\xi + \acute{\xi} - \nu - \lambda}. \quad (43)$$

Theorem 3. For $\xi, \acute{\xi}, \eta, \acute{\eta}, \lambda, \varpi \in \mathbb{C}, x > 0$ such that $Re(\lambda) > 0, Re(\varsigma) > -1, Re(\xi) > 0$ and $Re(\nu) > max\{0, Re(-\xi + \eta), Re(-\xi - \acute{\xi} - \acute{\eta} + \lambda)\}$, then prove the following formula

$$[D_{0^+}^{\xi,\acute{\xi},\eta,\acute{\eta},\lambda} t^{\nu-1} S_c^d(\sigma t^\alpha) R_{\varsigma,\tau}^{\varpi,\varepsilon}(\delta t^\mu)](x)$$

$$= x^{\xi + \acute{\xi} - \lambda + \nu - 1} \frac{\Gamma(\varepsilon)}{\Gamma(\varpi)} \sum_{k=0}^{[\frac{d}{c}]} \frac{(-c)_{dk}}{k!} G_{c,k}(\sigma x^\alpha)^k$$

$$\times {_4\psi_5}\left[\begin{array}{c} (\varpi, 1), (\nu + \alpha k + \mu), (-\eta + \xi + \nu + \alpha k + \mu), \\ (\xi + \acute{\xi} + \acute{\eta} - \lambda \nu + \alpha k + \mu); \\ (\varepsilon, 1), (\tau, \varsigma), (-\eta \nu + \alpha k + \mu), (\xi + \acute{\xi} - \lambda + \nu + \alpha k + \mu), \\ (\xi + \acute{\eta} - \lambda + \nu + \alpha k + \mu); \end{array} \delta x^\mu \right]. \quad (44)$$

Proof. By using (1) and (2) in the left hand side of (44) and representing it with $\mathfrak{U}$. After some simplification, we obtain

$$\mathfrak{U} = \sum_{k=0}^{[\frac{d}{c}]} \frac{(-c)_{dk}}{k!} G_{c,k}(\sigma t^\alpha)^k \sum_{n=0}^{\infty} \frac{(\varpi)_n}{(\varepsilon)_n \Gamma(\varsigma n + \tau)} \frac{(\delta t^\mu)^n}{n!}$$
$$\times [D_{0^+}^{\xi,\acute{\xi},\eta,\acute{\eta},\lambda} t^{\nu + \alpha k + \mu n - 1}](x). \quad (45)$$

Now, applying the integral Formula (42) to (45), we obtain

$$\mathfrak{U} = \sum_{k=0}^{[\frac{d}{c}]} \frac{(-c)_{dk}}{k!} G_{c,k} \sigma^k \sum_{n=0}^{\infty} \frac{(\varpi)_n}{(\varepsilon)_n \Gamma(\varsigma n + \tau)} \frac{(\delta)^n}{n!}$$

$$\times \Gamma\left[\begin{array}{c} \nu + \alpha k + \mu n, -\eta + \xi + \nu + \alpha k + \mu n, \xi + \acute{\xi} + \acute{\eta} - \lambda \nu + \alpha k + \mu n \\ -\eta \nu + \alpha k + \mu n, \xi + \acute{\xi} - \lambda + \nu + \alpha k + \mu n, \xi + \acute{\eta} - \lambda + \nu + \alpha k + \mu n \end{array}\right]$$

$$\times x^{\xi + \acute{\xi} - \lambda + \alpha k + \mu n - 1}, \quad (46)$$

which further implies

$$\mathfrak{U} = x^{\xi + \acute{\xi} - \lambda + \nu - 1} \frac{\Gamma(\varepsilon)}{\Gamma(\varpi)} \sum_{k=0}^{[\frac{d}{c}]} \frac{(-c)_{dk}}{k!} G_{c,k}(\sigma t^\alpha)^k$$

$$\times \sum_{n=0}^{\infty} \Gamma\left[\begin{array}{c} \varpi + n, \nu + \alpha k + \mu n, -\eta + \xi + \nu + \alpha k + \mu n, \\ \xi + \acute{\xi} + \acute{\eta} - \lambda \nu + \alpha k + \mu n \\ \varepsilon + n, \tau + \varsigma n, -\eta \nu + \alpha k + \mu n, \xi + \acute{\xi} - \lambda + \nu + \alpha k + \mu n, \\ \xi + \acute{\eta} - \lambda + \nu + \alpha k + \mu n \end{array}\right] \frac{(\delta x^\mu)^n}{n!}. \quad (47)$$

Now, by using definition (7), we arrive at required formula (44). □

Theorem 4. *For $\xi, \acute{\xi}, \eta, \acute{\eta}, \lambda, \omega \in \mathbb{C}, x > 0$ such that $Re(\lambda) > 0, Re(\varsigma) > -1, Re(\xi) > 0$ and $Re(\nu) > max\{0, Re(-\acute{\eta}),$
$Re(\acute{\xi} + \eta - \lambda), Re(\xi + \acute{\xi} - \lambda) + [Re(\lambda)] + 1\}$, then the following formula holds true*

$$[D_{0-}^{\xi,\acute{\xi},\eta,\acute{\eta},\lambda} t^{-\nu} S_c^d(\sigma t^\alpha) R_{\varsigma,\tau}^{\omega,\varepsilon}(\delta t^{-\mu})](x)$$

$$= x^{\xi+\acute{\xi}-\lambda-\nu} \frac{\Gamma(\varepsilon)}{\Gamma(\omega)} \sum_{k=0}^{[\frac{d}{c}]} \frac{(-c)_{dk}}{k!} G_{c,k}(\sigma x^\alpha)^k$$

$$\times {}_4\psi_5 \left[\begin{array}{c} (\omega,1), (\acute{\eta}+\nu+-\alpha k+\mu), (-\xi-\acute{\xi}+\lambda+\nu-\alpha k+\mu), \\ (\acute{\xi}-\eta+\lambda+\nu-\alpha k+\mu); \\ (\varepsilon,1), (\tau,\varsigma), (\nu-\alpha k+\mu n), (-\acute{\xi}+\acute{\eta}+\nu-\alpha k+\mu), \\ (-\xi-\acute{\xi}-\eta+\lambda+\nu-\alpha k+\mu); \end{array} \delta x^{-\mu} \right]. \quad (48)$$

Proof. By using (1) and (2) in the left hand side of (48) and representing it by $\mathfrak{E}$. After some simplification, we obtain

$$\mathfrak{E} = \sum_{k=0}^{[\frac{d}{c}]} \frac{(-c)_{dk}}{k!} G_{c,k}(\sigma)^k \sum_{n=0}^{\infty} \frac{(\omega)_n}{(\varepsilon)_n \Gamma(\varsigma n + \tau)} \frac{(\delta)^n}{n!}$$

$$\times [DI_{0+}^{\xi,\acute{\xi},\eta,\acute{\eta},\lambda} t^{-\nu+\alpha k-\mu n}](x). \quad (49)$$

Now, applying the integral Formula (43) to (49), we obtain

$$\mathfrak{V} = \sum_{k=0}^{[\frac{d}{c}]} \frac{(-c)_{dk}}{k!} G_{c,k}\sigma^k \sum_{n=0}^{\infty} \frac{(\omega)_n}{(\varepsilon)_n \Gamma(\varsigma n + \tau)} \frac{(\delta)^n}{n!}$$

$$\times \Gamma \left[\begin{array}{c} \acute{\eta}+\nu+-\alpha k+\mu n, -\xi-\acute{\xi}+\lambda+\nu-\alpha k+\mu n, \\ \acute{\xi}-\eta+\lambda+\nu-\alpha k+\mu n \\ \nu-\alpha k+\mu n, -\acute{\xi}+\acute{\eta}+\nu-\alpha k+\mu n, \\ -\xi-\acute{\xi}-\eta+\lambda+\nu-\alpha k+\mu n \end{array} \right]$$

$$\times x^{\xi+\acute{\xi}-\lambda-\nu+\alpha k-\mu n}, \quad (50)$$

which further implies

$$\mathfrak{V} = x^{\xi+\acute{\xi}-\lambda-\nu} \frac{\Gamma(\varepsilon)}{\Gamma(\omega)} \sum_{k=0}^{[\frac{d}{c}]} \frac{(-c)_{dk}}{k!} G_{c,k}(\sigma x^\alpha)^k$$

$$\times \sum_{n=0}^{\infty} \Gamma \left[\begin{array}{c} \omega+n, \acute{\eta}+\nu+-\alpha k+\mu n, -\xi-\acute{\xi}+\lambda+\nu-\alpha k+\mu n, \\ \acute{\xi}-\eta+\lambda+\nu-\alpha k+\mu n \\ \varepsilon+n, \tau+\varsigma n, \nu-\alpha k+\mu n, -\acute{\xi}+\acute{\eta}+\nu-\alpha k+\mu n, \\ -\xi-\acute{\xi}-\eta+\lambda+\nu-\alpha k+\mu n \end{array} \right] \frac{(\delta x^{-\mu})^n}{n!}. \quad (51)$$

Now, by using definition (7), we arrive at required formula (48). □

In the following corollaries, we look at some special cases involving the extended Wright function.

Corollary 7. *Assume that condition of Theorem 3 is fulfilled then using generalized Wright-type Function (3), the Formula (44) reduces to the following form.*

$$[D_{0+}^{\check{\zeta},\check{\xi},\eta,\acute{\eta},\lambda} t^{\nu-1} S_c^d(\sigma t^\alpha) R_{\varsigma,\tau}^{\varpi,\varepsilon}(-\delta t^\mu)](x)$$

$$= x^{\check{\zeta}+\check{\xi}-\lambda+\nu-1} \frac{\Gamma(\varepsilon)}{\Gamma(\varpi)} \sum_{k=0}^{\left[\frac{d}{c}\right]} \frac{(-c)_{dk}}{k!} G_{c,k}(\sigma x^\alpha)^k$$

$$\times {}_4\psi_5 \left[\begin{array}{c} (\varpi,1), (\nu+\alpha k+\mu), (-\eta+\check{\xi}+\nu+\alpha k+\mu), \\ (\check{\zeta}+\check{\xi}+\acute{\eta}-\lambda\nu+\alpha k+\mu); \\ (\varepsilon,1), (1-\varsigma,-\varsigma), (-\eta\nu+\alpha k+\mu), (\check{\zeta}+\check{\xi}-\lambda+\nu+\alpha k+\mu), \\ (\check{\xi}+\acute{\eta}-\lambda+\nu+\alpha k+\mu); 11 \end{array} \middle| -\delta x^\mu \right]. \quad (52)$$

Corollary 8. *Assume that condition of Theorem 4 is fulfilled then using generalized Wright-type Function (3), the Formula (48) reduces to the following form.*

$$[D_{0-}^{\check{\zeta},\check{\xi},\eta,\acute{\eta},\lambda} t^{-\nu} S_c^d(\sigma t^\alpha) R_{\varsigma,\tau}^{\varpi,\varepsilon}(\delta t^{-\mu})](x)$$

$$= x^{\check{\zeta}+\check{\xi}-\lambda-\nu} \frac{\Gamma(\varepsilon)}{\Gamma(\varpi)} \sum_{k=0}^{\left[\frac{d}{c}\right]} \frac{(-c)_{dk}}{k!} G_{c,k}(\sigma x^\alpha)^k$$

$$\times {}_4\psi_5 \left[\begin{array}{c} (\varpi,1), (\acute{\eta}+\nu+-\alpha k+\mu), (-\check{\zeta}-\check{\xi}+\lambda+\nu-\alpha k+\mu), \\ (\check{\xi}-\eta+\lambda+\nu-\alpha k+\mu); \\ (\varepsilon,1), (1-\varsigma,-\varsigma), (\nu-\alpha k+\mu n), (-\check{\xi}+\acute{\eta}+\nu-\alpha k+\mu), \\ (-\check{\zeta}-\check{\xi}-\eta+\lambda+\nu-\alpha k+\mu); \end{array} \middle| -\delta x^{-\mu} \right]. \quad (53)$$

Corollary 9. *Assume that condition of Theorem 3 is fulfilled and for $\varsigma \in (0,1)$ then using generalized Wright-type Function (4), the Formula (44) reduces to the following form.*

$$[D_{0+}^{\check{\zeta},\check{\xi},\eta,\acute{\eta},\lambda} t^{\nu-1} S_c^d(\sigma t^\alpha) F_{-\varsigma}^{\varpi,\varepsilon}(-\delta t^\mu)](x)$$

$$= x^{\check{\zeta}+\check{\xi}-\lambda+\nu-1} \frac{\Gamma(\varepsilon)}{\Gamma(\varpi)} \sum_{k=0}^{\left[\frac{d}{c}\right]} \frac{(-c)_{dk}}{k!} G_{c,k}(\sigma x^\alpha)^k$$

$$\times {}_4\psi_5 \left[\begin{array}{c} (\varpi,1), (\nu+\alpha k+\mu), (-\eta+\check{\xi}+\nu+\alpha k+\mu), \\ (\check{\zeta}+\check{\xi}+\acute{\eta}-\lambda\nu+\alpha k+\mu); \\ (\varepsilon,1), (0,-\varsigma), (-\eta\nu+\alpha k+\mu), (\check{\zeta}+\check{\xi}-\lambda+\nu+\alpha k+\mu), \\ (\check{\xi}+\acute{\eta}-\lambda+\nu+\alpha k+\mu); 11 \end{array} \middle| -\delta x^\mu \right]. \quad (54)$$

Corollary 10. *Assume that condition of Theorem 4 is fulfilled then using generalized Wright-type Function (4), the Formula (48) reduces to the following form.*

$$[D_{0-}^{\check{\zeta},\check{\xi},\eta,\acute{\eta},\lambda} t^{-\nu} S_c^d(\sigma t^\alpha) F_{-\varsigma}^{\varpi,\varepsilon}(\delta t^{-\mu})](x)$$

$$= x^{\check{\zeta}+\check{\xi}-\lambda-\nu} \frac{\Gamma(\varepsilon)}{\Gamma(\varpi)} \sum_{k=0}^{\left[\frac{d}{c}\right]} \frac{(-c)_{dk}}{k!} G_{c,k}(\sigma x^\alpha)^k$$

$$\times {}_4\psi_5 \left[\begin{array}{c} (\varpi,1), (\acute{\eta}+\nu+-\alpha k+\mu), (-\check{\zeta}-\check{\xi}+\lambda+\nu-\alpha k+\mu), \\ (\check{\xi}-\eta+\lambda+\nu-\alpha k+\mu); \\ (\varepsilon,1), (0,-\varsigma), (\nu-\alpha k+\mu n), (-\check{\xi}+\acute{\eta}+\nu-\alpha k+\mu), \\ (-\check{\zeta}-\check{\xi}-\eta+\lambda+\nu-\alpha k+\mu); \end{array} \middle| -\delta x^{-\mu} \right]. \quad (55)$$

Corollary 11. Assume that condition of Theorem 3 is fulfilled then using normalized Wright Function (5), the Formula (44) reduces to the following form.

$$\left[D_{0+}^{\zeta,\check{\zeta},\eta,\check{\eta},\lambda} t^{\nu-1} S_c^d(\sigma t^\alpha) R_{\varsigma,\tau}(\delta t^\mu)\right](x)$$

$$= x^{\zeta+\check{\zeta}-\lambda+\nu-1} \sum_{k=0}^{[\frac{d}{c}]} \frac{(-c)_{dk}}{k!} G_{c,k}(\sigma x^\alpha)^k$$

$$\times {}_4\psi_5 \left[\begin{array}{c} (\nu+\alpha k+\mu),(-\eta+\zeta+\nu+\alpha k+\mu), \\ (\zeta+\check{\zeta}+\check{\eta}-\lambda\nu+\alpha k+\mu); \\ (\tau,\varsigma),(-\eta\nu+\alpha k+\mu),(\zeta+\check{\zeta}-\lambda+\nu+\alpha k+\mu), \\ (\zeta+\check{\eta}-\lambda+\nu+\alpha k+\mu); \end{array} \delta x^\mu \right]. \tag{56}$$

Corollary 12. Assume that condition of Theorem 2 is fulfilled then using normalized Wright Function (5), the Formula (32) reduces to the following form.

$$\left[D_{0-}^{\zeta,\check{\zeta},\eta,\check{\eta},\lambda} t^{-\nu} S_c^d(\sigma t^\alpha) R_{\varsigma,\tau}(\delta t^{-\mu})\right](x)$$

$$= x^{\zeta+\check{\zeta}-\lambda-\nu} \sum_{k=0}^{[\frac{d}{c}]} \frac{(-c)_{dk}}{k!} G_{c,k}(\sigma x^\alpha)^k$$

$$\times {}_3\psi_4 \left[\begin{array}{c} (\check{\eta}+\nu+-\alpha k+\mu),(-\zeta-\check{\zeta}+\lambda+\nu-\alpha k+\mu), \\ (\check{\zeta}-\eta+\lambda+\nu-\alpha k+\mu); \\ (\tau,\varsigma),(\nu-\alpha k+\mu n),(-\check{\zeta}+\check{\eta}+\nu-\alpha k+\mu), \\ (-\zeta-\check{\zeta}-\eta+\lambda+\nu-\alpha k+\mu); \end{array} \delta x^{-\mu} \right]. \tag{57}$$

Remark 2. Following are the special cases of results discussed above.
(i) The formulas obtained reduce to formulas for Saigo's fractional differential operators (18) and (19) for $\check{\zeta} = 0$;
(ii) By substituting $\eta = -\zeta$ in Saigo's fractional differential operators, we obtain the formulas for Riemann–Liouville fractional differential operators (22) and (23).

4. Lavoie–Trottier Integral Formulas Involving Product of Srivastava's Polynomials and Extended Wright Function

In this section, we develop two extended integral formulas, involving the product of Srivastava polynomial (1) and extended Wright function (2).

Theorem 5. For $\zeta, \check{\zeta}, \eta, \check{\eta}, \lambda, \varpi \in \mathbb{C}$, $Re(\epsilon) > 0$, $Re(\omega) > 0$ and $y > 0$, the following formula holds true

$$\int_0^1 y^{\epsilon-1}(1-y)^{2\omega-1}\left(1-\frac{y}{3}\right)^{2\epsilon-1}\left(1-\frac{y}{4}\right)^{\omega-1} S_c^d\left(v\left(1-\frac{y}{4}\right)(1-y)^2\right)$$

$$\times R_{\varsigma,\tau}^{\varpi,\epsilon}\left(v\left(1-\frac{y}{4}\right)(1-y)^2\right) dy = \left(\frac{2}{3}\right)^{2\epsilon} \frac{\Gamma(\epsilon)\Gamma(\epsilon)}{\Gamma(\varpi)} \sum_{k=0}^{[\frac{d}{c}]} \frac{(-c)_{dk}}{k!} G_{c,k} y^k$$

$$\times {}_2\psi_3 \left[\begin{array}{c} (\varpi,1),(\omega+k,1); \\ (\epsilon,1),(\tau,\varsigma),(\epsilon+\omega+k,1); \end{array} v \right]. \tag{58}$$

Proof. By using (1) and (2) in left hand side of (58) and denoting it by $\mathfrak{H}$. After some simplification, we obtain

$$\mathfrak{H} = \sum_{k=0}^{[\frac{d}{c}]} \frac{(-c)_{dk}}{k!} G_{c,k} v^k \sum_{n=0}^{\infty} \frac{(\varpi)_n}{(\varepsilon)_n \Gamma(\varsigma n + \tau)} \frac{v^n}{n!}$$

$$\times \int_0^1 y^{\epsilon-1}(1-y)^{2(\omega+k+n)-1}\left(1-\frac{y}{3}\right)^{2\epsilon-1}\left(1-\frac{y}{4}\right)^{\omega+k+n} dy. \quad (59)$$

Now, applying the integral Formula (24) to (59), we obtain

$$\mathfrak{H} = \sum_{k=0}^{[\frac{d}{c}]} \frac{(-c)_{dk}}{k!} G_{c,k} v^k \sum_{n=0}^{\infty} \frac{(\varpi)_n}{(\varepsilon)_n \Gamma(\varsigma n + \tau)} \frac{v^n}{n!}$$

$$\times \left(\frac{2}{3}\right)^{2\epsilon} \frac{\Gamma(\epsilon)\Gamma(\omega+k+n)}{\Gamma(\epsilon+\omega+k+n)}, \quad (60)$$

which further implies

$$\mathfrak{H} = \left(\frac{2}{3}\right)^{2\epsilon} \frac{\Gamma(\epsilon)\Gamma(\epsilon)}{\Gamma(\varpi)} \sum_{k=0}^{[\frac{d}{c}]} \frac{(-c)_{dk}}{k!} G_{c,k} y^k$$

$$\times \sum_{n=0}^{\infty} \frac{\Gamma(\varpi+n)\Gamma(\omega+k+n)}{\Gamma(\epsilon+n)\Gamma(\tau+\varsigma n)\Gamma(\epsilon+\omega+k+n)} \frac{v^n}{n!}. \quad (61)$$

Now, by using definition (7), we arrive at the proof of the theorem. □

Theorem 6. *For $\xi, \acute{\xi}, \eta, \acute{\eta}, \lambda, \varpi \in \mathbb{C}$, $Re(\epsilon) > 0$, $Re(\omega) > 0$ and $y > 0$, the following formula holds true*

$$\int_0^1 y^{\epsilon-1}(1-y)^{2\omega-1}\left(1-\frac{y}{3}\right)^{2\epsilon-1}\left(1-\frac{y}{4}\right)^{\omega-1} S_c^d\left(vy\left(1-\frac{y}{4}\right)(1-y)^2\right)$$

$$\times R_{\varsigma,\tau}^{\varpi,\epsilon}\left(vy\left(1-\frac{y}{4}\right)(1-y)^2\right)dy = \left(\frac{2}{3}\right)^{2\epsilon} \frac{\Gamma(\epsilon)\Gamma(\epsilon)}{\Gamma(\varpi)} \sum_{k=0}^{[\frac{d}{c}]} \frac{(-c)_{dk}}{k!} G_{c,k}\left(\frac{4v}{9}\right)^k$$

$$\times {}_2\psi_3\left[\begin{array}{c}(\varpi,1),(\epsilon+k,1);\\ (\varepsilon,1),(\tau,\varsigma),(\epsilon+\omega+k,1);\end{array}\frac{4v}{9}\right]. \quad (62)$$

Proof. By using (1) and (2) in the left hand side of (62) and representing it by $\mathfrak{J}$. After some simplification, we obtain

$$\mathfrak{J} = \sum_{k=0}^{[\frac{d}{c}]} \frac{(-c)_{dk}}{k!} G_{c,k} v^k \sum_{n=0}^{\infty} \frac{(\varpi)_n}{(\varepsilon)_n \Gamma(\varsigma n + \tau)} \frac{v^n}{n!}$$

$$\times \int_0^1 y^{\epsilon+k+n-1}(1-y)^{2\omega-1}\left(1-\frac{y}{3}\right)^{2(\epsilon+k+n)-1}\left(1-\frac{y}{4}\right)^{\omega} dy. \quad (63)$$

Now, applying the integral Formula (24) to (63), we obtain

$$\mathfrak{I} = \sum_{k=0}^{[\frac{d}{c}]} \frac{(-c)_{dk}}{k!} G_{c,k} v^k \sum_{n=0}^{\infty} \frac{(\varpi)_n}{(\varepsilon)_n \Gamma(\varsigma n + \tau)} \frac{v^n}{n!}$$

$$\times \left(\frac{2}{3}\right)^{2(\varepsilon+k+n)} \frac{\Gamma(\omega)\Gamma(\varepsilon+k+n)}{\Gamma(\varepsilon+\omega+k+n)} \quad (64)$$

which further implies

$$\mathfrak{I} = \left(\frac{2}{3}\right)^{2\varepsilon} \frac{\Gamma(\omega)\Gamma(\varepsilon)}{\Gamma(\varpi)} \sum_{k=0}^{[\frac{d}{c}]} \frac{(-c)_{dk}}{k!} G_{c,k}^k \left(\frac{4v}{9}\right)^k$$

$$\times \sum_{n=0}^{\infty} \frac{\Gamma(\varpi+n)\Gamma(\omega+k+n)}{\Gamma(\varepsilon+n)\Gamma(\tau+\varsigma n)\Gamma(\varepsilon+\omega+k+n)} \frac{\left(\frac{4v}{9}\right)^n}{n!}. \quad (65)$$

Now, by using definition (7), we arrive at the proof of the theorem. □

In next corollaries, we discuss some special cases of extended Wright function.

Corollary 13. *Assume that condition of Theorem 5 is fulfilled then using generalized Wright-type Function (3), the Formula (58) reduces to the following form.*

$$\int_0^1 y^{\varepsilon-1}(1-y)^{2\omega-1}\left(1-\frac{y}{3}\right)^{2\varepsilon-1}\left(1-\frac{y}{4}\right)^{\omega-1} S_c^d\left(v\left(1-\frac{y}{4}\right)(1-y)^2\right)$$

$$\times M_{-\varsigma,1-\varsigma}^{\varpi,\varepsilon}\left(-v\left(1-\frac{y}{4}\right)(1-y)^2\right) dy = \left(\frac{2}{3}\right)^{2\varepsilon} \frac{\Gamma(\varepsilon)\Gamma(\varepsilon)}{\Gamma(\varpi)} \sum_{k=0}^{[\frac{d}{c}]} \frac{(-c)_{dk}}{k!} G_{c,k} y^k$$

$$\times {}_2\psi_3\left[\begin{array}{c}(\varpi,1),(\omega+k,1);\\(\varepsilon,1),(1-\varsigma,-\varsigma),(\varepsilon+\omega+k,1);\end{array} -v\right]. \quad (66)$$

Corollary 14. *Assume that condition of Theorem 6 is fulfilled then using generalized Wright-type Function (3), the Formula (62) reduces to the following form.*

$$\int_0^1 y^{\varepsilon-1}(1-y)^{2\omega-1}\left(1-\frac{y}{3}\right)^{2\varepsilon-1}\left(1-\frac{y}{4}\right)^{\omega-1} S_c^d\left(vy\left(1-\frac{y}{4}\right)(1-y)^2\right)$$

$$\times M_{-\varsigma,1-\varsigma}^{\varpi,\varepsilon}\left(-vy\left(1-\frac{y}{4}\right)(1-y)^2\right) dy = \left(\frac{2}{3}\right)^{2\varepsilon} \frac{\Gamma(\varepsilon)\Gamma(\varepsilon)}{\Gamma(\varpi)} \sum_{k=0}^{[\frac{d}{c}]} \frac{(-c)_{dk}}{k!} G_{c,k}\left(\frac{4v}{9}\right)^k$$

$$\times {}_2\psi_3\left[\begin{array}{c}(\varpi,1),(\varepsilon+k,1);\\(\varepsilon,1),(1-\varsigma,-\varsigma),(\varepsilon+\omega+k,1);\end{array} \frac{-4v}{9}\right]. \quad (67)$$

Corollary 15. *Assume that condition of Theorem 5 is fulfilled then using generalized Wright-type Function (4), the Formula (58) reduces to the following form.*

$$\int_0^1 y^{\epsilon-1}(1-y)^{2\omega-1}\left(1-\frac{y}{3}\right)^{2\epsilon-1}\left(1-\frac{y}{4}\right)^{\omega-1} S_c^d\left(v\left(1-\frac{y}{4}\right)(1-y)^2\right)$$
$$\times F_{-\varsigma}^{\varpi,\epsilon}\left(-v\left(1-\frac{y}{4}\right)(1-y)^2\right) dy = \left(\frac{2}{3}\right)^{2\epsilon} \frac{\Gamma(\epsilon)\Gamma(\epsilon)}{\Gamma(\varpi)} \sum_{k=0}^{[\frac{d}{c}]} \frac{(-c)_{dk}}{k!} G_{c,k} y^k$$
$$\times {}_2\psi_3\left[\begin{array}{c}(\varpi,1),(\omega+k,1);\\ (\epsilon,1),(0,-\varsigma),(\epsilon+\omega+k,1);\end{array} -v\right]. \quad (68)$$

Corollary 16. *Assume that condition of Theorem 6 is fulfilled then using generalized Wright-type Function (4), the Formula (62) reduces to the following form.*

$$\int_0^1 y^{\epsilon-1}(1-y)^{2\omega-1}\left(1-\frac{y}{3}\right)^{2\epsilon-1}\left(1-\frac{y}{4}\right)^{\omega-1} S_c^d\left(vy\left(1-\frac{y}{4}\right)(1-y)^2\right)$$
$$\times F_{-\varsigma}^{\varpi,\epsilon}\left(-vy\left(1-\frac{y}{4}\right)(1-y)^2\right) dy = \left(\frac{2}{3}\right)^{2\epsilon} \frac{\Gamma(\epsilon)\Gamma(\epsilon)}{\Gamma(\varpi)} \sum_{k=0}^{[\frac{d}{c}]} \frac{(-c)_{dk}}{k!} G_{c,k}\left(\frac{4v}{9}\right)^k$$
$$\times {}_2\psi_3\left[\begin{array}{c}(\varpi,1),(\epsilon+k,1);\\ (\epsilon,1),(0,-\varsigma),(\epsilon+\omega+k,1);\end{array} \frac{-4v}{9}\right]. \quad (69)$$

Corollary 17. *Assume that condition of Theorem 5 is fulfilled then using normalized Wright Function (5), the Formula (58) reduces to the following form.*

$$\int_0^1 y^{\epsilon-1}(1-y)^{2\omega-1}\left(1-\frac{y}{3}\right)^{2\epsilon-1}\left(1-\frac{y}{4}\right)^{\omega-1} S_c^d\left(v\left(1-\frac{y}{4}\right)(1-y)^2\right)$$
$$\times R_{\varsigma,\tau}\left(v\left(1-\frac{y}{4}\right)(1-y)^2\right) dy = \left(\frac{2}{3}\right)^{2\epsilon} \Gamma(\epsilon) \sum_{k=0}^{[\frac{d}{c}]} \frac{(-c)_{dk}}{k!} G_{c,k} y^k$$
$$\times {}_1\psi_2\left[\begin{array}{c}(\omega+k,1);\\ (\tau,\varsigma),(\epsilon+\omega+k,1);\end{array} v\right]. \quad (70)$$

Corollary 18. *Assume that condition of Theorem 6 is fulfilled then using normalized Wright Function (5), the Formula (62) reduces to the following form.*

$$\int_0^1 y^{\epsilon-1}(1-y)^{2\omega-1}\left(1-\frac{y}{3}\right)^{2\epsilon-1}\left(1-\frac{y}{4}\right)^{\omega-1} S_c^d\left(vy\left(1-\frac{y}{4}\right)(1-y)^2\right)$$
$$\times R_{\varsigma,\tau}\left(vy\left(1-\frac{y}{4}\right)(1-y)^2\right) dy = \left(\frac{2}{3}\right)^{2\epsilon} \Gamma(\epsilon) \sum_{k=0}^{[\frac{d}{c}]} \frac{(-c)_{dk}}{k!} G_{c,k}\left(\frac{4v}{9}\right)^k$$
$$\times {}_1\psi_2\left[\begin{array}{c}(\epsilon+k,1);\\ (\tau,\varsigma),(\epsilon+\omega+k,1);\end{array} \frac{4v}{9}\right]. \quad (71)$$

5. Oberhettinger Integral Formulas Involving Product of Srivastava's Polynomials and Extended Wright Function

The Oberhettinger's integral formulas involving the product of Srivastava's Polynomials (1) and extended Wright function (2) are established in this section.

$$\int_0^\infty y^{\phi-1}(y+b+\sqrt{y^2+2by})^{-\chi} dy = 2\chi b^{-\chi}\left(\frac{b}{2}\right)^\phi \frac{\Gamma(2\phi)\Gamma(\chi-\phi)}{\Gamma(1+\chi+\phi)}, \quad (72)$$

where $0 < Re(\phi) < Re(\chi)$.

Theorem 7. *For* $y > 0, \xi, \acute{\xi}, \eta, \acute{\eta}, \lambda, \omega \in \mathbb{C}, 0 < Re(\phi) < Re(\chi)$ *the following formula holds true*

$$\int_0^\infty y^{\phi-1}(y+b+\sqrt{y^2+2by})^{-\chi} S_c^d\left(\frac{v}{(y+b+\sqrt{y^2+2by})}\right)$$

$$\times R_{\varsigma,\tau}^{\omega,\varepsilon}\left(\frac{v}{(y+b+\sqrt{y^2+2by})}\right) dy$$

$$= b^{\phi-\chi} 2^{1-\phi} \frac{\Gamma(2\phi)\Gamma(\varepsilon)}{\Gamma(\omega)} \sum_{k=0}^{[\frac{d}{c}]} \frac{(-c)_{dk}}{k!} G_{c,k} \left(\frac{v}{b}\right)^k$$

$$\times {}_3\psi_4\left[\begin{array}{c}(\omega,1),(\chi+k+1,1),(\chi+\phi+k,1);\\(\varepsilon,1),(\tau,\varsigma),(\chi+k,1),(1+\chi+\phi+k,1);\end{array}\left|\frac{v}{b}\right.\right]. \quad (73)$$

Proof. By using (1) and (2) in the left hand side of (73) and representing it by $\mathfrak{U}$. After some simplification, we obtain

$$\mathfrak{U} = \sum_{k=0}^{[\frac{d}{c}]} \frac{(-c)_{dk}}{k!} G_{c,k} v^k \sum_{n=0}^\infty \frac{(\omega)_n}{(\varepsilon)_n \Gamma(\varsigma n+\tau)} \frac{v^n}{n!}$$

$$\times \int_0^\infty y^{\phi-1}(y+b+\sqrt{y^2+2by})^{-(\chi+k+n)} dy. \quad (74)$$

Now, applying the integral Formula (27) to (74), we obtain

$$\mathfrak{U} = \sum_{k=0}^{[\frac{d}{c}]} \frac{(-c)_{dk}}{k!} G_{c,k} v^k \sum_{n=0}^\infty \frac{(\omega)_n}{(\varepsilon)_n \Gamma(\varsigma n+\tau)} \frac{v^n}{n!}$$

$$\times 2(\chi+k+n) b^{-(\chi+k+n)} \left(\frac{b}{2}\right)^\phi \frac{\Gamma(2\phi)\Gamma(\chi-\phi+k+n)}{\Gamma(1+\chi+\phi+k+n)}, \quad (75)$$

which implies

$$\mathfrak{U} = b^{\phi-\chi} 2^{1-\phi} \frac{\Gamma(2\phi)\Gamma(\varepsilon)}{\Gamma(\omega)} \sum_{k=0}^{[\frac{d}{c}]} \frac{(-c)_{dk}}{k!} G_{c,k} (vb)^k$$

$$\times \frac{\Gamma(\omega+n)\Gamma(\chi+k+1+n)\Gamma(\chi-\phi+k+n)}{\Gamma(\varepsilon+n)\Gamma(\tau+\varsigma n)\Gamma(1+\chi+\phi+k+n)} \frac{\left(\frac{v}{b}\right)^n}{n!}. \quad (76)$$

In accordance with definition (7), we arrive at the formula (73). □

Theorem 8. *For* $y > 0, \xi, \acute{\xi}, \eta, \acute{\eta}, \lambda, \omega \in \mathbb{C}\ 0 < Re(\phi) < Re(\chi)$ *the following formula holds true*

$$\int_0^\infty y^{\phi-1}(y+b+\sqrt{y^2+2by})^{-\chi} S_c^d\left(\frac{vy}{(y+b+\sqrt{y^2+2by})}\right)$$

$$\times R_{\varsigma,\tau}^{\omega,\varepsilon}\left(\frac{vy}{(y+b+\sqrt{y^2+2by})}\right) dy$$

$$= b^{\phi-\chi} 2^{1-\phi} \frac{\Gamma(\chi-\phi)\Gamma(\varepsilon)}{\Gamma(\omega)} \sum_{k=0}^{[\frac{d}{c}]} \frac{(-c)_{dk}}{k!} G_{c,k} \left(\frac{v}{2}\right)^k$$

$$\times {}_3\psi_4\left[\begin{array}{c}(\omega,1),(\chi+k+1,1),(2\phi+2k,2);\\(\varepsilon,1),(\tau,\varsigma),(\chi+k,1),(1+\chi+\phi+2k,2);\end{array}\left|\frac{v}{2}\right.\right]. \quad (77)$$

Proof. By using (1) and (2) in the left hand side of (77) and representing it $\mathfrak{E}$. On some simplification, we obtain

$$\mathfrak{E} = \sum_{k=0}^{[\frac{d}{c}]} \frac{(-c)_{dk}}{k!} G_{c,k} v^k \sum_{n=0}^{\infty} \frac{(\omega)_n}{(\varepsilon)_n \Gamma(\varsigma n + \tau)} \frac{v^n}{n!}$$

$$\times \int_0^\infty y^{\phi+k+n-1}(y+b+\sqrt{y^2+2by})^{-(\chi+k+n)} dy. \tag{78}$$

Now, applying the integral Formula (27) to (78), we obtain

$$\mathfrak{E} = \sum_{k=0}^{[\frac{d}{c}]} \frac{(-c)_{dk}}{k!} G_{c,k} v^k \sum_{n=0}^{\infty} \frac{(\omega)_n}{(\varepsilon)_n \Gamma(\varsigma n + \tau)} \frac{v^n}{n!}$$

$$\times 2(\chi+k+n) b^{-(\chi+k+n)} \left(\frac{b}{2}\right)^{\phi+k+n} \frac{\Gamma(2\phi+2k+2n)\Gamma(\chi-\phi)}{\Gamma(1+\chi+\phi+2k+2n)}, \tag{79}$$

which implies

$$\mathfrak{E} = b^{\phi-\chi} 2^{1-\phi} \frac{\Gamma(\chi-\phi)\Gamma(\varepsilon)}{\Gamma(\omega)} \sum_{k=0}^{[\frac{d}{c}]} \frac{(-c)_{dk}}{k!} G_{c,k}\left(\frac{v}{2}\right)^k$$

$$\times \frac{\Gamma(\omega+n)\Gamma(\chi+k+1+n)\Gamma(2\phi+2k+2n)}{\Gamma(\varepsilon+n)\Gamma(\tau+\varsigma n)\Gamma(chi+k+n)\Gamma(1+\chi+\phi+2kk+2n)} \frac{\left(\frac{v}{2}\right)^n}{n!}. \tag{80}$$

In accordance with definition (7), we arrive at the formula (77). □

In the following corollaries, we present some special cases involving the extended Wright function.

Corollary 19. *Assume that condition of Theorem 7 is fulfilled then using generalized Wright-type Function (3), the Formula (73) reduces to the following form.*

$$\int_0^\infty y^{\phi-1}(y+b+\sqrt{y^2+2by})^{-\chi} S_c^d\left(\frac{v}{(y+b+\sqrt{y^2+2by})}\right)$$

$$\times M_{-\varsigma,1-\varsigma}^{\omega,\varepsilon}\left(\frac{-v}{(y+b+\sqrt{y^2+2by})}\right) dy$$

$$= b^{\phi-\chi} 2^{1-\phi} \frac{\Gamma(2\phi)\Gamma(\varepsilon)}{\Gamma(\omega)} \sum_{k=0}^{[\frac{d}{c}]} \frac{(-c)_{dk}}{k!} G_{c,k}\left(\frac{v}{b}\right)^k$$

$$\times {}_3\psi_4\left[\begin{array}{c}(\omega,1),(\chi+k+1,1),(\chi+\phi+k,1); \\ (\varepsilon,1),(1-\varsigma,-\varsigma),(\chi+k,1),(1+\chi+\phi+k,1);\end{array} \frac{-v}{b}\right]. \tag{81}$$

Corollary 20. *Assume that condition of Theorem 8 is fulfilled then using generalized Wright-type Function (3), the Formula (77) reduces to the following form.*

$$\int_0^\infty y^{\phi-1}(y+b+\sqrt{y^2+2by})^{-\chi} S_c^d\left(\frac{vy}{(y+b+\sqrt{y^2+2by})}\right)$$
$$\times M_{-\varsigma,1-\varsigma}^{\omega,\varepsilon}\left(\frac{-vy}{(y+b+\sqrt{y^2+2by})}\right)dy$$
$$= b^{\phi-\chi} 2^{1-\phi} \frac{\Gamma(\chi-\phi)\Gamma(\varepsilon)}{\Gamma(\omega)} \sum_{k=0}^{[\frac{d}{c}]} \frac{(-c)_{dk}}{k!} G_{c,k}\left(\frac{v}{2}\right)^k$$
$$\times {}_3\psi_4\left[\begin{array}{c}(\omega,1),(\chi+k+1,1),(2\phi+2k,2);\\ (\varepsilon,1),(1-\varsigma,-\varsigma),(\chi+k,1),(1+\chi+\phi+2k,2);\end{array}\frac{-v}{2}\right]. \quad (82)$$

Corollary 21. *Assume that condition of Theorem 7 is fulfilled then using generalized Wright-type Function (4), the Formula (73) reduces to the following form.*

$$\int_0^\infty y^{\phi-1}(y+b+\sqrt{y^2+2by})^{-\chi} S_c^d\left(\frac{v}{(y+b+\sqrt{y^2+2by})}\right)$$
$$\times F_{-\varsigma}^{\omega,\varepsilon}\left(\frac{-v}{(y+b+\sqrt{y^2+2by})}\right)dy$$
$$= b^{\phi-\chi} 2^{1-\phi} \frac{\Gamma(2\phi)\Gamma(\varepsilon)}{\Gamma(\omega)} \sum_{k=0}^{[\frac{d}{c}]} \frac{(-c)_{dk}}{k!} G_{c,k}\left(\frac{v}{b}\right)^k$$
$$\times {}_3\psi_4\left[\begin{array}{c}(\omega,1),(\chi+k+1,1),(\chi+\phi+k,1);\\ (\varepsilon,1),(0,-\varsigma),(\chi+k,1),(1+\chi+\phi+k,1);\end{array}\frac{-v}{b}\right]. \quad (83)$$

Corollary 22. *Assume that condition of Theorem 8 is fulfilled then using generalized Wright-type Function (4), the Formula (77) reduces to the following form.*

$$\int_0^\infty y^{\phi-1}(y+b+\sqrt{y^2+2by})^{-\chi} S_c^d\left(\frac{vy}{(y+b+\sqrt{y^2+2by})}\right)$$
$$\times F_{-\varsigma}^{\omega,\varepsilon}\left(\frac{-vy}{(y+b+\sqrt{y^2+2by})}\right)dy$$
$$= b^{\phi-\chi} 2^{1-\phi} \frac{\Gamma(\chi-\phi)\Gamma(\varepsilon)}{\Gamma(\omega)} \sum_{k=0}^{[\frac{d}{c}]} \frac{(-c)_{dk}}{k!} G_{c,k}\left(\frac{v}{2}\right)^k$$
$$\times {}_3\psi_4\left[\begin{array}{c}(\omega,1),(\chi+k+1,1),(2\phi+2k,2);\\ (\varepsilon,1),(0,-\varsigma),(\chi+k,1),(1+\chi+\phi+2k,2);\end{array}\frac{-v}{2}\right]. \quad (84)$$

Corollary 23. *Assume that condition of Theorem 7 is fulfilled then using generalized Wright-type Function (5), the Formula (73) reduces to the following form.*

$$\int_0^\infty y^{\phi-1}(y+b+\sqrt{y^2+2by})^{-\chi} S_c^d\left(\frac{v}{(y+b+\sqrt{y^2+2by})}\right)$$
$$\times R_{\varsigma,\tau}\left(\frac{-v}{(y+b+\sqrt{y^2+2by})}\right)dy$$
$$= b^{\phi-\chi} 2^{1-\phi} \Gamma(2\phi) \sum_{k=0}^{[\frac{d}{c}]} \frac{(-c)_{dk}}{k!} G_{c,k}\left(\frac{v}{b}\right)^k$$
$$\times {}_2\psi_3\left[\begin{array}{c}(\chi+k+1,1),(\chi+\phi+k,1);\\ (\tau,\varsigma),(\chi+k,1),(1+\chi+\phi+k,1);\end{array}\frac{-v}{b}\right]. \quad (85)$$

Corollary 24. *Assume that condition of Theorem 8 is fulfilled then using normalized Wright Function (5), the Formula (77) reduces to the following form.*

$$\int_0^\infty y^{\phi-1}(y+b+\sqrt{y^2+2by})^{-\chi} S_c^d\left(\frac{vy}{(y+b+\sqrt{y^2+2by})}\right)$$
$$\times R_{\varsigma,\tau}\left(\frac{vy}{(y+b+\sqrt{y^2+2by})}\right)dy$$
$$= b^{\phi-\chi} 2^{1-\phi}\Gamma(\chi-\phi) \sum_{k=0}^{[\frac{d}{c}]} \frac{(-c)_{dk}}{k!} G_{c,k} \left(\frac{v}{2}\right)^k$$
$$\times {}_2\psi_3\left[\begin{array}{c}(\chi+k+1,1),(2\phi+2k,2);\\(\tau,\varsigma),(\chi+k,1),(1+\chi+\phi+2k,2);\end{array}\frac{v}{2}\right]. \tag{86}$$

6. Conclusions

The present study is based on a well-known technique to explore certain general formulae involving special functions by skilfully utilizing the different integral and differential operators. In this article, we established new integral and differential formulas involving the product of Srivastava's polynomial and extended Wright function. The main consequences are presented in terms of Fox–Wright hypergeometric function. Unified integral representations of some of its special cases are also derived. The extended Wright function is related to Mittag–Leffler function (9), Meijer G-function (10) and Fox H-function (11), therefore all obtained results can be expressed in the form of these functions as well. Moreover, the general class of polynomials gives many known classical orthogonal polynomials as special cases for given suitable values for the coefficient $G_{c,k}$. The Hermite, Laguerre, Jacobi, and Konhauser polynomials are only a few examples. In continuation of this study, one can obtain the integral representation of more generalized special functions that have applicability in physics and engineering sciences.

Author Contributions: Conceptualization, S.M., G.R. and K.S.N.; methodology, S.M. and G.R.; software, S.N., G.R., Z.A.K. and K.S.N.; validation, S.M., G.R. and K.S.N.; formal analysis, Z.A.K.; investigation, S.N., S.M., G.R. and Z.A.K.; resources, K.S.N.; writing—original draft preparation, S.N., S.M., G.R., Z.A.K. and K.S.N.; writing—review and editing, G.R. and K.S.N.; funding acquisition, K.S.N. All authors have read and agreed to the published version of the manuscript.

Funding: This research received no external funding.

Institutional Review Board Statement: Not applicable.

Informed Consent Statement: Not applicable.

Data Availability Statement: Not applicable.

Conflicts of Interest: The authors declares that there is no conflict of interest regarding the publication of this paper.

References

1. Ali, S. On an interesting integral involving Gauss's hypergeometric function. *Adv. Comput. Sci. Appl.* **2012**, *1*, 244–246.
2. Choi, J.; Agarwal, P. Certain unified integrals involving a product of Bessel functions of first kind. *Honam Math. J.* **2013**, *35*, 667–677. [CrossRef]
3. Kabra, S.; Nagar, H.; Nisar, K.S.; Suthar, D.L. The Marichev-Saigo-Maeda fractional calculus operators pertaining to the generalized k-Struve function. *Appl. Math. Nonlinear Sci.* **2020**, *2*, 593–602. [CrossRef]
4. Sahin, R.; Yagc, O. Fractional calculus of the extended hypergeometric function. *Appl. Math. Nonlinear Sci.* **2020**, *5*, 396–384.
5. Samraiz, M.; Perveen, Z.; Rahman, G.; Nisar, K.S.; Kumar, D. On (k,s)-Hilfer Prabhakar Fractional Derivative with Applications in Mathematical Physics. *Front. Phys.* **2020**, *8*, 1–9. [CrossRef]
6. Ata, E.; Kiymaz, I.O. A study on certain properties of generalized special functions defined by Fox-Wright function. *Appl. Math. Nonlinear Sci.* **2020**, *5*, 147–162. [CrossRef]

7. Suthar, D.L.; Amsalu, H. Certain integrals associated with the generalized Bessel-Maitland function. *Appl. Appl. Math.* **2017**, *12*, 1002–1016.
8. Choi, J.; Agarwal, P.; Mathur, S.; Purohit, S.D. Certain new integral formulas involving the generalized Bessel functions. *Bull. Korean Math. Soc.* **2014**, *51*, 995–1003. [CrossRef]
9. Rakha, M.A.; Rathie, A.K.; Chaudhary, M.P.; Ali, S. On A new class of integrals involving hypergeometric function. *J. Inequal. Appl. Spec. Funct.* **2012**, *3*, 10–27. [CrossRef]
10. Samraiz, M.; Perveen, Z.; Abdeljawad, T.; Iqbal, S.; Naheed, S. On Certain Fractional Calculus Operators and Their Applications. *Phys. Scr.* **2020**, *95*, 115210. [CrossRef]
11. Saxena, R.K.; Ram, J.; Suthar, D.L. Fractional calculus of generalized Mittag-Leffler functions. *J. Indian Acad. Math.* **2009**, *31*, 165–172.
12. Suthar, D.L.; Andualem, M.; Debalkie, B. A study on generalized multivariable Mittag-Leffler function via generalized fractional calculus operators. *J. Math.* **2019**, *2019*, 1–7. [CrossRef]
13. Galué, L.; Kalla, S.L.; Kim, T.V. Composition of Erdelyi-Kober fractional operators. *Integral Transform. Spec. Funct.* **2000**, *9*, 185–196. [CrossRef]
14. Kiryakova, V. A brief story about the operators of the generalized fractional calculus. *Fract. Calc. Appl. Anal.* **2008**, *11*, 203–220.
15. Agarwal, P. Pathway fractional integral formulas involving Bessel function of the first kind. *Adv. Stud. Contemp. Math.* **2015**, *25*, 221–231.
16. Agarwal1, P.; Jain, S.; Agarwal, S.; Nagpal, M. On a new class of integrals involving Bessel functions of the first kind. *Commun. Numer. Anal.* **2014**, *2014*, cna-0021. [CrossRef]
17. Nisar, K.S.; Rahman, G.; Ghaffar, A.; Mubeen, S. New class of integrals involving extended Mittag-Lefflerer function. *J. Fract. Calc. Appl.* **2018**, *9*, 222–231.
18. Nisar, K.S.; Mondal, S.R. Certain unified integral formulas involving the generalized modified k-Bessel function of firrst kind. *Commun. Korean Math. Soc.* **2017**, *32*, 47–53.
19. Suthar, D.L.; Andualem, M. Integral Formulas Involving Product of Srivastava's Polynomials and Galu'e type Struve Functions. *Kyungpook Math. J.* **2019**, *59*, 725–734.
20. Suthar, D.L. Composition Formulae for the k-Fractional Calculus Operators Associated with k-Wright Function. *J. Math.* **2020**, *2020*, 5471715. [CrossRef]
21. Jangid, K.; Parmar, R.K.; Agarwal, R.; Purohit, S.D. Fractional calculusand integral transforms of general class of polynomialand incomplete Fox-Wright functions. *Adv. Differ. Equ.* **2020**, *2020*, 606. [CrossRef]
22. Shah, S.A.H.; Mubeen, S. Expressions of the Laguerre polynomial and some other special functions in terms of the generalized Meijer G-functions. *AIMS Math.* **2021**, *6*, 11631–11641. [CrossRef]
23. Shah, S.A.H.; Mubeen, S.; Rahman, G.; Younis, J. Relation of Some Known Functions in terms of Generalized Meijer G-Functions. *J. Math.* **2021**, *2021*, 7032459. [CrossRef]
24. Menaria, N.; Nisar, K.S.; Purohit, S.D. On a new class of integrals involving product of generalized Bessel function of the first kind and general class of polynomials. *Acta Univ. Apulensis* **2016**, *46*, 97–105.
25. Suthar, D.L.; Reddy, G.V.; Tsegaye, T. Unified Integrals Formulas Involving Product of Srivastava's Polynomials and Generalized Bessel-Maitland Function. *Int. J. Sci. Res.* **2017**, *6*, 708–710; ISSN 2277-8179.
26. Srivastava, H.M. A contour integral involving Fox's H-function. *Indian J. Math.* **1972**, *14*, 1–6.
27. El-Shahed, M.; Salem, A. An Extension of Wright Function and Its Properties. *J. Math.* **2015**, *2015*, 1–11. [CrossRef]
28. Wright, E.M. The asymptotic expansion of the generalized hypergeometric function. *J. Lond. Math. Soc.* **1935**, *10*, 286–293. [CrossRef]
29. Erdelyi, A. *Higher Transcendental Functions*; McGraw-Hill: New York, NY, USA; Toronto, ON, Canada; London, UK, 1953.
30. Marichev, O.I. Volterra equation of Mellin convolution type with a Horn function in the kernel. *Izvestiya Akademii Nauk BSSR. Seriya Fiziko-Matematicheskikh Nauk* **1974**, *1*, 128–129.
31. Saigo, M.; Maeda, N. More generalization of fractional calculus. In *Transform Methods and Special Functions*; Rusev, P., Dimovski, I., Kiryakova, V., Eds.; IMI-BAS: Sofia, Bulgaria, 1998; pp. 386–400.
32. Saigo, M. A remark on integral operators involving the Gauss hypergeometric functions. *Kyushu Univ.* **1978**, *11*, 135–143.
33. Samko, S.G.; Kilbas, A.A.; Maricheve, O.I. Fractional integrals and derivatives. In *Theory and Application*; Gordon and Breach Science Publishers: Yverdon, Switzerland, 1993.
34. Lavoie, J.L.; Trottier, G. On the sum of certain Appell's series. *Ganita* **1969**, *20*, 43–66.
35. Oberhettinger, F. *Tables of Mellin Transform*; Springer: New York, NY, USA, 1974.
36. Kataria, K.K.; Vellaisamy, P. Some fractional calculus results associated with the I-function. *Mathematiche (Catania)* **2015**, *70*, 173–190.

 *fractal and fractional*

Article

The Mittag-Leffler Function for Re-Evaluating the Chlorine Transport Model: Comparative Analysis

Abdulrahman F. Aljohani [1], Abdelhalim Ebaid [1,*], Ebrahem A. Algehyne [1], Yussri M. Mahrous [2], Carlo Cattani [3] and Hind K. Al-Jeaid [4]

[1] Computational & Analytical Mathematics and Their Applications Research Group, Department of Mathematics, Faculty of Science, University of Tabuk, Tabuk 71491, Saudi Arabia; a.f.aljohani@ut.edu.sa (A.F.A.); e.algehyne@ut.edu.sa (E.A.A.)
[2] Department of Studies and Basic Sciences, Faculty of Community, University of Tabuk, Tabuk 71491, Saudi Arabia; y.mahrous@ut.edu.sa
[3] Engineering School (DEIM), University of Tuscia, 01100 Viterbo, Italy; cattani@unitus.it
[4] Department of Mathematical Sciences, Umm Al-Qura University, Makkah 21955, Saudi Arabia; hkjeaid@uqu.edu.sa
* Correspondence: aebaid@ut.edu.sa

Abstract: This paper re-investigates the mathematical transport model of chlorine used as a water treatment model, when a variable order partial derivative is incorporated for describing the chlorine transport system. This model was introduced in the literature and governed by a fractional partial differential equation (FPDE) with prescribed boundary conditions. The obtained solution in the literature was based on implementing the Laplace transform (LT) combined with the method of residues and expressed in terms of regular exponential functions. However, the present analysis avoids such a method of residues, and thus a new analytical solution is introduced in this paper via Mittag-Leffler functions. Therefore, an effective approach is developed in this paper to solve the chlorine transport model with non-integer order derivative. In addition, our results are compared with several studies in the literature in case of integer-order derivative and the differences in results are explained.

Keywords: fractional partial differential equation; Mittag-Leffler function; boundary value problem; separation of variables; Laplace transform

1. Introduction

Water sciences is a growing field of research. The quality of water can be enhanced through suitable values of injection and maintaining residual chlorine in a network not by reducing chlorine. In industrial sciences, chlorine decay is not much more than that in the use of water networks operation and water quality control. This procedure is widely used in most countries to ensure the disinfection capacity of distributed water [1,2]. Therefore, the study of chlorine decay is of great importance due to its wide applications in engineering and industrial sciences [3]. Biswas et al. [4] formulated the standard model of chlorine transport in pipes. In addition, the standard model [4] (with integer-order derivative) has been re-analyzed utilizing different approximate methods [5,6]. Later, the author [7] generalized the standard model [4] by means of fractional calculus (FC). The dimensionless generalized model is governed by FPDE [7]:

$$ {}_0^C D_x^\alpha u(x,r) = \frac{A_0}{r}\frac{\partial}{\partial r}\left(\frac{1}{r}\frac{\partial u}{\partial r}\right) - A_1 u, \quad \alpha \in (0,1], \tag{1}$$

where α is the order of the fractional derivative in Caputo sense ${}_0^C D_x^\alpha u(x,r) = \frac{\partial^\alpha u}{\partial x^\alpha}$ and $u = u(x,r)$ is the chlorine concentration [7]. The model is subjected to the following boundary conditions (BCs) [7]:

$$u(0,r) = 1, \qquad 0 \leq r \leq 1, \qquad (2)$$

$$\frac{\partial}{\partial r}u(x,0) = 0, \qquad 0 \leq x \leq 1, \qquad (3)$$

$$\frac{\partial}{\partial r}u(x,1) + A_2 u(x,1) = 0, \qquad 0 \leq x \leq 1. \qquad (4)$$

Details of the parameters were addressed by the authors [4–7]. As $\alpha \to 1$, i.e., for classical partial derivative with respect to x, Biswas et al. [4], Yeh et al. [5], and Mahrous [6] obtained three different approximate solutions for the system (1)–(4). For $\alpha \in (0,1]$, the exact solution of the current model has been recently obtained in Ref. [7] through implementing the LT combined with the method of residues to determine the inverse LT of some expressions. However, our analysis avoids such a method of residues, and hence the inverse LT can be directly calculated in terms of the Mittag-Leffler functions. Moreover, useful and recent studies on the chlorine decay models are listed in Refs. [8–15]. Therefore, an effective approach is to be developed in this paper to resolve the Equations (1)–(4). The suggested approach is mainly based on the separation of variables method (SOV) combined with the LT. The SOV technique is used to convert the PDE (1) to a couple of ODEs via auxiliary parameter.

The LT method was widely applied to solve various models in physics and engineering such as diffusions process [16], fluid flow suspended with carbon-nanotubes [17], singular boundary value problems with applications [18,19], and the magnetohydrodynamics (MHD) convection over a flat plate [20]. In addition, the LT was successfully implemented to treat the Ambartsumian's model of interstellar brightness [21] (with ordinary derivative) and also in view of FC in Ref. [22]. Moreover, Handibag and Karande [23] applied the Laplace substitution method for solving PDEs involving mixed partial derivatives, while in Ref. [24], the same authors extended their idea to solve linear and nonlinear PDEs of nth order. In addition, the LT has been implemented to deal with a set of differential equations [25]. Additionally, the double LT was used by Dhunde and Waghmare [26] to solve nonlinear PDEs while the volterra integro-differential equations has been analyzed utilizing the triple LT by Mousa and Elzaki [27]. Very recently, Zhang and Nadeem [28] solved a set of nonlinear time-fractional differential equations by means of the LT. Besides, the solution in terms of the Mittag-Leffler functions for a class of first-order fractional initial value problems, using the LT, was introduced very recently by Ebaid and Al-Jeaid [29], while the geometric properties of the Mittag-Leffler functions were addressed by Srivastava et al. [30].

Therefore, the objective of this paper is to obtain the exact solution of the system (1)–(4) via the LT in a different and easier way than that one followed by Mahrous [7]. It will be shown that the present exact solution is of different physical meaning when compared with the corresponding results in Ref. [7]. Besides, the results will be discussed and interpreted. Finally, several comparisons are to be performed, and the differences in results will be explained.

2. The SOV Method

Based on the SOV method, we assume that

$$u(x,r) = \xi(x)\psi(r). \qquad (5)$$

Substituting (5) into (1) yields

$$\frac{1}{A_0 \xi(x)} {}_0^C D_x^\alpha \xi(x) + \frac{A_1}{A_0} = \frac{1}{r\psi(r)} \frac{d}{dr}\left(r \frac{d\psi(r)}{dr}\right), \qquad (6)$$

and accordingly we can write

$$\frac{1}{A_0 \xi(x)} {}_0^C D_x^\alpha \xi(x) + \frac{A_1}{A_0} = \frac{1}{\psi(r)}\left(\frac{d^2\psi(r)}{dr^2} + \frac{1}{r}\frac{d\psi(r)}{dr}\right) = \mu, \qquad (7)$$

where μ is an auxiliary parameter. From (7), we have the following FODE for $\xi(x)$:

$$\frac{1}{A_0\xi(x)}{}_0^C D_x^\alpha \xi(x) + \frac{A_1}{A_0} = \mu, \tag{8}$$

and the ODE for $\psi(r)$:

$$\frac{1}{\psi(r)}\left(\frac{d^2\psi(r)}{dr^2} + \frac{1}{r}\frac{d\psi(r)}{dr}\right) = \mu, \tag{9}$$

Following Biswas et al. [4], the equality $\mu = -\lambda^2$ ($\lambda > 0$) is used, hence, Equation (8) converts to

$$_0^C D_x^\alpha \xi(x) + \left(A_1 + A_0\lambda^2\right)\xi(x) = 0. \tag{10}$$

Additionally, Equation (9) becomes

$$\frac{d^2\psi(r)}{dr^2} + \frac{1}{r}\frac{d\psi(r)}{dr} + \lambda^2 \psi(r) = 0. \tag{11}$$

From the BC (3) and Equation (5), we obtain

$$\frac{d\psi(0)}{dr} = 0. \tag{12}$$

Additionally, the BC (4) and Equation (5) lead to

$$\frac{d\psi(1)}{dr} + A_2 \psi(1) = 0. \tag{13}$$

The solutions of Equations (10) and (11) will be provided in the following sections.

2.1. Solution of $\xi(x)$

Applying the LT on Equation (10) gives

$$s^\alpha \overline{\xi}(s) - s^{\alpha-1}\xi(0) + \left(A_1 + A_0\lambda^2\right)\overline{\xi}(s) = 0, \tag{14}$$

where $\overline{\xi}(s)$ is the LT of $\xi(x)$. Solving Equation (14) for $\overline{\xi}(s)$, we obtain

$$\overline{\xi}(s) = \frac{\xi(0)s^{\alpha-1}}{s^\alpha + (A_1 + A_0\lambda^2)}. \tag{15}$$

Applying the inverse LT on Equation (15) yields

$$\xi(x) = \xi(0)\mathcal{L}^{-1}\left(\frac{s^{\alpha-1}}{s^\alpha + (A_1 + A_0\lambda^2)}\right), \tag{16}$$

and hence

$$\xi(x) = \xi(0)E_\alpha\left(-(A_1 + A_0\lambda^2)x^\alpha\right), \tag{17}$$

where $E_\alpha(\cdot)$ is the Mittage-Leffler function of one parameter, where the equality ([22–29]) $\mathcal{L}^{-1}\left(\frac{s^{\alpha-\gamma}}{s^\alpha+\omega^2}\right) = x^{\gamma-1}E_{\alpha,\gamma}(-\omega^2 x^\alpha)$, $Re(s) > |\omega^2|^{\frac{1}{\alpha}}$ is applied to obtain Equation (17) when $\gamma = 1$ and $\omega^2 = A_1 + A_0\lambda^2$.

2.2. Solution of $\psi(r)$

The solution of Equation (11) is

$$\psi(r) = \delta_1 J_0(\lambda r) + \delta_2 Y_0(r), \tag{18}$$

where δ_1 and δ_2 are unknown constants. Besides, $J_0(\cdot)$ and $Y_0(\cdot)$ are Bessel functions. The physics of the present model require that $u(x,r)$ must be bounded at $r=0$. This implies that $\psi(r)$ must also be bounded at $r=0$, which leads to $\delta_2 = 0$, where $Y_0(r) \to \infty$ as $r \to 0$. Therefore, Equation (18) becomes

$$\psi(r) = \delta_1 J_0(\lambda r). \tag{19}$$

Implementing the property $J_0'(\lambda r) = -\lambda J_1(\lambda r)$, we have

$$\frac{d\psi(r)}{dr} = -\delta_1 \lambda J_1(\lambda r), \tag{20}$$

and hence

$$\frac{d\psi(0)}{dr} = -\delta_1 \lambda J_1(0) = 0. \tag{21}$$

Accordingly, the BC (12) is automatically satisfied since $J_1(0) = 0$. Applying the BC (13) yields

$$\frac{d\psi(1)}{dr} + A_2 \psi(1) = \delta_1(-\lambda J_1(\lambda) + A_2 J_0(\lambda)) = 0. \tag{22}$$

Under the condition $\delta_1 \neq 0$, Equation (22) yields

$$A_2 J_0(\lambda) - \lambda J_1(\lambda) = 0. \tag{23}$$

It should be noted here that Equation (23) has an infinite number of roots λ_n, so we can write

$$A_2 J_0(\lambda_n) - \lambda_n J_1(\lambda_n) = 0. \tag{24}$$

3. The Exact Solution $u(x,r)$

Substituting Equations (17) and (19) into Equation (5), we obtain $u(x,r)$ in the form:

$$u(x,r) = \xi(0)\delta_1 J_0(\lambda r) E_\alpha\left(-(A_1 + A_0 \lambda^2)x^\alpha\right), \tag{25}$$

or

$$u(x,r) = \sigma J_0(\lambda r) E_\alpha\left(-(A_1 + A_0 \lambda^2)x^\alpha\right), \tag{26}$$

where $\sigma = \xi(0)\delta_1$. Since Equation (24) has an infinite number of roots, then Equation (26) is given by the series:

$$u(x,r) = \sum_{n=1}^{\infty} \sigma_n J_0(\lambda_n r) E_\alpha\left(-(A_1 + A_0 \lambda_n^2)x^\alpha\right). \tag{27}$$

Applying the BC (2) on Equation (27) gives

$$1 = \sum_{n=1}^{\infty} \sigma_n J_0(\lambda_n r), \quad \text{where} \quad E_\alpha(0) = 1 \,\forall\, \alpha \in (0,1]. \tag{28}$$

Following Biswas et al. [4], we find that

$$\sigma_n = \frac{2 J_1(\lambda_n)}{\lambda_n \left(J_0^2(\lambda_n) + J_1^2(\lambda_n)\right)}. \tag{29}$$

Substituting (29) into (27) leads to

$$u(x,r) = 2\sum_{n=1}^{\infty} \frac{J_1(\lambda_n)J_0(\lambda_n r)E_\alpha\left(-(A_1+A_0\lambda_n^2)x^\alpha\right)}{\lambda_n\left(J_0^2(\lambda_n)+J_1^2(\lambda_n)\right)}. \tag{30}$$

Equation (24) implies $A_2 = \frac{\lambda_n J_1(\lambda_n)}{J_0(\lambda_n)}$, hence, Equation (30) is expressed as

$$u(x,r) = 2\sum_{n=1}^{\infty} \frac{\lambda_n J_1(\lambda_n)J_0(\lambda_n r)E_\alpha\left(-(A_1+A_0\lambda_n^2)x^\alpha\right)}{(A_2^2+\lambda_n^2)J_0^2(\lambda_n)}. \tag{31}$$

As $\alpha \to 1$, Equation (31) reduces to

$$u(x,r) = 2\sum_{n=1}^{\infty} \frac{\lambda_n J_1(\lambda_n)J_0(\lambda_n r)E_1\left(-(A_1+A_0\lambda_n^2)x\right)}{(A_2^2+\lambda_n^2)J_0^2(\lambda_n)}, \tag{32}$$

and therefore

$$u(x,r) = 2\sum_{n=1}^{\infty} \frac{\lambda_n J_1(\lambda_n)J_0(\lambda_n r)e^{-(A_1+A_0\lambda_n^2)x}}{(A_2^2+\lambda_n^2)J_0^2(\lambda_n)}, \tag{33}$$

which is identical to the solution obtained by Biswas et al. [4] for the chlorine decay model with classical x-partial derivative.

4. The Cup-Mixing Average Concentration

Following Biswas et al. [4], we define the dimensionless cup-mixing average concentration as

$$u_{av} = 2\int_0^1 u(x,r)\, rdr. \tag{34}$$

Substituting (31) into (34), yields

$$u_{av} = 2\sum_{n=1}^{\infty} \frac{\lambda_n J_1(\lambda_n)E_\alpha\left(-(A_1+A_0\lambda_n^2)x^\alpha\right)}{(A_2^2+\lambda_n^2)J_0^2(\lambda_n)}\int_0^1 rJ_0(\lambda_n r)dr, \tag{35}$$

or

$$u_{av} = 4\sum_{n=1}^{\infty} \frac{J_1^2(\lambda_n)}{(A_2^2+\lambda_n^2)J_0^2(\lambda_n)}E_\alpha\left(-(A_1+A_0\lambda_n^2)x^\alpha\right), \tag{36}$$

where the integral property $\int_0^1 rJ_0(\lambda_n r)dr = \frac{J_1(\lambda_n r)}{\lambda_n}$ is used. From Equation (24) and making use of $A_2 = \frac{\lambda_n J_1(\lambda_n)}{J_0(\lambda_n)}$, then

$$u_{av} = 4\sum_{n=1}^{\infty} \frac{A_2^2}{\lambda_n^2(A_2^2+\lambda_n^2)}E_\alpha\left(-(A_1+A_0\lambda_n^2)x^\alpha\right). \tag{37}$$

As $\alpha \to 1$, Equation (37) becomes

$$u_{av} = 4\sum_{n=1}^{\infty} \frac{A_2^2}{\lambda_n^2(A_2^2+\lambda_n^2)}e^{-(A_1+A_0\lambda_n^2)x}, \tag{38}$$

which agrees with the corresponding result in Ref. [4].

4.1. $A_2 \to \infty$ (The Pipe Walls Act as a Perfect Sink)

If the pipe walls act as a perfect sink, i.e., $A_2 \to \infty$ [4], then u_{av} is obtained from Equation (37) by

$$u_{av} = 4 \lim_{A_2 \to \infty} \left(\sum_{n=1}^{\infty} \frac{A_2^2}{\lambda_n^2 (A_2^2 + \lambda_n^2)} E_\alpha \left(-(A_1 + A_0 \lambda_n^2) x^\alpha \right) \right), \tag{39}$$

which gives

$$u_{av} = \sum_{n=1}^{\infty} \frac{4}{\lambda_n^2} E_\alpha \left(-(A_1 + A_0 \lambda_n^2) x^\alpha \right), \tag{40}$$

where λ_n's are the roots of $J_0(\lambda_n) = 0$. As $\alpha \to 1$, Equation (40) reduces to

$$u_{av} = \sum_{n=1}^{\infty} \frac{4}{\lambda_n^2} e^{-(A_1 + A_0 \lambda_n^2) x}, \tag{41}$$

which is the same result obtained in Ref. [4]. It may be important here to refer to that the series (39–41) are convergent for all positive values of the parameters A_0 and A_1. Such a point can be explained as follows. In Ref. [31] (see p. 9), it was mentioned that $0 < E_\alpha(-\Omega) \leq 1$ for $\Omega > 0$. Since the physical parameters A_0 and A_1, in addition to the roots λ_n, are always positive, then $0 < E_\alpha(-\Omega) = E_\alpha(-(A_1 + A_0\lambda_n^2)x^\alpha) \leq 1$ where $\Omega = (A_1 + A_0\lambda_n^2)x^\alpha > 0 \; \forall \; x \in [0,1], \alpha \in (0,1]$. Accordingly, we have from (39), (40) that $|u_{av}| \leq \sum_{n=1}^{\infty} \frac{4}{\lambda_n^2}$. To check the convergence of the series $\sum_{n=1}^{\infty} \frac{4}{\lambda_n^2}$, let $c_n = \frac{4}{\lambda_n^2}$, then $\lim_{n \to \infty} \left| \frac{c_{n+1}}{c_n} \right| = \lim_{n \to \infty} \left| \frac{\lambda_n^2}{\lambda_{n+1}^2} \right| \leq 1$, where $\lambda_n \leq \lambda_{n+1} \forall n \geq 1$. Similar proof can be easily shown for the series (41) and also for (37) and (38). Hence, the series (37–41) are convergent by the ratio test for all positive values of the parameters A_i, $i = 0, 1, 2$.

4.2. $A_2 \to 0$ (No Chlorine Consumption Takes Place at the Walls)

If $A_2 \to 0$ (the pipe walls are inert and no chlorine consumption takes place at the walls), then Equation (24) leads to $\lambda_n = 0$ or $J_1(\lambda_n) = 0$. The case $J_1(\lambda_n) = 0$ implies that $\sigma_n = 0$ (from Equation (29)), hence, trivial solution $u(x,r) = 0$ is obtained. The case $\lambda_n = 0$ transforms Equation (26) into the simple expression:

$$u(x,r) = \sigma E_\alpha(-A_1 x^\alpha). \tag{42}$$

Applying the BC (2) on Equation (40) gives $\sigma = 1$ and hence,

$$u(x,r) = E_\alpha(-A_1 x^\alpha). \tag{43}$$

According to (34), we obtain

$$u_{av} = 2 \int_0^1 E_\alpha(-A_1 x^\alpha) \, r dr = E_\alpha(-A_1 x^\alpha). \tag{44}$$

As $\alpha \to 1$, Equation (44) yields

$$u_{av} = e^{-A_1 x}, \tag{45}$$

which is the same expression obtained by Biswas et al. [4].

5. Results & Discussion

In this section, comparisons between the present results, as $\alpha \to 1$ (classical chlorine decay), and the corresponding ones in Refs. [4–6] are performed. Additionally, the comparisons between the present results and those obtained by Mahrous [7] are introduced for $\alpha \in (0,1]$ (fractional chlorine decay). In addition, the effect of the order of fractional derivative α on the variation of the cup-mixing average concentration u_{av} is discussed. Before doing so, we must have a clear picture about the nature of the roots of Equation (24). Here, the function $\phi(\lambda)$:

$$\phi(\lambda) = A_2 J_0(\lambda) - \lambda J_1(\lambda) = 0, \tag{46}$$

is supposed to facilitate the discussion.

5.1. Behavior of $\phi(\lambda)$

Behavior of $\phi(\lambda)$ is depicted in Figures 1–5 at various values of A_2. It is verified in all figures that there is an infinite number of roots. However, the roots are nearly identical for small $A_2 \in [0, 1)$ (Figure 1), except the first root. The thin curve (black, dashed) represents the function $\phi(\lambda) = -\lambda J_1(\lambda)$ ($A_2 = 0$). For $A_2 \in [1, 10)$, it can be conducted from Figure 2 that the roots, after the first seven ones, have approximately the same values. From Figure 3, it can be seen that the first two roots are nearly identical for $A_2 \in [10, 50)$, the rest of roots are different. For relatively higher values of $A_2 \in [50, 100)$ (Figure 4) and $A_2 \in [100, 1000)$ (Figure 5), the roots are nearly identical as shown from Figures 4 and 5. Although an infinite number of roots exist for the equation $\phi(\lambda) = 0$, Biswas et al. [4] considered certain approximate analytic formulas, using fitting data, for only the first three roots λ_1, λ_2 and λ_3 when deriving their results. Moreover, Yeh et al. [5] obtained an approximate formula for the first root and then they established their results. Furthermore, Mahrous [6] derived the first two roots and gave approximate analytic forms and then compared his results with Biswas et al. [4] and Yeh et al. [5]. In Table 1, the numeric values of the first three roots λ_1, λ_2 and λ_3 of Equation (24) are listed.

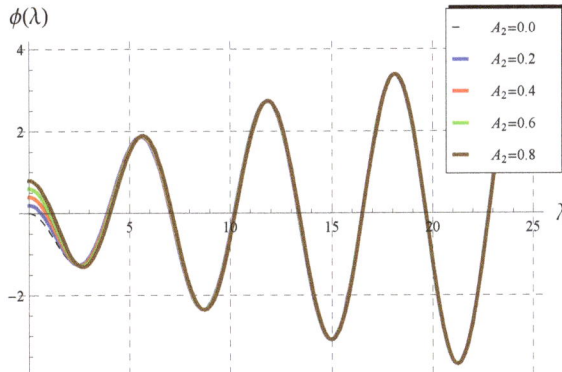

Figure 1. Behavior of $\phi(\lambda)$ vs. λ in the range $0 \leq A_2 < 1$. The thin curve (black, dashed) represents the function $\phi(\lambda) = -\lambda J_1(\lambda)$ ($A_2 = 0$).

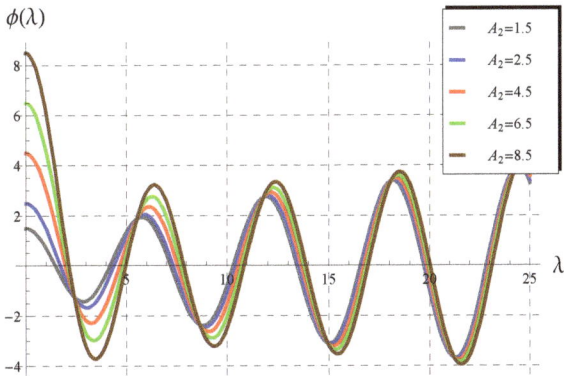

Figure 2. Behavior of $\phi(\lambda)$ vs. λ in the range $1 \leq A_2 < 10$.

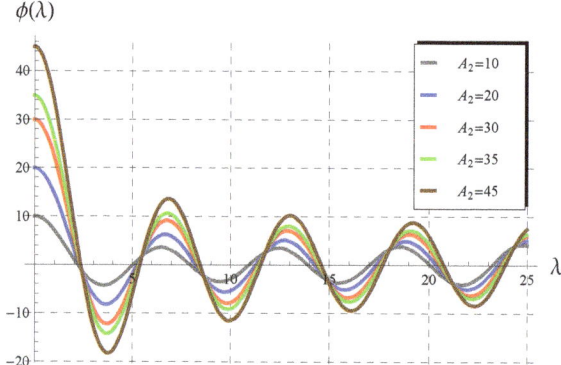

Figure 3. Behavior of $\phi(\lambda)$ vs. λ in the range $10 \leq A_2 < 50$.

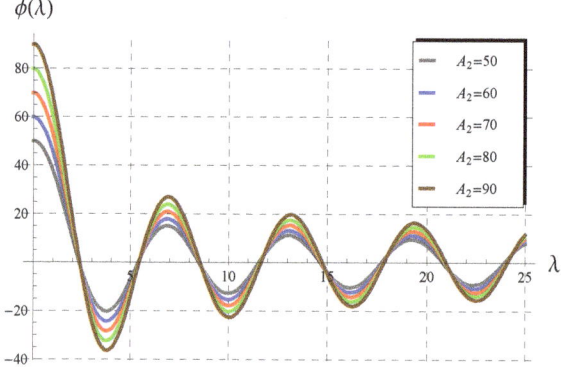

Figure 4. Behavior of $\phi(\lambda)$ vs. λ in the range $50 \leq A_2 < 100$.

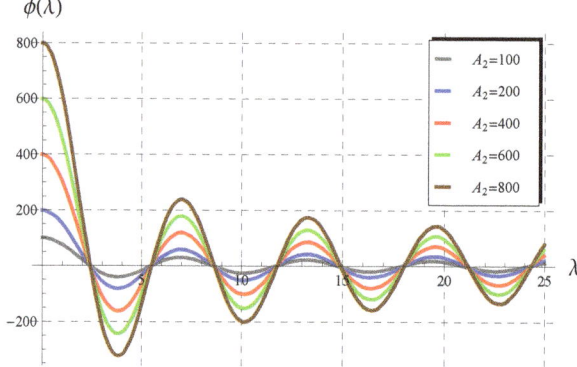

Figure 5. Behavior of $\phi(\lambda)$ vs. λ in the range $100 \leq A_2 < 1000$.

Table 1. The numerical values of λ_1, λ_2, and λ_3 of Equation (24) at different values of A_2.

A_2	λ_1	λ_2	λ_3
0.01	0.14125	3.83431	7.01701
0.1	0.44168	3.85771	7.02983
0.2	0.61698	3.88351	7.04403
0.5	0.94077	3.95937	7.08638
1	1.25578	4.07948	7.15580
2	1.59945	4.29096	7.28839
5	1.98981	4.71314	7.61771
10	2.17950	5.03321	7.95688
50	2.35724	5.41120	8.48399
100	2.38090	5.46521	8.56783

5.2. Experimental Values of A_2 (Wall Decay Rate)

Yeh et al. [5] mentioned that the values of A_2 are smaller than 0.1 according to the experimental studies for chlorine decay. Therefore, the comparison between the present results and those in the relevant literature are performed taking into account such experimental considerations, i.e., $A_2 < 0.1$. For this reason, Yeh et al. [5] considered only the first root λ_1 of Equation (24) and they obtained approximate expression $\lambda_1 = \sqrt{\frac{4A_2}{2+A_2}}$. Such λ_1 was obtained by Biswas et al. [4] as $\lambda_1 = 1.29861(A_2)^{0.477433}$, while Mahrous [6] derived $\lambda_1 = \sqrt{2\left(2 + A_2 - \sqrt{4 + A_2^2}\right)}$. The comparison between the present numerical values for λ_1, against A_2, and the above approaches is displayed in Figure 6. It is concluded from this figure that the values of Ref. [6] are the best when compared with the true numerical ones using Wolfram MATHEMATICA 12.

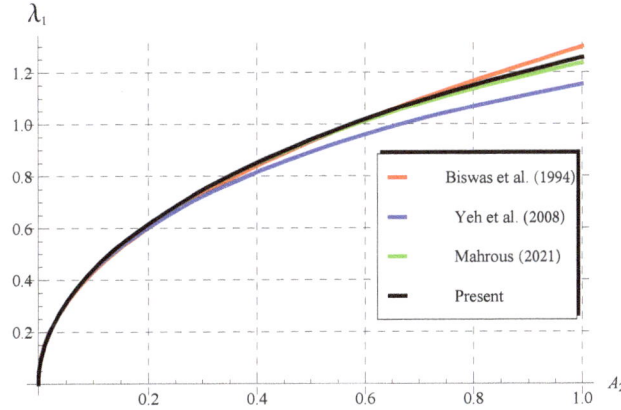

Figure 6. Comparisons between the present numerical and the published approximate values of the first root λ_1 vs. A_2 [4–6].

5.3. Comparisons as $\alpha \to 1$ (Classical Chlorine Decay)

In the case of $\alpha \to 1$, Biswas et al. [4] deduced the following approximation:

$$u_{av} = \frac{e^{-A_1 x}}{1+\epsilon}, \quad \epsilon = 2.4416 A_0 A_2 - 0.1559 A_0 A_2^2, \quad 0.01 \leq A_2 \leq 10, \quad (47)$$

as a consequence of the regression technique. In addition, Yeh et al. [5] showed that the u_{av} can be approximated as

$$u_{av} = \left(1 + \frac{2A_2}{4 + 2A_2 + A_2^2}\right) e^{-\left(A_1 + \frac{4A_0 A_2}{2+A_2}\right)x}, \quad 0 \leq A_2 < 0.1. \quad (48)$$

In addition, accurate approximate solution was obtained by Mahrous [6] in the form

$$u_{av} = 4A_2^2 \sum_{n=1}^{2} \frac{e^{-(A_1+A_0\lambda_n^2)x}}{\lambda_n^2(A_2^2+\lambda_n^2)}, \quad \lambda_{1,2} = \sqrt{2\left(2+A_2 \mp \sqrt{4+A_2^2}\right)}, \quad 0 \leq A_2 \leq 1. \quad (49)$$

In Figures 7–9, the present u_{av} (given in Equation (38)) is displayed versus A_1 (water decay rate), at the outlet $x = 1$ of a pipe, and compared with the above different approximations. The first three roots listed in Table 1 are used to conduct our results. For a fixed radial diffusivity value $A_0 = 1.4$, the comparisons are performed for $A_2 = 0.01$ (Figure 7), $A_2 = 0.1$ (Figure 8), and $A_2 = 0.5$ (Figure 9). It can be seen from Figures 7 and 8 that our results coincide with the published ones by Biswas et al. [4], Yeh et al. [5], and Mahrous [6]. However, the current results agree only with Mahrous [6] in Figure 9 when A_2 is slightly increased ($A_2 = 0.5$). This is because the value $A_2 = 0.5$ lies outside the range of validity addressed by Yeh et al. [5] ($0 \leq A_2 < 0.1$). Although $A_2 = 0.5$ lies inside the range of validity conducted by Biswas et al. [4] ($0.01 \leq A_2 \leq 10$), their estimated expression for the u_{av} deviate from our results and those of Mahrous [6]. Probably, the fitting data used by Biswas et al. [4] needs revisions in this case.

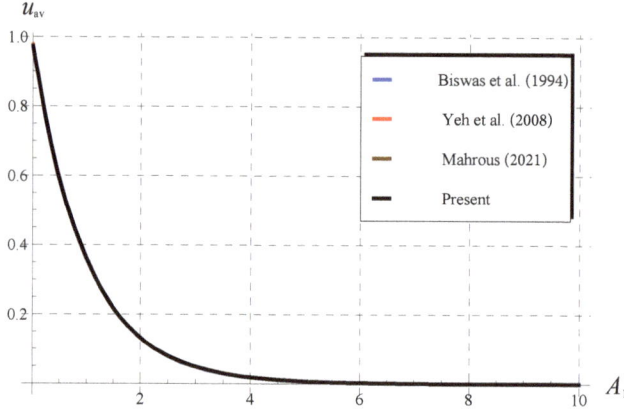

Figure 7. Comparisons between the present cup-mixing average concentration u_{av} and the corresponding ones in literature as $\alpha \to 1$ when $A_0 = 1.4$ and $A_2 = 0.01$ [4–6].

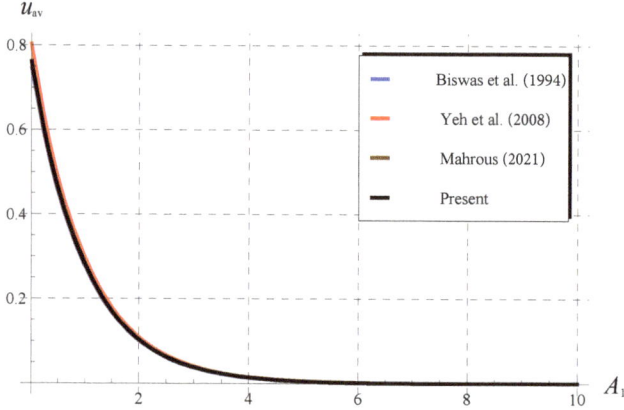

Figure 8. Comparisons between the present cup-mixing average concentration u_{av} and the corresponding ones in literature as $\alpha \to 1$ when $A_0 = 1.4$ and $A_2 = 0.1$ [4–6].

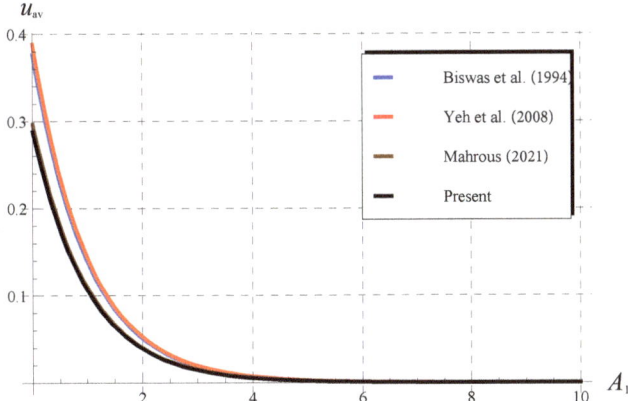

Figure 9. Comparisons between the present cup-mixing average concentration u_{av} and the corresponding ones in literature as $\alpha \to 1$ when $A_0 = 1.4$ and $A_2 = 0.5$ [4–6].

5.4. Comparisons for $\alpha \in (0, 1]$ (Fractional Chlorine Decay)

In this section, the behavior of the present cup-mixing average concentration u_{av} in the FC, given in Equation (37) and the the corresponding one in Ref. [7] are declared. For fractional chlorine decay, i.e., $\alpha \in (0,1]$, Mahrous [7] obtained the following exact solution for the u_{av}:

$$u_{av} = E_\alpha(-A_1 x^\alpha) - \frac{1}{\alpha} e^{(-A_1)^{1/\alpha} x} + 4 \sum_{n=1}^{\infty} \frac{A_2^2}{\lambda_n^2 (A_2^2 + \lambda_n^2)} e^{\left(-A_1 - A_0 \lambda_n^2\right)^{1/\alpha} x}. \tag{50}$$

For fixed $\alpha = 1/2$ and $A_0 = 1.4$, the comparisons between the two approaches in the FC are displayed through Figures 10–12 for the u_{av} at $A_2 = 0.01$ (Figure 10), $A_2 = 0.1$ (Figure 11), and $A_2 = 0.5$ (Figure 12). Here, it may be important to mention to that both of our approach and Ref. [7] use the same numeric values of the three roots in Table 1. However, a big difference in the behavior of u_{av} is detected. In all figures, the present u_{av} decreases with increasing A_2 in the whole domain, while the behavior of corresponding one in Ref. [7] is completely different. In addition, the u_{av} in Ref. [7] becomes negative in sub-domains of A_1, namely at the beginning. Moreover, the effect of α on the variation of u_{av} can be interpreted. Our curves in Figures 10–12 (black) at $\alpha = 1/2$ are always lower than those of Figures 7–9 (black) as $\alpha \to 1$. So, the u_{av} in the FC is of less amount than in classical calculus. As a final note on the comparisons made above, the present analysis agrees with the physical requirements of the problem, where the present u_{av} is always positive, i.e, unlike the negativity in Ref. [7]. In conclusion, the current analysis gives a clear picture and accurate solution of the chlorine decay in the FC.

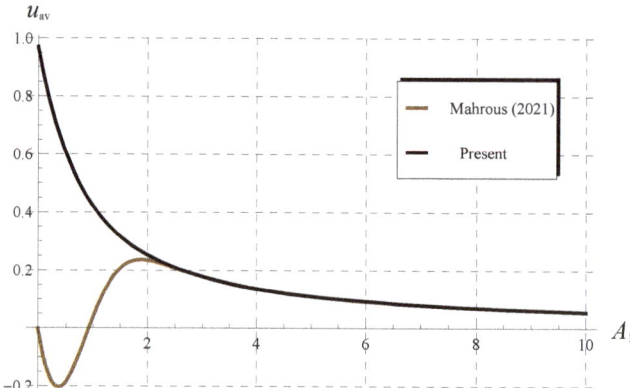

Figure 10. Comparisons between the present cup-mixing average concentration u_{av} and the corresponding ones in Mahrous [7] at $\alpha = \frac{1}{2}$ (fractional chlorine decay), $A_0 = 1.4$, and $A_2 = 0.01$.

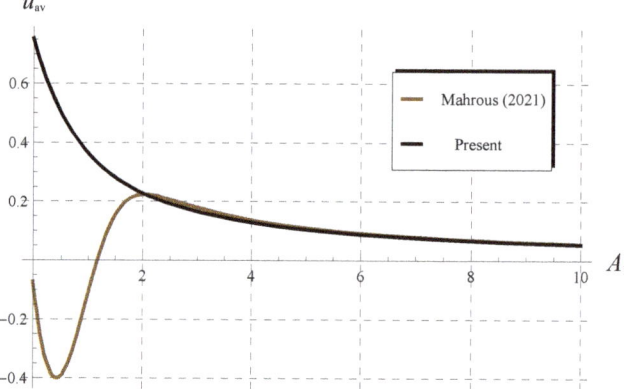

Figure 11. Comparisons between the present cup-mixing average concentration u_{av} and the corresponding ones in Mahrous [7] at $\alpha = \frac{1}{2}$ (fractional chlorine decay), $A_0 = 1.4$, and $A_2 = 0.1$.

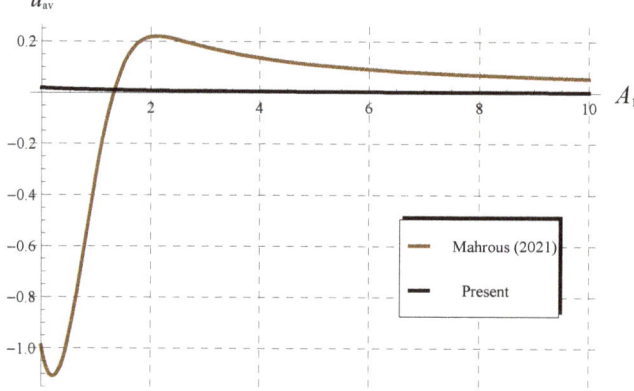

Figure 12. Comparisons between the present cup-mixing average concentration u_{av} and the corresponding ones in Mahrous [7] at $\alpha = \frac{1}{2}$ (fractional chlorine decay), $A_0 = 1.4$, and $A_2 = 0.5$.

6. Conclusions

The transport model of the chlorine concentration decay in the FC was investigated. The SOV method combined with the LT were applied to solve the current model. The dimensionless cup-mixing average concentration was obtained in closed form in terms of the Mittag-Leffler function. It was declared that the results reduce to the standard ones in the literature as the fractional order α tends to one. The obtained results were compared with several studies in the literature, and the difference in results is explained and interpreted in detail. In view of these comparisons, it can be concluded that the present analysis was effective to giving a clear picture and accurate solution for the chlorine transport model in the FC when compared with the previous solution in [7] for the same model.

Author Contributions: Conceptualization, A.F.A., A.E.; methodology, A.F.A., A.E. and H.K.A.-J.; software, E.A.A.; validation, C.C., A.F.A. and A.E.; resources, Y.M.M.; data curation, Y.M.M.; writing—original draft preparation, A.F.A., E.A.A. and A.E.; writing—review and editing, H.K.A.-J.; visualization, A.F.A., C.C., and Y.M.M.; supervision, A.F.A. and C.C.; project administration, A.F.A.; funding acquisition, A.F.A. All authors have read and agreed to the published version of the manuscript.

Funding: The authors extend their appreciation to the Deanship of Scientific Research at University of Tabuk for funding this work through Research Group No. S-0147-1441.

Institutional Review Board Statement: Not applicable.

Informed Consent Statement: Not applicable.

Data Availability Statement: Not applicable.

Conflicts of Interest: The authors declare no conflict of interest.

References

1. Clark, R.M.; Read, E.J.; Hoff, J.C. Analysis of Inactivation of Giardia Lamblia by Chlorine. *J. Environ. Eng.* **1989**, *115*, 80–90. [CrossRef]
2. LeChevallier, M.W.; Cawthon, C.D.; Lee, R.G. Inactivation of biofilm bacteria. *Appl. Environ. Microbiol.* **1988**, *54*, 2492–2499. [CrossRef] [PubMed]
3. Arnold, B.F.; Colford, J.M. Treating Water with Chlorine at Point-of-Use to Improve Water Quality and Reduce Child Diarrhea in Developing Countries: A Systematic Review and Meta-Analysis. *Am. J. Trop. Med. Hyg.* **2007**, *76*, 354–364. [CrossRef] [PubMed]
4. Biswas, P.; Lu, C.; Clark, R.M. A model for chlorine concentration decay in pipes. *Water Res.* **1993**, *27*, 1715–1724. [CrossRef]
5. Yeh, H.-D.; Wen, S.-B.; Chang, Y.-C.; Lu, C.-S. A new approximate solution for chlorine concentration decay in pipes. *Water Res.* **2008**, *42*, 2787–2795. [CrossRef]
6. Mahrous, Y.M. Accurate approximation for the chlorine transport in pipe. *Adv. Differ. Equ. Control Process.* **2021**, *25*, 115–126. [CrossRef]
7. Mahrous, Y.M. A possible generalized model of the chlorine concentration decay in pipes: Exact solution. *Int. Anal. Appl.* **2021**, *19*, 604–618. [CrossRef]
8. Monteiro, L.; Viegas, R.; Covas, D.; Menaia, J. Assessment of Current Models Ability to Describe Chlorine Decay and Appraisal of Water Spectroscopic Data as Model Inputs. *J. Environ. Eng.* **2017**, *143*, 04016071. [CrossRef]
9. Zhao, Y.; Yang, Y.J.; Shao, Y.; Neal, J.; Zhang, T. The dependence of chlorine decay and DBP formation kinetics on pipe flow properties in drinking water distribution. *Water Res.* **2018**, *141*, 32–45. watres.2018.04.048. [CrossRef]
10. Fisher, I.; Kastl, G.; Sathasivan, A. New Model of Chlorine-Wall Reaction for Simulating Chlorine Concentration in Drinking Water Distribution Systems. *Water Res.* **2017**, *125* (Suppl. C), 427–437. [CrossRef]
11. Ozdemir, O.; Buyruk, T. Effect of travel time and temperature on chlorine bulk decay in water supply pipes. *J. Environ. Eng.* **2018**, *144*, 04018002. [CrossRef]
12. Minaee, R.P.; Afsharnia, M.; Moghaddam, A.; Ebrahimi, A.A.; Askarishahi, M.; Mokhtari, M. Calibration of water quality model for distribution networks using genetic algorithm, particle swarm optimization, and hybrid methods. *MethodsX* **2019**, *6*, 540–548. [CrossRef] [PubMed]
13. Minaee, R.P.; Afsharnia, M.; Moghaddam, A.; Ebrahimi, A.A.; Askarishahi, M.; Mokhtari, M. Wall decay coefficient estimation in a real-life drinking water distribution network. *Water Resour. Manag.* **2019**, *33*, 1557–1569. [CrossRef]
14. Monteiro, L.; Carneiro, J.; Covas, D.I.C. Modelling chlorine wall decay in a full-scale water supply system. *Urban Water J.* **2020**, *17*, 754–762. [CrossRef]
15. Moghaddam, A.; Mokhtari, M.; Afsharnia, M.; Minaee, R.P. Simultaneous Hydraulic and Quality Model Calibration of a Real-World Water Distribution Network. *J. Water Resour. Plan. Manag.* **2020**, *146*, 06020007. [CrossRef]

16. Jakubowski, J.; Wisniewolski, M. On matching diffusions, Laplace transforms and partial differential equations. *Stoch. Proc. Appl.* **2015**, *125*, 3663–3690. [CrossRef]
17. Ebaid, A.; Sharif, M.A. Application of Laplace transform for the exact effect of a magnetic field on heat transfer of carbon-nanotubes suspended nanofluids. *Z. Naturforsch. A* **2015**, *70*, 471–475. [CrossRef]
18. Ebaid, A.; Wazwaz, A.M.; Alali, E.; Masaedeh, B. Hypergeometric series solution to a class of second-order boundary value problems via Laplace transform with applications to nanofluids. *Commun. Theor. Phys.* **2017**, *67*, 231. [CrossRef]
19. Ebaid, A.; Alali, E.; Saleh, H. The exact solution of a class of boundary value problems with polynomial coefficients and its applications on nanofluids. *J. Assoc. Arab Univ. Basi Appl. Sci.* **2017**, *24*, 156–159. [CrossRef]
20. Khaled, S.M. The exact effects of radiation and joule heating on magnetohydrodynamic Marangoni convection over a flat surface. *Therm. Sci.* **2018**, *22*, 63–72. [CrossRef]
21. Bakodah, H.O.; Ebaid, A. Exact solution of Ambartsumian delay differential equation and comparison with Daftardar-Gejji and Jafari approximate method. *Mathematics* **2018**, *6*, 331. [CrossRef]
22. Ebaid, A.; Cattani, C.; Juhani1, A.S.A.; El-Zahar, E.R. A novel exact solution for the fractional Ambartsumian equation. *Adv. Differ. Equ.* **2021**, *2021*, 88. [CrossRef]
23. Handibag, S.; Karande, B.D. Laplace substitution method for solving partial differential equations involving mixed partial derivatives. *Int. J. Comput. Eng. Res.* **2012**, *2*, 1049–1052.
24. Handibag, S.S.; Karande, B.D. Laplace substitution method for nth order linear and non-Linear PDE's involving mixed partial derivatives. *Int. Res. J. Eng. Technol.* **2015**, *2*, 378–388.
25. Pavani, P.V.; Priya, U.L.; Reddy, B.A. Solving differential equations by using Laplace transforms. *Int. J. Res. Anal. Rev.* **2018**, *5*, 1796–1799.
26. Dhunde, R.R.; Waghmare, G.L. Double Laplace iterative method for solving nonlinear partial differential equations. *New Trends Math. Sci.* **2019**, *7*, 138–149. [CrossRef]
27. Mousa, A.; Elzaki, T.M. Solution of volterra integro-differential equations by triple Laplace transform, Irish Interdiscip. *J. Sci. Res.* **2019**, *3*, 67–72.
28. Zhang, H.; Nadeem, M.; Rauf, A.; Hui, Z.G. A novel approach for the analytical solution of nonlinear time-fractional differential equations. *Int. J. Numer. Meth. Heat Fluid Flow* **2021**, *31*, 1069–1084. [CrossRef]
29. Ebaid, A.; Al-Jeaid, H.K. The Mittag–Leffler Functions for a Class of First-Order Fractional Initial Value Problems: Dual Solution via Riemann-Liouville Fractional Derivative. *Fractal Fract.* **2022**, *6*, 85. [CrossRef]
30. Srivastava, H.M.; Kumar, A.; Das, S.; Mehrez, K. Geometric Properties of a Certain Class of Mittag-Leffler-Type Functions. *Fractal Fract.* **2022**, *6*, 54. [CrossRef]
31. Alharb, W.; Hristova, S. New Series Solution of the Caputo Fractional Ambartsumian Delay Differential Equationation by Mittag-Leffler Functions. *Mathematics* **2021**, *9*, 157. [CrossRef]

fractal and fractional

Article

Hermite-Hadamard Fractional Integral Inequalities via Abel-Gontscharoff Green's Function

Yixia Li [1], Muhammad Samraiz [2], Ayesha Gul [2], Miguel Vivas-Cortez [3],* and Gauhar Rahman [4]

[1] College of Mathematics and Information Science, Xiangnan University, Chenzhou 423000, China; liyixia@xnu.edu.cn
[2] Department of Mathematics, University of Sargodha, Sargodha 40100, Pakistan; msamraiz@uos.edu.pk (M.S.); ayeshagul728@gmail.com (A.G.)
[3] Escuela de Ciencias Físicas y Matemáticas, Facultad de Ciencias Exactas y Naturales, Pontificia Universidad Católica del Ecuador, Av. 12 de Octubre 1076, Apartado, Quito 17-01-2184, Ecuador
[4] Department of Mathematics and Statistics, Hazara University, Mansehra 21300, Pakistan; gauhar55uom@gmail.com or drgauhar.rahman@hu.edu.pk
* Correspondence: mjvivas@puce.edu.ec

Abstract: The Hermite-Hadamard inequalities for κ-Riemann-Liouville fractional integrals (R-LFI) are presented in this study using a relatively novel approach based on Abel-Gontscharoff Green's function. In this new technique, we first established some integral identities. Such identities are used to obtain new results for monotonic functions whose second derivative is convex (concave) in absolute value. Some previously published inequalities are obtained as special cases of our main results. Various applications of our main consequences are also explored to special means and trapezoid-type formulae.

Keywords: Mittag-Leffler; Abel-Gontscharoff Green's function; Hermite-Hadamard inequalities; convex function; κ-Riemann-Liouville fractional integral

MSC: 26D15; 26D10; 26A33; 34B27

1. Introduction

Fractional calculus is a branch of mathematics that investigates the possibility of using a real or even a complex number as a differential and integral operator order. This theory has gained significant prominence in recent decades due to its wide applications in mathematical sciences. Samraiz et al. [1,2] explored some new fractional operators and their applications in mathematical physics. Tarasov [3] and Mainardi [4] explained, in detail, the history and applications of fractional calculus in mathematical economics and finance. The reader might also explore the literature [5–10] for further information on fractional calculus.

Convexity theory has a long history and has been the subject of significant research for more than a century. Many researchers have been interested in the various speculations, variants, and augmentations of convexity theory see, e.g., the books [11,12] and the articles [13,14]. This idea has aided the advancement of the concept of inequality prominently. Wu et al. [15] presented the Hermite-Hadamard inequalities for R-LFI, and Khan et al. [16] explored the inequalities by involving Green's function. This theory has been widely studied by various researchers [17–20].

Hermite (1822–1901) sent a letter to the journal Mathesis. An extract from that letter was published in Mathesis **3** (1883, p. 82). This inequality asserts that for a convex (concave) function F, we can write

$$F\left(\frac{\beta_1+\beta_2}{2}\right) \leq (\geq) \frac{1}{\beta_1-\beta_2}\int_{\beta_1}^{\beta_2} F(\tau)d\tau \leq (\geq) \frac{F(\beta_1)+F(\beta_2)}{2}$$

Definition 1 ([11]). *A function $F : I \to \mathbb{R}$ is said to be convex (concave) on I convex subset of $\mathbb{R}$ if the inequality*

$$F(\nu\beta_1 + (1-\nu)\beta_2) \leq (\geq)\nu F(\beta_1) + (1-\nu)F(\beta_2)$$

holds for all $0 \leq \nu \leq 1$ and $\beta_1, \beta_2 \in I$.

Definition 2 ([16]). *The left-sided and right-sided RL fractional integrals $I^\varrho_{\vartheta_1^+}F$ and $I^\varrho_{\vartheta_2^-}F$ of order $\varrho > 0$ on a finite interval $[\vartheta_1, \vartheta_2]$ are defined by*

$$I^\varrho_{\vartheta_1^+}F(\varsigma) = \frac{1}{\Gamma(\varrho)}\int_{\vartheta_1}^{\varsigma}(\varsigma-\nu)^{\varrho-1}F(\nu)d\nu, \quad \varsigma > \vartheta_1,$$

and

$$I^\varrho_{\vartheta_2^-}F(\varsigma) = \frac{1}{\Gamma(\varrho)}\int_{\varsigma}^{\vartheta_2}(\nu-\varsigma)^{\varrho-1}F(\nu)d\nu, \quad \varsigma < \vartheta_2,$$

respectively. The symbol $\Gamma(\varrho)$ represents the usual Euler's gamma function of ϱ defined by

$$\Gamma(\varrho) = \int_0^\infty \nu^{\varrho-1}e^{-\nu}d\nu. \tag{1}$$

Definition 3 ([15]). *The κ-fractional integrals of order ϱ with $\kappa > 0$ and $\vartheta \geq 0$ are defined as*

$$I^{\varrho,\kappa}_{\vartheta_1^+}F(\varsigma) = \frac{1}{\kappa\Gamma_\kappa(\varrho)}\int_{\vartheta_1}^{\varsigma}(\varsigma-\nu)^{\frac{\varrho}{\kappa}-1}F(\nu)d\nu, \quad \varsigma > \vartheta_1,$$

and

$$I^{\varrho,\kappa}_{\vartheta_2^-}F(\varsigma) = \frac{1}{\kappa\Gamma_\kappa(\varrho)}\int_{\varsigma}^{\vartheta_2}(\nu-\varsigma)^{\frac{\varrho}{\kappa}-1}F(\nu)d\nu, \quad \varsigma < \vartheta_2,$$

where $\Gamma_\kappa(\varrho)$ represents the κ-gamma function of ϱ defined by Diaz et al. in [21] with the following integral representation

$$\Gamma_\kappa(\varrho) = \int_0^\infty \nu^{\varrho-1}e^{\frac{-\nu^\kappa}{\kappa}}d\nu.$$

It is to be noted that $\kappa = 1$ gives the classical gamma function given in (1).

The Abel-Gontscharoff polynomial and theorem for the 'two-point right focal' problem are referenced in [16]. The Abel-Gontscharoff interpolating polynomial for the 'two-point' problem can be stated as

$$F(\varsigma) = F(\vartheta_1) + (\varsigma - \vartheta_1)F'(\vartheta_2) + \int_{\vartheta_1}^{\vartheta_2}G(\varsigma,\nu)F''(\nu)d\nu, \tag{2}$$

where $G(\varsigma, \nu)$ is Green's function for the two-point right focal problem.

Mehmood et al. in [22] introduced the following four functions by keeping Abel-Gontscharoff Green's function for the two-point right focal problem

$$G_1(\chi, \nu) = \begin{cases} \vartheta_1 - \nu, \vartheta_1 \leq \nu \leq \chi, \\ \vartheta_1 - \chi, \chi \leq \nu \leq \vartheta_2. \end{cases} \tag{3}$$

$$G_2(\chi, v) = \begin{cases} \chi - \vartheta_2, & \vartheta_1 \leq v \leq \chi, \\ v - \vartheta_2, & \chi \leq v \leq \vartheta_2. \end{cases} \quad (4)$$

$$G_3(\chi, v) = \begin{cases} \chi - \vartheta_1, & \vartheta_1 \leq v \leq \chi, \\ v - \vartheta_1, & \chi \leq v \leq \vartheta_2. \end{cases} \quad (5)$$

$$G_4(\chi, v) = \begin{cases} \vartheta_2 - v, & \vartheta_1 \leq v \leq \chi, \\ \vartheta_2 - \chi, & \chi \leq v \leq \vartheta_2. \end{cases} \quad (6)$$

Sarikaya et al. established in [20] the right and left R-LFI of the following Hermite-Hadamard type inequality.

Theorem 1. *Let* $F : [\vartheta_1, \vartheta_2] \to R$ *be a positive function with* $0 \leq \vartheta_1 < \vartheta_2$ *and* $F \in L[\vartheta_1, \vartheta_2]$. *If* F *is convex function on* $[\vartheta_1, \vartheta_2]$, *then the following inequalities for fractional integrals hold:*

$$F\left(\frac{\vartheta_1 + \vartheta_2}{2}\right) \leq \frac{\Gamma(\varrho + 1)}{2(\vartheta_2 - \vartheta_1)^\varrho} \left(I_{\vartheta_1^+}^\varrho F(\vartheta_2) + I_{\vartheta_2^-}^\varrho F(\vartheta_1)\right) \leq \frac{F(\vartheta_1) + F(\vartheta_2)}{2} \quad (7)$$

with $\varrho > 0$.

The motivation behind this study is to explore the Hermite-Hadamard inequalities using Green's functions presented above together with Abel-Gontscharoff interpolating polynomial corresponding to the choice $n = 2$.

2. Main Results

In this section, we establish Hermite-Hadamard inequalities for left generalized fractional integral via Abel-Gontscharoff Green's function for the two-point right focal problem (3).

Theorem 2. *Let* F *be a twice differentiable and convex function on* $[\vartheta_1, \vartheta_2]$ *that satisfying the relation given in* (2). *Then, the double inequality*

$$F\left(\frac{\varrho \vartheta_1 + \kappa \vartheta_2}{\varrho + \kappa}\right) \leq \frac{\Gamma_\kappa(\varrho + \kappa)}{(\vartheta_2 - \vartheta_1)^{\frac{\varrho}{\kappa}}} I_{\vartheta_1^+}^{\varrho, \kappa} F(\vartheta_2) \leq \frac{\varrho F(\vartheta_1) + \kappa F(\vartheta_2)}{\varrho + \kappa} \quad (8)$$

holds, where $\varrho, \kappa > 0$.

Proof. By making a substitution $\varsigma = \frac{\varrho \vartheta_1 + \kappa \vartheta_2}{\varrho + \kappa}$ in an Abel-Gontscharoff polynomial for the two-point right focal problem interpolating the polynomial presented by (2), we obtain

$$F\left(\frac{\varrho \vartheta_1 + \kappa \vartheta_2}{\varrho + \kappa}\right) = F(\vartheta_1) + \left(\frac{\kappa(\vartheta_2 - \vartheta_1)}{\varrho + \kappa}\right) F'(\vartheta_2)$$
$$+ \int_{\vartheta_1}^{\vartheta_2} G\left(\frac{\varrho \vartheta_1 + \kappa \vartheta_2}{\varrho + \kappa}, v\right) F''(v) dv. \quad (9)$$

Multiplying both sides of (2) by $\frac{\varrho(\vartheta_2 - \varsigma)^{\frac{\varrho}{\kappa} - 1}}{\kappa(\vartheta_2 - \vartheta_1)^{\frac{\varrho}{\kappa}}}$ and integrating with respect to ς, we obtain

$$\frac{\varrho}{\kappa(\vartheta_2 - \vartheta_1)^{\frac{\varrho}{\kappa}}} \int_{\vartheta_1}^{\vartheta_2} (\vartheta_2 - \varsigma)^{\frac{\varrho}{\kappa}} F(\varsigma) d\varsigma = \frac{\varrho}{\kappa(\vartheta_2 - \vartheta_1)^{\frac{\varrho}{\kappa}}} \left(\int_{\vartheta_1}^{\vartheta_2} (\vartheta_2 - \varsigma)^{\frac{\varrho}{\kappa} - 1} F(\vartheta_1) d\varsigma \right.$$
$$\left. + \int_{\vartheta_1}^{\vartheta_2} (\vartheta_2 - \varsigma)^{\frac{\varrho}{\kappa} - 1} (\varsigma - \vartheta_1) F'(\vartheta_2) d\varsigma + \int_{\vartheta_1}^{\vartheta_2} \int_{\vartheta_1}^{\vartheta_2} G(\varsigma, v)(\vartheta_2 - \varsigma)^{\frac{\varrho}{\kappa} - 1} F''(v) dv d\varsigma \right)$$

$$= \frac{\varrho}{\kappa(\vartheta_2-\vartheta_1)^{\frac{\varrho}{\kappa}}}\left(\frac{\kappa}{\varrho}F(\vartheta_1)(\vartheta_2-\vartheta_1)^{\frac{\varrho}{\kappa}}+\frac{\kappa^2 F'(\vartheta_2)(\vartheta_2-\vartheta_1)^{\frac{\varrho}{\kappa}+1}}{\varrho(\varrho+\kappa)}\right.$$
$$\left.+\int_{\vartheta_1}^{\vartheta_2}\int_{\vartheta_1}^{\vartheta_2}G(\varsigma,v)(\vartheta_2-\varsigma)^{\frac{\varrho}{\kappa}-1}F''(v)dvd\varsigma\right).$$

This can also be written as

$$\frac{\Gamma_\kappa(\varrho+\kappa)}{(\vartheta_2-\vartheta_1)^{\frac{\varrho}{\kappa}}}I^{\varrho,\kappa}_{\vartheta_1^+}F(\vartheta_2)=F(\vartheta_1)+\frac{\kappa F'(\vartheta_2)(\vartheta_2-\vartheta_1)}{(\varrho+\kappa)}$$
$$+\frac{\varrho}{\kappa(\vartheta_2-\vartheta_1)^{\frac{\varrho}{\kappa}}}\int_{\vartheta_1}^{\vartheta_2}\int_{\vartheta_1}^{\vartheta_2}G(\varsigma,v)(\vartheta_2-\varsigma)^{\frac{\varrho}{\kappa}-1}F''(v)dvd\varsigma. \tag{10}$$

Subtracting (10) from (9)

$$F\left(\frac{\varrho\vartheta_1+\kappa\vartheta_2}{\varrho+\kappa}\right)-\frac{\Gamma_\kappa(\varrho+\kappa)}{(\vartheta_2-\vartheta_1)^{\frac{\varrho}{\kappa}}}I^{\varrho,\kappa}_{\vartheta_1^+}F(\vartheta_2)=\int_{\vartheta_1}^{\vartheta_2}\left(G\left(\frac{\varrho\vartheta_1+\kappa\vartheta_2}{\varrho+\kappa},v\right)\right.$$
$$\left.-\frac{\varrho}{\kappa(\vartheta_2-\vartheta_1)^{\frac{\varrho}{\kappa}}}\int_{\vartheta_1}^{\vartheta_2}G(\varsigma,v)(\vartheta_2-\varsigma)^{\frac{\varrho}{\kappa}-1}d\varsigma\right)F''(v)dv. \tag{11}$$

Clearly,

$$\int_{\vartheta_1}^{\vartheta_2}G(\varsigma,v)(\vartheta_2-\varsigma)^{\frac{\varrho}{\kappa}-1}d\varsigma=\frac{\kappa^2}{\varrho(\varrho+\kappa)}\left((\vartheta_2-v)^{\frac{\varrho}{\kappa}+1}-(\vartheta_2-\vartheta_1)^{\frac{\varrho}{\kappa}+1}\right), \tag{12}$$

where

$$G\left(\frac{\varrho\vartheta_1+\kappa\vartheta_2}{\varrho+\kappa},v\right)=\begin{cases}\vartheta_1-v, & \vartheta_1\leq v\leq \frac{\varrho\vartheta_1+\kappa\vartheta_2}{\varrho+\kappa},\\ \frac{\kappa(\vartheta_1-\vartheta_2)}{\varrho+\kappa}, & \frac{\varrho\vartheta_1+\kappa\vartheta_2}{\varrho+\kappa}\leq v\leq \vartheta_2.\end{cases} \tag{13}$$

If $\vartheta_1\leq v\leq \frac{\varrho\vartheta_1+\kappa\vartheta_2}{\varrho+\kappa}$, then by utilizing (12) and (13), from (11), we obtain

$$G\left(\frac{\varrho\vartheta_1+\kappa\vartheta_2}{\varrho+\kappa},v\right)-\frac{\varrho}{\kappa(\vartheta_2-\vartheta_1)^{\frac{\varrho}{\kappa}}}\int_{\vartheta_1}^{\vartheta_2}G(\varsigma,v)(\vartheta_2-\varsigma)^{\frac{\varrho}{\kappa}-1}d\varsigma$$
$$=\vartheta_1-v-\frac{\kappa\left((\vartheta_2-v)^{\frac{\varrho}{\kappa}+1}-(\vartheta_2-\vartheta_1)^{\frac{\varrho}{\kappa}+1}\right)}{(\vartheta_2-\vartheta_1)^{\frac{\varrho}{\kappa}}(\varrho+\kappa)}.$$

Now, let

$$g(v)=\vartheta_1-v-\frac{\kappa\left((\vartheta_2-v)^{\frac{\varrho}{\kappa}+1}-(\vartheta_2-\vartheta_1)^{\frac{\varrho}{\kappa}+1}\right)}{(\vartheta_2-\vartheta_1)^{\frac{\varrho}{\kappa}}(\varrho+\kappa)},$$
$$g'(v)=-1+\frac{(\vartheta_2-v)^{\frac{\varrho}{\kappa}}}{(\vartheta_2-\vartheta_1)^{\frac{\varrho}{\kappa}}}\leq 0.$$

This proved that g is a decreasing function; therefore, we can write

$$G\left(\frac{\varrho\vartheta_1+\kappa\vartheta_2}{\varrho+\kappa},v\right) - \frac{\varrho}{\kappa(\vartheta_2-\vartheta_1)^{\frac{\varrho}{\kappa}}}\int_{\vartheta_1}^{\vartheta_2} G(\varsigma,v)(\vartheta_2-\varsigma)^{\frac{\varrho}{\kappa}-1}d\varsigma \leq 0. \qquad (14)$$

If $\frac{\varrho\vartheta_1+\kappa\vartheta_2}{\varrho+\kappa} \leq v \leq \vartheta_2$, then making use of (12) and (13), from (11), we obtain

$$G\left(\frac{\varrho\vartheta_1+\kappa\vartheta_2}{\varrho+\kappa},v\right) - \frac{\varrho}{\kappa(\vartheta_2-\vartheta_1)^{\frac{\varrho}{\kappa}}}\int_{\vartheta_1}^{\vartheta_2} G(\varsigma,v)(\vartheta_2-\varsigma)^{\frac{\varrho}{\kappa}-1}d\varsigma$$

$$= \frac{\kappa(\vartheta_1-\vartheta_2)}{\varrho+\kappa} - \frac{\varrho}{\kappa(\vartheta_2-\vartheta_1)^{\frac{\varrho}{\kappa}}}\left(\frac{\kappa^2\left((\vartheta_2-v)^{\frac{\varrho}{\kappa}+1}-(\vartheta_2-\vartheta_1)^{\frac{\varrho}{\kappa}+1}\right)}{\varrho(\varrho+\kappa)}\right)$$

$$= \frac{-\kappa(\vartheta_2-\vartheta_1)}{\varrho+\kappa} - \frac{\kappa\left((\vartheta_2-v)^{\frac{\varrho}{\kappa}+1}-(\vartheta_2-\vartheta_1)^{\frac{\varrho}{\kappa}+1}\right)}{(\vartheta_2-\vartheta_1)^{\frac{\varrho}{\kappa}}(\varrho+\kappa)}$$

$$= \frac{\kappa\left(-(\vartheta_2-\vartheta_1)(\vartheta_2-\vartheta_1)^{\frac{\varrho}{\kappa}}-(\vartheta_2-v)^{\frac{\varrho}{\kappa}+1}+(\vartheta_2-\vartheta_1)^{\frac{\varrho}{\kappa}+1}\right)}{(\vartheta_2-\vartheta_1)^{\frac{\varrho}{\kappa}}(\varrho+\kappa)}$$

$$= \frac{\kappa\left(-(\vartheta_2-\vartheta_1)^{\frac{\varrho}{\kappa}+1}-(\vartheta_2-v)^{\frac{\varrho}{\kappa}+1}+(\vartheta_2-\vartheta_1)^{\frac{\varrho}{\kappa}+1}\right)}{(\vartheta_2-\vartheta_1)^{\frac{\varrho}{\kappa}}(\varrho+\kappa)}$$

$$= \frac{-\kappa(\vartheta_2-v)^{\frac{\varrho}{\kappa}+1}}{(\vartheta_2-\vartheta_1)^{\frac{\varrho}{\kappa}}(\varrho+\kappa)} \leq 0. \qquad (15)$$

Since F is convex, therefore $F''(v) \geq 0$ and by using (14) and (15) in (11), we obtain

$$F\left(\frac{\varrho\vartheta_1+\kappa\vartheta_2}{\varrho+\kappa}\right) \leq \frac{\Gamma_\kappa(\varrho+\kappa)}{(\vartheta_2-\vartheta_1)^{\frac{\varrho}{\kappa}}} I_{\vartheta_1^+}^{\varrho,\kappa} F(\vartheta_2), \qquad (16)$$

which is the left half inequality of (8).

Next, we prove the right half inequality of (8). For this purpose, we choose $\varsigma = \vartheta_2$ in Equation (2), and we obtain

$$F(\vartheta_2) = F(\vartheta_1) + (\vartheta_2-\vartheta_1)F'(\vartheta_2) + \int_{\vartheta_1}^{\vartheta_2} G(\vartheta_2,v)F''(v)dv.$$

Adding $\frac{\varrho}{\kappa}F(\vartheta_1)$ on both sides and then dividing by $(\frac{\varrho}{\kappa}+1)$, we obtain

$$\frac{\varrho F(\vartheta_1)+\kappa F(\vartheta_2)}{\varrho+\kappa} = F(\vartheta_1) + \frac{\kappa(\vartheta_2-\vartheta_1)F'(\vartheta_2)}{\varrho+\kappa} + \frac{\kappa}{\varrho+\kappa}\int_{\vartheta_1}^{\vartheta_2} G(\vartheta_1,v)F''(v)dv. \qquad (17)$$

Subtracting (10) from (17), we have

$$\frac{\varrho F(\vartheta_1)+\kappa F(\vartheta_2)}{\varrho+\kappa} - \frac{\Gamma_\kappa(\varrho+\kappa)}{(\vartheta_2-\vartheta_1)^{\frac{\varrho}{\kappa}}} I_{\vartheta_1^+}^{\varrho,\kappa} F(\vartheta_2) = \int_{\vartheta_1}^{\vartheta_2}\left(\frac{\kappa G(\vartheta_2,v)}{\varrho+\kappa}\right.$$

$$\left. - \frac{\varrho}{\kappa(\vartheta_2-\vartheta_1)^{\frac{\varrho}{\kappa}}}\int_{\vartheta_1}^{\vartheta_2} G(\varsigma,v)(\vartheta_2-\varsigma)^{\frac{\varrho}{\kappa}-1}d\varsigma\right) F''(v)dv. \qquad (18)$$

Using the value of Green's function $(\vartheta_1 - v)$ for $\vartheta_1 \leq v \leq \vartheta_2$ and Equation (12), we can write

$$\frac{\kappa G(\vartheta_2, v)}{\varrho + \kappa} - \frac{\varrho}{\kappa(\vartheta_2 - \vartheta_1)^{\frac{\varrho}{\kappa}}} \left(\frac{\kappa^2 \left((\vartheta_2 - v)^{\frac{\varrho}{\kappa}+1} - (\vartheta_2 - \vartheta_1)^{\frac{\varrho}{\kappa}+1} \right)}{\varrho(\varrho + \kappa)} \right)$$

$$= \frac{\kappa(\vartheta_1 - v)}{\varrho + \kappa} - \frac{\kappa \left((\vartheta_2 - v)^{\frac{\varrho}{\kappa}+1} - (\vartheta_2 - \vartheta_1)^{\frac{\varrho}{\kappa}+1} \right)}{(\vartheta_2 - \vartheta_1)^{\frac{\varrho}{\kappa}}(\varrho + \kappa)}$$

$$= \frac{\kappa \left((\vartheta_1 - v)(\vartheta_2 - \vartheta_1)^{\frac{\varrho}{\kappa}} - (\vartheta_2 - v)^{\frac{\varrho}{\kappa}+1} + (\vartheta_2 - \vartheta_1)^{\frac{\varrho}{\kappa}+1} \right)}{(\vartheta_2 - \vartheta_1)^{\frac{\varrho}{\kappa}}(\varrho + \kappa)}$$

$$= \frac{\kappa \left((\vartheta_2 - v)(\vartheta_2 - \vartheta_1)^{\frac{\varrho}{\kappa}} - (\vartheta_2 - v)^{\frac{\varrho}{\kappa}+1} \right)}{(\vartheta_2 - \vartheta_1)^{\frac{\varrho}{\kappa}}(\varrho + \kappa)} \geq 0. \quad (19)$$

Now, using the convexity of F and (18) in (19), we obtain

$$\frac{\varrho F(\vartheta_1) + \kappa F(\vartheta_2)}{\varrho + \kappa} \geq \frac{\Gamma_\kappa(\varrho + \kappa)}{(\vartheta_2 - \vartheta_1)^{\frac{\varrho}{\kappa}}} I_{\vartheta_1^+}^{\varrho,\kappa} F(\vartheta_2). \quad (20)$$

Finally, by combining (16) and (20), we arrive at required result. □

The following remark proved the generalization of Theorem 2.

Remark 1. *Substituting $\kappa = 1$ in inequality (8), we find the following results presented in ([16], Theorem 2.2).*

$$F\left(\frac{\varrho \vartheta_1 + \vartheta_2}{\varrho + 1} \right) \leq \frac{\Gamma(\varrho + 1)}{(\vartheta_2 - \vartheta_1)^\varrho} I_{\vartheta_1^+}^\varrho F(\vartheta_2) \leq \frac{\varrho F(\vartheta_1) + F(\vartheta_2)}{\varrho + 1}.$$

In next result, we consider the absolute value of difference presented in (18) and utilizing (19) along with additional conditions on F.

Theorem 3. *Let F be a twice differentiable function on $[\vartheta_1, \vartheta_2]$ and $\varrho, \kappa > 0$. Then, we have the following inequalities*

(i) If $|F''|$ is an increasing function, then

$$\left| \frac{\varrho F(\vartheta_1) + \kappa F(\vartheta_2)}{\varrho + \kappa} - \frac{\Gamma_\kappa(\varrho + \kappa)}{(\vartheta_2 - \vartheta_1)^{\frac{\varrho}{\kappa}}} I_{\vartheta_1^+}^{\varrho,\kappa} F(\vartheta_2) \right| \leq \frac{\varrho \kappa \left| F''(\vartheta_2) \right| (\vartheta_2 - \vartheta_1)^2}{2(\varrho + \kappa)(\varrho + 2\kappa)}. \quad (21)$$

(ii) If $|F''|$ is decreasing function, then

$$\left| \frac{\varrho F(\vartheta_1) + \kappa F(\vartheta_2)}{\varrho + \kappa} - \frac{\Gamma_\kappa(\varrho + \kappa)}{(\vartheta_2 - \vartheta_1)^{\frac{\varrho}{\kappa}}} I_{\vartheta_1^+}^{\varrho,\kappa} F(\vartheta_2) \right| \leq \frac{\varrho \kappa \left| F''(\vartheta_1) \right| (\vartheta_2 - \vartheta_1)^2}{2(\varrho + \kappa)(\varrho + 2\kappa)}. \quad (22)$$

(iii) If $|F''|$ is a convex function, then

$$\left| \frac{\varrho F(\vartheta_1) + \kappa F(\vartheta_2)}{\varrho + \kappa} - \frac{\Gamma_\kappa(\varrho + \kappa)}{(\vartheta_2 - \vartheta_1)^{\frac{\varrho}{\kappa}}} I_{\vartheta_1^+}^{\varrho,\kappa} F(\vartheta_2) \right| \leq \frac{\max \left(\left|F''(\vartheta_1)\right|, \left|F''(\vartheta_2)\right| \right) \varrho \kappa (\vartheta_2 - \vartheta_1)^2}{2(\varrho + \kappa)(\varrho + 2\kappa)}. \quad (23)$$

Proof. (i) From (18) and (19), we can write

$$\left|\frac{\varrho F(\vartheta_1)+\kappa F(\vartheta_2)}{\varrho+\kappa}-\frac{\Gamma_\kappa(\varrho+\kappa)}{(\vartheta_2-\vartheta_1)^{\frac{\varrho}{\kappa}}}I^{\varrho,\kappa}_{\vartheta_1^+}F(\vartheta_2)\right|$$

$$\leq \frac{\kappa}{(\vartheta_2-\vartheta_1)^{\frac{\varrho}{\kappa}}(\varrho+\kappa)}\int_{\vartheta_1}^{\vartheta_2}\left((\vartheta_2-v)(\vartheta_2-\vartheta_1)^{\frac{\varrho}{\kappa}}-(\vartheta_2-v)^{\frac{\varrho}{\kappa}+1}\right)F''(v)dv. \qquad (24)$$

Since $(\vartheta_2-\vartheta_1)^{\frac{\varrho}{\kappa}}(\vartheta_2-v)-(\vartheta_2-v)^{\frac{\varrho}{\kappa}+1}\geq 0$ and $|F''|$ is an increasing function, this implies

$$\left|\frac{\varrho F(\vartheta_1)+\kappa F(\vartheta_2)}{\varrho+\kappa}-\frac{\Gamma_\kappa(\varrho+\kappa)}{(\vartheta_2-\vartheta_1)^{\frac{\varrho}{\kappa}}}I^{\varrho,\kappa}_{\vartheta_1^+}F(\vartheta_2)\right|$$

$$\leq \frac{\kappa|F''(\vartheta_2)|}{(\vartheta_2-\vartheta_1)^{\frac{\varrho}{\kappa}}(\varrho+\kappa)}\int_{\vartheta_1}^{\vartheta_2}\left((\vartheta_2-v)(\vartheta_2-\vartheta_1)^{\frac{\varrho}{\kappa}}-(\vartheta_2-v)^{\frac{\varrho}{\kappa}+1}\right)dv$$

$$= \frac{\kappa|F''(\vartheta_2)|}{(\vartheta_2-\vartheta_1)^{\frac{\varrho}{\kappa}}(\varrho+\kappa)}\left(\frac{(\vartheta_2-\vartheta_1)^{\frac{\varrho}{\kappa}}(\vartheta_2-\vartheta_1)^2}{2}-\frac{\kappa(\vartheta_2-\vartheta_1)^{\frac{\varrho}{\kappa}+2}}{\varrho+2\kappa}\right)$$

$$= \frac{\kappa|F''(\vartheta_2)|\left(\varrho(\vartheta_2-\vartheta_1)^2\right)}{2(\varrho+\kappa)(\varrho+2\kappa)},$$

which is inequality (21).

(ii) Again, using (18) and (19) and by following the same procedure as in case (i), we obtain

$$\left|\frac{\varrho F(\vartheta_1)+\kappa F(\vartheta_2)}{\varrho+\kappa}-\frac{\Gamma_\kappa(\varrho+\kappa)}{(\vartheta_2-\vartheta_1)^{\frac{\varrho}{\kappa}}}I^{\varrho,\kappa}_{\vartheta_1^+}F(\vartheta_2)\right|\leq \frac{\kappa|F''(\vartheta_1)|\varrho(\vartheta_2-\vartheta_1)^2}{2(\varrho+\kappa)(\varrho+2\kappa)}.$$

(iii) By using (21) and (22) and the fact that F'' is bounded above by $\max\left(|F''(\vartheta_1)|,|F''(\vartheta_2)|\right)$ being a convex function on the interval $(\vartheta_1,\vartheta_2)$, we find

$$\left|\frac{\varrho F(\vartheta_1)+\kappa F(\vartheta_2)}{\varrho+\kappa}-\frac{\Gamma_\kappa(\varrho+\kappa)}{(\vartheta_2-\vartheta_1)^{\frac{\varrho}{\kappa}}}I^{\varrho,\kappa}_{\vartheta_1^+}F(\vartheta_2)\right|\leq \frac{\kappa\max\left|F''(\vartheta_1)|,|F''(\vartheta_2)|\right|\varrho(\vartheta_2-\vartheta_1)^2}{2(\varrho+\kappa)(\varrho+2\kappa)}.$$

□

The following remark relates the above theorem with the published results in [16].

Remark 2. *With a choice $\kappa=1$ in inequalities (21)–(23), we find the following results presented in ([16], Theorem 2.3).*

$$\left|\frac{\varrho F(\vartheta_1)+F(\vartheta_2)}{\varrho+1}-\frac{\Gamma(\varrho+1)}{(\vartheta_2-\vartheta_1)^{\varrho}}I^{\varrho}_{\vartheta_1^+}F(\vartheta_2)\right|\leq \frac{|F''(\vartheta_2)|\varrho(\vartheta_2-\vartheta_1)^2}{2(\varrho+1)(\varrho+2)},$$

$$\left|\frac{\varrho F(\vartheta_1)+F(\vartheta_2)}{\varrho+1}-\frac{\Gamma(\varrho+1)}{(\vartheta_2-\vartheta_1)^{\varrho}}I^{\varrho}_{\vartheta_1^+}F(\vartheta_2)\right|\leq \frac{|F''(\vartheta_1)|\varrho(\vartheta_2-\vartheta_1)^2}{2(\varrho+1)(\varrho+2)},$$

$$\left|\frac{\varrho F(\vartheta_1) + F(\vartheta_2)}{\varrho + 1} - \frac{\Gamma(\varrho+1)}{(\vartheta_2 - \vartheta_1)^\varrho} I^\varrho_{\vartheta_1^+} F(\vartheta_2)\right| \leq \frac{\max\left(\left|F''(\vartheta_1)\right|, \left|F''(\vartheta_2)\right|\right) \varrho(\vartheta_2 - \vartheta_1)^2}{2(\varrho+1)(\varrho+2)}.$$

Using Green's function from (13) and some additional features on F, we obtain the following theorem.

Theorem 4. *Let F be a twice differentiable function on $[\vartheta_1, \vartheta_2]$ and $\varrho, \kappa > 0$. Then, the following statements holds.*

(i) *If $|F''|$ is an increasing function, then*

$$\left|F\left(\frac{\varrho\vartheta_1 + \kappa\vartheta_2}{\varrho + \kappa}\right) - \frac{\Gamma_\kappa(\varrho + \kappa)}{(\vartheta_2 - \vartheta_1)^{\frac{\varrho}{\kappa}}} I^{\varrho,\kappa}_{\vartheta_1^+} F(\vartheta_2)\right|$$
$$\leq \frac{\kappa(\kappa)^{\frac{\varrho}{\kappa}+3}(\vartheta_2 - \vartheta_1)^2}{(\varrho + \kappa)^{\frac{\varrho}{\kappa}+3}(\varrho + 2\kappa)}\left(\left|F''\left(\frac{\varrho\vartheta_1 + \kappa\vartheta_2}{\varrho + \kappa}\right)\right|\right.$$
$$\left.\left(\frac{\varrho\left((\varrho + \kappa)^{\frac{\varrho}{\kappa}+1} - 2(\varrho)^{\frac{\varrho}{\kappa}+1}\right)}{2\kappa(\kappa)^{\frac{\varrho}{\kappa}+1}}\right) + \left|F''(\vartheta_2)\right|\left(\frac{\varrho}{\kappa}\right)^{\frac{\varrho}{\kappa}+2}\right).$$

(ii) *If $|F''|$ is a decreasing function, then*

$$\left|F\left(\frac{\varrho\vartheta_1 + \kappa\vartheta_2}{\varrho + \kappa}\right) - \frac{\Gamma_\kappa(\varrho + \kappa)}{(\vartheta_2 - \vartheta_1)^{\frac{\varrho}{\kappa}}} I^{\varrho,\kappa}_{\vartheta_1^+} F(\vartheta_2)\right|$$
$$\leq \frac{\kappa(\kappa)^{\frac{\varrho}{\kappa}+3}(\vartheta_2 - \vartheta_1)^2}{(\varrho + \kappa)^{\frac{\varrho}{\kappa}+3}(\varrho + 2\kappa)}\left(\left|F''(\vartheta_1)\right|\right.$$
$$\left.\left(\frac{\varrho\left((\varrho + \kappa)^{\frac{\varrho}{\kappa}+1} - 2(\varrho)^{\frac{\varrho}{\kappa}+1}\right)}{2\kappa(\kappa)^{\frac{\varrho}{\kappa}+1}}\right) + \left|F''\left(\frac{\varrho\vartheta_1 + \kappa\vartheta_2}{\varrho + \kappa}\right)\right|\left(\frac{\varrho}{\kappa}\right)^{\frac{\varrho}{\kappa}+2}\right).$$

(iii) *If $|F''|$ is a convex function, then*

$$\left|F\left(\frac{\varrho\vartheta_1 + \kappa\vartheta_2}{\varrho + \kappa}\right) - \frac{\Gamma_\kappa(\varrho + \kappa)}{(\vartheta_2 - \vartheta_1)^{\frac{\varrho}{\kappa}}} I^{\varrho,\kappa}_{\vartheta_1^+} F(\vartheta_2)\right| \leq \frac{\kappa(\kappa)^{\frac{\varrho}{\kappa}+3}(\vartheta_2 - \vartheta_1)^2}{(\varrho + \kappa)^{\frac{\varrho}{\kappa}+3}(\varrho + 2\kappa)}$$
$$\times \left(\max\left(\left|F''(\vartheta_1)\right|, \left|F''\left(\frac{\varrho\vartheta_1 + \kappa\vartheta_2}{\varrho + \kappa}\right)\right|\right)\right)\left(\frac{\varrho\left((\varrho + \kappa)^{\frac{\varrho}{\kappa}+1} - 2(\varrho)^{\frac{\varrho}{\kappa}+1}\right)}{2\kappa(\kappa)^{\frac{\varrho}{\kappa}+1}}\right)$$
$$+ \max\left(\left|F''\left(\frac{\varrho\vartheta_1 + \kappa\vartheta_2}{\varrho + \kappa}\right)\right|, \left|F''(\vartheta_2)\right|\right)\left(\frac{\varrho}{\kappa}\right)^{\frac{\varrho}{\kappa}+2}.$$

Proof. (i) By using (11)–(13), we can write

$$F\left(\frac{\varrho\vartheta_1+\kappa\vartheta_2}{\varrho+\kappa}\right)-\frac{\Gamma_\kappa(\varrho+\kappa)}{(\vartheta_2-\vartheta_1)^{\frac{\varrho}{\kappa}}}I_{\vartheta_1^+}^{\varrho,\kappa}F(\vartheta_2) = \int_{\vartheta_1}^{\frac{\varrho\vartheta_1+\kappa\vartheta_2}{\varrho+\kappa}}\left(G\left(\frac{\varrho\vartheta_1+\kappa\vartheta_2}{\varrho+\kappa},v\right)\right.$$

$$-\frac{\varrho}{\kappa(\vartheta_2-\vartheta_1)^{\frac{\varrho}{\kappa}}}\int_{\vartheta_1}^{\vartheta_2}G(\varsigma,v)(\vartheta_2-\varsigma)^{\frac{\varrho}{\kappa}-1}d\varsigma\right)F''(v)dv+\int_{\frac{\varrho\vartheta_1+\kappa\vartheta_2}{\varrho+\kappa}}^{\vartheta_2}\left(G\left(\frac{\varrho\vartheta_1+\kappa\vartheta_2}{\varrho+\kappa},v\right)\right.$$

$$-\frac{\varrho}{\kappa(\vartheta_2-\vartheta_1)^{\frac{\varrho}{\kappa}}}\int_{\vartheta_1}^{\vartheta_2}G(\varsigma,v)(\vartheta_2-\varsigma)^{\frac{\varrho}{\kappa}-1}d\varsigma\bigg)F''(v)dv$$

$$=-\frac{\kappa}{(\varrho+\kappa)(\vartheta_2-\vartheta_1)^{\frac{\varrho}{\kappa}}}\left(\int_{\vartheta_1}^{\frac{\varrho\vartheta_1+\kappa\vartheta_2}{\varrho+\kappa}}\left((\vartheta_2-v)^{\frac{\varrho}{\kappa}+1}-(\vartheta_2-\vartheta_1)^{\frac{\varrho}{\kappa}+1}-\left(\frac{\varrho+\kappa}{\kappa}\right)\right.\right.$$

$$\times(\vartheta_2-\vartheta_1)^{\frac{\varrho}{\kappa}}(\vartheta_1-v)\bigg)F''(v)dv+\int_{\frac{\varrho\vartheta_1+\kappa\vartheta_2}{\varrho+\kappa}}^{\vartheta_2}(\vartheta_2-v)^{\frac{\varrho}{\kappa}+1}F''(v)dv\bigg). \quad (25)$$

This can also be written as

$$\left|F\left(\frac{\varrho\vartheta_1+\kappa\vartheta_2}{\varrho+\kappa}\right)-\frac{\Gamma_\kappa(\varrho+\kappa)}{(\vartheta_2-\vartheta_1)^{\frac{\varrho}{\kappa}}}I_{\vartheta_1^+}^{\varrho,\kappa}F(\vartheta_2)\right|$$

$$\leq\frac{\kappa}{(\varrho+\kappa)(\vartheta_2-\vartheta_1)^{\frac{\varrho}{\kappa}}}\Bigg(\int_{\vartheta_1}^{\frac{\varrho\vartheta_1+\kappa\vartheta_2}{\varrho+\kappa}}\bigg|(\vartheta_2-v)^{\frac{\varrho}{\kappa}+1}-(\vartheta_2-\vartheta_1)^{\frac{\varrho}{\kappa}+1}$$

$$-\left(\frac{\varrho+\kappa}{\kappa}\right)(\vartheta_2-\vartheta_1)^{\frac{\varrho}{\kappa}}(\vartheta_1-v)\bigg|\big|F''(v)\big|dv$$

$$+\int_{\frac{\varrho\vartheta_1+\kappa\vartheta_2}{\varrho+\kappa}}^{\vartheta_2}(\vartheta_2-v)^{\frac{\varrho}{\kappa}+1}\big|F''(v)\big|dv\Bigg)$$

$$\leq\frac{\kappa}{(\varrho+\kappa)(\vartheta_2-\vartheta_1)^{\frac{\varrho}{\kappa}}}\left(\left|F''\left(\frac{\varrho\vartheta_1+\kappa\vartheta_2}{\varrho+\kappa}\right)\right|\int_{\vartheta_1}^{\frac{\varrho\vartheta_1+\kappa\vartheta_2}{\varrho+\kappa}}(\vartheta_2-v)^{\frac{\varrho}{\kappa}+1}\right.$$

$$-(\vartheta_2-\vartheta_1)^{\frac{\varrho}{\kappa}+1}-\left(\frac{\varrho+\kappa}{\kappa}\right)(\vartheta_2-\vartheta_1)^{\frac{\varrho}{\kappa}}(\vartheta_1-v)dv$$

$$+\big|F''(\vartheta_2)\big|\int_{\frac{\varrho\vartheta_1+\kappa\vartheta_2}{\varrho+\kappa}}^{\vartheta_2}(\vartheta_2-v)^{\frac{\varrho}{\kappa}+1}dv\Bigg)$$

$$=\frac{\kappa}{(\varrho+\kappa)(\vartheta_2-\vartheta_1)^{\frac{\varrho}{\kappa}}}\left(\left|F''\left(\frac{\varrho\vartheta_1+\kappa\vartheta_2}{\varrho+\kappa}\right)\right|(\vartheta_2-\vartheta_1)^{\frac{\varrho}{\kappa}+2}\right.$$

$$\times\left(\frac{-\kappa(\varrho)^{\frac{\varrho}{\kappa}+2}}{(\varrho+\kappa)^{\frac{\varrho}{\kappa}+2}(\varrho+2\kappa)}+\frac{\kappa}{\varrho+2\kappa}-\frac{\kappa}{\varrho+\kappa}+\frac{\kappa}{2(\varrho+\kappa)}\right)$$

$$+\big|F''(\vartheta_2)\big|(\vartheta_2-\vartheta_1)^{\frac{\varrho}{\kappa}+2}\left(\frac{\kappa(\varrho)^{\frac{\varrho}{\kappa}+2}}{(\varrho+\kappa)^{\frac{\varrho}{\kappa}+2}(\varrho+2\kappa)}\right)\Bigg)$$

$$= \frac{\kappa(\kappa)^{\frac{\varrho}{\kappa}+3}(\vartheta_2-\vartheta_1)^2}{(\varrho+\kappa)^{\frac{\varrho}{\kappa}+3}(\varrho+2\kappa)} \left(\left|F''\left(\frac{\varrho\vartheta_1+\kappa\vartheta_2}{\varrho+\kappa}\right)\right|\right.$$

$$\left.\times \left(\frac{\varrho\left((\varrho+\kappa)^{\frac{\varrho}{\kappa}+1}-2(\varrho)^{\frac{\varrho}{\kappa}+1}\right)}{2\kappa(\kappa)^{\frac{\varrho}{\kappa}+1}}\right) + \left|F''(\vartheta_2)\right|\left(\frac{\varrho}{\kappa}\right)^{\frac{\varrho}{\kappa}+2}\right).$$

Part (ii) can be proved by the same procedure as above.

(iii) Since

$$\left|F\left(\frac{\varrho\vartheta_1+\kappa\vartheta_2}{\varrho+\kappa}\right) - \frac{\Gamma_\kappa(\varrho+\kappa)}{(\vartheta_2-\vartheta_1)^{\frac{\varrho}{\kappa}}} I^{\varrho,\kappa}_{\vartheta_1^+} F(\vartheta_2)\right|$$

$$\leq \frac{\kappa}{(\varrho+\kappa)(\vartheta_2-\vartheta_1)^{\frac{\varrho}{\kappa}}} \left(\left(\int_{\vartheta_1}^{\frac{\varrho\vartheta_1+\kappa\vartheta_2}{\varrho+\kappa}} (\vartheta_2-v)^{\frac{\varrho}{\kappa}+1}\right.\right.$$

$$\left.\left. - (\vartheta_2-\vartheta_1)^{\frac{\varrho}{\kappa}+1} - \left(\frac{\varrho}{\kappa}+1\right)(\vartheta_2-\vartheta_1)^{\frac{\varrho}{\kappa}}(\vartheta_1-v)\right)|F''(v)|dv\right.$$

$$\left. + \int_{\frac{\varrho\vartheta_1+\kappa\vartheta_2}{\varrho+\kappa}}^{\vartheta_2} (\vartheta_2-v)^{\frac{\varrho}{\kappa}+1}|F''(v)|dv\right).$$

Since every convex function F defined on an interval $[\vartheta_1,\vartheta_2]$ is bounded above by $\max\{F(\vartheta_1),F(\vartheta_2)\}$. Therefore, we have

$$\left|F\left(\frac{\varrho\vartheta_1+\kappa\vartheta_2}{\varrho+\kappa}\right) - \frac{\Gamma_\kappa(\varrho+\kappa)}{(\vartheta_2-\vartheta_1)^{\frac{\varrho}{\kappa}}} I^{\varrho,\kappa}_{\vartheta_1^+} F(\vartheta_2)\right|$$

$$\leq \frac{\kappa}{(\varrho+\kappa)(\vartheta_2-\vartheta_1)^{\frac{\varrho}{\kappa}}} \left(\max\left(\left|F''(\vartheta_1)\right|,\left|F''\left(\frac{\varrho\vartheta_1+\kappa\vartheta_2}{\varrho+\kappa}\right)\right|\right)\right.$$

$$\times \int_{\vartheta_1}^{\frac{\varrho\vartheta_1+\kappa\vartheta_2}{\varrho+\kappa}} \left((\vartheta_2-v)^{\frac{\varrho}{\kappa}+1} - (\vartheta_2-\vartheta_1)^{\frac{\varrho}{\kappa}+1} - \left(\frac{\varrho}{\kappa}+1\right)\right.$$

$$\left.\times (\vartheta_2-\vartheta_1)^{\frac{\varrho}{\kappa}}(\vartheta_1-v)\right)dv + \max\left(\left|F''\left(\frac{\varrho\vartheta_1+\kappa\vartheta_2}{\varrho+\kappa}\right)\right|,\left|F''(\vartheta_2)\right|\right)$$

$$\left.\times \int_{\frac{\varrho\vartheta_1+\kappa\vartheta_2}{\varrho+\kappa}}^{\vartheta_2} (\vartheta_2-v)^{\frac{\varrho}{\kappa}+1}dv\right)$$

$$= \frac{\kappa(\kappa)^{\frac{\varrho}{\kappa}+3}(\vartheta_2-\vartheta_1)^2}{(\varrho+\kappa)^{\frac{\varrho}{\kappa}+3}(\varrho+2\kappa)} \left(\max\left(\left|F''(\vartheta_1)\right|,\left|F''\left(\frac{\varrho\vartheta_1+\kappa\vartheta_2}{\varrho+\kappa}\right)\right|\right)\right.$$

$$\times \left(\frac{\varrho\left((\varrho+\kappa)^{\frac{\varrho}{\kappa}+1}-2(\varrho)^{\frac{\varrho}{\kappa}+1}\right)}{2\kappa(\kappa)^{\frac{\varrho}{\kappa}+1}}\right) + \max\left(\left|F''\left(\frac{\varrho\vartheta_1+\kappa\vartheta_2}{\varrho+\kappa}\right)\right|,\right.$$

$$\left.\left.\left|F''(\vartheta_2)\right|\right)\left(\frac{\varrho}{\kappa}\right)^{\frac{\varrho}{\kappa}+2}\right),$$

which is the desired inequality. □

Remark 3. *Let $\kappa=1$, and we obtain the following results presented in ([16], Theorem 2.5).*

The following theorem involves the change of variables in Theorem 4.

Theorem 5. *Let F be a twice differentiable and $|F''|$ be a convex function on $[\vartheta_1, \vartheta_2]$. Then, the inequality*

$$\left| F\left(\frac{\varrho\vartheta_1 + \kappa\vartheta_2}{\varrho + \kappa}\right) - \frac{\Gamma_\kappa(\varrho + \kappa)}{(\vartheta_2 - \vartheta_1)^{\frac{\varrho}{\kappa}}} I^{\varrho,\kappa}_{\vartheta_1^+} F(\vartheta_2) \right|$$

$$\leq \frac{\kappa^3(\vartheta_2 - \vartheta_1)^2}{6(\varrho + \kappa)^3(\varrho + 3\kappa)} \left(|F''(\vartheta_1)| \left(9\left(\frac{\varrho^2}{\kappa}\right) + 23\varrho + 12\kappa\right) \right.$$

$$+ \left. |F''(\vartheta_2)| \frac{(7(\varrho^2) + 17\varrho\kappa + 12\kappa^2)}{\varrho + 2\kappa} \right).$$

holds for any $\varrho, \kappa > 0$.

Proof. Equation (25) can be written as

$$F\left(\frac{\varrho\vartheta_1 + \kappa\vartheta_2}{\varrho + \kappa}\right) - \frac{\Gamma_\kappa(\varrho + \kappa)}{(\vartheta_2 - \vartheta_1)^{\frac{\varrho}{\kappa}}} I^{\varrho,\kappa}_{\vartheta_1^+} F(\vartheta_2)$$

$$= \frac{\kappa}{(\varrho + \kappa)(\vartheta_2 - \vartheta_1)^{\frac{\varrho}{\kappa}}} \left(\int_{\vartheta_1}^{\frac{\varrho\vartheta_1 + \kappa\vartheta_2}{\varrho + \kappa}} \left(\left(\frac{\varrho + \kappa}{\kappa}\right)(\vartheta_2 - \vartheta_1)^{\frac{\varrho}{\kappa}} \right. \right.$$

$$\times (\vartheta_1 - v) - (\vartheta_2 - v)^{\frac{\varrho}{\kappa}+1} + (\vartheta_2 - \vartheta_1)^{\frac{\varrho}{\kappa}+1} \Big) F''(v) dv$$

$$- \int_{\frac{\varrho\vartheta_1 + \kappa\vartheta_2}{\varrho + \kappa}}^{\vartheta_2} (\vartheta_2 - v)^{\frac{\varrho}{\kappa}+1} F''(v) dv \bigg).$$

Let $\tau \in [0, 1]$ and $v = \tau\vartheta_1 + (1 - \tau)\vartheta_2$, then

$$F\left(\frac{\varrho\vartheta_1 + \kappa\vartheta_2}{\varrho + \kappa}\right) - \frac{\Gamma_\kappa(\varrho + \kappa)}{(\vartheta_2 - \vartheta_1)^{\frac{\varrho}{\kappa}}} I^{\varrho,\kappa}_{\vartheta_1^+} F(\vartheta_2)$$

$$= \frac{\kappa}{(\varrho + \kappa)(\vartheta_2 - \vartheta_1)^{\frac{\varrho}{\kappa}}} \left(\int_1^{\frac{\varrho}{\varrho+\kappa}} \left(\left(\frac{\varrho}{\kappa} + 1\right)(\vartheta_2 - \vartheta_1)^{\frac{\varrho}{\kappa}} \left(\vartheta_1 - \tau\vartheta_1 \right. \right.\right.$$

$$- (1-\tau)\vartheta_2 \bigg) - \bigg(\vartheta_2 - \tau\vartheta_1 - (1-\tau)\vartheta_2\bigg)^{\frac{\varrho}{\kappa}+1}$$

$$+ (\vartheta_2 - \vartheta_1)^{\frac{\varrho}{\kappa}+1} \bigg) F''\bigg(\tau\vartheta_1 + (1-\tau)\vartheta_2\bigg)\bigg(\vartheta_1 - \vartheta_2\bigg) d\tau$$

$$- \int_{\frac{\varrho}{\varrho+\kappa}}^0 \bigg(\vartheta_2 - \tau\vartheta_1 - (1-\tau)\vartheta_2\bigg)^{\frac{\varrho}{\kappa}+1} F''\bigg(\tau\vartheta_1 + (1-\tau)\vartheta_2\bigg)$$

$$\times \bigg(\vartheta_1 - \vartheta_2\bigg) d\tau \bigg)$$

$$= -\frac{\kappa(\vartheta_2 - \vartheta_1)^{\frac{\varrho}{\kappa}+2}}{(\varrho + \kappa)(\vartheta_2 - \vartheta_1)^{\frac{\varrho}{\kappa}}} \left(\int_1^{\frac{\varrho}{\varrho+\kappa}} \left(-(1-\tau)\left(\frac{\varrho}{\kappa} + 1\right) + 1 \right) \right. \tag{26}$$

$$\times F''\Big(\tau\vartheta_1+(1-\tau)\vartheta_2\Big)d\tau - \int_1^{\frac{\varrho}{\varrho+\kappa}} \tau^{\frac{\varrho}{\kappa}+1} F''\Big(\tau\vartheta_1+(1-\tau)\vartheta_2\Big)d\tau$$

$$+ \int_0^{\frac{\varrho}{\varrho+\kappa}} \tau^{\frac{\varrho}{\kappa}+1} F''\Big(\tau\vartheta_1+(1-\tau)\vartheta_2\Big)d\tau \Bigg)$$

$$= \frac{\kappa(\vartheta_2-\vartheta_1)^2}{(\varrho+\kappa)} \Bigg(\int_{\frac{\varrho}{\varrho+\kappa}}^1 \bigg(\Big(\frac{\varrho}{\kappa}\Big)\tau+\tau-\frac{\varrho}{\kappa}\bigg) F''\Big(\tau\vartheta_1+(1-\tau)\vartheta_2\Big)d\tau$$

$$- \int_0^1 \tau^{\frac{\varrho}{\kappa}+1} F''\Big(\tau\vartheta_1+(1-\tau)\vartheta_2\Big)d\tau \Bigg).$$

Taking the absolute on both sides and using the convexity of $|F''|$, we obtain

$$\Bigg| F\Big(\frac{\varrho\vartheta_1+\kappa\vartheta_2}{\varrho+\kappa}\Big) - \frac{\Gamma_\kappa(\varrho+\kappa)}{(\vartheta_2-\vartheta_1)^{\frac{\varrho}{\kappa}}} I_{\vartheta_1^+}^{\varrho,\kappa} F(\vartheta_2) \Bigg|$$

$$\leq \frac{\kappa(\vartheta_2-\vartheta_1)^2}{(\varrho+\kappa)} \Bigg(\int_{\frac{\varrho}{\varrho+\kappa}}^1 \bigg(\Big(\frac{\varrho}{\kappa}\Big)\tau+\tau-\frac{\varrho}{\kappa}\bigg)$$

$$\times \Big| F''\Big(\tau\vartheta_1+(1-\tau)\vartheta_2\Big)\Big| d\tau$$

$$+ \int_0^1 \tau^{\frac{\varrho}{\kappa}+1} \Big| F''\Big(\tau\vartheta_1+(1-\tau)\vartheta_2\Big)\Big| d\tau \Bigg)$$

$$\leq \frac{\kappa(\vartheta_2-\vartheta_1)^2}{(\varrho+\kappa)} \Bigg(\int_{\frac{\varrho}{\varrho+\kappa}}^1 \bigg(\tau\Big(\frac{\varrho}{\kappa}+1\Big)-\frac{\varrho}{\kappa}\bigg)$$

$$\times \Big(\tau\big|F''(\vartheta_1)\big|+(1-\tau)\big|F''(\vartheta_2)\big|\Big) d\tau$$

$$+ \int_0^1 \tau^{\frac{\varrho}{\kappa}+1} \Big(\tau\big|F''(\vartheta_1)\big|+\big(1-\tau\big)\big|F''(\vartheta_2)\big|\Big) d\tau \Bigg)$$

$$\leq \frac{\kappa(\vartheta_2-\vartheta_1)^2}{(\varrho+\kappa)} \Bigg(\frac{\kappa\big|F''(\vartheta_1)\big|\big(3\varrho+2\kappa\big)}{6(\varrho+\kappa)^2} + \frac{\kappa^2\big|F''(\vartheta_2)\big|}{6(\varrho+\kappa)^2}$$

$$+ \frac{\kappa\big|F''(\vartheta_1)\big|}{\varrho+3\kappa} + \frac{\kappa^2\big|F''(\vartheta_2)\big|}{(\varrho+2\kappa)(\varrho+3\kappa)} \Bigg)$$

$$\leq \frac{\kappa^3(\vartheta_2-\vartheta_1)^2}{6(\varrho+\kappa)^3(\varrho+3\kappa)} \Bigg(\big|F''(\vartheta_1)\big|\Big(9\Big(\frac{\varrho^2}{\kappa}\Big)+23\varrho+12\kappa\Big)$$

$$+ \frac{\big|F''(\vartheta_2)\big|\big(7(\varrho^2)+17\varrho\kappa+12\kappa^2\big)}{\varrho+2\kappa} \Bigg).$$

Hence, the proof is done. □

Remark 4. By setting $\kappa = 1$, we obtain the following result presented in ([16], Theorem 2.7).

$$\left| F\left(\frac{\varrho\vartheta_1 + \vartheta_2}{\varrho + 1}\right) - \frac{\Gamma(\varrho+1)}{(\vartheta_2 - \vartheta_1)^\varrho} I^\varrho_{\vartheta_1^+} F(\vartheta_2) \right|$$
$$\leq \frac{(\vartheta_2 - \vartheta_1)^2}{6(\varrho+1)^3(\varrho+3)} \left(|F''(\vartheta_1)|(9(\varrho^2) + 23\varrho + 12) \right.$$
$$+ \left. |F''(\vartheta_2)| \frac{(7(\varrho^2) + 17\varrho + 12)}{\varrho+2} \right).$$

Theorem 6. Let F be a twice differentiable and $|F''|$ be a convex function on $[\vartheta_1, \vartheta_2]$. Then, the inequality

$$\left| \frac{\varrho F(\vartheta_1) + \kappa F(\vartheta_2)}{\varrho + \kappa} - \frac{\Gamma_\kappa(\varrho+\kappa)}{(\vartheta_2-\vartheta_1)^{\frac{\varrho}{\kappa}}} I^{\varrho,\kappa}_{\vartheta_1^+} F(\vartheta_2) \right|$$
$$\leq \frac{\kappa \varrho (\vartheta_2 - \vartheta_1)^2}{3(\varrho+\kappa)(\varrho+3\kappa)} \left(|F''(\vartheta_1)| + |F''(\vartheta_2)| \frac{(\varrho+5\kappa)}{2(\varrho+2\kappa)} \right)$$

holds for any $\varrho, \kappa > 0$.

Proof. From Equation (18), we can write

$$\frac{\varrho F(\vartheta_1) + \kappa F(\vartheta_2)}{\varrho + \kappa} - \frac{\Gamma_\kappa(\varrho+\kappa)}{(\vartheta_2-\vartheta_1)^{\frac{\varrho}{\kappa}}} I^{\varrho,\kappa}_{\vartheta_1^+} F(\vartheta_2)$$
$$= \frac{\kappa}{(\vartheta_2-\vartheta_1)^{\frac{\varrho}{\kappa}}(\varrho+\kappa)} \int_{\vartheta_1}^{\vartheta_2} \left((\vartheta_2 - v)(\vartheta_2 - \vartheta_1)^{\frac{\varrho}{\kappa}} \right.$$
$$- \left. (\vartheta_2 - v)^{\frac{\varrho}{\kappa}+1} \right) F''(v) dv.$$

For $\tau \in [0,1]$, substituting $v = \tau\vartheta_1 + (1-\tau)\vartheta_2$, we obtain

$$\frac{\varrho F(\vartheta_1) + \kappa F(\vartheta_2)}{\varrho + \kappa} - \frac{\Gamma_\kappa(\varrho+\kappa)}{(\vartheta_2-\vartheta_1)^{\frac{\varrho}{\kappa}}} I^{\varrho,\kappa}_{\vartheta_1^+} F(\vartheta_2)$$
$$= \frac{\kappa(\vartheta_2-\vartheta_1)}{(\vartheta_2-\vartheta_1)^{\frac{\varrho}{\kappa}}(\varrho+\kappa)} \int_0^1 \left((\vartheta_2-\vartheta_1)^{\frac{\varrho}{\kappa}} (\tau(\vartheta_2-\vartheta_1)) \right.$$
$$- \left. (\tau(\vartheta_2-\vartheta_1)^{\frac{\varrho}{\kappa}+1}) \right) F''\left(\tau\vartheta_1 + (1-\tau)\vartheta_2\right) d\tau,$$
$$= \frac{\kappa(\vartheta_2-\vartheta_1)^2}{\varrho+\kappa} \int_0^1 \left(\tau - \tau^{\frac{\varrho}{\kappa}+1}\right) F''\left(\tau\vartheta_1 + (1-\tau)\vartheta_2\right) d\tau. \qquad (27)$$

By using the convexity of $|F''|$, we obtain

$$\left| \frac{\varrho F(\vartheta_1) + \kappa F(\vartheta_2)}{\varrho + \kappa} - \frac{\Gamma_\kappa(\varrho+\kappa)}{(\vartheta_2-\vartheta_1)^{\frac{\varrho}{\kappa}}} I^{\varrho,\kappa}_{\vartheta_1^+} F(\vartheta_2) \right|$$
$$\leq \frac{\kappa(\vartheta_2-\vartheta_1)^2}{\varrho+\kappa} \int_0^1 \left(\tau - \tau^{\frac{\varrho}{\kappa}+1}\right) \left| F''\left(\tau\vartheta_1 + (1-\tau)\vartheta_2\right) \right| d\tau,$$

$$\leq \frac{\kappa(\vartheta_2 - \vartheta_1)^2}{\varrho + \kappa} \int_0^1 \left(\tau - \tau^{\frac{\varrho}{\kappa}+1}\right)\left(\tau\left|F''(\vartheta_1)\right| + (1-\tau)\left|F''(\vartheta_2)\right|\right)d\tau,$$

$$= \frac{\kappa(\vartheta_2 - \vartheta_1)^2}{\varrho + \kappa}\left(\left|F''(\vartheta_1)\right|\left(\frac{\varrho}{3(\varrho + 3\kappa)}\right)\right.$$

$$+ \left|F''(\vartheta_2)\right|\left(\frac{(\varrho^2 + 5\varrho\kappa)}{6(\varrho + 2\kappa)(\varrho + 3\kappa)}\right)\right)$$

$$= \frac{\varrho\kappa(\vartheta_2 - \vartheta_1)^2}{3(\varrho + \kappa)(\varrho + 3\kappa)}\left(\left|F''(\vartheta_1)\right| + \left|F''(\vartheta_2)\right|\frac{(\varrho + 5\kappa)}{2(\varrho + 2\kappa)}\right).$$

This completes the proof. □

Remark 5. *Corresponding to the choice* $\kappa = 1$, *in Theorem 6, we obtain the following result explored in ([16], Theorem 2.9)*

$$\left|\frac{\varrho F(\vartheta_1) + F(\vartheta_2)}{\varrho + 1} - \frac{\Gamma(\varrho + 1)}{(\vartheta_2 - \vartheta_1)^{\varrho}} I^{\varrho}_{\vartheta_1^+} F(\vartheta_2)\right|$$

$$\leq \frac{\varrho(\vartheta_2 - \vartheta_1)^2}{3(\varrho + 1)(\varrho + 3)}\left(\left|F''(\vartheta_1)\right| + \left|F''(\vartheta_2)\right|\frac{(\varrho + 5)}{2(\varrho + 2)}\right).$$

The next theorem is a combination of Equation (26) given in Theorem 5 and the well-known Jensen's inequality.

Theorem 7. *Let F be a twice differentiable and $|F''|$ be a concave function on $[\vartheta_1, \vartheta_2]$. Then, the inequality*

$$\left|F\left(\frac{\varrho\vartheta_1 + \kappa\vartheta_2}{\varrho + \kappa}\right) - \frac{\Gamma_{\kappa}(\varrho + \kappa)}{(\vartheta_2 - \vartheta_1)^{\frac{\varrho}{\kappa}}} I^{\varrho,\kappa}_{\vartheta_1^+} F(\vartheta_2)\right|$$

$$\leq \frac{\kappa(\vartheta_2 - \vartheta_1)^2}{\varrho + \kappa}\left(\frac{\kappa}{2(\varrho + \kappa)}\left|F''\left(\frac{3\varrho\vartheta_1 + 2\vartheta_1\kappa + \vartheta_2\kappa}{3(\varrho + \kappa)}\right)\right|\right.$$

$$+ \frac{\kappa}{\varrho + 2\kappa}\left|F''\left(\frac{\varrho\vartheta_1 + 2\vartheta_1\kappa + \vartheta_2\kappa}{\varrho + 3\kappa}\right)\right|\right)$$

holds for any $\varrho, \kappa > 0$.

Proof. Equation (26) can be rewritten in the following way

$$F\left(\frac{\varrho\vartheta_1 + \kappa\vartheta_2}{\varrho + \kappa}\right) - \frac{\Gamma_{\kappa}(\varrho + \kappa)}{(\vartheta_2 - \vartheta_1)^{\frac{\varrho}{\kappa}}} I^{\varrho,\kappa}_{\vartheta_1^+} F(\vartheta_2)$$

$$= \frac{\kappa(\vartheta_2 - \vartheta_1)^2}{\varrho + \kappa}\left(\int_{\frac{\varrho}{\varrho+\kappa}}^{1} \left(\tau\left(\frac{\varrho}{\kappa} + 1\right) - \frac{\varrho}{\kappa}\right) F''\left(\tau\vartheta_1 + (1-\tau)\vartheta_2\right) d\tau\right.$$

$$\left. - \int_0^1 \tau^{\frac{\varrho}{\kappa}+1} F''\left(\tau\vartheta_1 + (1-\tau)\vartheta_2\right) d\tau\right).$$

By using the condition of absolute value and then Jensen's integral inequality, we find

$$\left| F\left(\frac{\varrho\vartheta_1 + \kappa\vartheta_2}{\varrho + \kappa}\right) - \frac{\Gamma_\kappa(\varrho + \kappa)}{(\vartheta_2 - \vartheta_1)^{\frac{\varrho}{\kappa}}} I_{\vartheta_1^+}^{\varrho,\kappa} F(\vartheta_2) \right|$$

$$\leq \frac{\kappa(\vartheta_2 - \vartheta_1)^2}{\varrho + \kappa} \left[\int_{\frac{\varrho}{\varrho+\kappa}}^{1} \left(\tau\left(\frac{\varrho}{\kappa}+1\right) - \frac{\varrho}{\kappa}\right) d\tau \right.$$

$$\times \left| F''\left(\frac{\int_{\frac{\varrho}{\varrho+\kappa}}^{1} \left(\tau(\frac{\varrho}{\kappa}+1) - \frac{\varrho}{\kappa}\right)\left(\tau\vartheta_1 + (1-\tau)\vartheta_2\right) d\tau}{\int_{\frac{\varrho}{\varrho+\kappa}}^{1} \left(\tau(\frac{\varrho}{\kappa}+1) - \frac{\varrho}{\kappa}\right) d\tau} \right) \right|$$

$$+ \int_0^1 t^{\frac{\varrho}{\kappa}+1} \left| F''\left(\frac{\int_0^1 \tau^{\frac{\varrho}{\kappa}+1}\left(\tau\vartheta_1 + (1-\tau)\vartheta_2\right) d\tau}{\int_0^1 \tau^{\frac{\varrho}{\kappa}+1} d\tau} \right) \right|$$

$$= \frac{\kappa(\vartheta_2 - \vartheta_1)^2}{\varrho + \kappa} \left(\frac{\kappa}{2(\varrho + \kappa)} \left| F''\left(\frac{3\varrho\vartheta_1 + 2\vartheta_1\kappa + \vartheta_2\kappa}{3(\varrho + \kappa)} \right) \right| \right.$$

$$+ \frac{\kappa}{\varrho + 2\kappa} \left| F''\left(\frac{\varrho\vartheta_1 + 2\vartheta_1\kappa + \vartheta_2\kappa}{\varrho + 3\kappa} \right) \right| \right).$$

□

Remark 6. *Letting* $\kappa = 1$ *in Theorem 7 gives the following result presented in ([16], Theorem 2.11).*

$$\left| F\left(\frac{\varrho\vartheta_1 + \vartheta_2}{\varrho + 1}\right) - \frac{\Gamma(\varrho + 1)}{(\vartheta_2 - \vartheta_1)^\varrho} I_{\vartheta_1^+}^\varrho F(\vartheta_2) \right|$$

$$\leq \frac{(\vartheta_2 - \vartheta_1)^2}{\varrho + 1} \left(\frac{1}{2(\varrho + 1)} \left| F''\left(\frac{3\varrho\vartheta_1 + 2\vartheta_1 + \vartheta_2}{3(\varrho + 1)} \right) \right| \right.$$

$$+ \frac{1}{\varrho + 2} \left| F''\left(\frac{\varrho\vartheta_1 + 2\vartheta_1 + \vartheta_2}{\varrho + 3} \right) \right| \right).$$

Theorem 8. *Let* F *be a twice differentiable and* $|F''|$ *be a concave function on* $[\vartheta_1, \vartheta_2]$. *Then, for any* $\varrho, \kappa > 0$, *we have the inequality*

$$\left| \frac{\varrho F(\vartheta_1) + \kappa F(\vartheta_2)}{\varrho + \kappa} - \frac{\Gamma_\kappa(\varrho + \kappa)}{(\vartheta_2 - \vartheta_1)^{\frac{\varrho}{\kappa}}} I_{\vartheta_1^+}^{\varrho,\kappa} f(\vartheta_2) \right|$$

$$\leq \frac{\kappa\varrho(\vartheta_2 - \vartheta_1)^2}{2(\varrho + \kappa)(\varrho + 2\kappa)} \left| F''\left(\frac{2\vartheta_1\varrho + \varrho\vartheta_2 + 4\vartheta_1\kappa + 5\vartheta_2\kappa}{3(\varrho + 3\kappa)} \right) \right|.$$

Proof. Equation (27) can also be expressed by the following relation.

$$\frac{\varrho F(\vartheta_1) + \kappa F(\vartheta_2)}{\varrho + \kappa} - \frac{\Gamma_\kappa(\varrho + \kappa)}{(\vartheta_2 - \vartheta_1)^{\frac{\varrho}{\kappa}}} I_{\vartheta_1^+}^{\varrho,\kappa} F(\vartheta_2)$$

$$= \frac{\kappa(\vartheta_2 - \vartheta_1)^2}{(\varrho + \kappa)} \int_0^1 \left(\tau - \tau^{\frac{\varrho}{\kappa}+1}\right) F''\left(\tau\vartheta_1 + (1-\tau)\vartheta_2\right) d\tau,$$

By using the condition of absolute value and then Jensen's integral inequality, we find

$$\left| \frac{\varrho F(\vartheta_1) + \kappa F(\vartheta_2)}{\varrho + \kappa} - \frac{\Gamma_\kappa(\varrho + \kappa)}{(\vartheta_2 - \vartheta_1)^{\frac{\varrho}{\kappa}}} I_{\vartheta_1^+}^{\varrho,\kappa} F(\vartheta_2) \right|$$

$$\leq \frac{\kappa(\vartheta_2 - \vartheta_1)^2}{(\varrho + \kappa)} \int_0^1 \left(\tau - \tau^{\frac{\varrho}{\kappa}+1} \right) d\tau \left| F'' \left(\frac{\int_0^1 \left(\tau - \tau^{\frac{\varrho}{\kappa}+1} \right) \left(\tau \vartheta_1 + (1 - \tau) \vartheta_2 \right) d\tau}{\int_0^1 \left(\tau - \tau^{\frac{\varrho}{\kappa}+1} \right) d\tau} \right) \right|$$

$$= \frac{\kappa \varrho (\vartheta_2 - \vartheta_1)^2}{2(\varrho + \kappa)(\varrho + 2\kappa)} \left| F'' \left(\frac{2\varrho \vartheta_1 + \varrho \vartheta_2 + 4\kappa \vartheta_1 + 5\kappa \vartheta_2}{3(\varrho + 3\kappa)} \right) \right|.$$

Hence, the desired result is proven. □

Remark 7. *If we choose $\kappa = 1$ in Theorem 8, we obtain the following result presented in ([16], Theorem 2.13).*

$$\left| \frac{\varrho F(\vartheta_1) + F(\vartheta_2)}{\varrho + 1} - \frac{\Gamma(\varrho + 1)}{(\vartheta_2 - \vartheta_1)^\varrho} I_{\vartheta_1^+}^{\varrho} F(\vartheta_2) \right|$$

$$\leq \frac{\varrho(\vartheta_2 - \vartheta_1)^2}{2(\varrho + 1)(\varrho + 2)} \left| F'' \left(\frac{2\vartheta_1 \varrho + \varrho \vartheta_2 + 4\vartheta_1 + 5\vartheta_2}{3(\varrho + 3)} \right) \right|.$$

3. Some Applications to Special Means

(i) The arithmetic mean:

$$A = A(\vartheta_1, \vartheta_2) = \frac{\vartheta_1 + \vartheta_2}{2}, \quad \vartheta_1, \vartheta_2 > 0. \tag{28}$$

(ii) The logarithmic mean:

$$L(\vartheta_1, \vartheta_2) = \frac{\vartheta_2 - \vartheta_1}{\ln \vartheta_2 - \ln \vartheta_1}, \quad \vartheta_1 \neq \vartheta_2 \; \vartheta_1, \vartheta_2 > 0.$$

(iii) The generalized logarithmic mean:

$$L_n(\vartheta_1, \vartheta_2) = \left(\frac{\vartheta_2^{n+1} - \vartheta_1^{n+1}}{(n+1)(\vartheta_2 - \vartheta_1)} \right)^{\frac{1}{n}}, \quad n \in \mathbb{Z} \setminus \{-1, 0\}, \; \vartheta_1 \neq \vartheta_2, \; \vartheta_1, \vartheta_2 > 0. \tag{29}$$

Proposition 1. *Let $\vartheta_1, \vartheta_2 \in \Re^+$, $\vartheta_1 < \vartheta_2$, then we have the following inequalities.*

$$\left| A(e^{\vartheta_1}, e^{\vartheta_2}) - L(e^{\vartheta_1}, e^{\vartheta_2}) \right| \leq \frac{e^{\vartheta_2}(\vartheta_2 - \vartheta_1)^2}{12},$$

$$\left| A(e^{\vartheta_1}, e^{\vartheta_2}) - L(e^{\vartheta_1}, e^{\vartheta_2}) \right| \leq \frac{e^{\vartheta_1}(\vartheta_2 - \vartheta_1)^2}{12},$$

$$\left| A(e^{\vartheta_1}, e^{\vartheta_2}) - L(e^{\vartheta_1}, e^{\vartheta_2}) \right| \leq \frac{\max\left(e^{\vartheta_1}, e^{\vartheta_2}\right)(\vartheta_2 - \vartheta_1)^2}{12}$$

and
$$\left| A(e^{\vartheta_1}, e^{\vartheta_2}) - L(e^{\vartheta_1}, e^{\vartheta_2}) \right| \leq \frac{(\vartheta_2 - \vartheta_1)^2 (e^{\vartheta_1} + e^{\vartheta_2})}{24}.$$

Proof. Using Theorem 3 and making some simplification, we can write

$$\left| \frac{\varrho F(\vartheta_1) + \kappa F(\vartheta_2)}{\varrho + \kappa} - \frac{\varrho}{\kappa(\vartheta_2 - \vartheta_1)^{\frac{\varrho}{\kappa}}} \int_{\vartheta_1}^{\vartheta_2} (\vartheta_2 - \varsigma)^{\frac{\varrho}{\kappa}-1} F(\varsigma) d\varsigma \right| \leq \frac{\varrho \kappa \left| F''(\vartheta_2) \right| (\vartheta_2 - \vartheta_1)^2}{2(\varrho + \kappa)(\varrho + 2\kappa)}.$$

By substituting $\varrho = \kappa$, $F(\vartheta) = e^{\vartheta}$ and using simple calculation, we obtain

$$\left| \frac{\kappa(e^{\vartheta_1}) + \kappa(e^{\vartheta_2})}{\kappa + \kappa} - \frac{\kappa}{\kappa(\vartheta_2 - \vartheta_1)^{\frac{K}{\kappa}}} \int_{\vartheta_1}^{\vartheta_2} (\vartheta_2 - \varsigma)^{\frac{K}{\kappa}-1} (e^{\varsigma}) d\varsigma \right| \frac{\left| e^{\vartheta_2} (\vartheta_2 - \vartheta_1) \right|^2}{12}.$$

This can also be written as

$$\left| \frac{e^{\vartheta_1} + e^{\vartheta_2}}{2} - \frac{e^{\vartheta_2} - e^{\vartheta_1}}{(\vartheta_2 - \vartheta_1)} \right| \leq \frac{e^{\vartheta_2} (\vartheta_2 - \vartheta_1)^2}{12}.$$

Now, making use of (28) and (29), we arrive at the result

$$\left| A(e^{\vartheta_1}, e^{\vartheta_2}) - L(e^{\vartheta_1}, e^{\vartheta_2}) \right| \leq \frac{|e^{\vartheta_2}|(\vartheta_2 - \vartheta_1)^2}{12}.$$

By using the same procedure in part (ii) and part (iii) of Theorem 3 and Theorem 6, we find the remaining inequalities. □

Proposition 2. Let $\vartheta_1, \vartheta_2 \in \Re^+$, $\vartheta_1 < \vartheta_2$, then the inequalities

$$\left| A(\vartheta_1^n, \vartheta_2^n) - L_n^n(\vartheta_1, \vartheta_2) \right| \leq \frac{(\vartheta_2 - \vartheta_1)^2}{24} \left(|n(n-1)| \left(\vartheta_1^{n-2} + \vartheta_2^{n-2} \right) \right),$$

$$\left| A(\vartheta_1^n, \vartheta_2^n) - L_n^n(\vartheta_1, \vartheta_2) \right| \leq \frac{(\vartheta_2 - \vartheta_1)^2 |n(n-1)| \vartheta_2^{n-2}}{12},$$

$$\left| A(\vartheta_1^n, \vartheta_2^n) - L_n^n(\vartheta_1, \vartheta_2) \right| \leq \frac{(\vartheta_2 - \vartheta_1)^2 |n(n-1)| \vartheta_1^{n-2}}{12}$$

and

$$\left| A(\vartheta_1^n, \vartheta_2^n) - L_n^n(\vartheta_1, \vartheta_2) \right| \leq \frac{\max \left(|n(n-1)| \vartheta_1^{n-2}, |n(n-1)| \vartheta_2^{n-2} \right) (\vartheta_2 - \vartheta_1)^2}{12}.$$

are true for $n \in \mathbb{Z}$ with $|n(n-1)| \geq 2$.

Proof. Using Theorem 6 and making some simplification, we obtain

$$\left| \frac{\varrho F(\vartheta_1) + \kappa F(\vartheta_2)}{\varrho + \kappa} - \frac{\varrho}{\kappa(\vartheta_2 - \vartheta_1)^{\frac{\varrho}{\kappa}}} \int_{\vartheta_1}^{\vartheta_2} (\vartheta_2 - \varsigma)^{\frac{\varrho}{\kappa} - 1} F(\varsigma) d\varsigma \right|$$

$$\leq \frac{\kappa \varrho(\vartheta_2 - \vartheta_1)^2}{3(\varrho + \kappa)(\varrho + 3\kappa)} \left(|F''(\vartheta_1)| + |F''(\vartheta_2)| \frac{(\varrho + 5\kappa)}{2(\varrho + 2\kappa)} \right)$$

By substituting $\varrho = \kappa$ and $F(\vartheta) = \vartheta^n$, where $\vartheta > 0$ and $|n(n-1)| \geq 2$, we obtain

$$\left| \frac{\vartheta_1^n + \vartheta_2^n}{2} - \frac{1}{(\vartheta_2 - \vartheta_1)} \int_{\vartheta_1}^{\vartheta_2} (\varsigma)^n d\varsigma \right| \leq \frac{(\vartheta_2 - \vartheta_1)^2}{24} \left(|n(n-1)| \vartheta_1^{n-2} + |n(n-1)| \vartheta_2^{n-2} \right),$$

$$\left| \frac{(\vartheta_1^n) + (\vartheta_2^n)}{2} - \frac{\vartheta_2^{n+1} - \vartheta_1^{n+1}}{(n+1)(\vartheta_2 - \vartheta_1)} \right| \leq \frac{(\vartheta_2 - \vartheta_1)^2}{24} \left(|n(n-1)| \left(\vartheta_1^{n-2} + \vartheta_2^{n-2} \right) \right).$$

Now, by using Equations (28) and (29), we obtain the desired result. Similarly by using the same polynomial function in Theorem 3, we obtain the rest of the inequalities. □

4. Conclusions

The bounds of various functions are studied in optimization theory—a branch of mathematics. The innovative fractional Hermite-Hadamard type inequalities established in this research are based on functions whose second order derivatives with absolute values are convex (concave). A new technique is used to explore the main results by involving Green's function and Abel-Gontscharoff interpolating polynomials for two-point problems with a combination of κ-R-LFI. Jensen's inequality is capably utilized with wide applications in optimization theory. Some applications of our main findings are presented to special means. This study motivates the researchers to establish the various Hermite-Hadamard inequalities by using the other Green's functions G_2, G_3, and G_4 with more general fractional operators.

Author Contributions: Conceptualization, Y.L., M.S. and A.G.; methodology, M.S. and A.G.; software, M.S., G.R. and M.V.-C.; validation, Y.L., M.V.-C. and G.R.; formal analysis, M.S. and A.G.; investigation, Y.L., M.S. and A.G.; resources, M.S., G.R. and M.V.-C.; data curation, Y.L., M.S. and G.R.; writing—original draft preparation, Y.L., M.S. and A.G.; writing—review and editing, M.V.-C. and G.R.; visualization, Y.L. and M.V.-C.; supervision, M.S., M.V.-C. and G.R.; project administration, Y.L. and M.S.; funding acquisition, M.V.-C. All authors have read and agreed to the published version of the manuscript.

Funding: The work was supported by the key Scientific Research Projects of Hunan Provincial Department of Education in 2021 (grant number: 21A0526).

Institutional Review Board Statement: Not applicable.

Informed Consent Statement: Not applicable.

Data Availability Statement: Not applicable.

Acknowledgments: The first author thank to the key Scientific Research Projects of Hunan Provincial Department of Education.

Conflicts of Interest: The authors declare that they have no conflict of interest.

References

1. Samraiz, M.; Perveen, Z.; Abdeljawad, T.; Iqbal, S.; Naheed, S. On Certain Fractional Calculus Operators and Their Applications in Mathematical Physics. *Phys. Scr.* **2020**, *95*, 115210. [CrossRef]
2. Samraiz, M.; Perveen, Z.; Rahman, G.; Nisar, K.S.; Kumar, D. On (k,s)-Hilfer Prabhakar Fractional Derivative with Applications in Mathematical Physics. *Front. Phys.* **2020**, *8*, 309. [CrossRef]
3. Tarasov, V.E. On History of Mathematical Economics, Application of Fractional Calculus. *Mathematics* **2019**, *7*, 509. [CrossRef]
4. Mainardi, F. On the Advent of Fractional Calculus in Econophysics via Continuous-Time Random Walk. *Mathematics* **2020**, *8*, 641. [CrossRef]
5. Samraiz, M.; Umer, M.; Kashuri, A.; Abdeljawad, T.; Iqbal, S.; Mlaiki, N. On Weighted (k,s)-Riemann-Liouville Fractional Operators and Solution of Fractional Kinetic Equation. *Fractal. Frac.* **2021**, *5*, 118. [CrossRef]
6. Johansyah, M.D.; Supriatna, A.K.; Rusyaman, E.; Saputra, J. Application of fractional differential equation in economic growth model: A systematic review approach. *AIMS Math.* **2021**, *6*, 10266–10280. [CrossRef]
7. Baleanu, D.; Agarwal, R.P. Fractional calculus in the sky. *Adv. Differ. Equ.* **2021**, *2021*, 117. [CrossRef]
8. Sweilam, N.H.; Al-Mekhlafi, S.M.; Assiri, T.; Atangana, A. Optimal control for cancer treatment mathematical model using Atangana–Baleanu–Caputo fractional derivative. *Adv. Differ. Equ.* **2020**, *2020*, 334. [CrossRef]
9. Khan, T.U.; Khan, M.A. Generalized conformable fractional operator. *J. Comput. Appl. Math.* **2019**, *346*, 378–389. [CrossRef]
10. Jarad, F.; Ugurlu, E.; Abdeljawad, T.; Baleanu, D. On a new class of fractional operators. *Adv. Differ. Equ.* **2017**, *2017*, 247. [CrossRef]
11. Niculescu, C.P.; Persson, L.E. *Convex Functions and Their Applications: A Contemporary Approach*; CMC Books in Mathematics; Springer: New York, NY, USA, 2004.
12. Pečarić, J.; Proschan, F.; Tong, Y.L. *Convex Functions, Partial Orderings and Statistical Application*; Acadmic Press: New York, NY, USA, 1992.
13. Varosanec, S. On h-convexity. *J. Math. Anal. Appl.* **2007**, *326*, 303–311. [CrossRef]
14. Hudzik, H.; Maligranda, L. Some remarks on s-convex functions. *Aequationes Math.* **1994**, *48*, 100–111. [CrossRef]
15. Wu, S.; Iqbal, S.; Aamir, M.; Samraiz, M. On some Hermite-Hadamard inequalities involving k-fractional operators. *J. Inequal. Appl.* **2021**, *2021*, 32. [CrossRef]
16. Khan, M.A.; Iqbal, A.; Suleman, M.; Chu, Y.-M. The right Riemann-Liouville fractional Hermite-Hadamard type inequalities derived from Green's function. *AIP Adv.* **2020**, *10*, 045032.
17. Sharma, N.; Singh, S.K.; Mishra, S.K.; Hamdi, A. Hermite–Hadamard-type inequalities for interval-valued preinvex functions via Riemann–Liouville fractional integrals. *J. Inequal. Appl.* **2020**, *2020*, 591. [CrossRef]
18. Chen, H.; Katugampola, U.N. Hermite-Hadamard and Hermite-Hadamard-Fejer type inequalities for generalized fractional integrals. *J. Math. Anal. Appl.* **2017**, *446*, 1274–1291. [CrossRef]
19. Set, E.; Ozdemir, M.O.; Dragomir, S.S. On the Hadamard-type of inequalities involving several kinds of convexity. *J. Inequal. Appl.* **2010**, *2010*, 286845. [CrossRef]
20. Sarikaya, M.Z.; Set, E.; Yaldiz, H.; Basak, N. Hermite-Hadamard inequalities for fractional integrals and related fractional inequalities. *Math. Comput. Model.* **2013**, *57*, 2403–2407. [CrossRef]
21. Diaz, R.; Pariguan, E. On hypergeometric functions and Pochhammer k-symbol. *Divulg Math.* **2007**, *15*, 179–192.
22. Mehmood, N.; Agarwal, R.P.; Butt, S.I.; Pecaric, J.E. New generalizations of Popoviciu-type inequalities via new Green's functions and Montgomery identity. *J. Inequal. Appl.* **2017**, *2017*, 1353. [CrossRef] [PubMed]

fractal and fractional

Article
Dirichlet Averages of Generalized Mittag-Leffler Type Function

Dinesh Kumar [1], Jeta Ram [2] and Junesang Choi [3,*]

[1] Department of Applied Sciences, College of Agriculture-Jodhpur, Agriculture University Jodhpur, Jodhpur 342304, India; dinesh_dino03@yahoo.com
[2] Department of Mathematics and Statistics, Jai Narain Vyas University, Jodhpur 342005, India; bishnoi_jr@yahoo.com
[3] Department of Mathematics, Dongguk University, Gyeongju 38066, Korea
* Correspondence: junesang@dongguk.ac.kr; Tel.: +82-010-6525-2262

Abstract: Since Gösta Magus Mittag-Leffler introduced the so-called Mittag-Leffler function in 1903 and studied its features in five subsequent notes, passing the first half of the 20th century during which the majority of scientists remained almost unaware of the function, the Mittag-Leffler function and its various extensions (referred to as Mittag-Leffler type functions) have been researched and applied to a wide range of problems in physics, biology, chemistry, and engineering. In the context of fractional calculus, Mittag-Leffler type functions have been widely studied. Since Carlson established the notion of Dirichlet average and its different variations, these averages have been explored and used in a variety of fields. This paper aims to investigate the Dirichlet and modified Dirichlet averages of the *R*-function (an extended Mittag-Leffler type function), which are provided in terms of Riemann-Liouville integrals and hypergeometric functions of several variables. Principal findings in this article are (possibly) applicable. This article concludes by addressing an open problem.

Keywords: Dirichlet averages; *B*-splines; dirichlet splines; Riemann–Liouville fractional integrals; hypergeometric functions of one and several variables; generalized Mittag-Leffler type function; Srivastava–Daoust generalized Lauricella hypergeometric function

MSC: 26A33; 33C20; 33E12; 33E20

1. Introduction and Preliminaries

The Mittag-Leffler function $E_\alpha(z)$ (see [1])

$$E_\alpha(z) = \sum_{\ell=0}^{\infty} \frac{z^\ell}{\Gamma(\alpha \ell + 1)} \quad (\Re(\alpha) > 0), \tag{1}$$

Γ being the familiar Gamma function (see, for example, Section 1.1 in [2]), is named after the eminent Swedish mathematician Gösta Magus Mittag-Leffler (1846–1927), who explored its features in 1902–1905 in five notes (consult, for instance, [1]) related to his summation technique for divergent series (see also Chapter 1, [3]). Because $\Gamma(\ell+1) = \ell!$ ($\ell \in \mathbb{N}_0$) and therefore $E_1(z) = e^z$, this function gives a straightforward extension of the exponential function. Here and elsewhere, let $\mathbb{N}, \mathbb{Z}_0^-, \mathbb{R}, \mathbb{R}^+$, and $\mathbb{C}$ be the sets of positive integers, non-positive integers, real numbers, positive real numbers, and complex numbers, respectively, and put $\mathbb{N}_0 := \mathbb{N} \cup \{0\}$. Passing the first half of the 20th century during which the majority of scientists remained almost unaware of the function, the Mittag-Leffler function and its various extensions (referred to as Mittag-Leffler type functions) have been studied and applied to a wide range of problems in physics, biology, chemistry, engineering, etc. This function's most significant features are described in Chapter XVIII [4], which is dedicated to so-called miscellaneous functions. The Mittag-Leffler function was categorized as miscellaneous because it was not until the 1960s that it was discovered as belonging to a

broader class of higher transcendental functions known as Fox H-functions, thus the term "miscellaneous" (consult, for instance, [5]). In reality, this class was not well-established until Fox's landmark study (see [6]). The simplest (and most crucial for applications) extension of the Mittag-Leffler function, notably the two-parametric Mittag-Leffler function

$$E_{\alpha,\beta}(z) = \sum_{\ell=0}^{\infty} \frac{z^\ell}{\Gamma(\alpha\ell + \beta)} \quad (\alpha, \beta \in \mathbb{C}, \Re(\alpha) > 0) \tag{2}$$

was separately studied by Humbert and Agarwal in 1953 (see, for example, [7]) and by Dzherbashyan in 1954 (see, for example, [8]). However, it first appeared formally in Wiman's article [9]. Prabhakar [10] introduced the following three-parametric Mittag-Leffler function:

$$E_{\alpha,\beta}^{\gamma}(z) = \sum_{\ell=0}^{\infty} \frac{(\gamma)_\ell}{\ell!\, \Gamma(\alpha\ell + \beta)} z^\ell \quad (\alpha, \beta, \gamma \in \mathbb{C}, \Re(\alpha) > 0, \Re(\gamma) > 0), \tag{3}$$

where $(\lambda)_\nu$ denotes the Pochhammer symbol defined (for $\lambda, \nu \in \mathbb{C}$) by

$$(\lambda)_\nu := \frac{\Gamma(\lambda + \nu)}{\Gamma(\lambda)} = \begin{cases} 1 & (\nu = 0;\ \lambda \in \mathbb{C} \setminus \{0\}) \\ \lambda(\lambda+1)\cdots(\lambda + n - 1) & (\nu = n \in \mathbb{N};\ \lambda \in \mathbb{C}), \end{cases} \tag{4}$$

it being accepted conventionally that $(0)_0 := 1$. This Function (3) is being used for a variety of applicable issues. Scientists, engineers, and statisticians recognize the significance of the aforementioned H-function due to its great potential for applications in several scientific and technical domains. In addition to the Mittag-Leffler Functions (1)–(3), the H-function includes a variety of functions (see, for example, [5]). Among several monographs on the H-function, monograph [5] discusses the theory of the H-function with a focus on its applications. The H-function (or Fox's H-function [6]) is defined by means of a Mellin–Barnes type integral in the following manner (consult also [5]):

$$\begin{aligned} H(z) = H_{p,q}^{m,n}(z) &= H_{p,q}^{m,n}\!\left[z \,\bigg|\, \begin{matrix}(a_p, \alpha_p) \\ (b_q, \beta_q)\end{matrix}\right] \\ &= H_{p,q}^{m,n}\!\left[z \,\bigg|\, \begin{matrix}(a_1,\alpha_1), \ldots, (a_p, \alpha_p) \\ (b_1,\beta_1), \ldots, (b_q, \beta_q)\end{matrix}\right] = \frac{1}{2\pi\omega} \int_{\mathfrak{L}} \Omega(s)\, z^{-s}\, ds, \end{aligned} \tag{5}$$

where $\omega = \sqrt{-1}$, and

$$\Omega(s) := \frac{\prod\limits_{j=1}^{m} \Gamma(b_j + \beta_j s) \cdot \prod\limits_{j=1}^{n} \Gamma(1 - a_j - \alpha_j s)}{\prod\limits_{j=m+1}^{q} \Gamma(1 - b_j - \beta_j s) \cdot \prod\limits_{j=n+1}^{p} \Gamma(a_j + \alpha_j s)}. \tag{6}$$

We also assume the following: $z^{-s} = \exp[-s\{\ln|z| + i \arg z\}]$, where $\ln|z|$ is the natural logarithm, and $\eta < \arg z < \eta + 2\pi$ for some $\eta \in \mathbb{R}$. The integration path $\mathfrak{L} = \mathfrak{L}_{i\gamma\infty}(\gamma \in \mathbb{R})$ extends from $\gamma - i\infty$ to $\gamma + i\infty$ with indentations, if necessary, so that the poles of $\Gamma(1 - a_j - \alpha_j s)$ $(1 \leq j \leq n \in \mathbb{N}_0)$ can be separated from those of $\Gamma(b_j + \beta_j s)$ $(1 \leq j \leq m \in \mathbb{N}_0)$ and has no those poles on it. The parameters $p, q \in \mathbb{N}_0$ satisfy the conditions $0 \leq n \leq p$, $0 \leq m \leq q$; the parameters $\alpha_j, \beta_j \in \mathbb{R}^+$ and $a_j, b_j \in \mathbb{C}$. The empty product in (6) (and elsewhere) is (as usual) understood to be unity.

For the existence conditions of the H-function, one may refer to Appendix F.4 [3], Section 1.2 [5]. Here it is recalled that the three-parametric Mittag-Leffler function (Prabhakar function) (3) is represented by the following Mellin–Barnes integral (see p.10, Example 1.5 in [5]):

$$E_{\alpha,\beta}^{\gamma}(z) = \frac{1}{2\pi\omega\,\Gamma(\gamma)} \int_{\xi-\omega\infty}^{\xi+\omega\infty} \frac{\Gamma(s)\,\Gamma(\gamma-s)}{\Gamma(\beta-\alpha s)} (-z)^{-s}\,ds \qquad (7)$$

$$\left(|\arg z| < 2\pi,\ \xi \in \mathbb{R}\ \text{(fixed)},\ \alpha \in \mathbb{R}^+,\ \Re(\beta) > 0,\ \gamma \in \mathbb{C}\setminus\mathbb{Z}_0^-\right).$$

We find from (5) and (7) that

$$E_{\alpha,\beta}^{\gamma}(z) = \frac{1}{\Gamma(\gamma)} H_{1,2}^{1,1}\left[-z\ \middle|\ \begin{matrix}(1-\gamma,1)\\(0,1),\ (1-\beta,\alpha)\end{matrix}\right]. \qquad (8)$$

Using (8) in the relation $E_{\alpha,\beta}^1(z) = E_{\alpha,\beta}(z)$, we get (consult, for example, p.9, Equation (1.50) in [5])

$$E_{\alpha,\beta}(z) = H_{1,2}^{1,1}\left[-z\ \middle|\ \begin{matrix}(0,1)\\(0,1),\ (1-\beta,\alpha)\end{matrix}\right]. \qquad (9)$$

Indeed, the Mittag-Leffler type functions in association with the fractional calculus have been actively researched (see, for example, [11,12]).

Carlson developed the notion of the Dirichlet average in his work [13] (see also [14–18]). Carlson also provided a full and thorough analysis of the numerous varieties of Dirichlet averages. A function's so-called Dirichlet average is the integral mean of the function with regard to the Dirichlet measure. Subsequently and more recently, this study topic has been explored in publications such as [19–28]. Neuman and Van Fleet [19] defined Dirichlet averages of multivariate functions and demonstrated their recurrence formula. Daiya and Kumar [20] researched the double Dirichlet averages of S-functions. Saxena and Daiya [29] proposed and explored the S-functions. Kilbas and Kattuveettill [22] investigated Dirichlet averages of the three-parametric Mittag-Leffler Function (3), whose representations are provided in terms of the Riemann–Liouville fractional integrals and the hypergeometric functions with multiple variables. Saxena et al. [25] explored Dirichlet averages of the generalized multi-index Mittag-Leffler functions (see, for instance, [30]), whose representations are expressed in terms of Riemann–Liouville integrals and hypergeometric functions of several variables. Using Riemann–Liouville fractional integral operators, Vyas [31] investigated the solution of the Euler–Darboux equation in terms of Dirichlet averages of boundary conditions on Hölder space and weighted Hölder spaces of continuous functions. For further Dirichlet averages in connection with fractional calculus, one may consult [21,24,32–36]. These Dirichlet averages were used in a number of studies, in particular, Dirichlet splines (see [19]), B-splines (see [18,23]), and Stolarsky means (see [37]).

In this work, we propose to investigate the Dirichlet and modified Dirichlet averages of the R-function (an extended Mittag-Leffler type function) (see, for details, Section 2). Main results stated in this paper, which are presented in terms of Riemann–Liouville integrals and hypergeometric functions of several variables, are (potentially) useful.

Let Ω be a convex set in $\mathbb{C}$ and $z := (z_1,\ldots,z_n) \in \Omega^n$ ($n \in \mathbb{N}\setminus\{1\}$). Suppose that f is a measurable function on Ω. Then the general Dirichlet average of the function f is defined as follows (see [15]):

$$F(b;z) = \int_{E_{n-1}} f(u \circ z)\,d\mu_b(u), \qquad (10)$$

where b and u denote the arrays of n parameters $b_1,\ldots,b_n$ and $u_1,\ldots,u_n$, respectively, and $d\mu_b(u)$ is the Dirichlet measure defined by

$$d\mu_b(u) = \frac{1}{B(b)} u_1^{b_1-1}\cdots u_{n-1}^{b_{n-1}-1}(1-u_1-\cdots-u_{n-1})^{b_n-1}\,du_1\cdots du_{n-1}, \qquad (11)$$

and E_{n-1} is the Euclidean simplex in $\mathbb{R}^{n-1}$ ($n \in \mathbb{N}\setminus\{1,2\}$) given by

$$E_{n-1} = \{(u_1,\ldots,u_{n-1}) : u_j \geq 0\ (j \in \overline{1,n-1}),\ u_1 + \cdots + u_{n-1} \leq 1\}, \qquad (12)$$

and $B(b)$ is the multivariate Beta-function defined by

$$B(b) := \frac{\Gamma(b_1)\cdots\Gamma(b_n)}{\Gamma(b_1+\cdots+b_n)} \quad (\Re(b_j) > 0 \; (j \in \overline{1,n})),$$

and

$$u \circ z := \sum_{j=1}^{n-1} u_j z_j + (1 - u_1 - \cdots - u_{n-1}) z_n.$$

Here and throughout this paper, the notation $\overline{1,p} := \{1, \ldots, p\}$ ($p \in \mathbb{N}$) is used. The special case of (11) when $n = 2$ reduces to the following form:

$$d\mu_{\beta,\beta'}(u) = \frac{\Gamma(\beta+\beta')}{\Gamma(\beta)\Gamma(\beta')} u^{\beta-1} \, du. \tag{13}$$

Carlson [15] investigated the average (10) for the function $f(z) = z^k$ ($k \in \mathbb{R}$) in the following form:

$$R_k(b; z) = \int_{E_{n-1}} (u \circ z)^k \, d\mu_b, \tag{14}$$

whose special case $n = 2$ was given as follows (see [13,15]):

$$R_k(\beta, \beta'; x, y) = \frac{1}{B(\beta, \beta')} \int_0^1 [ux + (1-u)y]^k \, u^{\beta-1}(1-u)^{\beta'-1} \, du, \tag{15}$$

where $\beta, \beta' \in \mathbb{C}$ with $\min\{\Re(\beta), \Re(\beta')\} > 0$, and $x, y \in \mathbb{R}$, $B(\beta, \beta')$ is the familiar Beta function (consult, for instance, Chapter 1, [2]).

The Riemann–Liouville fractional integral of a function f is defined as follows (consult, for instance, (p. 69) [38]): For $\alpha \in \mathbb{C}$ with $\Re(\alpha) > 0$ and $a \in \mathbb{R}$,

$$(I_{a+}^{\alpha} f)(x) = \frac{1}{\Gamma(\alpha)} \int_a^x (x-t)^{\alpha-1} f(t) \, dt \quad (x > a). \tag{16}$$

The Srivastava–Daoust generalization $F_{C:D^{(1)};\ldots;D^{(n)}}^{A:B^{(1)};\ldots;B^{(n)}}$ of the Lauricella hypergeometric function F_D in n variables is defined by (see (p. 454) [39]; see also (p. 37) [40], (p. 209) [5])

$$F_{C:D^{(1)};\ldots;D^{(n)}}^{A:B^{(1)};\ldots;B^{(n)}} \left(\begin{array}{c} [(a) : \theta^{(1)}, \ldots, \theta^{(n)}] : \; [(b^{(1)}) : \varphi^{(1)}]; \ldots; [(b^{(n)}) : \varphi^{(n)}]; \\ [(c) : \psi^{(1)}, \ldots, \psi^{(n)}] : \; [(d^{(1)}) : \delta^{(1)}]; \ldots; [(d^{(n)}) : \delta^{(n)}]; \end{array} x_1, \ldots, x_n \right)$$

$$= \sum_{m_1,\ldots,m_n=0}^{\infty} \frac{\prod_{j=1}^{A}(a_j)_{m_1\theta_j^{(1)}+\cdots+m_n\theta_j^{(n)}} \prod_{j=1}^{B^{(1)}}(b_j^{(1)})_{m_1\varphi_j^{(1)}} \cdots \prod_{j=1}^{B^{(n)}}(b_j^{(n)})_{m_n\varphi_j^{(n)}}}{\prod_{j=1}^{C}(c_j)_{m_1\psi_j^{(1)}+\cdots+m_n\psi_j^{(n)}} \prod_{j=1}^{D^{(1)}}(d_j^{(1)})_{m_1\delta_j^{(1)}} \cdots \prod_{j=1}^{D^{(n)}}(d_j^{(n)})_{m_n\delta_j^{(n)}}}$$

$$\times \frac{x_1^{m_1}}{m_1!} \cdots \frac{x_n^{m_n}}{m_n!}, \tag{17}$$

where the coefficients, for all $k \in \overline{1,n}$,

$$\theta_j^{(k)} \; (j \in \overline{1,A}); \; \varphi_j^{(k)} \; \left(j \in \overline{1, B^{(k)}}\right); \; \psi_j^{(k)} \; (j \in \overline{1,C}); \; \delta_j^{(k)} \; \left(j \in \overline{1, D^{(k)}}\right)$$

are real and positive, and (a) abbreviates the array of A parameters $a_1, \ldots, a_A$, $\left(b^{(k)}\right)$ abbreviates the array of $B^{(k)}$ parameters $b_j^{(k)}$ $\left(j \in \overline{1, B^{(k)}}\right)$ for all $k \in \overline{1,n}$, with similar interpretations for (c) and $\left(d^{(k)}\right)$ ($k \in \overline{1,n}$); et cetera.

One may refer to Srivastava and Daoust [41] for the specific convergence requirements of the multiple series (17).

2. The Generalized Mittag-Leffler Type Function (the R-Function)

The R-function, which Kumar and Kumar [42] proposed and Kumar and Purohit [43] studied, is defined as follows:

$$ {}_p^\kappa R_q^{\alpha,\beta;\gamma}(z) = {}_p^\kappa R_q^{\alpha,\beta;\gamma}(a_1,\ldots,a_p;b_1,\ldots,b_q;z) = \sum_{n=0}^\infty \frac{\prod_{j=1}^p (a_j)_n}{\prod_{j=1}^q (b_j)_n} \frac{(\gamma)_{\kappa n}}{\Gamma(\alpha n + \beta)} \frac{z^n}{n!} \quad (18)$$

$$(\alpha,\, \beta,\, \gamma \in \mathbb{C};\; \Re(\alpha) > \max\{0, \Re(\kappa) - 1\};\; \Re(\kappa) > 0),$$

where $(a_j)_n$ $(j \in \overline{1,p})$ and $(b_j)_n$ $(j \in \overline{1,q})$ are the Pochhammer symbols in (4). The series (18) is defined when

$$b_j \in \mathbb{C} \setminus \mathbb{Z}_0^- \quad (j \in \overline{1,q}). \quad (19)$$

If any parameter a_j is a negative integer or zero, then the series (18) terminates to become a polynomial in z.

Assuming that none of the numerator parameters is zero or a negative integer (otherwise the question of convergence will not arise) and with the restriction given by (19), the ${}_p^\kappa R_q^{\alpha,\beta;\gamma}$ series in (18)

(i) converges for $|z| < \infty$, if $p < q + 1$,
(ii) converges for $|z| < 1$, if $p = q + 1$, and
(iii) diverges for all $z \in \mathbb{C} \setminus \{0\}$ if $p > q + 1$.

Furthermore, if we set

$$\omega := \sum_{j=1}^q b_j - \sum_{j=1}^p a_j, \quad (20)$$

then it is seen that the ${}_p^\kappa R_q^{\alpha,\beta;\gamma}$ series in (18), with $p = q + 1$, is

(a) absolutely convergent for $|z| = 1$, if $\Re(\omega) > 0$,
(b) conditionally convergent for $|z| = 1$ $(z \neq 1)$, if $-1 < \Re(\omega) \leqq 0$, and
(c) divergent for $|z| = 1$, if $\Re(\omega) \leqq -1$.

Remark 1. *The R-function in (18) is general enough to include, as its special cases, such functions as (for example) the generalized Mittag-Leffler function $E_{\alpha,\beta}^{\gamma,\kappa}(z)$ introduced by Srivastava and Tomovski [44]:*

$$ {}_1^\kappa R_1^{\alpha,\beta;\gamma}(z) = \sum_{n=0}^\infty \frac{(\gamma)_{\kappa n}}{\Gamma(\alpha n + \beta)} \frac{z^n}{n!} = E_{\alpha,\beta}^{\gamma,\kappa}(z) \quad (21)$$

as well as the Mittag-Leffler function $E_\alpha(z)$ (see [1]):

$$ {}_1^1 R_1^{\alpha,1;1}(z) = \sum_{n=0}^\infty \frac{z^n}{\Gamma(\alpha n + 1)} = E_\alpha(z). \quad (22)$$

3. Bivariate Dirichlet Averages

The Dirichlet average of the generalized Mittag-Leffner type Function (18) is denoted and defined as follows:

$$ {}_p^\kappa \mathcal{M}_q^{\alpha,\delta;\gamma} \left[(\beta, \beta'; x, y) \right]_{(b)_{1,q}}^{(a)_{1,p}} $$
$$:= \int_{E_1} \left[{}_p^\kappa R_q^{\alpha,\delta;\gamma}(a_1,\ldots,a_p;b_1,\ldots,b_q;(u \circ z)) \right] d\mu_{\beta,\beta'}(u), \quad (23)$$

where $(a)_{1,n}$ and $(b)_{1,n}$ $(n \in \mathbb{N})$ denote the horizontal arrays $a_1,\ldots,a_n$ and $b_1,\ldots,b_n$, respectively; $z = (x,y) \in \mathbb{R}^2$ and $\min\{\Re(\beta), \Re(\beta')\} > 0$. In fact, it is shown that the Dirichlet average of the R-function (18) is stated in terms of the Riemann–Liouville fractional integrals (16) claimed by Theorems 1 and 2.

Theorem 1. *Let $z, \alpha, \beta, \beta', \delta, \gamma, \kappa \in \mathbb{C}$ such that $\Re(\alpha) > \max\{0, \Re(\kappa) - 1\}$ and $\min\{\Re(\kappa), \Re(\beta), \Re(\beta')\} > 0$. Also let $x, y \in \mathbb{R}$ with $x > y$ and $I_{0+}^{\beta'}$ be the Riemann–Liouville fractional integral given in (16). Then the Dirichlet average of the generalized Mittag-Leffler type function (18) is given by the following formula:*

$$\,_p^\kappa\mathcal{M}_q^{\alpha,\delta;\gamma}\left[(\beta,\beta';x,y)\right]_{(b)_{1,q}}^{(a)_{1,p}} = \frac{\Gamma(\beta+\beta')}{\Gamma(\beta)(x-y)^{\beta+\beta'-1}}\left(I_{0+}^{\beta'}f\right)(x-y), \qquad (24)$$

where the function f is given by

$$f(t) = t^{\beta-1}\,_p^\kappa R_q^{\alpha,\delta;\gamma}(a_1,\ldots,a_p;b_1,\ldots,b_q;y+t). \qquad (25)$$

Proof. With the aid of (10) to (13), by applying the R-function (18) to (23), we find that

$$\begin{aligned}\mathcal{D}_1 &:= \,_p^\kappa\mathcal{M}_q^{\alpha,\delta;\gamma}\left[(\beta,\beta';x,y)\right]_{(b)_{1,q}}^{(a)_{1,p}} \\ &= \frac{1}{B(\beta,\beta')}\int_0^1 u^{\beta-1}(1-u)^{\beta'-1}\sum_{n=0}^\infty \frac{\prod_{j=1}^p(a_j)_n}{\prod_{j=1}^q(b_j)_n}\frac{(\gamma)_{\kappa n}[y+u(x-y)]^n}{\Gamma(\alpha n+\delta)\,n!}du.\end{aligned} \qquad (26)$$

By changing the order of integration and summation, which is verified under the stated conditions, we get

$$\mathcal{D}_1 = \frac{1}{B(\beta,\beta')}\sum_{n=0}^\infty \frac{\prod_{j=1}^p(a_j)_n}{\prod_{j=1}^q(b_j)_n}\frac{(\gamma)_{\kappa n}}{\Gamma(\alpha n+\delta)\,n!}\int_0^1 u^{\beta-1}(1-u)^{\beta'-1}[y+u(x-y)]^n du.$$

Setting $t := u(x-y)$, we find that

$$\begin{aligned}\mathcal{D}_1 &= \frac{\Gamma(\beta+\beta')}{\Gamma(\beta)\Gamma(\beta')}\sum_{n=0}^\infty \frac{\prod_{j=1}^p(a_j)_n}{\prod_{j=1}^q(b_j)_n}\frac{(\gamma)_{\kappa n}}{\Gamma(\alpha n+\delta)\,n!}\left(\frac{1}{x-y}\right)^{\beta+\beta'-1} \\ &\quad \times \int_0^{x-y} t^{\beta-1}(x-y-t)^{\beta'-1}(y+t)^n dt \\ &= \left(\frac{1}{x-y}\right)^{\beta+\beta'-1}\frac{\Gamma(\beta+\beta')}{\Gamma(\beta)} \\ &\quad \times \left[\frac{1}{\Gamma(\beta')}\int_0^{x-y}\left\{\sum_{n=0}^\infty \frac{\prod_{j=1}^p(a_j)_n}{\prod_{j=1}^q(b_j)_n}\frac{(\gamma)_{\kappa n}(y+t)^n}{\Gamma(\alpha n+\delta)\,n!}\right\}t^{\beta-1}(x-y-t)^{\beta'-1}dt\right].\end{aligned}$$

Then, using (16) and (18), we arrive at the desired result in (24). This completes the proof. □

We take into account the following modification to the Dirichlet average in (23):

$$\begin{aligned}\,_p^{\kappa,\lambda}\mathcal{M}_q^{\alpha,\delta;\gamma}&\left[(\beta,\beta';x,y)\right]_{(b)_{1,q}}^{(a)_{1,p}} \\ &= \int_{E_1}(u\circ z)^{\lambda-1}\left[\,_p^\kappa R_q^{\alpha,\delta;\gamma}(a_1,\ldots,a_p;b_1,\ldots,b_q;(u\circ z)^\gamma)\right]d\mu_{\beta,\beta'}(u),\end{aligned} \qquad (27)$$

where $\lambda \in \mathbb{C}$ with $\Re(\lambda) > 0$ and $z = (x,y)$.

Theorem 2. Let $z, \alpha, \beta, \beta', \delta, \gamma \in \mathbb{C}$ with $\min\{\Re(\beta), \Re(\beta')\} > 0$ and $\kappa \in \mathbb{N}$. Furthermore, let $x, y \in \mathbb{R}$ with $x > y$ and the convergence conditions of the R-function be satisfied. Then the following formula holds true: For $\Re(\lambda) > 0$,

$$\,_p^{\kappa,\lambda}\mathcal{M}_q^{\alpha,\delta;\gamma}\left[(\beta,\beta';x,y)\right]_{(b)_{1,q}}^{(a)_{1,p}} = \frac{\Gamma(\beta+\beta')}{\Gamma(\beta)(x-y)^{\beta+\beta'-1}}\left(I_{y+}^{\beta'}g\right)(x), \tag{28}$$

where the function g is given by

$$g(t) = t^{\lambda-1}(t-y)^{\beta-1}\,_p^{\kappa}R_q^{\alpha,\delta;\gamma}(a_1,\ldots,a_p;b_1,\ldots,b_q;t^\gamma). \tag{29}$$

Proof. With the aid of (10)–(13), by applying the R-function (18)–(27), we find that

$$\mathcal{D}_2 := \,_p^{\kappa,\lambda}\mathcal{M}_q^{\alpha,\delta;\gamma}\left[(\beta,\beta';x,y)\right]_{(b)_{1,q}}^{(a)_{1,p}}$$

$$= \frac{1}{B(\beta,\beta')}\int_0^1 u^{\beta-1}(1-u)^{\beta'-1}[y+u(x-y)]^{\lambda-1}$$

$$\times \sum_{n=0}^{\infty}\frac{\prod_{j=1}^p(a_j)_n}{\prod_{j=1}^q(b_j)_n}\frac{(\gamma)_{\kappa n}[y+u(x-y)]^{n\gamma}}{n!\Gamma(\alpha n+\delta)}\,du$$

$$= \frac{\Gamma(\beta+\beta')}{\Gamma(\beta)\Gamma(\beta')}\sum_{n=0}^{\infty}\frac{\prod_{j=1}^p(a_j)_n}{\prod_{j=1}^q(b_j)_n}\frac{(\gamma)_{\kappa n}}{n!\Gamma(\alpha n+\delta)}$$

$$\times \int_0^1 u^{\beta-1}(1-u)^{\beta'-1}[y+u(x-y)]^{n\gamma+\lambda-1}du.$$

Then, setting $t := y + u(x-y)$, we obtain

$$\mathcal{D}_2 = \frac{\Gamma(\beta+\beta')}{\Gamma(\beta)\Gamma(\beta')}\sum_{n=0}^{\infty}\frac{\prod_{j=1}^p(a_j)_n}{\prod_{j=1}^q(b_j)_n}\frac{(\gamma)_{\kappa n}}{\Gamma(\alpha n+\delta)\,n!}\left(\frac{1}{x-y}\right)^{\beta+\beta'-1}$$

$$\times \int_y^x t^{n\gamma+\lambda-1}(t-y)^{\beta-1}(x-t)^{\beta'-1}dt$$

$$= \left(\frac{1}{x-y}\right)^{\beta+\beta'-1}\frac{\Gamma(\beta+\beta')}{\Gamma(\beta)}$$

$$\times \left[\frac{1}{\Gamma(\beta')}\int_y^x\left\{\sum_{n=0}^{\infty}\frac{\prod_{j=1}^p(a_j)_n}{\prod_{j=1}^q(b_j)_n}\frac{(\gamma)_{\kappa n}t^{n\gamma}}{\Gamma(\alpha n+\delta)\,n!}\right\}t^{\lambda-1}(t-y)^{\beta-1}(x-t)^{\beta'-1}dt\right].$$

Finally, using (16), we are led to the desired result (28). This complete the proof. $\square$

4. Dirichlet Average Expressed in Terms of Srivastava–Daoust Function

This section discusses an alternative formulation of the modified Dirichlet averages of the R-function.

Theorem 3. Let $\beta, \beta', \delta, \lambda \in \mathbb{C}$ with $\min\{\Re(\beta), \Re(\beta'), \Re(\lambda)\} > 0$ and $x, y, \kappa, \alpha, \gamma \in \mathbb{R}$ with $x > y$ and $\min\{\kappa, \alpha, -\gamma\} > 0$. The convergence conditions of the R-function are supposed to be satisfied. Then the following formula holds true:

$$\,_p^{\kappa,\lambda}\mathcal{M}_q^{\alpha,\delta;\gamma}\left[(\beta,\beta';x,y)\right]_{(b)_{1,q}}^{(a)_{1,p}}$$

$$= \frac{y^{\lambda-1}}{\Gamma(\delta)}F_{0:q+2;1}^{1:p+1;1}\left(\begin{array}{c}[1-\kappa:-\gamma,1]:\\ \underline{\qquad}:\end{array}\begin{array}{c}[(a),\gamma:1_{(p)},\kappa];\\ [(b),\delta,1-\kappa:1_{(q)},\alpha,-\gamma];\end{array}\right. \tag{30}$$

$$\left.\begin{array}{c}[\beta:1];\\ [\beta+\beta':1];\end{array}y^\gamma,1-\frac{x}{y}\right),$$

where $a_{(\ell)}$, here and throughout this paper, abbreviates the array of ℓ times repetition of the same parameter a's, $a, \ldots, a$, and (a) and (b) abbreviate the arrays of p and q parameters $a_1, \ldots, a_p$ and $b_1, \ldots, b_q$, respectively.

Proof. In view of (28) and (15), we have

$$\mathcal{D}_3 := {}_p^{\kappa,\lambda}\mathcal{M}_q^{\alpha,\delta,\gamma}\left[(\beta,\beta';x,y)\right]_{(b)_{1,q}}^{(a)_{1,p}}$$

$$= \frac{1}{B(\beta,\beta')}\int_0^1 u^{\beta-1}(1-u)^{\beta'-1}[y+u(x-y)]^{\lambda-1}$$

$$\times \sum_{n=0}^{\infty} \frac{\prod_{j=1}^{p}(a_j)_n}{\prod_{j=1}^{q}(b_j)_n} \frac{(\gamma)_{\kappa n}[y+u(x-y)]^{n\gamma}}{n!\Gamma(\alpha n+\delta)}\, du.$$

Exchanging the order of integral and summation and using the generalized binomial series

$$(1-z)^{-a} = \sum_{n=0}^{\infty} \frac{(a)_n z^n}{n!} \quad (|z|<1;\ a\in\mathbb{C})$$

and the Beta function, we obtain

$$\mathcal{D}_3 := \frac{\Gamma(\beta+\beta')}{\Gamma(\beta)\Gamma(\beta')}\sum_{n=0}^{\infty} \frac{\prod_{j=1}^{p}(a_j)_n}{\prod_{j=1}^{q}(b_j)_n} \frac{(\gamma)_{\kappa n} y^{n\gamma+\lambda-1}}{n!\Gamma(\alpha n+\delta)}$$

$$\times \int_0^1 u^{\beta-1}(1-u)^{\beta'-1}\left[1-\left(1-\frac{x}{y}\right)u\right]^{n\gamma+\lambda-1} du$$

$$= y^{\lambda-1}\sum_{n=0}^{\infty} \frac{\prod_{j=1}^{p}(a_j)_n}{\prod_{j=1}^{q}(b_j)_n} \frac{(\gamma)_{\kappa n} y^{n\gamma}}{n!\Gamma(\alpha n+\delta)}\, {}_2F_1\!\left(\begin{array}{c}\beta,1-\gamma n-\lambda;\\ \beta+\beta';\end{array} 1-\frac{x}{y}\right).$$

Applying $\Gamma(\lambda+\nu) = \Gamma(\lambda)(\lambda)_\nu$ $(\lambda, \nu \in \mathbb{C})$ and

$$(1-\lambda-\gamma n)_r = \frac{\Gamma(1-\lambda-\gamma n+r)}{\Gamma(1-\lambda-\gamma n)} = \frac{(1-\lambda)_{-\gamma n+r}}{(1-\lambda)_{-\gamma n}},$$

we find

$$\mathcal{D}_3 = \frac{y^{\lambda-1}}{\Gamma(\delta)}\sum_{n=0}^{\infty}\sum_{r=0}^{\infty} \frac{\prod_{j=1}^{p}(a_j)_n(\gamma)_{\kappa n}(1-\kappa)_{-\gamma n+r}(\beta)_r}{\prod_{j=1}^{q}(b_j)_n(\delta)_{\alpha n}(1-\kappa)_{-\gamma n}(\beta+\beta')_r}\frac{(y^\gamma)^n \left(1-\frac{x}{y}\right)^r}{n!\, r!},$$

which, in view of (17), leads to the right-hand side of (30). This completes the proof. □

5. Multivariate Dirichlet Averages

Consider the Dirichlet average (23) and its modification (27) where $(z) := (z_1, \ldots, z_n) \in \mathbb{C}^n$ and $d_1, \ldots, d_n$ are parameters. Our finding is predicated on the following basic premise in Lemma 1 (see [22]).

Lemma 1. *Let $d_j, r_j \in \mathbb{C}$ $(j \in \overline{1,n};\ n \in \mathbb{N})$ such that $\min\{\Re(d_j), \Re(r_j)\} > -1$. Furthermore, let E_{n-1} denote the Euclidean simplex in (12) and $d\mu_d(u)$ stand for the Dirichlet measure in (11). Then the following formula holds true:*

$$\int_{E_{n-1}} u_1^{r_1}\cdots u_{n-1}^{r_{n-1}}(1-u_1-\cdots-u_{n-1})^{r_n}\, d\mu_d(u) = \frac{(d_1)_{r_1}\cdots(d_n)_{r_n}}{(d_1+\cdots+d_n)_{r_1+\cdots+r_n}} \qquad (31)$$

(see Equation (52) [22]).

The Lauricella function F_D defined for complex parameters $d = (d_1, \ldots, d_n) \in \mathbb{C}^n$ is defined as follows (consult, for example, Section 1.4 in [40]):

$$F_D(a, (d); c; z) = \sum_{m_1, \ldots, m_n = 0}^{\infty} \frac{(a)_{m_1 + \cdots + m_n} (d_1)_{m_1} \cdots (d_n)_{m_n}}{(c)_{m_1 + \cdots + m_n}} \frac{z_1^{m_1} \cdots z_n^{m_n}}{m_1! \cdots m_n!}. \tag{32}$$

The series (32) converges for all variables inside unit circle $\max_{1 \le j \le n} |z_j| < 1$.

Here we investigate the following Dirichlet average:

$$\begin{aligned}
&{}_p^{\kappa,\eta} \mathcal{M}_q^{\alpha,\delta;\gamma} [(d); (1-z)]_{(b)_{1,q}}^{(a)_{1,p}} \\
&= \int_{E_{n-1}} (1 - u \circ z)^{\eta - 1} \left[{}_p^{\kappa} R_q^{\alpha,\delta;\gamma} (a_1, \ldots, a_p; b_1, \ldots, b_q; (1 - u \circ z)^{\gamma}) \right] d\mu_d(u).
\end{aligned} \tag{33}$$

We also need the following multinomial expansion:

$$(1 - z_1 - \cdots - z_n)^{\rho} = \sum_{r_1, \ldots, r_n = 0}^{\infty} (-\rho)_{r_1 + \cdots + r_n} \frac{z_1^{r_1} \cdots z_n^{r_n}}{r_1! \cdots r_n!} \quad (|z_1 + \cdots + z_n| < 1). \tag{34}$$

Theorem 4. Let $\kappa, \alpha, \gamma \in \mathbb{R}$ with $\min\{\kappa, \alpha, \gamma\} > 0$ and $\delta, \eta, d_j, z_j \in \mathbb{C}$ with $\Re(\eta) > 0$ and $\Re(d_j) > 0$ $(j \in \overline{1,n})$. Convergence conditions of the R-function are assumed to be satisfied. Then the following result holds true:

$$\begin{aligned}
&{}_p^{\kappa,\eta} \mathcal{M}_q^{\alpha,\delta;\gamma} [d_1, \ldots, d_n; 1 - z_1, \ldots, 1 - z_n]_{(b)_{1,q}}^{(a)_{1,p}} \\
&= \frac{1}{\Gamma(\delta)} F_{2:q+1;0;\ldots;0}^{0:p+2;1;\ldots;1} \left(\begin{array}{c} \underline{\qquad} \\ [\eta, \sum_{j=1}^{n} d_j : \theta^{(1)}, \theta^{(2)}] : \end{array} \begin{array}{c} [(a), \gamma, \eta : 1_{(p)}, \kappa, \gamma]; \\ [(b), \delta : 1_{(q)}, \alpha]; \end{array} \right. \\
&\qquad \left. \begin{array}{c} [d_1 : 1]; \quad \ldots; \quad [d_n : 1]; \\ \underline{\qquad} ; \quad \ldots; \quad \underline{\qquad} ; \end{array} \quad 1, -z_1, \ldots, -z_n \right),
\end{aligned} \tag{35}$$

where (a) and (b) abbreviate the arrays of p and q parameters $a_1, \ldots, a_p$ and $b_1, \ldots, b_q$, respectively, $\theta^{(1)}$ and $\theta^{(2)}$ abbreviate the arrays of $n + 1$ parameters γ, $(-1)_{(n)}$ and 0, $1_{(n)}$, respectively.

Proof. Considering the multivariate Dirichlet average (33), we have

$$\begin{aligned}
\mathcal{D}_4 &:= {}_p^{\kappa,\eta} \mathcal{M}_q^{\alpha,\delta;\gamma} [(d); (1-z)]_{(b)_{1,q}}^{(a)_{1,p}} \\
&= \int_{E_{n-1}} (1 - u \circ z)^{\eta - 1} \sum_{n=0}^{\infty} \frac{\prod_{j=1}^{p} (a_j)_n (1 - u \circ z)^{\gamma n} (\gamma)_{\kappa n}}{\prod_{j=1}^{q} (b_j)_n \Gamma(\alpha n + \delta) n!} d\mu_d(u) \\
&= \sum_{n=0}^{\infty} \frac{\prod_{j=1}^{p} (a_j)_n (\gamma)_{\kappa n}}{\prod_{j=1}^{q} (b_j)_n \Gamma(\alpha n + \delta) n!} \int_{E_{n-1}} (1 - u \circ z)^{\gamma n + \eta - 1} d\mu_d(u).
\end{aligned}$$

Applying Lemma 1 and the polynomial expansion (34), and assuming $|u_1 z_1 + \cdots + u_n z_n| < 1$, we arrive at

$$\mathcal{D}_4 = \sum_{n=0}^{\infty} \frac{\prod_{j=1}^{p}(a_j)_n (\gamma)_{\kappa n}}{\prod_{j=1}^{q}(b_j)_n \Gamma(\alpha n + \delta) n!} \sum_{r_1,\ldots,r_n=0}^{\infty} (1-\gamma n - \eta)_{r_1+\cdots+r_n} \frac{z_1^{r_1}\cdots z_n^{r_n}}{r_1!\cdots r_n!}$$
$$\times \int_{E_{n-1}} u_1^{r_1}\cdots u_n^{r_n} (1-u_1-\cdots-u_{n-1})^{r_n} d\mu_d(u)$$
$$= \sum_{n=0}^{\infty} \frac{\prod_{j=1}^{p}(a_j)_n (\gamma)_{\kappa n}}{\prod_{j=1}^{q}(b_j)_n \Gamma(\alpha n + \delta) n!}$$
$$\times \sum_{r_1,\ldots,r_n=0}^{\infty} \frac{(1-\gamma n - \eta)_{r_1+\cdots+r_n} (d_1)_{r_1}\cdots(d_n)_{r_n}}{(d_1+\cdots+d_n)_{r_1+\cdots+r_n}} \frac{z_1^{r_1}\cdots z_n^{r_n}}{r_1!\cdots r_n!}.$$

The n-fold inner sum (with respect to $r_1,\cdots,r_n$) forms a Lauricella $F_D^{(n)}$ function in n variables (see, for instance, (p. 33) [40]), we have

$$\mathcal{D}_4 = \sum_{n=0}^{\infty} \frac{\prod_{j=1}^{p}(a_j)_n (\gamma)_{\kappa n}}{\prod_{j=1}^{q}(b_j)_n \Gamma(\alpha n + \delta) n!}$$
$$\times F_D^{(n)}[1-\gamma n - \eta; d_1,\ldots,d_n; d_1+\cdots+d_n; z_1,\ldots,z_n].$$

Using $\Gamma(\delta + \alpha n) = \Gamma(\delta)(\delta)_{\alpha n}$ and

$$(1-\gamma n - \eta)_{r_1+\cdots+r_n} = (-1)^{r_1+\cdots+r_n} \frac{(\eta)_{\gamma n}}{(\eta)_{\gamma n - r_1 - \cdots - r_n}},$$

we obtain

$$\mathcal{D}_4 = \frac{1}{\Gamma(\delta)} \sum_{n,r_1,\ldots,r_n=0}^{\infty} \frac{\left(\prod_{j=1}^{p}(a_j)_n\right)(\gamma)_{\kappa n}(\eta)_{\gamma n}(d_1)_{r_1}\cdots(d_n)_{r_n}}{\left(\prod_{j=1}^{q}(b_j)_n\right)(\delta)_{\alpha n}(\eta)_{\gamma n - r_1 - \cdots - r_n}(d_1+\cdots+d_n)_{r_1+\cdots+r_n}}$$
$$\times \frac{(-z_1)^{r_1}\cdots(-z_n)^{r_n}}{n!\, r_1!\cdots r_n!},$$

which, in view of (17), is easily seen to yield the expression of the right-hand side of (35). □

6. Concluding Remarks

The Dirichlet and modified Dirichlet averages of the R-function in (18) (a generalized Mittag-Leffler type function) were explored. In Theorems 1 and 2, the bivariate Dirichlet averages of the R-function (18) were expressed in terms of the Riemann–Liouville fractional integrals whose kernel functions are products of some elementary functions and the R-function (18). In Theorem 3, the bivariate Dirichlet average of the R-function (18) (see Theorem 2) was shown to be expressed in terms of the Srivastava–Daoust generalization (17) of the Lauricella hypergeometric function. In Theorem 4, the multivariate Dirichlet average of the R-function (18) was proven to be expressed in terms of the Srivastava–Daoust generalization (17) of the Lauricella hypergeometric function. The main results in Theorems 1–4 are believed to be useful.

The Mittag-Leffler function $E_\alpha(z)$ in (1), the two-parametric Mittag-Leffler function $E_{\alpha,\beta}(z)$ in (2), the three-parametric Mittag-Leffler function $E_{\alpha,\beta}^\gamma(z)$ in (3), and the R-function in (18) are obviously contained as special cases in the well-known Fox–Wright function ${}_p\Psi_q$ (see, for details, p. 21 [40]; see also p. 56 [38]). Because the R-function in (18) is of general character, all results in Theorems 1–4 are seen to be able to yield a large number of particular instances. The following corollary demonstrates just a particular instance of Theorem 1:

Corollary 1. *Let the conditions in Theorem 1 be satisfied and set $p = q = 1$ and $a_j = b_j = 1$ in (24). Then the Dirichlet average for the generalized Mittag-Leffler function holds true:*

$${}_1^\kappa\mathcal{M}_1^{\alpha,\delta;\gamma}[(\beta,\beta';x,y)] = \frac{\Gamma(\beta+\beta')}{\Gamma(\beta)(x-y)^{\beta+\beta'-1}}\left\{I_{0+}^{\beta'}\left(t^{\beta-1}E_{\alpha,\delta}^{\gamma,\kappa}(y+t)\right)\right\}(x-y), \quad (36)$$

where $E_{\alpha,\delta}^{\gamma,\kappa}$ is given in (21).

As with the H-function of the single variable in (5), the H-function of multiple variables is generated using multiple contour integrals of the Mellin–Barnes type (see pp. 205–207, Appendix A.1 in [5]). This article concludes with the questions posed: Like (8),

- Express (possibly) the Srivastava–Daoust generalization (17) of the Lauricella hypergeometric function in terms of the multivariate H-function;
- Express (possibly) the right members of Theorems 3 and 4 in terms of the multivariate H-function.

Author Contributions: Writing—original draft, D.K., J.R. and J.C.; writing—review and editing, D.K. and J.C. All authors have read and agreed to the published version of the manuscript.

Funding: The third-named author was supported by the Basic Science Research Program through the National Research Foundation of Korea (NRF) funded by the Ministry of Education (NRF-2020R111A1A01052440).

Institutional Review Board Statement: Not applicable.

Informed Consent Statement: Not applicable.

Acknowledgments: The authors are grateful for the valuable and encouraging comments made by the anonymous referees, which helped to improve this paper. The first and third named authors are dedicating this paper to Professor Jeta Ram who passed away in November of 2020.

Conflicts of Interest: The authors declare no conflict of interest.

References

1. Mittag-Leffler, G.M. Sur la nouvelle fonction $E_\alpha(x)$. *C. R. Acad. Sci. Paris* **1903**, *137*, 554–558.
2. Srivastava, H.M.; Choi, J. *Zeta and q-Zeta Functions and Associated Series and Integrals*; Elsevier Science Publishers: Amsterdam, The Netherland; London, UK; New York, NY, USA, 2012.
3. Gorenflo, R.; Kilbas, A.A.; Mainardi, F.; Rogosin, S. *Mittag-Leffler Functions, Related Topics and Applications*, 2nd ed.; Springer: Berlin/Heidelberg, Germany, 2020.
4. Erdélyi, A.; Magnus, W.; Oberhettinger, F.; Tricomi, F.G. *Higher Transcendental Functions*; McGraw-Hill Book Company: New York, NY, USA; Toronto, ON, Canada; London, UK, 1955; Volume 3.
5. Mathai, A.M.; Saxena, R.K.; Haubold, H.J. *The H-Function: Theory and Applications*; Springer: Dordrecht, The Netherlands; New York, NY, USA, 2010.
6. Fox, C. The G and H functions as symmetrical Fourier kernels. *Thans. Am. Math. Soc.* **1961**, *98*, 395–429. [CrossRef]
7. Humbert, P. Quelques résultats relatifs à la fonction de Mittag-Leffler. *C. R. Acad. Sci. Paris* **1953**, *236*, 1467–1468.
8. Dzherbashian, M.M. On integral representation of functions continuous on given rays (generalization of the Fourier integrals). *Izvestija Akad. Nauk SSSR Ser. Mat.* **1954**, *18*, 427–448. (In Russian)
9. Wiman, A. Über den fundamentalsatz der theorie der funkntionen $E_\alpha(x)$. *Acta Math.* **1905**, *29*, 191–201. [CrossRef]
10. Prabhakar, T.R. A singular integral equation with a generalized Mittag-Leffler function in the kernel. *Yokohama Math. J.* **1971**, *19*, 7–15.
11. Kilbas, A.A.; Saigo, M.; Saxena, R.K. Generalized Mittag-Leffler functions and generalized fractional calculus operators. *Integral Transf. Spec. Funct.* **2004**, *15*, 31–49. [CrossRef]
12. Shukla, A.K.; Prajapati, J.C. On a generalization of Mittag-Leffler function and its properties. *J. Math. Anal. Appl.* **2007**, *336*, 797–811. [CrossRef]
13. Carlson, B.C. *Special Functions of Applied Mathematics*; Academic Press: New York, NY, USA, 1977.
14. Carlson, B.C. Lauricella's hypergeometric function F_D. *J. Math. Anal. Appl.* **1963**, *7*, 452–470. [CrossRef]
15. Carlson, B.C. A connection between elementary and higher transcendental functions. *SIAM J. Appl. Math.* **1969**, *17*, 116–148. [CrossRef]
16. Carlson, B.C. Invariance of an integral average of a logarithm. *Amer. Math. Mon.* **1975**, *82*, 379–382. [CrossRef]
17. Carlson, B.C. Dirichlet Averages of $x^t \log x$. *SIAM J. Math. Anal.* **1987**, *18*, 550–565. [CrossRef]

18. Carlson, B.C. *B*-splines, hypergeometric functions and Dirichlet average. *J. Approx. Theory* **1991**, *67*, 311–325. [CrossRef]
19. Neuman, E.; Fleet, P.J.V. Moments of Dirichlet splines and their applications to hypergeometric functions. *J. Comput. Appl. Math.* **1994**, *53*, 225–241. [CrossRef]
20. Daiya, J.; Kumar, D. *S*-function associated with fractional derivative and double Dirichlet average. *AIMS Math.* **2020**, *5*, 1372–1382. [CrossRef]
21. Vyas, D.N.; Banerji, P.K.; Saigo, M. On Dirichlet average and fractional integral of a general claqss of polynomials. *J. Fract. Calc.* **1994**, *6*, 61–64.
22. Kilbas, A.A.; Kattuveettill, A. Representations of Dirichlet averages of generalized Mittag-Leffler function via fractional integrals and special functions. *Frac. Calc. Appl. Anal.* **2008**, *11*, 471–492.
23. Massopust, P.; Forster, B. Multivariate complex *B*-splines and Dirichlet averages. *J. Approx. Theory* **2010**, *162*, 252–269. [CrossRef]
24. Gupta, S.C.; Agrawal, B.M. Double Dirichlet averages and fractional derivatives. *Ganita Sandesh* **1991**, *5*, 47–53.
25. Saxena, R.K.; Pogány, T.K.; Ram, J.; Daiya, J. Dirichlet averages of generalized multi-index Mittag-Leffler functions. *Armen. J. Math.* **2010**, *3*, 174–187.
26. Dickey, J.M. Multiple hypergeometric functions: Probabilistic interpretations and statistical uses. *J. Amer. Statist. Assoc.* **1983**, *78*, 628–637. [CrossRef]
27. Vyas, D.N. Some results on hypergeometric functions suggested by Dirichlet averages. *J. Indian Acad. Math.* **2011**, *33*, 705–715.
28. Ahmad, F.; Jain, D.K.; Jain, A.; Ahmad, A. Dirichlet averages of Wright-type hypergeometric function. *Inter. J. Discrete Math.* **2017**, *2*, 6–9.
29. Saxena, R.K.; Daiya, J. Integral transforms of the *S*-function. *Le Math.* **2015**, *70*, 147–159.
30. Saxena, R.K.; Nishimoto, K. *N*–fractional calculus of generalized Mittag–Leffler functions. *J. Fract. Calc.* **2010**, *37*, 43–52.
31. Vyas, D.N. Dirichlet averages, fractional integral operators and solution of Euler-Darboux equation on Hölder spaces. *Appl. Math.* **2016**, *7*, 69827. [CrossRef]
32. Deora, Y.; Banerji, P.K. Double Dirichlet average of e^x using fractional derivative. *J. Fract. Calc.* **1993**, *3*, 81–86.
33. Deora, Y.; Banerji, P.K. Triple Dirichlet average and fractional derivative. *Rev. Téc. Fac. Ing. Univ. Zulia* **1993**, *16*, 157–161.
34. Deora, Y.; Banerji, P.K.; Saigo, M. Fractional integral and Dirichlet averages. *J. Frac. Calc.* **1994**, *6*, 55–59.
35. Deora, Y.; Banerji, P.K. An application of fractional calculus to the solution of Euler-Darboux equation in terms of Dirichlet averages. *J. Fract. Calc.* **1994**, *5*, 91–94.
36. Ram, C.; Choudhary, P.; Gehlot, K.S. Representation of Dirichlet average of *K*-series via fractional integrals and special functions. *Internat. J. Math. Appl.* **2013**, *1*, 1–11.
37. Simić, S.; Bin-Mohsin, B. Stolarsky means in many variables. *Mathematics* **2020**, *8*, 1320. [CrossRef]
38. Kilbas, A.A.; Srivastava, H.M.; Trujillo, J.J. *Theory and Applications of Fractional Differential Equations*, North-Holland Mathematical Studies; Elsevier (North-Holland) Science Publishers: Amsterdam, The Netherlands; London, UK; New York, NY, USA, 2006; Volume 204.
39. Srivastava, H.M.; Daoust, M.C. Certain generalized Neumann expansions associated with the Kampé de Fériet function. *Nederl. Akad. Wetensch. Proc. Ser. A Indag. Math.* **1969**, *72*, 449–457.
40. Srivastava, H.M.; Karlsson, P.W. *Multiple Gaussian Hypergeometric Series*; Halsted Press (Ellis Horwood Limited): Chichester, UK; John Wiley and Sons: New York, NY, USA, 1985.
41. Srivastava, H.M.; Daoust, M.C. A note on convergence of Kempé de Fériet double hypergeometric series. *Math. Nachr.* **1972**, *53*, 151–157. [CrossRef]
42. Kumar, D.; Kumar, S. Fractional calculus of the generalized Mittag-Leffler type function. *Int. Sch. Res. Notices* **2014**, *2014*, 907432. [CrossRef] [PubMed]
43. Kumar, D.; Purohit, S.D. Fractional differintegral operators of the generalized Mittag-Leffler type function. *Malaya J. Mat.* **2014**, *2*, 419–425.
44. Srivastava, H.M.; Tomovski, Ž. Fractional calculus with an integral operator containing a generalized Mittag-Leffler function in the kernel. *Appl. Math. Comput.* **2009**, *211*, 198–210. [CrossRef]

Article

Fractional Integral Inequalities of Hermite–Hadamard Type for $(h, g; m)$-Convex Functions with Extended Mittag-Leffler Function

Maja Andrić

Faculty of Civil Engineering, Architecture and Geodesy, University of Split, Matice hrvatske 15, 21000 Split, Croatia; maja.andric@gradst.hr

Abstract: Several fractional integral inequalities of the Hermite–Hadamard type are presented for the class of $(h, g; m)$-convex functions. Applied fractional integral operators contain extended generalized Mittag-Leffler functions as their kernel, thus enabling new fractional integral inequalities that extend and generalize the known results. As an application, the upper bounds of fractional integral operators for $(h, g; m)$-convex functions are given.

Keywords: fractional calculus; Mittag-Leffler function; convex function; Hermite–Hadamard inequality

Citation: Andrić, M. Fractional Integral Inequalities of Hermite–Hadamard Type for $(h, g; m)$-Convex Functions with Extended Mittag-Leffler Function. *Fractal Fract.* **2022**, *6*, 301. https://doi.org/10.3390/fractalfract6060301

Academic Editor: Ahmed I. Zayed

Received: 5 May 2022
Accepted: 28 May 2022
Published: 29 May 2022

Publisher's Note: MDPI stays neutral with regard to jurisdictional claims in published maps and institutional affiliations.

Copyright: © 2022 by the author. Licensee MDPI, Basel, Switzerland. This article is an open access article distributed under the terms and conditions of the Creative Commons Attribution (CC BY) license (https://creativecommons.org/licenses/by/4.0/).

1. Introduction

In recent years, in the field of applied sciences, fractional calculus has been used with different boundary conditions to develop mathematical models relating to real-world problems. This significant interest in the theory of fractional calculus has been stimulated by many of its applications, especially in the various fields of physics and engineering.

Inequalities involving integrals of functions and their derivatives are of great importance in mathematical analysis and its applications. Inequalities containing fractional derivatives have applications in regard to fractional differential equations, especially in establishing the uniqueness of the solutions of initial value problems and their upper bounds. This kind of application motivated the researchers towards the theory of integral inequalities, with the aim of extending and generalizing classical inequalities using different fractional integral operators.

The motivation for this research on Hermite–Hadamard-type integral inequalities was provided by recent studies on these inequalities for different types of integral operators (see [1–8]) and different classes of convexity (see [9–17]). The famous Hermite–Hadamard inequality provides an estimate of the (integral) mean value of a continuous convex function.

Theorem 1 (The Hermite–Hadamard inequality). *Let $f : [a, b] \to \mathbb{R}$ be a continuous convex function. Then*

$$f\left(\frac{a+b}{2}\right) \leq \frac{1}{b-a} \int_a^b f(x)\, dx \leq \frac{f(a)+f(b)}{2}.$$

Its fractional version, involving Riemann–Liouville fractional integrals, is given in [18].

Theorem 2 ([18]). *Let $f : [a, b] \to \mathbb{R}$ be a convex function with $f \in L_1[a, b]$. Then for $\sigma > 0$*

$$f\left(\frac{a+b}{2}\right) \leq \frac{\Gamma(\sigma+1)}{2(b-a)^\sigma} \left[J_{a+}^\sigma f(b) + J_{b-}^\sigma f(a)\right] \leq \frac{f(a)+f(b)}{2}.$$

Recall that the left-sided and the right-sided Riemann-Liouville fractional integrals of order $\sigma > 0$ are defined as in [19] for $f \in L_1[a, b]$ with

$$J_{a+}^{\sigma} f(x) = \frac{1}{\Gamma(\sigma)} \int_a^x (x-t)^{\sigma-1} f(t)\, dt, \quad x \in (a,b], \tag{1}$$

$$J_{b-}^{\sigma} f(x) = \frac{1}{\Gamma(\sigma)} \int_x^b (t-x)^{\sigma-1} f(t)\, dt, \quad x \in [a,b). \tag{2}$$

Our aim is to prove Hermite–Hadamard's inequality in more general settings, and for this we need an extended generalized Mittag-Leffler function with its fractional integral operators and a class of $(h, g; m)$-convex functions.

The paper is structured as follows. In Section 2, we give present preliminary results and definitions that will be used in this paper. In Section 3, several Hermite–Hadamard-type inequalities for $(h, g; m)$-convex functions using fractional integral operators are presented. Furthermore, several properties and identities of these operators are given. As an application, in Section 4 we derive the upper bounds of fractional integral operators involving $(h, g; m)$-convex functions. In the last section, Section 5, we present the conclusions of this research.

2. Preliminaries

2.1. An Extended Generalized form of the Mittag-Leffler Function

The Mittag-Leffler function

$$E_\rho(z) = \sum_{n=0}^{\infty} \frac{z^n}{\Gamma(\rho n + 1)} \quad (z \in \mathbb{C}, \Re(\rho) > 0)$$

with its generalizations appears as a solution of fractional differential or integral equations. The first generalization for two parameters was carried out by Wiman [8]:

$$E_{\rho,\sigma}(z) = \sum_{n=0}^{\infty} \frac{z^n}{\Gamma(\rho n + \sigma)}, \quad (z, \rho, \sigma \in \mathbb{C}, \Re(\rho) > 0), \tag{3}$$

after which Prabhakar defined the Mittag-Leffler function of three parameters [3]:

$$E_{\rho,\sigma}^{\delta}(z) = \sum_{n=0}^{\infty} \frac{(\delta)_n}{\Gamma(\rho n + \sigma)} \frac{z^n}{n!}, \quad (z, \rho, \sigma, \delta \in \mathbb{C}, \Re(\rho) > 0). \tag{4}$$

Recently we presented in [1] (see also [2]) an extended generalized form of the Mittag-Leffler function $E_{\rho,\sigma,\tau}^{\delta,c,v,r}(z;p)$:

Definition 1 ([1])**.** *Let $\rho, \sigma, \tau, \delta, c \in \mathbb{C}, \Re(\rho), \Re(\sigma), \Re(\tau) > 0, \Re(c) > \Re(\delta) > 0$ with $p \geq 0$, $r > 0$ and $0 < q \leq r + \Re(\rho)$. Then the extended generalized Mittag-Leffler function $E_{\rho,\sigma,\tau}^{\delta,c,q,r}(z;p)$ is defined by*

$$E_{\rho,\sigma,\tau}^{\delta,c,q,r}(z;p) = \sum_{n=0}^{\infty} \frac{B_p(\delta + nq, c - \delta)}{B(\delta, c - \delta)} \frac{(c)_{nq}}{\Gamma(\rho n + \sigma)} \frac{z^n}{(\tau)_{nr}}. \tag{5}$$

Note, we use the generalized Pochhammer symbol $(c)_{nq} = \frac{\Gamma(c+nq)}{\Gamma(c)}$ and an extended beta function $B_p(x,y) = \int_0^1 t^{x-1}(1-t)^{y-1} e^{-\frac{p}{t(1-t)}} dt$, where $\Re(x), \Re(y), \Re(p) > 0$.

Remark 1. *Several generalizations of the Mittag-Leffler function can be obtained for different parameter choices. For instance, the function (5) is reduced to*

(i) *the Salim-Faraj function $E_{\rho,\sigma,\tau}^{\delta,\tau,q}(z)$ for $p = 0$ [5],*

(ii) *the Rahman function $E_{\rho,\sigma}^{\delta,q,c}(z;p)$ for $\tau = r = 1$ [4],*

(iii) *the Shukla–Prajapati function $E_{\rho,\sigma}^{\delta,q}(z)$ for $p = 0$ and $\tau = r = 1$ [6],*

(iv) *the Prabhakar function $E_{\rho,\sigma}^{\delta}(z)$ for $p = 0$ and $\tau = r = q = 1$ [3],*

(v) the Wiman function $E_{\rho,\sigma}(z)$ for $p = 0$ and $\tau = r = q = \delta = 1$ [8],
(vi) the Mittag-Leffler function $E_\rho(z)$ for $p = 0$, $\tau = r = q = \delta = 1$ and $\sigma = 1$.

Next we have corresponding fractional integral operators, the left-sided $\varepsilon_{a^+,\rho,\sigma,\tau}^{\omega,\delta,c,q,r} f$ and the right-sided $\varepsilon_{b^-,\rho,\sigma,\tau}^{\omega,\delta,c,q,r} f$, where the kernel is a function $E_{\rho,\sigma,\tau}^{\delta,c,q,r}(z;p)$:

Definition 2 ([1]). *Let $\omega, \rho, \sigma, \tau, \delta, c \in \mathbb{C}$, $\Re(\rho), \Re(\sigma), \Re(\tau) > 0$, $\Re(c) > \Re(\delta) > 0$ with $p \geq 0$, $r > 0$ and $0 < q \leq r + \Re(\rho)$. Let $f \in L_1[a,b]$ and $x \in [a,b]$. Then the left-sided and the right-sided generalized fractional integral operators $\varepsilon_{a^+,\rho,\sigma,\tau}^{\omega,\delta,c,q,r} f$ and $\varepsilon_{b^-,\rho,\sigma,\tau}^{\omega,\delta,c,q,r} f$ are defined by*

$$\left(\varepsilon_{a^+,\rho,\sigma,\tau}^{\omega,\delta,c,q,r} f\right)(x;p) = \int_a^x (x-t)^{\sigma-1} E_{\rho,\sigma,\tau}^{\delta,c,q,r}(\omega(x-t)^\rho;p) f(t) dt, \tag{6}$$

$$\left(\varepsilon_{b^-,\rho,\sigma,\tau}^{\omega,\delta,c,q,r} f\right)(x;p) = \int_x^b (t-x)^{\sigma-1} E_{\rho,\sigma,\tau}^{\delta,c,q,r}(\omega(t-x)^\rho;p) f(t) dt. \tag{7}$$

Remark 2. *If we apply different parameter choices, then* (6) *is a generalization of*
(i) *the Salim-Faraj fractional integral operator $\varepsilon_{a^+,\rho,\sigma,\tau}^{\omega,\delta,q,r} f(x)$ for $p = 0$ [5],*
(ii) *the Rahman fractional integral operator $\varepsilon_{a^+,\rho,\sigma}^{\omega,\delta,q,c} f(x;p)$ for $\tau = r = 1$ [4],*
(iii) *the Srivastava–Tomovski fractional integral operator $\varepsilon_{a^+,\rho,\sigma}^{\omega,\delta,q} f(x)$ for $p = 0$ and $\tau = r = 1$ [7],*
(iv) *the Prabhakar fractional integral operator $\varepsilon(\rho,\sigma;\delta;\omega) f(x)$ for $p = 0$ and $\tau = r = q = 1$ [3],*
(v) *the left-sided Riemann–Liouville fractional integral $J_{a^+}^\sigma f(x)$ for $p = \omega = 0$, that is,* (1).

We listed reductions for the left-sided fractional integral operator, whereas the analogs are valid for the right-sided.

More details on this generalized form of the Mittag-Leffler function and its fractional integral operators can be found in [1,2]. Here are some results we will use in this study:

Theorem 3 ([1]). *If $\alpha, \omega, \rho, \sigma, \tau, \delta, c \in \mathbb{C}$, $\Re(\rho), \Re(\sigma), \Re(\tau) > 0$, $\Re(c) > \Re(\delta) > 0$ with $p \geq 0$, $r > 0$ and $0 < q \leq r + \Re(\rho)$, then for power functions $(t-a)^{\alpha-1}$ and $(b-t)^{\alpha-1}$ follow*

$$\left(\varepsilon_{a^+,\rho,\sigma,\tau}^{\omega,\delta,c,q,r}(t-a)^{\alpha-1}\right)(x;p) = \Gamma(\alpha)(x-a)^{\alpha+\sigma-1} E_{\rho,\sigma+\alpha,\tau}^{\delta,c,q,r}(\omega(x-a)^\rho;p), \tag{8}$$

$$\left(\varepsilon_{b^-,\rho,\sigma,\tau}^{\omega,\delta,c,q,r}(b-t)^{\alpha-1}\right)(x;p) = \Gamma(\alpha)(b-x)^{\alpha+\sigma-1} E_{\rho,\sigma+\alpha,\tau}^{\delta,c,q,r}(\omega(b-x)^\rho;p). \tag{9}$$

If we set $a = 0$ and $x = 1$ in (8), or $b = 1$ and $x = 0$ in (9), then we obtain the following corollary.

Corollary 1 ([1]). *If $\alpha, \omega, \rho, \sigma, \tau, \delta, c \in \mathbb{C}$, $\Re(\rho), \Re(\sigma), \Re(\tau) > 0$, $\Re(c) > \Re(\delta) > 0$ with $p \geq 0$, $r > 0$ and $0 < q \leq r + \Re(\rho)$, then*

$$\frac{1}{\Gamma(\alpha)} \int_0^1 t^{\alpha-1}(1-t)^{\sigma-1} E_{\rho,\sigma,\tau}^{\delta,c,q,r}(\omega(1-t)^\rho;p) dt = E_{\rho,\sigma+\alpha,\tau}^{\delta,c,q,r}(\omega;p).$$

Setting $\alpha = 1$ in Theorem 3, we obtain following identities for the constant function:

Corollary 2 ([2]). *Let the assumptions of Theorem 3 hold with $\alpha = 1$. Then*

$$\left(\varepsilon^{\omega,\delta,c,q,r}_{a^+,\rho,\sigma,\tau}1\right)(x;p) = (x-a)^\sigma E^{\delta,c,q,r}_{\rho,\sigma+1,\tau}(\omega(x-a)^\rho;p), \tag{10}$$

$$\left(\varepsilon^{\omega,\delta,c,q,r}_{b^-,\rho,\sigma,\tau}1\right)(x;p) = (b-x)^\sigma E^{\delta,c,q,r}_{\rho,\sigma+1,\tau}(\omega(b-x)^\rho;p). \tag{11}$$

In this paper, we will use simplified notation to avoid a complicated manuscript form:

$$E(z;p) := E^{\delta,c,q,r}_{\rho,\sigma,\tau}(z;p)$$

and

$$(\boldsymbol{\varepsilon}^\omega_{a^+}f)(x;p) := \left(\varepsilon^{\omega,\delta,c,q,r}_{a^+,\rho,\sigma,\tau}f\right)(x;p),$$

$$(\boldsymbol{\varepsilon}^\omega_{b^-}f)(x;p) := \left(\varepsilon^{\omega,\delta,c,q,r}_{b^-,\rho,\sigma,\tau}f\right)(x;p).$$

Of course, the conditions on all parameters $\rho, \sigma, \tau, \omega, \delta, c, q, r$ are essential and will be added to all theorems.

2.2. A Class of $(h,g;m)$-Convex Functions

Another direction for the generalization of the Hermite–Hadamard inequality is the use of different classes of convexity. For this we need a class of $(h,g;m)$-convex functions, the properties of which were recently presented in [14]:

Definition 3 ([14]). *Let h be a nonnegative function on $J \subseteq \mathbb{R}$, $(0,1) \subseteq J$, $h \not\equiv 0$ and let g be a positive function on $I \subseteq \mathbb{R}$. Furthermore, let $m \in (0,1]$. A function $f : I \to \mathbb{R}$ is said to be an $(h,g;m)$-convex function if it is nonnegative and if*

$$f(tx + m(1-t)y) \leq h(t)f(x)g(x) + m\,h(1-t)f(y)g(y) \tag{12}$$

holds for all $x, y \in I$ and all $t \in (0,1)$.

If (12) holds in the reversed sense, then f is said to be an $(h,g;m)$-concave function.

This class unifies a certain range of convexity, enabling generalizations of known results. For different choices of functions h, g and parameter m, a class of $(h,g;m)$-convex functions is reduced to a class of P-functions [15], h-convex functions [17], m-convex functions [16], $(h-m)$-convex functions [11], (s,m)-Godunova–Levin functions of the second kind [10], exponentially s-convex functions in the second sense [9], etc. For example, if we set $h(t) = t^s$, $s \in (0,1]$, $g(x) = e^{-\beta x}$, $\beta \in \mathbb{R}$, then we obtain a class defined in [13]:

A function $f : I \subset \mathbb{R} \to \mathbb{R}$ is called exponentially (s,m)-convex in the second sense if the following inequality holds

$$f(tx + m(1-t)y) \leq \frac{t^s}{e^{\beta x}}f(x) + \frac{(1-t)^s}{e^{\beta y}}mf(y) \tag{13}$$

for all $x, y \in I$ and all $t \in [0,1]$, where $\beta \in \mathbb{R}$, $s, m \in (0,1]$.

Next we need the Hermite–Hadamard inequality for $(h,g;m)$-convex functions:

Theorem 4 ([14]). *Let f be a nonnegative $(h,g;m)$-convex function on $[0,\infty)$ where h is a nonnegative function on $J \subseteq \mathbb{R}$, $(0,1) \subseteq J$, $h \not\equiv 0$, g is a positive function on $[0,\infty)$ and $m \in (0,1]$. If $f,g,h \in L_1[a,b]$, where $0 \leq a < b < \infty$, then the following inequalities hold*

$$
\begin{aligned}
f\left(\frac{a+b}{2}\right) &\leq \frac{h\left(\frac{1}{2}\right)}{b-a}\int_a^b\left[f(x)g(x)+mf\left(\frac{x}{m}\right)g\left(\frac{x}{m}\right)\right]dx \\
&\leq \frac{h\left(\frac{1}{2}\right)f(a)g(a)}{b-a}\int_a^b h\left(\frac{b-x}{b-a}\right)g(x)\,dx \\
&\quad + \frac{mh\left(\frac{1}{2}\right)f\left(\frac{b}{m}\right)g\left(\frac{b}{m}\right)}{b-a}\int_a^b h\left(\frac{x-a}{b-a}\right)g(x)\,dx \\
&\quad + \frac{mh\left(\frac{1}{2}\right)f\left(\frac{a}{m}\right)g\left(\frac{a}{m}\right)}{b-a}\int_a^b h\left(\frac{b-x}{b-a}\right)g\left(\frac{x}{m}\right)dx \\
&\quad + \frac{m^2 h\left(\frac{1}{2}\right)f\left(\frac{b}{m^2}\right)g\left(\frac{b}{m^2}\right)}{b-a}\int_a^b h\left(\frac{x-a}{b-a}\right)g\left(\frac{x}{m}\right)dx. \quad (14)
\end{aligned}
$$

3. Fractional Integral Inequalities of the Hermite–Hadamard Type for $(h,g;m)$-Convex Functions

The Hermite–Hadamard inequality for $(h,g;m)$-convex functions is obtained in [14], where some special results are pointed out and several known inequalities are improved upon. In [12], the article that followed, a few more inequalities of the Hermite–Hadamard type are presented. Here we will obtain their fractional generalizations, using (5)–(7), that is, the extended generalized Mittag-Leffler function E with fractional integral operators $\varepsilon_{a^+}^{\omega} f$ and $\varepsilon_{b^-}^{\omega} f$ in the real domain.

In this section, it is necessary to introduce the following conditions on the parameters and the interval $[a,b]$:

Assumption 1. *Let $\omega \in \mathbb{R}$, $\rho,\sigma,\tau > 0$, $c > \delta > 0$ with $p \geq 0$ and $0 < q \leq r+\rho$. Furthermore, let $0 \leq a < b < \infty$.*

We start with the left side, i.e., the first Hermite–Hadamard fractional integral inequality for $(h,g;m)$-convex functions involving the extended generalized Mittag-Leffler function.

Theorem 5. *Let Assumption 1 hold. Let f be a nonnegative $(h,g;m)$-convex function on $[0,\infty)$, where h is a nonnegative function on $J \subseteq \mathbb{R}$, $(0,1) \subseteq J$, $h \not\equiv 0$, g is a positive function on $[0,\infty)$ and $m \in (0,1]$. If $f,g \in L_1[a,\frac{b}{m}]$, then the following inequality holds*

$$f\left(\frac{a+b}{2}\right)(\varepsilon_{a^+}^{\bar{\omega}} 1)(b;p) \leq h\left(\tfrac{1}{2}\right)\left[(\varepsilon_{a^+}^{\bar{\omega}} fg)(b;p) + m^{\sigma+1}(\varepsilon_{\frac{b}{m}^-}^{\bar{\bar{\omega}}} fg)(\tfrac{a}{m};p)\right], \quad (15)$$

where

$$\bar{\omega} = \frac{\omega}{(b-a)^\rho}, \quad \bar{\bar{\omega}} = \frac{m^\rho \omega}{(b-a)^\rho}. \quad (16)$$

Proof. Let f be an $(h,g;m)$-convex function on $[0,\infty)$, $m \in (0,1]$. Then for $t = \frac{1}{2}$ we have

$$f\left(\frac{x+my}{2}\right) \leq h\left(\tfrac{1}{2}\right)f(x)g(x) + mh\left(\tfrac{1}{2}\right)f(y)g(y).$$

Choosing $y \equiv \frac{y}{m}$ we obtain

$$f\left(\frac{x+y}{2}\right) \leq h\left(\tfrac{1}{2}\right)\left[f(x)g(x) + mf\left(\tfrac{y}{m}\right)g\left(\tfrac{y}{m}\right)\right].$$

Let $x = ta + (1-t)b$ and $y = (1-t)a + tb$. Then

$$\begin{aligned} f\left(\frac{a+b}{2}\right) &\leq h\left(\tfrac{1}{2}\right)\Big[f(ta+(1-t)b)\,g(ta+(1-t)b) \\ &\quad + mf\left((1-t)\tfrac{a}{m}+t\tfrac{b}{m}\right)g\left((1-t)\tfrac{a}{m}+t\tfrac{b}{m}\right)\Big]. \end{aligned}$$

In the following step we will need to multiply both sides of the above inequality by $t^{\sigma-1}\mathbf{E}(\omega t^\rho; p)$ and integrate on $[0,1]$ with respect to the variable t, which gives us

$$\begin{aligned} & f\left(\frac{a+b}{2}\right) \int_0^1 t^{\sigma-1}\mathbf{E}(\omega t^\rho;p)\,dt \\ &\leq h\left(\tfrac{1}{2}\right)\int_0^1 f(ta+(1-t)b)\,g(ta+(1-t)b)\,t^{\sigma-1}\mathbf{E}(\omega t^\rho;p)\,dt \\ &\quad + mh\left(\tfrac{1}{2}\right)\int_0^1 f\left((1-t)\tfrac{a}{m}+t\tfrac{b}{m}\right)g\left((1-t)\tfrac{a}{m}+t\tfrac{b}{m}\right) t^{\sigma-1}\mathbf{E}(\omega t^\rho;p)\,dt. \end{aligned}$$

With substitutions $u = ta + (1-t)b$ and $v = (1-t)\tfrac{a}{m} + t\tfrac{b}{m}$ we obtain

$$\begin{aligned} & \frac{1}{(b-a)^\sigma} f\left(\frac{a+b}{2}\right) \int_a^b (b-u)^{\sigma-1}\mathbf{E}(\overline{\omega}(b-u)^\rho;p)\,du \\ &\leq \frac{h\left(\tfrac{1}{2}\right)}{(b-a)^\sigma}\int_a^b f(u)g(u)(b-u)^{\sigma-1}\mathbf{E}(\overline{\omega}(b-u)^\rho;p)\,du \\ &\quad + \frac{m^{\sigma+1} h\left(\tfrac{1}{2}\right)}{(b-a)^\sigma}\int_{\tfrac{a}{m}}^{\tfrac{b}{m}} f(v)g(v)\left(v-\tfrac{a}{m}\right)^{\sigma-1}\mathbf{E}(\overline{\overline{\omega}}(v-\tfrac{a}{m})^\rho;p)\,dv. \end{aligned}$$

Since $m \in (0,1]$, then $a \leq a/m$, $b \leq b/m$ and $[a,b] \subset [a,\tfrac{b}{m}]$. Therefore, the condition $f,g \in L_1[a,\tfrac{b}{m}]$ is stated in this theorem. The above inequality can be written as

$$\begin{aligned} & \frac{1}{(b-a)^\sigma} f\left(\frac{a+b}{2}\right)(\boldsymbol{\varepsilon}_{a+}^{\overline{\omega}}1)(b;p) \\ &\leq \frac{h\left(\tfrac{1}{2}\right)}{(b-a)^\sigma}\left[(\boldsymbol{\varepsilon}_{a+}^{\overline{\omega}}fg)(b;p) + m^{\sigma+1}(\boldsymbol{\varepsilon}_{\tfrac{b}{m}-}^{\overline{\overline{\omega}}}fg)(\tfrac{a}{m};p)\right]. \end{aligned}$$

Note that with Corollary 2 we can obtain the constant $(\boldsymbol{\varepsilon}_{a+}^{\overline{\omega}}1)(b;p)$. This completes the proof. □

Next we have the second Hermite–Hadamard fractional integral inequality.

Theorem 6. *Let the assumptions of Theorem 5 hold with $f, g, h \in L_1[a, \tfrac{b}{m}]$. Then*

$$(e^{\overline{\omega}}_{a^+}fg)(b;p) + m^{\sigma+1}(e^{\overline{\overline{\omega}}}_{\frac{b-}{m}}fg)(\tfrac{a}{m};p)$$

$$\leq f(a)g(a)\int_a^b h\left(\frac{b-x}{b-a}\right)g(x)(b-x)^{\sigma-1}\mathbf{E}(\overline{\omega}(b-x)^\rho;p)dx$$

$$+mf\left(\frac{b}{m}\right)g\left(\frac{b}{m}\right)\int_a^b h\left(\frac{x-a}{b-a}\right)g(x)(b-x)^{\sigma-1}\mathbf{E}(\overline{\omega}(b-x)^\rho;p)dx$$

$$+m^{\sigma+1}f\left(\frac{a}{m}\right)g\left(\frac{a}{m}\right)\int_{\frac{a}{m}}^{\frac{b}{m}} h\left(\frac{b-mx}{b-a}\right)g(x)(x-\tfrac{a}{m})^{\sigma-1}\mathbf{E}(\overline{\overline{\omega}}(x-\tfrac{a}{m})^\rho;p)dx$$

$$+m^{\sigma+2}f\left(\frac{b}{m^2}\right)g\left(\frac{b}{m^2}\right)\int_{\frac{a}{m}}^{\frac{b}{m}} h\left(\frac{mx-a}{b-a}\right)g(x)(x-\tfrac{a}{m})^{\sigma-1}\mathbf{E}(\overline{\overline{\omega}}(x-\tfrac{a}{m})^\rho;p)dx, \qquad (17)$$

where $\overline{\omega}$ and $\overline{\overline{\omega}}$ are defined by (16).

Proof. Due to the $(h,g;m)$-convexity of f we have

$$f(ta+(1-t)b) \leq h(t)f(a)g(a) + mh(1-t)f\left(\frac{b}{m}\right)g\left(\frac{b}{m}\right).$$

Multiplying both sides of above inequality by $g(ta+(1-t)b)t^{\sigma-1}\mathbf{E}(\omega t^\rho;p)$ and integrating on $[0,1]$ with respect to the variable t, we obtain

$$\int_0^1 f(ta+(1-t)b)g(ta+(1-t)b)t^{\sigma-1}\mathbf{E}(\omega t^\rho;p)dt$$

$$\leq f(a)g(a)\int_0^1 h(t)g(ta+(1-t)b)t^{\sigma-1}\mathbf{E}(\omega t^\rho;p)dt$$

$$+mf\left(\frac{b}{m}\right)g\left(\frac{b}{m}\right)\int_0^1 h(1-t)g(ta+(1-t)b)t^{\sigma-1}\mathbf{E}(\omega t^\rho;p)dt.$$

With the substitution $u=ta+(1-t)b$ we obtain

$$\frac{1}{(b-a)^\sigma}\int_a^b f(u)g(u)(b-u)^{\sigma-1}\mathbf{E}(\overline{\omega}(b-u)^\rho;p)du$$

$$\leq \frac{f(a)g(a)}{(b-a)^\sigma}\int_a^b h\left(\frac{b-u}{b-a}\right)g(u)(b-u)^{\sigma-1}\mathbf{E}(\overline{\omega}(b-u)^\rho;p)du$$

$$+\frac{mf\left(\frac{b}{m}\right)g\left(\frac{b}{m}\right)}{(b-a)^\sigma}\int_a^b h\left(\frac{u-a}{b-a}\right)g(u)(b-u)^{\sigma-1}\mathbf{E}(\overline{\omega}(b-u)^\rho;p)du,$$

that is

$$(e^{\overline{\omega}}_{a^+}fg)(b;p)$$

$$\leq f(a)g(a)\int_a^b h\left(\frac{b-u}{b-a}\right)g(u)(b-u)^{\sigma-1}\mathbf{E}(\overline{\omega}(b-u)^\rho;p)du$$

$$+mf\left(\frac{b}{m}\right)g\left(\frac{b}{m}\right)\int_a^b h\left(\frac{u-a}{b-a}\right)g(u)(b-u)^{\sigma-1}\mathbf{E}(\overline{\omega}(b-u)^\rho;p)du. \qquad (18)$$

Again, due to the $(h,g;m)$-convexity of f we have

$$f\left((1-t)\frac{a}{m}+t\frac{b}{m}\right) \leq h(1-t)f\left(\frac{a}{m}\right)g\left(\frac{a}{m}\right) + mh(t)f\left(\frac{b}{m^2}\right)g\left(\frac{b}{m^2}\right).$$

Multiplying both sides of above inequality by $g\left((1-t)\frac{a}{m}+t\frac{b}{m}\right)t^{\sigma-1}E(\omega t^\rho;p)$ and integrating on $[0,1]$ with respect to the variable t, we obtain

$$\int_0^1 f\left((1-t)\frac{a}{m}+t\frac{b}{m}\right)g\left((1-t)\frac{a}{m}+t\frac{b}{m}\right)t^{\sigma-1}E(\omega t^\rho;p)dt$$
$$\leq f\left(\frac{a}{m}\right)g\left(\frac{a}{m}\right)\int_0^1 h(1-t)g\left((1-t)\frac{a}{m}+t\frac{b}{m}\right)t^{\sigma-1}E(\omega t^\rho;p)dt$$
$$+mf\left(\frac{b}{m^2}\right)g\left(\frac{b}{m^2}\right)\int_0^1 h(t)g\left((1-t)\frac{a}{m}+t\frac{b}{m}\right)t^{\sigma-1}E(\omega t^\rho;p)dt.$$

With the substitution $v=(1-t)\frac{a}{m}+t\frac{b}{m}$ we obtain

$$\frac{m^\sigma}{(b-a)^\sigma}\int_{\frac{a}{m}}^{\frac{b}{m}} f(v)g(v)(v-\tfrac{a}{m})^{\sigma-1}E(\overline{\overline{\omega}}(v-\tfrac{a}{m})^\rho;p)dv$$
$$\leq \frac{m^\sigma f\left(\frac{a}{m}\right)g\left(\frac{a}{m}\right)}{(b-a)^\sigma}\int_{\frac{a}{m}}^{\frac{b}{m}} h\left(\frac{b-mv}{b-a}\right)g(v)(v-\tfrac{a}{m})^{\sigma-1}E(\overline{\overline{\omega}}(v-\tfrac{a}{m})^\rho;p)dv$$
$$+\frac{m^{\sigma+1} f\left(\frac{b}{m^2}\right)g\left(\frac{b}{m^2}\right)}{(b-a)^\sigma}\int_{\frac{a}{m}}^{\frac{b}{m}} h\left(\frac{vm-a}{b-a}\right)g(v)(v-\tfrac{a}{m})^{\sigma-1}E(\overline{\overline{\omega}}(v-\tfrac{a}{m})^\rho;p)dv,$$

that is

$$(\boldsymbol{\varepsilon}_{\frac{b^-}{m}}^{\overline{\overline{\omega}}} fg)(\tfrac{a}{m};p)$$
$$\leq f\left(\frac{a}{m}\right)g\left(\frac{a}{m}\right)\int_{\frac{a}{m}}^{\frac{b}{m}} h\left(\frac{b-mv}{b-a}\right)g(v)(v-\tfrac{a}{m})^{\sigma-1}E(\overline{\overline{\omega}}(v-\tfrac{a}{m})^\rho;p)dv$$
$$+mf\left(\frac{b}{m^2}\right)g\left(\frac{b}{m^2}\right)\int_{\frac{a}{m}}^{\frac{b}{m}} h\left(\frac{vm-a}{b-a}\right)g(v)(v-\tfrac{a}{m})^{\sigma-1}E(\overline{\overline{\omega}}(v-\tfrac{a}{m})^\rho;p)dv. \quad (19)$$

Inequality (17) now follows from (18) and (19). □

In the following we derive fractional integral inequalities of Hermite–Hadamard type for different types of convexity, and state several corollaries, using special functions for h and/or g, and the parameter m. The first consequence of Theorems 5 and 6 obtained via the setting $g \equiv 1$ (i.e., $g(x) = 1$) is the Hermite–Hadamard fractional integral inequality for $(h-m)$-convex functions given in ([20], Theorem 2.1):

Corollary 3. Let Assumption 1 hold. Let f be a nonnegative $(h-m)$-convex function on $[0,\infty)$ where h is a nonnegative function on $J \subseteq \mathbb{R}$, $(0,1) \subseteq J$, $h \not\equiv 0$ and $m \in (0,1]$. If $f \in L_1[a,\frac{b}{m}]$ and $h \in L_1[0,1]$, then following inequalities hold

$$f\left(\frac{a+b}{2}\right)(\boldsymbol{\varepsilon}_{a^+}^{\overline{\omega}} 1)(b;p) \leq h\left(\tfrac{1}{2}\right)\left[(\boldsymbol{\varepsilon}_{a^+}^{\overline{\omega}} f)(b;p)+m^{\sigma+1}(\boldsymbol{\varepsilon}_{\frac{b^-}{m}}^{\overline{\overline{\omega}}} f)(\tfrac{a}{m};p)\right]$$
$$\leq h\left(\tfrac{1}{2}\right)(b-a)^\sigma\left\{\left[f(a)+m^2 f\left(\frac{b}{m^2}\right)\right](\boldsymbol{\varepsilon}_{1^-}^\omega h)(0;p)\right.$$
$$\left.+\left[mf\left(\frac{a}{m}\right)+mf\left(\frac{b}{m}\right)\right](\boldsymbol{\varepsilon}_{0^+}^\omega h)(1;p)\right\}, \quad (20)$$

where $\overline{\omega}$ and $\overline{\overline{\omega}}$ are defined by (16).

Proof. First we use substitutions $t = \frac{b-x}{b-a}$ and $z = \frac{mx-a}{b-a}$ in Theorem 6, after which we apply identities

$$\int_0^1 h(t) t^{\sigma-1} E(\omega t^\rho; p) dt = (\varepsilon_{1-}^\omega h)(0; p) \quad (21)$$

and

$$\int_0^1 h(1-t) t^{\sigma-1} E(\omega t^\rho; p) dt$$
$$= \int_0^1 h(z)(1-z)^{\sigma-1} E(\omega(1-z)^\rho; p) dz = (\varepsilon_{0+}^\omega h)(1; p). \quad (22)$$

The result now follows from the above and Theorem 5. □

By setting the function $g \equiv 1$ and the parameter $m = 1$, the previous result is reduced to the Hermite–Hadamard fractional integral inequality for h-convex functions:

Corollary 4. *Let Assumption 1 hold. Let f be a nonnegative h-convex function on $[0, \infty)$ where h is a nonnegative function on $J \subseteq \mathbb{R}$, $(0,1) \subseteq J$, $h \not\equiv 0$. If $f \in L_1[a, \frac{b}{m}]$ and $h \in L_1[0,1]$, then the following inequalities hold*

$$f\left(\frac{a+b}{2}\right)(\varepsilon_{a+}^{\overline{\omega}} 1)(b; p)$$
$$\leq h\left(\tfrac{1}{2}\right)\left[(\varepsilon_{a+}^{\overline{\omega}} f)(b; p) + (\varepsilon_{b-}^{\overline{\omega}} f)(a; p)\right]$$
$$\leq h\left(\tfrac{1}{2}\right)(b-a)^\sigma [f(a) + f(b)]\left[(\varepsilon_{1-}^\omega h)(0; p) + (\varepsilon_{0+}^\omega h)(1; p)\right], \quad (23)$$

where $\overline{\omega}$ is defined by (16).

In the following, we set the function $h \equiv \text{id}$, the identity function. With $g \equiv 1$ we obtain the Hermite–Hadamard fractional integral inequality for m-convex functions from ([21], Theorem 3.1):

Corollary 5. *Let Assumption 1 hold. Let f be a nonnegative m-convex function on $[0, \infty)$ with $m \in (0, 1]$. If $f \in L_1[a, \frac{b}{m}]$, then the following inequalities hold*

$$f\left(\frac{a+b}{2}\right)(\varepsilon_{a+}^{\overline{\omega}} 1)(b; p) \leq \frac{1}{2}\left[(\varepsilon_{a+}^{\overline{\omega}} f)(b; p) + m^{\sigma+1}(\varepsilon_{\frac{b}{m}-}^{\overline{\overline{\omega}}} f)(\tfrac{a}{m}; p)\right]$$
$$\leq \frac{(b-a)^\sigma}{2}\left\{\left[f(a) + m^2 f\left(\frac{b}{m^2}\right)\right](\varepsilon_{1-}^\omega \text{id})(0; p)\right.$$
$$\left. + \left[mf\left(\frac{a}{m}\right) + mf\left(\frac{b}{m}\right)\right](\varepsilon_{0+}^\omega \text{id})(1; p)\right\}, \quad (24)$$

where $\overline{\omega}$ and $\overline{\overline{\omega}}$ are defined by (16).

The Hermite–Hadamard fractional integral inequality for convex functions is given in ([21], Theorem 2.1). Here it is a merely a consequence for $h \equiv \text{id}$, $g \equiv 1$ and $m = 1$:

Corollary 6. *Let Assumption 1 hold. Let f be a nonnegative convex function on $[0, \infty)$. If $f \in L_1[a, b]$, then the following inequalities hold*

$$f\left(\frac{a+b}{2}\right)(\varepsilon_{a+}^{\overline{\omega}} 1)(b; p) \leq \frac{1}{2}\left[(\varepsilon_{a+}^{\overline{\omega}} f)(b; p) + (\varepsilon_{b-}^{\overline{\omega}} f)(a; p)\right]$$
$$\leq \frac{f(a) + f(b)}{2}(\varepsilon_{a+}^{\overline{\omega}} 1)(b; p), \quad (25)$$

where $\overline{\omega}$ is defined by (16).

Proof. Here we use

$$(\varepsilon_{0+}^{\omega} \text{ id})(1;p) + (\varepsilon_{1-}^{\omega} \text{ id})(0;p)$$
$$= \int_0^1 t^{\sigma} \mathbf{E}(\omega t^{\rho}; p) dt + \int_0^1 t(1-t)^{\sigma-1} \mathbf{E}(\omega(1-t)^{\rho}; p) dt$$
$$= \int_0^1 (1-t)^{\sigma} \mathbf{E}(\omega(1-t)^{\rho}; p) dt + \int_0^1 t(1-t)^{\sigma-1} \mathbf{E}(\omega(1-t)^{\rho}; p) dt$$
$$= \int_0^1 (1-t)^{\sigma-1} \mathbf{E}(\omega(1-t)^{\rho}; p) dt$$
$$= (\varepsilon_{0+}^{\omega} 1)(1;p) = \frac{1}{(b-a)^{\sigma}} (\varepsilon_{a+}^{\bar{\omega}} 1)(b;p).$$

□

We have presented several Hermite–Hadamard-type inequalities for the $(h,g;m)$-convex function using fractional integral operators, where the kernel is an extended generalized Mittag-Leffler function. If we apply different parameter choices, as in Remark 2, then we obtain corresponding inequalities for different fractional operators.

Several Properties of Fractional Integral Operators $\varepsilon_{a+}^{\omega} f$ and $\varepsilon_{b-}^{\omega} f$

At the end of this section we give several results for fractional integral operators.

Proposition 1. *Let $\omega, \rho, \sigma, \tau, \delta, c \in \mathbb{C}$, $\Re(\rho), \Re(\sigma), \Re(\tau) > 0$, $\Re(c) > \Re(\delta) > 0$ with $p \geq 0$, $r > 0$ and $0 < q \leq r + \Re(\rho)$.*

(i) *If the function $f \in L_1[a,b]$ is symmetric about $\frac{a+b}{2}$, then*

$$(\varepsilon_{a+}^{\omega} f)(b;p) = (\varepsilon_{b-}^{\omega} f)(a;p). \tag{26}$$

In particular,

$$(\varepsilon_{a+}^{\omega} 1)(b;p) = (\varepsilon_{b-}^{\omega} 1)(a;p). \tag{27}$$

(ii) *Furthermore,*

$$\left(\varepsilon_{a+}^{\omega} (t-a)^{\alpha-1}\right)(b;p) = (\varepsilon_{b-}^{\omega} (b-t)^{\alpha-1})(a;p), \tag{28}$$

$$\left(\varepsilon_{a+}^{\omega} (b-t)^{\alpha-1}\right)(b;p) = (\varepsilon_{b-}^{\omega} (t-a)^{\alpha-1})(a;p). \tag{29}$$

In particular,

$$\left(\varepsilon_{0+}^{\omega} t^{\alpha-1}\right)(1;p) = (\varepsilon_{1-}^{\omega} (1-t)^{\alpha-1})(0;p), \tag{30}$$

$$\left(\varepsilon_{0+}^{\omega} (1-t)^{\alpha-1}\right)(1;p) = (\varepsilon_{1-}^{\omega} t^{\alpha-1})(0;p). \tag{31}$$

Proof. (i) If the function f is symmetric about $\frac{a+b}{2}$, i.e., $f(t) = f(a+b-t)$ for all $t \in [a,b]$, then, substituting $z = a + b - t$, Equation (26) easily follows:

$$(\varepsilon_{a+}^{\omega} f)(b;p) = \int_a^b (b-t)^{\sigma-1} \mathbf{E}(\omega(b-t)^{\rho}; p) f(t) dt$$
$$= \int_a^b (z-a)^{\sigma-1} \mathbf{E}(\omega(z-a)^{\rho}; p) f(a+b-z) dz$$
$$= \int_a^b (z-a)^{\sigma-1} \mathbf{E}(\omega(z-a)^{\rho}; p) f(z) dz = (\varepsilon_{b-}^{\omega} f)(a;p).$$

Note that (27) also follows directly from Corollary 2 if we set $x = b$ in (10) and $x = a$ in (11).

(ii) Equations (28) and (29) follow with the substitution $z = a + b - t$. Furthermore, (28) follows directly from Theorem 3 if we set $x = b$ in (8) and $x = a$ in (9). The final two equations are obtained for $a = 0$ and $b = 1$. □

Remark 3. *To obtain the Hermite–Hadamard inequality for convex functions involving Riemann–Liouville fractional integrals, given in Theorem 2, first we need to set $p = \omega = 0$ in (5)*

$$E(z;0) = \sum_{n=0}^{\infty} \frac{(\delta)_{nq}}{\Gamma(\rho n + \sigma)} \frac{z^n}{(\tau)_{nr}}.$$

Since $\mathbf{E}(0;0) = E_{\rho,\sigma,\tau}^{\delta,c,q,r}(0;0) = \frac{1}{\Gamma(\sigma)}$, setting $p = \omega = 0$ in (6) we obtain Riemann–Liouville fractional integrals

$$(\pmb{\varepsilon}_{a+}^{0} f)(x;0) = \frac{1}{\Gamma(\sigma)} \int_{a}^{x} (x-t)^{\sigma-1} f(t)\, dt = J_{a+}^{\sigma} f(x),$$

$$(\pmb{\varepsilon}_{b-}^{0} f)(x;0) = \frac{1}{\Gamma(\sigma)} \int_{x}^{b} (t-x)^{\sigma-1} f(t)\, dt = J_{b-}^{\sigma} f(x).$$

Note that a direct consequence of Theorem 3 is

$$(\pmb{\varepsilon}_{0+}^{\omega}\ \text{id})(1;p) = E_{\rho,\sigma+2,\tau}^{\delta,c,q,r}(\omega;p). \tag{32}$$

For the reader's convenience, we will directly prove this:

$$\begin{aligned}
(\pmb{\varepsilon}_{0+}^{\omega}\ \text{id})(1;p) &= \int_0^1 t(1-t)^{\sigma-1} E(\omega(1-t)^{\rho};p) dt \\
&= \int_0^1 t(1-t)^{\sigma-1} \sum_{n=0}^{\infty} \frac{B_p(\delta+nq,c-\delta)}{B(\delta,c-\delta)} \frac{(c)_{nq}}{\Gamma(\rho n + \sigma)} \frac{\omega^n (1-t)^{n\rho}}{(\tau)_{nr}} dt \\
&= \sum_{n=0}^{\infty} \frac{B_p(\delta+nq,c-\delta)}{B(\delta,c-\delta)} \frac{(c)_{nq}}{\Gamma(\rho n + \sigma)} \frac{\omega^n}{(\tau)_{nr}} \int_0^1 t(1-t)^{n\rho+\sigma-1} dt \\
&= \sum_{n=0}^{\infty} \frac{B_p(\delta+nq,c-\delta)}{B(\delta,c-\delta)} \frac{(c)_{nq}}{\Gamma(\rho n + \sigma)} \frac{\omega^n}{(\tau)_{nr}} B(2, n\rho+\sigma) \\
&= \sum_{n=0}^{\infty} \frac{B_p(\delta+nq,c-\delta)}{B(\delta,c-\delta)} \frac{(c)_{nq}}{\Gamma(\rho n + \sigma)} \frac{\omega^n}{(\tau)_{nr}} \frac{\Gamma(2)\Gamma(n\rho+\sigma)}{\Gamma(2+n\rho+\sigma)} \\
&= \sum_{n=0}^{\infty} \frac{B_p(\delta+nq,c-\delta)}{B(\delta,c-\delta)} \frac{(c)_{nq}}{\Gamma(\rho n + (\sigma+2))} \frac{\omega^n}{(\tau)_{nr}} \\
&= E_{\rho,\sigma+2,\tau}^{\delta,c,q,r}(\omega;p).
\end{aligned}$$

Hence,

$$\left(\pmb{\varepsilon}_{0+}^{0}\ \text{id}\right)(1;0) = \frac{1}{\Gamma(\sigma+2)}$$

and

$$(\pmb{\varepsilon}_{1-}^{0}\ \text{id})(0;0) = \int_0^1 t^{\sigma} E(0;p) dt = \frac{1}{(\sigma+1)\Gamma(\sigma)}, \tag{33}$$

from which follows

$$\left(\pmb{\varepsilon}_{0+}^{0}\ \text{id}\right)(1;0) + (\pmb{\varepsilon}_{1-}^{0}\ \text{id})(0;0) = \frac{1}{\Gamma(\sigma+1)}.$$

Finally, if we set $h(x) = x$, $g \equiv 1$, $m = 1$ and $p = \omega = 0$, then Theorems 5 and 6 are reduced to Theorem 2.

4. Applications: Bounds of Fractional Integral Operators for $(h, g; m)$-Convex Functions

As an application, in this section we obtain the upper bounds of fractional integral operators for $(h, g; m)$-convex functions.

Assumption 2. *Let $\omega \in \mathbb{R}$, $\rho, \sigma, \tau > 0$, $c > \delta > 0$ with $p \geq 0$ and $0 < q \leq r + \rho$. Let f be a nonnegative $(h, g; m)$-convex function on $[0, \infty)$ where h is a nonnegative function on $J \subseteq \mathbb{R}$, $(0,1) \subseteq J$, $h \not\equiv 0$, g is a positive function on $[0, \infty)$, and $m \in (0, 1]$. Furthermore, let $0 \leq a < b < \infty$.*

Theorem 7. *Let Assumption 2 hold. If $f, g \in L_1[a, b]$ and $h \in L_1[0, 1]$, then for $x \in [a, b]$ the following inequality holds*

$$\frac{1}{(x-a)^\sigma}(\boldsymbol{\varepsilon}_{a^+}^{\omega_a} f)(x; p) \leq f(a)g(a)(\boldsymbol{\varepsilon}_{1^-}^{\omega} h)(0; p) + mf\left(\frac{x}{m}\right)g\left(\frac{x}{m}\right)(\boldsymbol{\varepsilon}_{0^+}^{\omega} h)(1; p). \tag{34}$$

where

$$\omega_a = \frac{\omega}{(x-a)^\rho}. \tag{35}$$

Proof. Let f be an $(h, g; m)$-convex function on $[0, \infty)$, $x \in [a, b]$, $m \in (0, 1]$ and $t \in (0, 1)$. Then, similarly to Theorem 6, we use

$$f(ta + (1-t)x) \leq h(t)f(a)g(a) + mh(1-t)f\left(\frac{x}{m}\right)g\left(\frac{x}{m}\right).$$

Multiplying both sides of the above inequality by $t^{\sigma-1}\boldsymbol{E}(\omega t^\rho; p)$ and integrating on $[0, 1]$ with respect to the variable t, we obtain

$$\int_0^1 f(ta + (1-t)x)t^{\sigma-1}\boldsymbol{E}(\omega t^\rho; p)dt$$
$$\leq f(a)g(a)\int_0^1 h(t)\,t^{\sigma-1}\boldsymbol{E}(\omega t^\rho; p)dt + mf\left(\frac{x}{m}\right)g\left(\frac{x}{m}\right)\int_0^1 h(1-t)t^{\sigma-1}\boldsymbol{E}(\omega t^\rho; p)dt.$$

With the substitution $u = ta + (1-t)x$ and identities (21), (22), we obtain the inequality (34). □

Theorem 8. *Let Assumption 2 hold. If $f, g \in L_1[a, b]$ and $h \in L_1[0, 1]$, then for $x \in [a, b]$ the following inequality holds*

$$\frac{1}{(b-x)^\sigma}(\boldsymbol{\varepsilon}_{b^-}^{\omega_b} f)(x; p) \leq f(b)g(b)(\boldsymbol{\varepsilon}_{1^-}^{\omega} h)(0; p) + mf\left(\frac{x}{m}\right)g\left(\frac{x}{m}\right)(\boldsymbol{\varepsilon}_{0^+}^{\omega} h)(1; p), \tag{36}$$

where

$$\omega_b = \frac{\omega}{(b-x)^\rho}. \tag{37}$$

Proof. Using

$$f(tb + (1-t)x) \leq h(t)f(b)g(b) + mh(1-t)f\left(\frac{x}{m}\right)g\left(\frac{x}{m}\right),$$

the proof follows analogously to that of Theorem 7. □

From the two previous theorems we can directly obtain the following result.

Corollary 7. Let Assumption 2 hold. If $f, g \in L_1[a, b]$ and $h \in L_1[0, 1]$, then for $x \in [a, b]$ the following inequality holds

$$\frac{1}{(x-a)^\sigma}(\boldsymbol{\varepsilon}_{a^+}^{\omega_a} f)(x; p) + \frac{1}{(b-x)^\sigma}(\boldsymbol{\varepsilon}_{b^-}^{\omega_b} f)(x; p)$$
$$\leq [f(a)g(a) + f(b)g(b)](\boldsymbol{\varepsilon}_{1^-}^\omega h)(0; p) + 2m f\left(\frac{x}{m}\right) g\left(\frac{x}{m}\right)(\boldsymbol{\varepsilon}_{0^+}^\omega h)(1; p). \quad (38)$$

where ω_a and ω_b are defined by (35) and (37).

If we set $x = b$ in Theorem 7 and $x = a$ in Theorem 8, then we obtain the next fractional integral inequality of the Hermite–Hadamard type.

Theorem 9. Let Assumption 2 hold. If $f, g, h \in L_1[a, b]$, then the following inequalities hold

$$\frac{1}{(b-a)^\sigma}\left[(\boldsymbol{\varepsilon}_{a^+}^{\overline{\omega}} f)(b; p) + (\boldsymbol{\varepsilon}_{b^-}^{\overline{\omega}} f)(a; p)\right]$$
$$\leq [f(a)g(a) + f(b)g(b)](\boldsymbol{\varepsilon}_{1^-}^\omega h)(0; p)$$
$$+ m\left[f\left(\frac{a}{m}\right)g\left(\frac{a}{m}\right) + f\left(\frac{b}{m}\right)g\left(\frac{b}{m}\right)\right](\boldsymbol{\varepsilon}_{0^+}^\omega h)(1; p), \quad (39)$$

where $\overline{\omega}$ is defined by (16).

In the following we will extend our interval to $[ma, b]$. Since $m \in (0, 1]$, then $ma \leq a$, $mb \leq b$, and $[a, b] \subset [ma, b]$.

Theorem 10. Let Assumption 2 hold. If $f, g \in L_1[ma, b]$ and $h \in L_1[0, 1]$, then the following inequality holds

$$\frac{1}{(mb-a)^\sigma}\left[(\boldsymbol{\varepsilon}_{a^+}^{\omega_1} f)(mb; p) + \left(\boldsymbol{\varepsilon}_{mb^-}^{\omega_1} f\right)(a; p)\right]$$
$$+ \frac{1}{(b-ma)^\sigma}\left[(\boldsymbol{\varepsilon}_{b^-}^{\omega_2} f)(ma; p) + (\boldsymbol{\varepsilon}_{ma^+}^{\omega_2} f)(b; p)\right]$$
$$\leq (m+1)[f(a)g(a) + f(b)g(b)][(\boldsymbol{\varepsilon}_{1^-}^\omega h)(0; p) + (\boldsymbol{\varepsilon}_{0^+}^\omega h)(1; p)], \quad (40)$$

where

$$\omega_1 = \frac{\omega}{(mb-a)^\rho}, \quad \omega_2 = \frac{\omega}{(b-ma)^\rho}. \quad (41)$$

Proof. Let f be an $(h, g; m)$-convex function on $[0, \infty)$, $m \in (0, 1]$ and $t \in (0, 1)$. Then

$$f(ta + m(1-t)b) \leq h(t)f(a)g(a) + mh(1-t)f(b)g(b),$$
$$f((1-t)a + mtb) \leq h(1-t)f(a)g(a) + mh(t)f(b)g(b)$$

and

$$f(tb + m(1-t)a) \leq h(t)f(b)g(b) + mh(1-t)f(a)g(a),$$
$$f((1-t)b + mta) \leq h(1-t)f(b)g(b) + mh(t)f(a)g(a).$$

First we add the above inequalities, i.e.,

$$f(ta + m(1-t)b) + f((1-t)a + mtb)$$
$$+ f(tb + m(1-t)a) + f((1-t)b + mta)$$
$$\leq (m+1)[f(a)g(a) + f(b)g(b)]h(t)$$
$$+ (m+1)[f(a)g(a) + f(b)g(b)]h(1-t).$$

Then we use multiplication by $t^{\sigma-1}\mathbf{E}(\omega t^\rho; p)$ and integration on $[0,1]$ with respect to the variable t to obtain

$$\int_0^1 f(ta + m(1-t)b)t^{\sigma-1}\mathbf{E}(\omega t^\rho; p)dt$$
$$+ \int_0^1 f((1-t)a + mtb)t^{\sigma-1}\mathbf{E}(\omega t^\rho; p)dt$$
$$+ \int_0^1 f(tb + m(1-t)a)t^{\sigma-1}\mathbf{E}(\omega t^\rho; p)dt$$
$$+ \int_0^1 f((1-t)b + mta)t^{\sigma-1}\mathbf{E}(\omega t^\rho; p)dt$$
$$\leq (m+1)[f(a)g(a) + f(b)g(b)]\int_0^1 h(t)\, t^{\sigma-1}\mathbf{E}(\omega t^\rho; p)dt$$
$$+ (m+1)[f(a)g(a) + f(b)g(b)]\int_0^1 h(1-t)t^{\sigma-1}\mathbf{E}(\omega t^\rho; p)dt.$$

For the left side of the inequality we need several substitutions. For instance, if we set $u = ta + m(1-t)b$, then we get

$$\int_0^1 f(ta + m(1-t)b)t^{\sigma-1}\mathbf{E}(\omega t^\rho; p)dt$$
$$= \frac{1}{(mb-a)^\sigma}\int_a^{mb} f(u)(mb-u)^{\sigma-1}\mathbf{E}(\tfrac{\omega}{(mb-a)^\rho}(mb-u)^\rho; p)du.$$

Hence,

$$\frac{1}{(mb-a)^\sigma}\int_a^{mb} f(u)(mb-u)^{\sigma-1}\mathbf{E}(\tfrac{\omega}{(mb-a)^\rho}(mb-u)^\rho; p)du$$
$$+ \frac{1}{(mb-a)^\sigma}\int_a^{mb} f(u)(u-a)^{\sigma-1}\mathbf{E}(\tfrac{\omega}{(mb-a)^\rho}(u-a)^\rho; p)du$$
$$+ \frac{1}{(b-ma)^\sigma}\int_{ma}^b f(u)(u-ma)^{\sigma-1}\mathbf{E}(\tfrac{\omega}{(b-ma)^\rho}(u-ma)^\rho; p)du$$
$$+ \frac{1}{(b-ma)^\sigma}\int_{ma}^b f(u)(b-u)^{\sigma-1}\mathbf{E}(\tfrac{\omega}{(b-ma)^\rho}(b-u)^\rho; p)du$$
$$\leq (m+1)[f(a)g(a) + f(b)g(b)](\boldsymbol{\varepsilon}_{1-}^\omega h)(0; p)$$
$$+ (m+1)[f(a)g(a) + f(b)g(b)](\boldsymbol{\varepsilon}_{0+}^\omega h)(1; p),$$

that is

$$\frac{1}{(mb-a)^\sigma}\left(\boldsymbol{\varepsilon}_{a+}^{\frac{\omega}{(mb-a)^\rho}} f\right)(mb; p) + \frac{1}{(mb-a)^\sigma}\left(\boldsymbol{\varepsilon}_{mb-}^{\frac{\omega}{(mb-a)^\rho}} f\right)(a; p)$$
$$+ \frac{1}{(b-ma)^\sigma}\left(\boldsymbol{\varepsilon}_{b-}^{\frac{\omega}{(b-ma)^\rho}} f\right)(ma; p) + \frac{1}{(b-ma)^\sigma}\left(\boldsymbol{\varepsilon}_{ma+}^{\frac{\omega}{(b-ma)^\rho}} f\right)(b; p)$$
$$\leq (m+1)[f(a)g(a) + f(b)g(b)]\left[(\boldsymbol{\varepsilon}_{1-}^\omega h)(0; p) + (\boldsymbol{\varepsilon}_{0+}^\omega h)(1; p)\right].$$

This provides the require inequality. □

Remark 4. *With an extended generalized Mittag-Leffler function from Definition 1 and a class of $(h, g; m)$-convex functions as in Definition 3, for different parameters p, τ, r, q, ω and for different choices of functions h, g and parameter m, we obtain corresponding upper bounds of different fractional operators for different classes of convexity.*

5. Conclusions

This research was on Hermite–Hadamard-type inequalities existing in a more general setting. We used a fractional integral operator containing an extended generalized Mittag-Leffler function in the kernel, and obtained Hermite–Hadamard fractional integral inequalities for a class of $(h, g; m)$-convex functions. Furthermore, we presented the upper bounds of the fractional integral operators for $(h, g; m)$-convex functions. The obtained results generalize and extend the corresponding inequalities for different classes of convex functions.

Funding: This research received no external funding.

Institutional Review Board Statement: Not applicable.

Informed Consent Statement: Not applicable.

Data Availability Statement: Not applicable.

Acknowledgments: This research was partially supported through project KK.01.1.1.02.0027 (a project co-financed by the Croatian Government and the European Union through the European Regional Development Fund—the Competitiveness and Cohesion Operational Programme).

Conflicts of Interest: The author declares no conflict of interest.

References

1. Andrić, M.; Farid, G.; Pečarić, J. A further extension of Mittag-Leffler function. *Fract. Calc. Appl. Anal.* **2018**, *21*, 1377–1395. [CrossRef]
2. Andrić, M.; Farid, G.; Pečarić, J. Analytical Inequalities for Fractional Calculus Operators and the Mittag-Leffler Function, Applications of integral operators containing an extended generalized Mittag-Leffler function in the kernel. *Ser. Monogr. Inequal.* **2021**, *20*, 272.
3. Prabhakar, T.R. A singular integral equation with a generalized Mittag Leffler function in the kernel. *Yokohama Math. J.* **1971**, *19*, 7–15.
4. Rahman, G.; Baleanu, D.; Qurashi, M.A.; Purohit, S.D.; Mubeen, S.; Arshad, M. The extended Mittag-Leffler function via fractional calculus. *J. Nonlinear Sci. Appl.* **2017**, *10*, 4244–4253. [CrossRef]
5. Salim, T.O.; Faraj, A.W. A generalization of Mittag-Leffler function and integral operator associated with fractional calculus. *J. Fract. Calc. Appl.* **2012**, *3*, 1–13.
6. Shukla, A.K.; Prajapati, J.C. On a generalization of Mittag-Leffler function and its properties. *J. Math. Anal. Appl.* **2007**, *336*, 797–811. [CrossRef]
7. Srivastava, H.M.; Tomovski, Z. Fractional calculus with an integral operator containing a generalized Mittag-Leffler function in the kernel. *Appl. Math. Comput.* **2009**, *211*, 198–210. [CrossRef]
8. Wiman, A. Über die Nullstellen der Funktionen $E_\alpha(x)$. *Acta Math.* **1905**, *29*, 217–234. [CrossRef]
9. Mehreen N.; Anwar, M. Hermite–Hadamard type inequalities for exponentially p-convex functions and exponentially s-convex functions in the second sense with applications. *J. Inequal. Appl.* **2019**, *2019*, 92. [CrossRef]
10. Noor, M.A.; Noor K.I.; Awan, M.U. Fractional Ostrowski inequalities for (s, m)-Godunova-Levin functions. *Facta Univ. Ser. Math. Inform.* **2015**, *30*, 489–499.
11. Özdemir, M.E.; Akdemir, A.O.; Set, E. On $(h - m)$-convexity and Hadamard-type inequalities. *Transylv. J. Math. Mech.* **2016**, *8*, 51–58.
12. Andrić, M. Fejér type inequalities for $(h, g; m)$-convex functions. *TWMS J. Pure Appl. Math.* **2021**, accepted.
13. Qiang, X.; Farid, G.; Pečarić, J.; Akbar, S.B. Generalized fractional integral inequalities for exponentially (s, m)-convex functions. *J. Inequal. Appl.* **2020**, *2020*, 70. [CrossRef]
14. Andrić, M.; Pečarić, J. On $(h, g; m)$-convexity and the Hermite-Hadamard inequality. *J. Convex Anal.* **2022**, *29*, 257–268.
15. Dragomir, S.S.; Pečarić, J.E.; Persson, L.E. Some inequalities of Hadamard type. *Soochow J. Math.* **1995**, *21*, 335–341.
16. Toader, G.H. Some generalizations of the convexity. *Proc. Colloq. Approx. Optim.* **1984**, 329–338.
17. Varošanec, S. On h-convexity. *J. Math. Anal. Appl.* **2007**, *326*, 303–311. [CrossRef]
18. Sarikaya, M.Z.; Set, E.; Yaldiz H.; Basak N. Hermite-Hadamard's inequalities for fractional integrals and related fractional inequalities. *Math. Comp. Model.* **2013**, *57*, 2403–2407. [CrossRef]
19. Kilbas, A.A.; Srivastava, H.M.; Trujillo, J.J. Theory and Applications of Fractional Differential Equations. In *Mathematics Studies*; Elsevier: Amsterdam, The Netherlands, 2006; Volume 204.
20. Kang, S.M.; Farid, G.; Nazeer, W.; Mehmood, S. $(h - m)$-convex functions and associated fractional Hadamard and Fejér-Hadamard inequalities via an extended generalized Mittag-Leffler function. *J. Inequal. Appl.* **2019**, *2019*, 78. [CrossRef]
21. Kang, S.M.; Farid, G.; Nazeer, W.; Tariq, B. Hadamard and Fejér-Hadamard inequalities for extended generalized fractional integrals involving special functions. *J. Inequal. Appl.* **2018**, *2018*, 119. [CrossRef] [PubMed]

fractal and fractional

Article

A New Fractional Poisson Process Governed by a Recursive Fractional Differential Equation

Zhehao Zhang

Department of Financial and Actuarial Mathematics, Xi'an Jiaotong-Liverpool University, Suzhou 215123, China; zhehao.zhang@xjtlu.edu.cn

Abstract: This paper proposes a new fractional Poisson process through a recursive fractional differential governing equation. Unlike the homogeneous Poisson process, the Caputo derivative on the probability distribution of k jumps with respect to time is linked to all probability distribution functions of j jumps, where j is a non-negative integer less than or equal to k. The distribution functions of arrival times are derived, while the inter-arrival times are no longer independent and identically distributed. Further, this new fractional Poisson process can be interpreted as a homogeneous Poisson process whose natural time flow has been randomized, and the underlying time randomizing process has been studied. Finally, the conditional distribution of the kth order statistic from random number samples, counted by this fractional Poisson process, is also discussed.

Keywords: fractional differential equations; Mittag–Leffler functions; Fox H function; subordinator and inverse stable subordinator; Lamperti law; order statistic

1. Introduction

Since the inter-arrival times of a Poisson process being independent and exponentially distributed are not supported by real data (see [1,2] and references therein), the fractional Poisson processes have received various attention. There are several different approaches to this concept. Jumarie [3] studies the fractional version of the Poisson process through the fractional master equation. Laskin [4] modifies the differential equation governing the probability distribution function of a homogeneous Poisson process through the Riemann–Liouville fractional derivative.

Another approach, followed by [5] is to generalize the inter-arrival times of a homogeneous Poisson process through the Mittag–Leffler distribution (see [6]). Later, Reference [7] shows that this fractional version is a true renewal process, without the independent and stationary increments.

If we denote the homogeneous Poisson process as $\{N_t\}_{t \geq 0}$ with intensity λ, where $\lambda > 0$, and $\frac{\partial^\beta}{\partial t^\beta}$ as the Caputo fractional derivative, where $\beta \in (0,1)$, i.e.,

$$\frac{\partial^\beta}{\partial t^\beta} f(t) = \frac{1}{\Gamma(1-\alpha)} \int_0^t (t-s)^{-\alpha} \left(\frac{d}{ds} f(s) \right) ds,$$

then one fractional method, proposed by [8] and denoted as $\{M_t^\beta\}_{t \geq 0}$, is to generalize the probability distribution function of N_t from

$$\frac{\partial}{\partial t} \mathbb{P}(N_t = k) = -\lambda (\mathbb{P}(N_t = k) - \mathbb{P}(N_t = k-1)), \quad k \in \mathbb{N}_0, \qquad (1)$$

to

$$\frac{\partial^\beta}{\partial t^\beta} \mathbb{P}(M_t^\beta = k) = -\lambda \left(\mathbb{P}(M_t^\beta = k) - \mathbb{P}(M_t^\beta = k-1) \right), \quad k \in \mathbb{N}_0, \qquad (2)$$

i.e., it is the time being fictionalized from a calculus point of view. More interestingly, if we consider the inverse β-stable subordinator $\{E_t^\beta\}_{t\geq 0}$, where $\beta \in (0,1)$, i.e.,

$$\mathbb{E}[e^{-qE_t^\beta}] = E_\beta(-qt^\beta), \tag{3}$$

where

$$E_\beta(z) = \sum_{j=0}^{\infty} \frac{z^j}{\Gamma(\beta j + 1)}, \quad \beta \in \mathbb{C}, \Re(\beta) > 0$$

is the Mittag–Leffler function of one variable, and assume that $\{N_t\}_{t\geq 0}$ and $\{E_t^\beta\}_{t\geq 0}$ are independent, then

$$M_t^\beta \stackrel{d}{=} N_{E_t^\beta}, \tag{4}$$

i.e., it is the time being randomized from a probability point of view. Beghin and Orsingher [9], Meerschaert et al. [10] prove that $\{M_t^\beta\}_{t\geq 0}$ is still a renewal process with inter-arrival times being independent and identically Mittag–Leffler distributed random variables, and study the case where Equation (2) is generalized to the nth order differential equation. The probability distribution function of $\{M_t^\beta\}_{t\geq 0}$ is,

$$\mathbb{P}\left(M_t^\beta = k\right) = \left(\lambda t^\beta\right)^k E_{\beta,\beta k+1}^{k+1}\left(-\lambda t^\beta\right), \quad k \in \mathbb{N}_0,$$

where

$$E_{\beta,\gamma}^\delta(z) = \sum_{k=0}^{\infty} \frac{(\delta)_k}{\Gamma(\beta k + \gamma)} \frac{z^k}{k!}, \quad \beta, \gamma, \delta \in \mathbb{C}, \Re(\beta) > 0$$

is the Mittag–Leffler function of three variables. Later, [11] describe the non-homogeneous version of this fractional Poisson process through its non-local governing equation. This fractional Poisson process has been applied in various fields. We refer to [12] for its applications in the transport of charged carriers, and [13] for its applications in risk theory.

Another type of fractional Poisson process, proposed by [14,15], is constructed through the integral representation, by replacing the Gaussian measure in the definition of fractional Brownian motion with the Poisson counting measure. This fractional version displays long range dependence, has a fatter tail than the Gaussian process, and converges to fractional Brownian motion in distribution. Wang et al. [16] study the non-homogeneous versions of this fractional process.

This paper defines a new fractional Poisson process, denoted as $\{N_t^\beta\}_{t\geq 0}$, through a governing equation, which generalizes Equation (1) by connecting $\mathbb{P}(N_t^\beta = k)$ to $\mathbb{P}(N_t^\beta = j)$ for all $j \leq k$ through Caputo fractional derivative, i.e.,

$$\frac{\partial^\beta}{\partial t^\beta}\mathbb{P}\left(N_t^\beta = k\right) = -\lambda^\beta \sum_{j=0}^{k} \frac{(-\beta)_j}{j!} \mathbb{P}\left(N_t^\beta = k-j\right), \quad j,k \in \mathbb{N}_0.$$

Since $\mathbb{P}(N_t^\beta = k) = 0$ for $k \notin \mathbb{N}_0$, then the upper bound of the summation on the right hand side can be extended to infinity. Thus, the fractional differentiation on the time of the probability distribution function is related to the probabilities of all possible values this new process could take. Particularly, when $\beta = 1$, the above equation goes back to Equation (1).

We first study the probability properties of this fractional Poisson process. Later, we find this fractional process can be interpreted as a homogeneous Poisson process whose natural time flow has been randomized, and the underlying time process at time one follows a Lamperti distribution. The transforms of this underlying time process have also been studied. Finally, we discuss the order statistics counted by this fractional Poisson process.

2. Main Results

Theorem 1. *Let* $\{N_t^\beta\}_{t\geq 0}$ *be a fractional Poisson process with parameter* $\lambda > 0$ *and* $\beta \in (0,1)$, *which satisfies the governing equation*

$$\frac{\partial^\beta}{\partial t^\beta}\mathbb{P}\left(N_t^\beta = k\right) = -\lambda^\beta \sum_{j=0}^{k} \frac{(-\beta)_j}{j!}\mathbb{P}\left(N_t^\beta = k-j\right), \quad j,k \in \mathbb{N}_0, \tag{5}$$

where $\mathbb{P}(N_t^\beta = k) = 0$ *for* $k \notin \mathbb{N}_0$. *Then the probability distribution function of this process is*

$$\mathbb{P}\left(N_t^\beta = k\right) = \frac{(-1)^k}{k!} E_{\beta, 1-k}\left(-\lambda^\beta t^\beta\right), \quad k \in \mathbb{N}_0, \tag{6}$$

where

$$E_{\beta,\gamma}(z) = \sum_{j=0}^{\infty} \frac{z^j}{\Gamma(\beta j + \gamma)}, \quad \beta, \gamma \in \mathbb{C}, \Re(\beta) > 0,$$

is the Mittag–Leffler function of two variables, and the probability density function of its arrival times $\{T_k\}_{k \in \mathbb{N}}$ *is*

$$f_{T_k}(t) = \frac{(-1)^{k+1}}{\Gamma(k)} \lambda(\lambda t)^{\beta-1} E_{\beta, \beta+1-k}\left(-\lambda^\beta t^\beta\right), \quad k \in \mathbb{N}. \tag{7}$$

Proof. From the definition, we may write the right hand side of the governing equation into an infinite series,

$$\frac{\partial^\beta}{\partial t^\beta}\mathbb{P}\left(N_t^\beta = k\right) = -\lambda^\beta \sum_{j=0}^{\infty} \frac{(-\beta)_j}{j!}\mathbb{P}\left(N_t^\beta = k-j\right),$$

and therefore the Laplace transform of N_t^β is

$$\frac{\partial^\beta}{\partial t^\beta}\mathbb{E}\left[e^{-qN_t^\beta}\right] = -\lambda^\beta \sum_{j=0}^{\infty} \frac{(-\beta)_j}{j!}\mathbb{E}\left[e^{-q\left(N_t^\beta + j\right)}\right]$$

$$= -\lambda^\beta \sum_{j=0}^{\infty} \frac{(-\beta)_j}{j!} e^{-qj} \mathbb{E}\left[e^{-qN_t^\beta}\right]$$

$$= -\lambda^\beta \left(1 - e^{-q}\right)^\beta \mathbb{E}\left[e^{-qN_t^\beta}\right].$$

Taking Laplace transform from t to s gives

$$s^\beta \mathcal{L}_s\left\{\mathbb{E}\left[e^{-qN_t^\beta}\right]\right\} - s^{\beta-1}\mathbb{E}\left[e^{-qN_0^\beta}\right] = -\lambda^\beta\left(1 - e^{-q}\right)^\beta \mathcal{L}_s\left\{\mathbb{E}\left[e^{-qN_t^\beta}\right]\right\}$$

$$\mathcal{L}_s\left\{\mathbb{E}\left[e^{-qN_t^\beta}\right]\right\} = \frac{s^{\beta-1}}{s^\beta + \lambda^\beta(1 - e^{-q})^\beta}.$$

This leads to

$$\mathbb{E}\left[e^{-qN_t^\beta}\right] = E_{\beta,1}\left(-\lambda^\beta(1-e^{-q})^\beta t^\beta\right) \qquad (8)$$

$$= \sum_{n=0}^{\infty} \frac{(-\lambda^\beta t^\beta)^n}{\Gamma(\beta n+1)}(1-e^{-q})^{\beta n}$$

$$= \sum_{k=0}^{\infty} e^{-qk} \sum_{n=0}^{\infty} \frac{(-\lambda^\beta t^\beta)^n}{\Gamma(\beta n+1)} \frac{(-\beta n)_k}{k!},$$

and therefore

$$\mathbb{P}\left(N_t^\beta = k\right) = \frac{1}{k!} \sum_{n=0}^{\infty} \frac{(-\lambda^\beta t^\beta)^n (-\beta n)_k}{\Gamma(\beta n+1)}$$

$$= \frac{1}{k!} \sum_{n=0}^{\infty} \frac{(-1)^k(-\lambda^\beta t^\beta)^n}{\Gamma(\beta n+1-k)}$$

$$= \frac{(-1)^k}{k!} E_{\beta,1-k}\left(-\lambda^\beta t^\beta\right).$$

For the arrival times $\{T_k\}_{k\in\mathbb{N}}$, since $\{N_t^\beta \geq k\}$ and $\{T_k \leq t\}$ are equivalent, then we have

$$\mathbb{P}(T_k \leq t) = \sum_{j=k}^{\infty} \mathbb{P}\left(N_t^\beta = j\right)$$

$$= \sum_{j=k}^{\infty} \frac{(-1)^j}{j!} E_{\beta,1-j}\left(-\lambda^\beta t^\beta\right)$$

$$= \sum_{n=0}^{\infty} \frac{(-\lambda^\beta t^\beta)^n}{\Gamma(\beta n+1)} \sum_{j=k}^{\infty} \frac{1}{j!} \frac{\Gamma(-\beta n+j)}{\Gamma(-\beta n)}$$

$$= \sum_{n=0}^{\infty} \frac{(-\lambda^\beta t^\beta)^n}{\Gamma(\beta n+1)} \frac{(-\beta n)_k}{k!} {}_2F_1(1, k-n\beta, 1+k, 1)$$

$$= \sum_{n=1}^{\infty} \frac{(-\lambda^\beta t^\beta)^n}{\Gamma(\beta n+1)} \frac{(-\beta n)_k}{k!} \frac{\Gamma(1+k)\Gamma(\beta n)}{\Gamma(k)\Gamma(1+\beta n)}$$

$$= \frac{(-1)^k}{\Gamma(k)} \sum_{n=1}^{\infty} \frac{(-\lambda^\beta t^\beta)^n}{\Gamma(\beta n+1-k)} \frac{1}{\beta n},$$

where ${}_2F_1(\alpha,\beta,\gamma,z)$ is the Hypergeometric function; see [17] (Chapter 9). The proof is completed after differentiating it with respect to t once. □

Remark 1. 1. When $\beta = 1$,

$$\mathbb{E}\left[e^{-qN_t^\beta}\right] = \sum_{n=0}^{\infty} \frac{(-\lambda(1-e^{-q})t)^n}{\Gamma(n+1)} = \exp\{\lambda t(e^{-q}-1)\}$$

$$\mathbb{P}\left(N_t^\beta = k\right) = \frac{1}{k!} \sum_{n=0}^{\infty} \frac{(-1)^k(-\lambda t)^n}{\Gamma(n+1-k)} = \frac{1}{k!} \sum_{n=k}^{\infty} \frac{(-1)^k(-\lambda^\beta t^\beta)^n}{\Gamma(n+1-k)} = \frac{(\lambda t)^k}{k!} e^{-\lambda t},$$

which goes back to a homogeneous Poisson process.

2. Since $\beta \in (0,1)$, the integral representation of the Mittag–Leffler function remains [18] (Lemma 2.2.2), and we have

$$\mathbb{P}\left(N_t^\beta = k\right) = \frac{(-1)^k}{k!} \frac{1}{2\pi i} \int_{c-i\infty}^{c+i\infty} \frac{\Gamma(s)\Gamma(1-s)}{\Gamma(1-k-\beta s)} \left(\lambda^\beta t^\beta\right)^s ds$$

$$= \frac{(-1)^k}{k!} H_{1,2}^{1,1}\left[\lambda^\beta t^\beta \middle| \begin{array}{c} (0,1) \\ (0,1) \quad (k,\beta) \end{array}\right],$$

where $H_{p,q}^{m,n}\left[z \middle| \begin{array}{c} (a_i, A_i)_{1,p} \\ (b_j, B_j)_{1,q} \end{array}\right]$ is the Fox's H function. The convergence of this contour integral can be checked by [19] (Equation (1.13)) The integral representation, by closing the contour in two different directions, leads to

$$E_{\beta,\gamma}(z) = \sum_{j=0}^{\infty} \frac{z^j}{\Gamma(\beta j + \gamma)} = -\sum_{j=1}^{\infty} \frac{z^{-j}}{\Gamma(\gamma - \beta j)},$$

which determines the asymptotic behavior of these probability functions.

3. Equation (8) indicates that

$$\mathbb{E}\left[e^{-qN_{t_1+t_2}^\beta}\right] = E_{\beta,1}\left(-\lambda^\beta(1-e^{-q})^\beta(t_1+t_2)^\beta\right)$$

$$\neq E_{\beta,1}\left(-\lambda^\beta(1-e^{-q})^\beta t_1^\beta\right) E_{\beta,1}\left(-\lambda^\beta(1-e^{-q})^\beta t_2^\beta\right)$$

$$= \mathbb{E}\left[e^{-qN_{t_1}^\beta}\right] \mathbb{E}\left[e^{-qN_{t_2}^\beta}\right],$$

i.e., $\{N_t^\beta\}_{t \geq 0}$ no longer possesses independent increments and therefore loses the lack of memory property of the homogeneous Poisson process.

4. Since

$$\frac{d}{dq}\mathbb{E}\left[e^{-qN_t^\beta}\right] = -e^{-q}(1-e^{-q})^{\beta-1} t^\beta \lambda^\beta E_{\beta,\beta}\left(-\lambda^\beta(1-e^{-q})^\beta t^\beta\right),$$

which tends to ∞ as q tends to 0, then $\mathbb{E}\left[\left(N_t^\beta\right)^n\right]$ does not exist for all $n \in \mathbb{N}$, unlike $\mathbb{E}\left[\left(M_t^\beta\right)^n\right]$.

Given Equation (6), Equation (5) can be verified directly.

$$\frac{\partial^\beta}{\partial t^\beta}\mathbb{P}\left(N_t^\beta = k\right) = \frac{1}{k!} \sum_{n=1}^{\infty} \frac{(-1)^k(-\lambda^\beta)^n}{\Gamma(\beta n + 1 - k)} \frac{\Gamma(\beta n + 1)}{\Gamma(\beta n - \beta + 1)} t^{\beta(n-1)}$$

$$= -\lambda^\beta \frac{1}{k!} \sum_{n=0}^{\infty} \frac{(-\lambda^\beta)^n}{\Gamma(\beta n + 1)} (-\beta n - \beta)_k t^{\beta n}$$

$$= -\lambda^\beta \frac{1}{k!} \sum_{n=0}^{\infty} \frac{(-\lambda^\beta)^n}{\Gamma(\beta n + 1)} \sum_{j=0}^{k} \binom{k}{j} (-\beta)_j (-\beta n)_{k-j} t^{\beta n}$$

$$= -\lambda^\beta \sum_{j=0}^{k} \frac{k!}{j!(k-j)!} (-\beta)_j \frac{1}{k!} \sum_{n=0}^{\infty} \frac{(-\lambda^\beta)^n}{\Gamma(\beta n + 1)} \frac{(-1)^{k-j}\Gamma(\beta n + 1)}{\Gamma(\beta n + 1 - (k-j))} t^{\beta n}$$

$$= -\lambda^\beta \sum_{j=0}^{k} \frac{(-\beta)_j}{j!} \frac{1}{(k-j)!} \sum_{n=0}^{\infty} \frac{(-1)^{k-j}(-\lambda^\beta)^n}{\Gamma(\beta n + 1 - (k-j))} t^{\beta n}$$

$$= -\lambda^\beta \sum_{j=0}^{k} \frac{(-\beta)_j}{j!} \mathbb{P}\left(N_t^\beta = k\right).$$

We present a few numerical examples of $\mathbb{P}\left(N_t^\beta = k\right)$ and $f_{T_k}(t)$ from Figures 1–6.

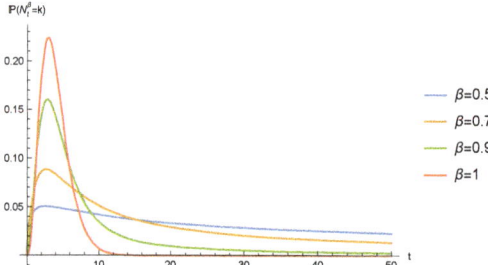

Figure 1. $\mathbb{P}\left(N_t^\beta = 3\right)$ with $\lambda = 1$.

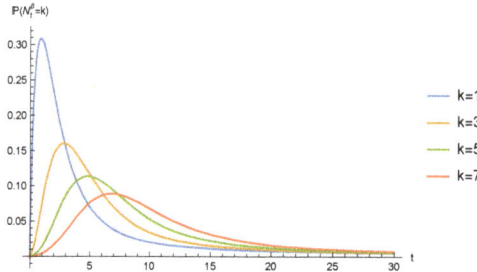

Figure 2. $\mathbb{P}\left(N_t^{0.9} = k\right)$ with $\lambda = 1$.

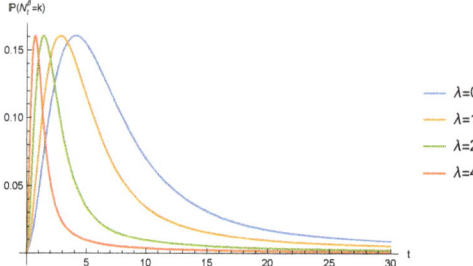

Figure 3. $\mathbb{P}\left(N_t^{0.9} = 3\right)$.

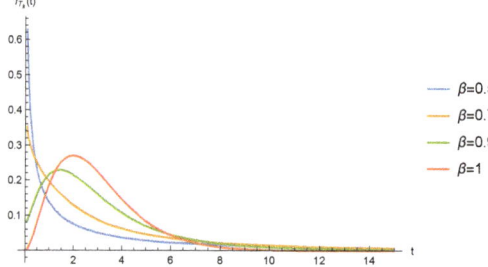

Figure 4. $f_{T_3}(t)$ with $\lambda = 1$.

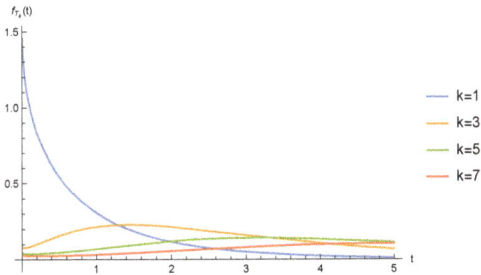

Figure 5. $f_{T_k}(t)$ with $\lambda = 1$ and $\beta = 0.9$.

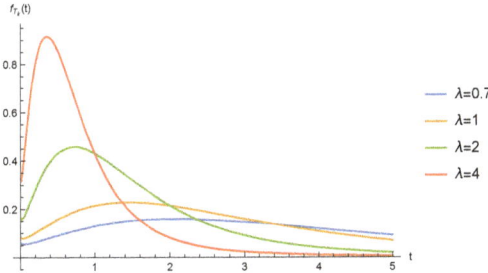

Figure 6. $f_{T_3}(t)$ with $\beta = 0.9$.

The Laplace transform of T_k is

$$\mathbb{E}\left[e^{-qT_k}\right] = q\mathcal{L}_q\{\mathbb{P}(T_k \leq t)\} = \frac{(-1)^k}{\Gamma(k)} \sum_{n=1}^{\infty} \frac{\Gamma(\beta n)}{\Gamma(\beta n + 1 - k)} \left(-\frac{\lambda^\beta}{q^\beta}\right)^n, \qquad (9)$$

which allows us to determine whether $\{N_t^\beta\}_{t \geq 0}$ is still a renewal process. For $k = 1$,

$$\mathbb{E}\left[e^{-qT_1}\right] = \frac{\lambda^\beta}{q^\beta + \lambda^\beta} = \mathbb{E}\left[e^{-q\tau_1}\right],$$

i.e., τ_1 is Mittag–Leffler distributed with survival function

$$\mathbb{P}(\tau_1 > t) = E_{\beta,1}\left(-\lambda^\beta t^\beta\right).$$

For $k = 2$,

$$\mathbb{E}\left[e^{-qT_2}\right] = \frac{\lambda^\beta \left(q^\beta(1-\beta) + \lambda^\beta\right)}{(q^\beta + \lambda^\beta)^2} = \left(\frac{\lambda^\beta}{q^\beta + \lambda^\beta}\right)^2 + (1-\beta)\frac{\lambda^\beta q^\beta}{(q^\beta + \lambda^\beta)^2}.$$

If τ_1 and τ_2 are independent, then

$$\mathbb{E}\left[e^{-q\tau_2}\right] = \frac{\mathbb{E}\left[e^{-qT_2}\right]}{\mathbb{E}\left[e^{-qT_1}\right]} = \frac{\left(\frac{\lambda^\beta}{q^\beta + \lambda^\beta}\right)^2 + (1-\beta)\left(\frac{\lambda^\beta q^\beta}{(q^\beta + \lambda^\beta)^2}\right)}{\frac{\lambda^\beta}{q^\beta + \lambda^\beta}} = \beta \frac{\lambda^\beta}{q^\beta + \lambda^\beta} + (1-\beta),$$

which implies that τ_2 is a mixture of Mittag–Leffler distributed random variable with probability β and a mass point at zero with probability $1 - \beta$. So, if τ_1 and τ_2 are independent, then they are not equal in distribution and the mass point at zero implies the multiple jumps at one time. Particularly, when $\beta = 1$, Equation (9) turns out to be

$$\mathbb{E}\left[e^{-qT_k}\right] = \frac{(-1)^k}{\Gamma(k)} \sum_{n=k}^{\infty} \frac{\Gamma(n)}{\Gamma(n+1-k)}\left(-\frac{\lambda}{q}\right)^n = \left(\frac{\lambda}{q+\lambda}\right)^k,$$

which is the Laplace transform of a gamma distribution and goes back to a homogeneous Poisson process.

If we denote a β-stable subordinator as $\{D_t^\beta\}_{t \geq 0}$, i.e.,

$$\mathbb{E}\left[e^{-qD_t^\beta}\right] = e^{-tq^\beta},$$

then based on the definition,

$$E_t^\beta = \inf\left\{u \geq 0 : D_u^\beta > t\right\}, \quad t \geq 0.$$

$\{E_t^\beta\}_{t \geq 0}$ is non-decreasing and its sample paths are almost surely continuous if $\{D_t^\beta\}_{t \geq 0}$ is strictly increasing. From Equation (3), it can be seen that $\{E_t^\beta\}_{t \geq 0}$ possesses non-Markovian with non-stationary and non-independent increments. The probability functions of $\{D_t^\beta\}_{t \geq 0}$ and $\{E_t^\beta\}_{t \geq 0}$ are usually in complicated forms, and [20] find the densities of the product, quotient, and power of them in terms of the Fox's H function. Since

$$\mathbb{E}\left[\left(D_t^\beta\right)^\rho\right] = t^{\frac{\rho}{\beta}} \frac{\Gamma(1-\frac{\rho}{\beta})}{\Gamma(1-\rho)}, \quad \Re(\rho) < \beta,$$

then for two independent β-stable processes $\{D_{1,t}^\beta\}_{t \geq 0}$ and $\{D_{2,t}^\beta\}_{t \geq 0}$, we have

$$\mathbb{E}\left[\left(\frac{D_{1,t}^\beta}{D_{2,t}^\beta}\right)^\rho\right] = \frac{\Gamma(1-\frac{\rho}{\beta})\Gamma(1+\frac{\rho}{\beta})}{\Gamma(1-\rho)\Gamma(1+\rho)}, \quad \Re(\rho) \in (-\beta, \beta). \tag{10}$$

If we denote $L = \left(\frac{D_{1,t}^\beta}{D_{2,t}^\beta}\right)^\beta$, then L is a Lamperti random variable and its probability density function with respect to the Lebesgue measure on $\mathbb{R}$ is

$$f_L(x) = \frac{\sin(\pi\beta)}{\pi\beta} \frac{1}{x^2 + 2x\cos(\pi\beta) + 1}, \quad x > 0. \tag{11}$$

See [21,22] for a detailed discussion on the Lamperti law and the stable law. Meanwhile, the Mellin transform of $\{E_t^\beta\}_{t \geq 0}$ is

$$\mathbb{E}\left[\left(E_t^\beta\right)^\rho\right] = t^{\beta\rho} \frac{\Gamma(1+\rho)}{\Gamma(1+\beta\rho)}.$$

From [23], for $\rho \in (0,1)$, the Mellin transform and the Laplace transform of a positive random variable can be connected through

$$\mathbb{E}[X^\rho] = \frac{\rho}{\Gamma(1-\rho)} \int_0^\infty q^{-\rho-1}\left(1 - \mathbb{E}\left[e^{-qX}\right]\right)dq, \quad \rho \in (0,1).$$

Replacing X with E_t^β gives

$$\int_0^\infty q^{-\rho-1}\left(1 - E_{\beta,1}\left(-qt^\beta\right)\right)dq = t^{\beta\rho} \frac{\Gamma(\rho)\Gamma(1-\rho)}{\Gamma(1+\beta\rho)}. \tag{12}$$

The next theorem gives a parallel result to distributional equality Equation 4.

Theorem 2. *Let $\{N_t^\beta\}_{t\geq 0}$ be a fractional Poisson process with parameter $\lambda > 0$ and $\beta \in (0,1)$. If $\{U_t^\beta\}_{t\geq 0}$ is a non-negative process such that*

$$f_{U_t^\beta}(x) = \frac{1}{\beta}\frac{1}{t} H_{2,2}^{1,1}\left[\frac{x}{t} \left| \begin{array}{cc} \left(1-\frac{1}{\beta},\frac{1}{\beta}\right) & (0,1) \\ \left(1-\frac{1}{\beta},\frac{1}{\beta}\right) & (0,1) \end{array} \right.\right], \quad x > 0, \tag{13}$$

and independent with $\{N_t^\beta\}_{t\geq 0}$, then

$$N_t^\beta \stackrel{d}{=} N_{U_t^\beta}, \tag{14}$$

Proof. If Equation (14) is true, then

$$\mathbb{E}\left[e^{-qN_t^\beta}\right] = \mathbb{E}\left[\mathbb{E}\left[e^{-qN_t^\beta}|U_t^\beta\right]\right] = \mathbb{E}\left[e^{-\lambda(1-e^{-q})U_t^\beta}\right] = E_{\beta,1}\left(-\lambda^\beta(1-e^{-q})^\beta t^\beta\right),$$

which gives

$$\mathbb{E}\left[e^{-qU_t^\beta}\right] = E_{\beta,1}\left(-q^\beta t^\beta\right). \tag{15}$$

Applying Equation (12) gives

$$\begin{aligned}
\mathbb{E}\left[\left(U_t^\beta\right)^\rho\right] &= \frac{\rho}{\Gamma(1-\rho)} \int_0^\infty q^{-\rho-1}\left(1 - E_{\beta,1}\left(-q^\beta t^\beta\right)\right) dq \\
&= \frac{\rho}{\Gamma(1-\rho)} \frac{1}{\beta} \int_0^\infty s^{-\frac{\rho}{\beta}-1}\left(1 - E_{\beta,1}\left(-st^\beta\right)\right) ds \\
&= \frac{\rho}{\Gamma(1-\rho)} \frac{1}{\beta} t^{\beta\frac{\rho}{\beta}} \frac{\Gamma\left(\frac{\rho}{\beta}\right)\Gamma\left(1-\frac{\rho}{\beta}\right)}{\Gamma\left(1+\beta\frac{\rho}{\beta}\right)} \\
&= t^\rho \frac{\Gamma\left(1-\frac{\rho}{\beta}\right)\Gamma\left(1+\frac{\rho}{\beta}\right)}{\Gamma(1-\rho)\Gamma(1+\rho)}, \quad \Re(\rho) \in (-\beta, \beta).
\end{aligned} \tag{16}$$

Finally, applying the inverse Mellin transform gives

$$\begin{aligned}
f_{U_t^\beta}(x) &= \frac{1}{2\pi i} \int_{c-i\infty}^{c+i\infty} \mathbb{E}\left[\left(U_t^\beta\right)^\rho\right] x^{-1-\rho} d\rho \\
&= \frac{1}{\beta}\frac{1}{2\pi i} \int_{c-i\infty}^{c+i\infty} t^\rho \frac{\Gamma\left(-\frac{\rho}{\beta}\right)\Gamma\left(1+\frac{\rho}{\beta}\right)}{\Gamma(-\rho)\Gamma(1+\rho)} x^{-1-\rho} d\rho \\
&= \frac{1}{\beta}\frac{1}{t}\frac{1}{2\pi i} \int_{c_1-i\infty}^{c_1+i\infty} \frac{\Gamma\left(\frac{1}{\beta}-\frac{1}{\beta}s\right)\Gamma\left(1-\frac{1}{\beta}+\frac{1}{\beta}s\right)}{\Gamma(1-s)\Gamma(s)}\left(\frac{x}{t}\right)^{-s} ds \\
&= \frac{1}{\beta}\frac{1}{t} H_{2,2}^{1,1}\left[\frac{x}{t} \left| \begin{array}{cc} \left(1-\frac{1}{\beta},\frac{1}{\beta}\right) & (0,1) \\ \left(1-\frac{1}{\beta},\frac{1}{\beta}\right) & (0,1) \end{array} \right.\right].
\end{aligned}$$

□

Remark 2. 1. From Equation (15), $\{U_t^\beta\}_{t\geq 0}$ does not possess independent and stationary increments.

2. Comparing Equation (10) and Equation (16), it can be seen that

$$U_t^\beta \stackrel{d}{=} t \frac{D_{1,t}^\beta}{D_{2,t}^\beta} \stackrel{d}{=} tL^{\frac{1}{\beta}} \stackrel{d}{=} t\left(\frac{E_{1,t}^\beta}{E_{2,t}^\beta}\right)^{\frac{1}{\beta}},$$

which leads to a simpler form of Equation (13) after a change of variable in Equation (11),

$$f_{U_t^\beta}(x) = \frac{t^\beta}{\pi} \frac{x^{\beta-1}\sin(\pi\beta)}{x^{2\beta} + 2x^\beta t^\beta \cos(\pi\beta) + t^{2\beta}}, \quad x > 0. \tag{17}$$

Meanwhile, this expression can be seen from Equation (13) directly,

$$\begin{aligned}
f_{U_t^\beta}(x) &= \frac{1}{\beta}\frac{1}{t} H_{2,2}^{1,1}\left[\frac{x}{t}\Bigg|\begin{array}{cc}\left(1-\frac{1}{\beta},\frac{1}{\beta}\right) & (0,1) \\ \left(1-\frac{1}{\beta},\frac{1}{\beta}\right) & (0,1)\end{array}\right] \\
&= \frac{1}{t}\frac{1}{2\pi i}\int_{c-i\infty}^{c+i\infty} \frac{\Gamma\left(\frac{1}{\beta}-y\right)\Gamma\left(1-\frac{1}{\beta}+y\right)}{\Gamma(1-\beta y)\Gamma(\beta y)}\left(\frac{x}{t}\right)^{-\beta y} dy \\
&= \frac{1}{t}\sum_{n=0}^\infty \lim_{y \to \frac{1}{\beta}+n}\left(y-\frac{1}{\beta}-n\right)\Gamma\left(\frac{1}{\beta}-y\right)\frac{\Gamma\left(1-\frac{1}{\beta}+y\right)}{\Gamma(1-\beta y)\Gamma(\beta y)}\left(\frac{x}{t}\right)^{-\beta y} \\
&= \frac{1}{t}\sum_{n=0}^\infty \frac{(-1)^n}{\Gamma(-\beta n)\Gamma(1+\beta n)}\left(\frac{t}{x}\right)^{(1+\beta n)} \\
&= -\frac{1}{t}\sum_{n=0}^\infty (-1)^n \frac{\sin(n\pi\beta)}{\pi}\left(\frac{t}{x}\right)^{(1+\beta n)} \\
&= -\frac{1}{\pi t}\left(\frac{t}{x}\right)\sum_{n=0}^\infty \sin(n\pi\beta)\left(-\left(\frac{x}{t}\right)^\beta\right)^{-n} \\
&= \frac{1}{\pi}\frac{1}{x}\frac{\left(\frac{x}{t}\right)^\beta \sin(\pi\beta)}{\left(\frac{x}{t}\right)^{2\beta} + 2\left(\frac{x}{t}\right)^\beta \cos(\pi\beta) + 1} \\
&= \frac{t^\beta}{\pi}\frac{x^{\beta-1}\sin(\pi\beta)}{x^{2\beta} + 2x^\beta t^\beta \cos(\pi\beta) + t^{2\beta}}.
\end{aligned}$$

With this simplified expression, we can see that when $t \to 0$, $f_{U_t^\beta}(x) \to 0$ and when $x \to 0$, $f_{U_t^\beta}(x) \to \infty$.

3. From Equation (16), the Mellin transform only exists for $\Re(\rho) \in (-\beta, \beta)$, and therefore U_t^β does not have the first moment for $t > 0$. This fits our observation in Theorem 1 that N_t^β does not have the first moment for $t > 0$ either.

We give a few numerical examples of $f_{U_t^\beta}(x)$ from Figures 7–10. In the first two figures, there is a clear sign that the density functions approach to infinity as the variable x tends to zero. In the last two figures, the density functions approach to zero as the variable t tends to zero. These behaviors fit the theoretical analysis on Equation (17).

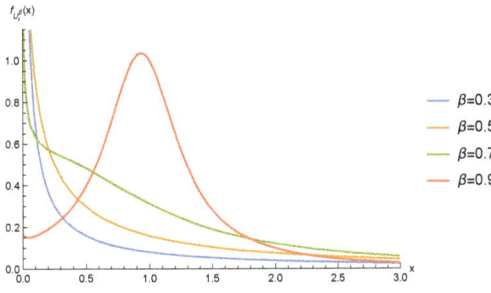

Figure 7. $f_{U_1^\beta}(x)$.

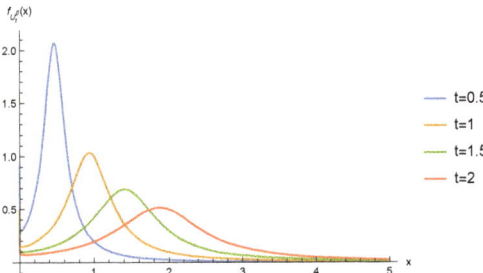

Figure 8. $f_{U_t^{0.9}}(x)$.

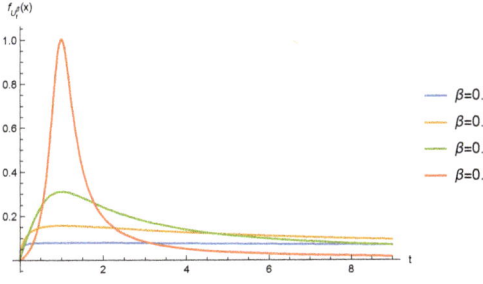

Figure 9. $f_{U_1^\beta}(x)$.

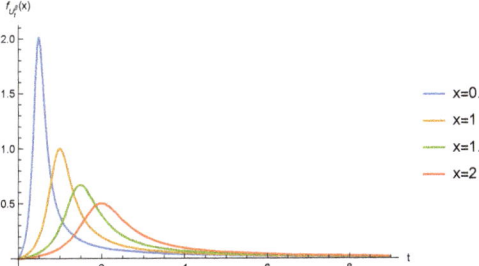

Figure 10. $f_{U_t^{0.9}}(x)$.

Equation (13) can be checked through the Laplace transform directly

$$\mathbb{E}\left[e^{-qU_t^\beta}\right] = \frac{1}{\beta}\frac{1}{t}tH_{2,3}^{2,1}\left[qt\left|\begin{matrix}\left(0,\frac{1}{\beta}\right) & (0,1) \\ (0,1) & \left(0,\frac{1}{\beta}\right) & (0,1)\end{matrix}\right.\right]$$

$$= \frac{1}{\beta}H_{1,2}^{1,1}\left[qt\left|\begin{matrix}\left(0,\frac{1}{\beta}\right) \\ \left(0,\frac{1}{\beta}\right) & (0,1)\end{matrix}\right.\right]$$

$$= H_{1,2}^{1,1}\left[(qt)^\beta\left|\begin{matrix}(0,1) \\ (0,1) & (0,\beta)\end{matrix}\right.\right]$$

$$= E_{\beta,1}\left(-(qt)^\beta\right).$$

With the distributional equality Equation (14), Equation (6) can be calculated directly,

$$\mathbb{P}\left(N_t^\beta = k\right) = \mathbb{P}\left(N_{U_t^\beta} = k\right)$$

$$= \int_0^\infty \frac{(\lambda x)^k}{k!}e^{-\lambda x}\mathbb{P}\left(U_t^\beta \in dx\right)$$

$$= \int_0^\infty \frac{(\lambda x)^k}{k!}e^{-\lambda x}\frac{1}{\beta}\frac{1}{t}\frac{1}{2\pi i}\int_{c-i\infty}^{c+i\infty}\frac{\Gamma\left(\frac{1}{\beta}-\frac{1}{\beta}s\right)\Gamma\left(1-\frac{1}{\beta}+\frac{1}{\beta}s\right)}{\Gamma(1-s)\Gamma(s)}\left(\frac{x}{t}\right)^{-s}dsdx$$

$$= \frac{1}{k!}\frac{1}{\beta}\frac{1}{2\pi i}\int_{c_1-i\infty}^{c_1+i\infty}\frac{\Gamma\left(\frac{1}{\beta}u\right)\Gamma\left(1-\frac{1}{\beta}u\right)\Gamma(k+u)}{\Gamma(u)\Gamma(1-u)}(\lambda t)^{-u}du$$

$$= \frac{1}{k!}\frac{1}{\beta}\frac{1}{2\pi i}\int_{c_1-i\infty}^{c_1+i\infty}\frac{\Gamma\left(\frac{1}{\beta}u\right)\Gamma\left(1-\frac{1}{\beta}u\right)}{\Gamma(1-u)}(-1)^k\frac{\Gamma(-u+1)}{\Gamma(-u+1-k)}(\lambda t)^{-u}du$$

$$= \frac{(-1)^k}{k!}\frac{1}{\beta}\frac{1}{2\pi i}\int_{c_1-i\infty}^{c_1+i\infty}\frac{\Gamma\left(\frac{1}{\beta}u\right)\Gamma\left(1-\frac{1}{\beta}u\right)}{\Gamma(1-k-u)}(\lambda t)^{-u}du$$

$$= \frac{(-1)^k}{k!}\frac{1}{2\pi i}\int_{c_2-i\infty}^{c_2+i\infty}\frac{\Gamma(s)\Gamma(1-s)}{\Gamma(1-k-\beta s)}(\lambda t)^{-\beta s}ds$$

$$= \frac{(-1)^k}{k!}E_{\beta,1-k}\left(-\lambda^\beta t^\beta\right).$$

Corollary 1. *The Laplace and Mellin transforms of the density function of $\{U_t^\beta\}_{t\geq 0}$, where $\beta \in (0,1)$, with respect to the time variable are*

$$\int_0^\infty e^{-st}f_{U_t^\beta}(x)dt = H_{1,2}^{1,1}\left[(sx)^\beta\left|\begin{matrix}\left(1-\frac{1}{\beta},1\right) \\ \left(1-\frac{1}{\beta},1\right) & (0,\beta)\end{matrix}\right.\right], \quad x > 0,$$

and

$$\int_0^\infty t^\rho f_{U_t^\beta}(x)dt = x^\rho\frac{\Gamma\left(1-\frac{\rho+1}{\beta}\right)\Gamma\left(1+\frac{\rho+1}{\beta}\right)}{\Gamma(2+\rho)\Gamma(-\rho)}, \quad x > 0.$$

Proof. We first rewrite the density function of U_t^β as

$$f_{U_t^\beta}(x) = \frac{1}{\beta t} H_{2,2}^{1,1}\left[\frac{x}{t} \middle| \begin{array}{cc} \left(1-\frac{1}{\beta},\frac{1}{\beta}\right) & (0,1) \\ \left(1-\frac{1}{\beta},\frac{1}{\beta}\right) & (0,1) \end{array}\right]$$

$$= \frac{1}{\beta x} H_{2,2}^{1,1}\left[\frac{x}{t} \middle| \begin{array}{cc} \left(1,\frac{1}{\beta}\right) & (1,1) \\ \left(1,\frac{1}{\beta}\right) & (1,1) \end{array}\right]$$

$$= \frac{1}{\beta x} H_{2,2}^{1,1}\left[\frac{t}{x} \middle| \begin{array}{cc} \left(0,\frac{1}{\beta}\right) & (0,1) \\ \left(0,\frac{1}{\beta}\right) & (0,1) \end{array}\right].$$

Then, the Laplace transform is

$$\int_0^\infty e^{-st} f_{U_t^\beta}(x) dt = \int_0^\infty e^{-st} \frac{1}{\beta x} H_{2,2}^{1,1}\left[\frac{t}{x} \middle| \begin{array}{cc} \left(0,\frac{1}{\beta}\right) & (0,1) \\ \left(0,\frac{1}{\beta}\right) & (0,1) \end{array}\right] dt$$

$$= \frac{1}{\beta x} x H_{2,3}^{2,1}\left[sx \middle| \begin{array}{cc} \left(1-\frac{1}{\beta},\frac{1}{\beta}\right) & (0,1) \\ (0,1) & \left(1-\frac{1}{\beta},\frac{1}{\beta}\right) & (0,1) \end{array}\right]$$

$$= \frac{1}{\beta} H_{1,2}^{1,1}\left[sx \middle| \begin{array}{c} \left(1-\frac{1}{\beta},\frac{1}{\beta}\right) \\ \left(1-\frac{1}{\beta},\frac{1}{\beta}\right) & (0,1) \end{array}\right]$$

$$= H_{1,2}^{1,1}\left[(sx)^\beta \middle| \begin{array}{c} \left(1-\frac{1}{\beta},1\right) \\ \left(1-\frac{1}{\beta},1\right) & (0,\beta) \end{array}\right],$$

and the Mellin transform is

$$\int_0^\infty t^{s-1} f_{U_t^\beta}(x) dt = \int_0^\infty t^{s-1} \frac{1}{\beta x} H_{2,2}^{1,1}\left[\frac{t}{x} \middle| \begin{array}{cc} \left(0,\frac{1}{\beta}\right) & (0,1) \\ \left(0,\frac{1}{\beta}\right) & (0,1) \end{array}\right] dt$$

$$= \frac{x^s}{\beta x} \frac{\Gamma\left(1-\frac{s}{\beta}\right)\Gamma\left(\frac{s}{\beta}\right)}{\Gamma(s)\Gamma(1-s)}$$

$$= x^{s-1} \frac{\Gamma\left(1-\frac{s}{\beta}\right)\Gamma\left(1+\frac{s}{\beta}\right)}{\Gamma(1+s)\Gamma(1-s)},$$

which completes the proof. □

Let $(X_1, X_2, \ldots, X_n)$ be a series of n independent and identically distributed random variables with probability density function f_X. Denote $\left(X_{(1)}, X_{(2)}, \ldots, X_{(n)}\right)$ as the order statistics of this series if $X_{(1)} \leq \ldots \leq X_{(k)} \leq \ldots \leq X_{(n)}$, and $X_{(k)}^{N_t^\beta}$ as the kth order statistic from N_t^β samples, for $k \in \{1, \ldots, N_t^\beta\}$. The following result is an application of $\{N_t^\beta\}_{t \geq 0}$ on the order statistics, which shows that the probability of $\{X_{(k)}^{N_t^\beta} \mid N_t^\beta \geq k\}$ can be expressed as the ratio of probabilities of a fractional Poisson process.

Theorem 3. *Let $\{N_t^\beta\}_{t \geq 0}$ be a fractional Poisson process with parameter $\lambda > 0$, $\beta \in (0,1)$ and $\{X_i\}_{i \in \mathbb{N}}$ be a sequence of i.i.d. random variables with probability distribution function F_X,*

$$\mathbb{P}\left(X_{(k)}^{N_t^\beta} < x \mid N_t^\beta \geq k\right) = \frac{\mathbb{P}\left(\tilde{N}_t^\beta \geq k\right)}{\mathbb{P}\left(N_t^\beta \geq k\right)}, \quad k \in \mathbb{N},$$

where $\{\tilde{N}_t^\beta\}_{t\geq 0}$ is a fractional Poisson process with parameter $\lambda F_X > 0$ and $\beta \in (0,1)$.

Proof. By the conditional probability law,

$$\mathbb{P}\left(X_{(k)}^{N_t^\beta} < x \mid N_t^\beta \geq k\right) = \frac{\sum_{n=k}^\infty \mathbb{P}\left(X_{(k)}^{N_t^\beta} < x, N_t^\beta = n\right)}{\mathbb{P}\left(N_t^\beta \geq k\right)}$$

$$= \frac{\sum_{n=k}^\infty \mathbb{P}\left(X_{(k)}^{N_t^\beta} < x \mid N_t^\beta = n\right)\mathbb{P}\left(N_t^\beta = n\right)}{\mathbb{P}\left(N_t^\beta \geq k\right)}.$$

The numerator could be computed as

$$\sum_{n=k}^\infty \mathbb{P}\left(X_{(k)}^{N_t^\beta} < x \mid N_t^\beta = n\right)\mathbb{P}\left(N_t^\beta = n\right)$$

$$= \sum_{n=k}^\infty \sum_{j=k}^n \binom{n}{j} F_X^j(x)(1-F_X(x))^{n-j}\frac{(-1)^n}{n!}\sum_{m=0}^\infty \frac{(-\lambda^\beta t^\beta)^m}{\Gamma(\beta m+1-n)}$$

$$= \sum_{j=k}^\infty \sum_{m=0}^\infty \sum_{n=j}^\infty \frac{1}{(n-j)!j!} F_X^j(x)(1-F_X(x))^{n-j}(-1)^n\frac{(-\lambda^\beta t^\beta)^m}{\Gamma(\beta m+1-n)}$$

$$= \sum_{j=k}^\infty \frac{F_X^j(x)}{j!}\sum_{m=0}^\infty (-\lambda^\beta t^\beta)^m \sum_{n=0}^\infty \frac{(1-F_X(x))^n}{n!}\frac{(-1)^{n+j}}{\Gamma(\beta m+1-n-j)}$$

$$= \sum_{j=k}^\infty \frac{F_X^j(x)}{j!}\sum_{m=0}^\infty (-\lambda^\beta t^\beta)^m \frac{(-1)^j F_X^{\beta m-j}(x)}{\Gamma(\beta m+1-j)}$$

$$= \sum_{j=k}^\infty \frac{(-1)^j}{j!}\sum_{m=0}^\infty \frac{\left(-\lambda^\beta F_X^\beta(x)t^\beta\right)^m}{\Gamma(\beta m+1-j)}$$

$$= \mathbb{P}(\tilde{N}_t^\beta \geq k).$$

This completes the proof. □

3. Conclusions and Future Work

In this paper, we discuss a new fractional Poisson process governed by a recursive fractional differential governing equation. The probability distribution function and the Laplace transform of this process are derived. Moreover, this process is equivalent in distribution with a homogeneous Poisson process whose natural time flow is randomized by a Lamperti process. Finally, order statistic from random number samples counted by this fractional Poisson process is studied.

Further research may focus on investigating the distribution properties of the inter-arrival times, generalizing from the first-order differential equation to the nth-order differential equation, and the applications in the risk theory, e.g., the discounted sum counted by this point process.

Funding: This research was supported by The Natural Science Foundation of the Jiangsu Higher Education Institutions of China program [No. 20KJB110019] and the Research Development Fund of Xi'an-Jiaotong Liverpool University [No. RDF-19-02-19].

Acknowledgments: The author thanks the editor and two anonymous reviewers for the valuable comments and suggestions to improve the quality of this paper and clarify the presentation.

Conflicts of Interest: The author declares no conflict of interest.

References

1. Scalas, E.; Gorenflo, R.; Luckock, H.; Mainardi, F.; Mantelli, M.; Raberto, M. Anomalous waiting times in high-frequency financial data. *Quant. Financ.* **2004**, *4*, 695–702. [CrossRef]
2. Kerss, A.; Leonenko, N.N.; Sikorskii, A. Fractional Skellam processes with applications to finance. *Fract. Calc. Appl. Anal. Int. J. Theory Appl.* **2014**, *17*, 532. [CrossRef]
3. Jumarie, G. Fractional master equation: Non-standard analysis and Liouville–Riemann derivative. *Chaos, Solitons Fractals* **2001**, *12*, 2577–2587. [CrossRef]
4. Laskin, N. Fractional Poisson process. *Commun. Nonlinear Sci. Numer. Simul.* **2003**, *8*, 201–213. [CrossRef]
5. Repin, O.N.; Saichev, A.I. Fractional Poisson Law. *Radiophys. Quantum Electron.* **2000**, *43*, 738–741. [CrossRef]
6. Pillai, R.N. On Mittag-Leffler functions and related distributions. *Ann. Inst. Stat. Math.* **1990**, *42*, 157–161. [CrossRef]
7. Mainardi, F.; Gorenflo, R.; Vivoli, A. Beyond the Poisson renewal process: A tutorial survey. *J. Comput. Appl. Math.* **2007**, *205*, 725–735. [CrossRef]
8. Beghin, L.; Orsingher, E. Fractional poisson processes and related planar random motions. *Electron. J. Probab.* **2009**, *14*, 1790–1826. [CrossRef]
9. Beghin, L.; Orsingher, E. Poisson-type processes governed by fractional and higher-order recursive differential equations. *Electron. J. Probab.* **2010**, *15*, 684–709. [CrossRef]
10. Meerschaert, M.; Nane, E.; Vellaisamy, P. The Fractional Poisson Process and the Inverse Stable Subordinator. *Electron. J. Probab.* **2011**, *16*, 1600–1620. [CrossRef]
11. Leonenko, N.; Scalas, E.; Trinh, M. The fractional non-homogeneous Poisson process. *Stat. Probab. Lett.* **2017**, *120*, 147–156. [CrossRef]
12. Uchaikin, V.V.; Sibatov, R.T. Fractional theory for transport in disordered semiconductors. *Commun. Nonlin. Sci. Numer. Simul.* **2008**, *13*, 715–727. [CrossRef]
13. Kumar, A.; Leonenko, N.; Pichler, A. Fractional risk process in insurance. *Math. Financ. Econ.* **2020**, *14*, 43. [CrossRef]
14. Wang, X.T.; Wen, Z.X. Poisson fractional processes. *Chaos Solitons Fract.* **2003**, *18*, 169–177. [CrossRef]
15. Wang, X.T.; Wen, Z.X.; Zhang, S.Y. Fractional Poisson process (II). *Chaos Solitons Fract.* **2006**, *28*, 143–147. [CrossRef]
16. Wang, X.T.; Zhang, S.Y.; Fan, S. Nonhomogeneous fractional Poisson processes. *Chaos Solitons Fract.* **2007**, *31*, 236–241. [CrossRef]
17. Lebedev, N.N. *Special Functions and Their Applications*; Dover Publications: New York, NY, USA, 1972.
18. Mathai, A.M.; Haubold, H.J. *Special Functions for Applied Scientists*; Springer: New York, NY, USA, 2008. [CrossRef]
19. Mathai, A.; Saxena, R.K.; Haubold, H.J. *The H-Function*; Springer: New York, NY, USA, 2010. [CrossRef]
20. Kataria, K.; Vellaisamy, P. On densities of the product, quotient and power of independent subordinators. *J. Math. Anal. Appl.* **2018**, *462*, 1627–1643. [CrossRef]
21. James, L.F. Lamperti-type laws. *Ann. Appl. Probab.* **2010**, *20*, 1303–1340. [CrossRef]
22. Devroye, L.; James, L. On simulation and properties of the stable law. *Stat. Methods Appl.* **2014**, *23*, 307–343. [CrossRef]
23. Lin, G.D. On the Mittag-Leffler distributions. *J. Stat. Plan. Inference* **1998**, *74*, 1–9. [CrossRef]

MDPI
St. Alban-Anlage 66
4052 Basel
Switzerland
Tel. +41 61 683 77 34
Fax +41 61 302 89 18
www.mdpi.com

Fractal and Fractional Editorial Office
E-mail: fractalfract@mdpi.com
www.mdpi.com/journal/fractalfract

www.ingramcontent.com/pod-product-compliance
Lightning Source LLC
LaVergne TN
LVHW070449100526
838202LV00014B/1693